Principles and Practice

Spencer L. Shorte • Friedrich Frischknecht
Editors

Imaging Cellular and Molecular Biological Functions

With 138 Figures, 82 in color and 13 Tables

Springer

Dr. S. L. Shorte
Plateforme d'Imagerie Dynamique
PFID-Imagopole
Insitut Pasteur
25-28 rue du Docteur Roux
F-75015 Paris
France
sshorte@pasteur.fr

Dr. Friedrich Frischknecht
Department of Parasitology
Hygiene Institute
Heidelberg University Medical School
INF 324
69120 Heidelberg
freddy.frischknecht@med.uni-heidelberg.de

Cover illustration: The image shows artistic rendering of three-dimensional image series reconstructions from two different points of view using confocal axial tomography by micro-rotation; for detailed description see: Renaud O., Heintzmann R., Saez-Cirion A., Schnelle T., Mueller T. and Shorte S.L.: *A system and methodology for high-content visual screening of individual intact living cells in suspension Proc. Of SPIE Vol. 6441, 64410Q (2007) "Imaging Manipulation, and Analysis of Biomolecules, Cells, and Tissues V", Ed. Daniel L. Farkas, Robert C. Leif, Dan V. Nicolau*

ISBN-13: 978-3-540-71330-2 e-ISBN-13: 978-3–540–71331–9

Library of Congress Control Number: 2007929272

Springer-Verlag is a part of Springer Science+Business Media

springer.com

Editor: Dr. Sabine Schreck, Heidelberg
Desk Editor: Anette Lindqvist, Heidelberg
Production: SPi
Typesetting: SPi
Cover Design: WMX Design Heidelberg

Printed on acid-free paper 39/3152-HM 5 4 3 2 1 0

Preface

Among many biological scientists "imaging" has come to be considered a buzzword necessary to help win funding, or to make dull conclusions sexy; and if you are one of these people, this book will certainly be of great utility to you too! Notwithstanding these less laudable needs, over 100 years since the first movies of microscopic life were recorded on cinematographic film, imaging in the biological sciences has matured into something resembling, arguably, an emerging discipline. Today it is common to find universities offering young students in biology courses entitled "Imaging and photonics", "Bioimaging", "Digital imaging", "Fluorescent and luminescent probes", and "Bioinformatics and image analysis". So, for a growing number of biological (and biomedical) research groups, departments, institutes, and companies "imaging sciences" are becoming an essential area of resource investment and focus. Nonetheless, there is a lack of clear definition; and it remains for the community to agree whether "imaging" is merely a collection of technologies and methods, or a scientific research discipline in itself. This book does not presume to answer this question nor to define "imaging as a science". Rather we hope to provide the reader with an informative and up-to-date methods volume delineating the broader context of this discourse.

Imaging cellular and molecular biological functions offers a unique selection of essays by leading experts describing methods, techniques, and experimental considerations for imaging biological processes. The first of three sections lays out a series of comprehensive chapters serving as foundations that reinforce the fundamental aspects of digital imaging as a tool (from hardware to software). Specifically, two chapters cover from the formation of digital images within the imaging-microscope setup to their subsequent processing and treatment for analysis. This is accompanied by a "how to" concerning establishing and running state-of-the-art imaging facilities, providing a unique and valuable insight into what is rarely considered from a practical and systematic point of view. Finally, the first section leaves us with a detailed reflection on the important caveat concerning data management that arises as soon as we begin to address the enormous data flood produced by imaging applications; and a possible open-source "community/ stakeholder" driven solution is proposed therein.

A critical applications area for imaging molecular and cellular biological processes is the study of spatiotemporal dynamics of macromolecular interactions (e.g. protein–protein, protein–DNA, protein–lipid, etc.). So, the second section focuses on selected methodological topics around this theme, including in-depth principles for practice concerning those methods that are rapidly becoming routine, including the application of (1) Förster resonance energy transfer (FRET), which can be used to quantitatively measure molecular interactions, (2) fluorescence recovery after photobleaching (FRAP), providing a quantitative measure for diffusion and/or transportation of molecules, and (3) fluorescence correlation spectroscopy (FCS), which enables the direct quantitative determination of molecular kinetics and biochemical parameters (concentrations, affinities, etc) of fluorescently labelled molecules in vitro and inside living cells. Among the chapters covering these subjects there is a degree of overlap that arises naturally, and this is expected to help the reader to fully grasp these sophisticated methods in all their diversity of application, and most important their complementary nature. Finally, the second section contains a definitive commentary on the topic of signal colocalisation, which must be one of the most widely used, but often poorly applied imaging methods, and which impacts almost every aspect of what we try to do with imaging microscopes. To close this section, one chapter describes the transition from collecting imaging data to interpretation in the context of hypothesis-driven experimentation that uses modelling as a means to test data-model validity. The approach uses fluorescence speckle microscopy to illustrate the principles for *in silico* modelling as a means to help validate data interpretation. However, the general arguments for *in silico* biology that uses in situ data parameters are easily extended to other types of imaging experiments, and remain perhaps one of the most exciting and unique advantages offered by the state of the art in imaging.

Finally, the third section presents detailed applications examples using both basic and advanced methods chosen chiefly because of their special tendency in each case to inspire readers to create and customise their own "imaging solutions", rather than to reproduce recipes. The philosophy of this volume is to provide readers with a means to enter into imaging with the confidence to construct methods and design novel experimental solutions using these powerful approaches as tools to answer their own specific questions. Indeed, good molecular and cellular imaging is not solely about recipes; rather, like its distant cousins in other sophisticated biotechnology methods areas (e.g. molecular biology), it is a mixture of scientific utility, empirical accuracy, and careful interpretation. Towards these ends the third section aims to impart to the reader exactly how in imaging it is especially true that "necessity breeds invention". To these ends we examine in detail diverse paradigms, including in vivo imaging of molecular beacons in single cells, tracking living parasites inside intact living animals, mapping of three-dimensional dynamics of amoeboid motility, and the spatiotemporal kinetics of intracellular protein signalling cascades in living cells. Further there is a deep reflection on the cutting edge of the much overlooked area of single-cell microbiology, where new imaging methods are opening unexpected avenues of study. Finally, three chapters describe utilities, general methods, and experimental design consideration for automated high-content analysis, with a view to applications using functional assays, and their optimisation for high throughput.

While this work is clearly neither complete in describing currently available methods, microscopes, and processing packages, nor a monograph, we hope it provides more than a readable collection. Aiming at the biologist, chemist, engineer, medical researcher, and physicist alike, at all levels, including student, researcher, principal investigator, commercial scientist, and provost, we hope to share with you some of our enthusiasm for this area of research, and to provide you with a book that will serve as more than an eventual table-prop.

April 2007

Freddy Frischknecht and Spencer Shorte
Heidelberg, Paris

Contents

Contributors

Amino, R.
Department of Biochemistry, Federal University of Sao Paulo, Rua Tres de Maio 100, 04044-020 Sao Paolo, S.P., Brazil

Anderson, K.I.
Beatson Cancer Research Institute, Switchback Road, Garscube Estate, Glasgow G61 1BD, UK

Athelogou, M.
Definiens AG, Trappentreustr 1, 80339 Munich, Germany

Baatz, M.
Definiens AG, Trappentreustr 1, 80339 Munich, Germany

Baines, I.
Max Planck Institute of Molecular Cell Biology and Genetics, Pfotenhauerstrasse 108, 01307 Dresden, Germany

Barsacchi, R.
Max Planck Institute of Molecular Cell Biology and Genetics, Pfotenhauerstrasse 108, 01307 Dresden, Germany

Binnig, G.
Definiens AG, Trappentreustr 1, 80339 Munich, Germany

Brehm-Stecher, B.F.
Department of Food Science & Human Nutrition, Iowa State University, Ames, IA 50011, USA

Cabal, G.
Department of Cell Biology of Infection, Institut Pasteur, 25–28 Rue du Dr Roux, 75015 Paris, France

Cathcart, J.A.
Optical Microscopy, National High Magnetic Field Laboratory, The Florida State University, Tallahassee, FL 32310, USA

Davidson, M.W.
Optical Microscopy, National High Magnetic Field Laboratory, and Department of Biological Science, The Florida State University, Tallahassee, FL 32310, USA

Danuser, G.
Department of Cell Biology, The Scripps Research Institute, 10550 North Torrey Pines Road, Mail Drop CB 167, La Jolla, CA 92037, USA

Dorn, J.F.
Department of Cell Biology, The Scripps Research Institute, 10550 North Torrey Pines Road, Mail Drop CB 167, La Jolla, CA 92037, USA

Enninga, J.
Department of Cell Biology of Infection, Institut Pasteur, 25–28 Rue du Dr Roux, 75015 Paris, France

Fava, E.
Technology Development Studio, Max Planck Institute of Molecular Cell Biology and Genetics, Pfotenhauerstrasse 108, 01307 Dresden, Germany

Franke-Fayard, B.
Department of Parasitology, Leiden University Medical Center, 2300 RC Leiden, The Netherlands

Frischknecht, F.
Department of Parasitology, Hygiene Institute, University of Heidelberg Medical School, Im Neuenheimer Feld 324, 69120 Heidelberg, Germany

Gradl, G.
Evotec Technologies GmbH, Schnackenburgallee 114, 22525 Hamburg, Germany

Griffin, J.D.
Optical Microscopy, National High Magnetic Field Laboratory, The Florida State University, Tallahassee, FL 32310, USA

Hazelwood, K.L.
Optical Microscopy, National High Magnetic Field Laboratory, The Florida State University, Tallahassee, FL 32310, USA

Hinnah, C.
Evotec Technologies GmbH, Schnackenburgallee 114, 22525 Hamburg, Germany

Hoppe, A.
Department of Microbiology and Immunology, University of Michigan Medical School, Ann Arbor, Michigan 48109-0620, USA

Janse, C.
Department of Parasitology, Leiden University Medical Center, 2300 RC Leiden, The Netherlands

Jaqaman, K.
Department of Cell Biology, The Scripps Research Institute, 10550 North Torrey Pines Road, Mail Drop CB 167, La Jolla, CA 92037, USA

Kaestner, L.
Institute for Molecular Cell Biology, Medical Faculty Building 61, Saarland University, 66421 Homburg/Saar, Germany

Kirsch, A.
Evotec Technologies GmbH, Schnackenburgallee 114, 22525 Hamburg, Germany

Krausz, E.
Technology Development Studio, Max Planck Institute of Molecular Cell Biology and Genetics, Pfotenhauerstrasse 108, 01307 Dresden, Germany

Kuhl, S.
W.M. Keck Dynamic Image Analysis Facility, Department of Biological Sciences, The University of Iowa, Iowa City, IA 52242, USA

Li, D.
Institut National de la Santé et de la Recherche Médicale (INSERM) U603, 75006 Paris, France, Centre National de la Recherche Scientifique (CNRS) UMR 8154, 75006 Paris, France, and Laboratory of Neurophysiology & New Microscopies, Université Paris Descartes, 75006 Paris, France

Lipp, L.
Institute for Molecular Cell Biology, Medical Faculty Building 61, Saarland University, 66421 Homburg/Saar, Germany

Mhlanga, M.M.
Department of Cell Biology of Infection, Institut Pasteur, 25–28 Rue du Dr Roux, 75015 Paris, France

Meijering, E.
Biomedical Imaging Group Rotterdam, Departments of Medical Informatics and Radiology, Erasmus MC – University Medical Center Rotterdam, 3000 DR Rotterdam, The Netherlands

Ménard, R.
Unité de Biologie et Génétique du Paludisme, Department of Parasitology, Institut Pasteur, 25–28 Rue du Dr Roux, 75015 Paris, France

Müller, J.
Evotec Technologies GmbH, Schnackenburgallee 114, 22525 Hamburg, Germany

Nojima, D.
Evotec Technologies GmbH, Schnackenburgallee 114, 22525 Hamburg, Germany

Oheim, M.
Institut National de la Santé et de la Recherche Médicale (INSERM) U603, 75006 Paris, France, Centre National de la Recherche Scientifique (CNRS) UMR 8154, 75006 Paris, France, and Laboratory of Neurophysiology & New Microscopies, Université Paris Descartes, 75006 Paris, France

Olenych, S.G.
Optical Microscopy, National High Magnetic Field Laboratory, The Florida State University, Tallahassee, FL 32310, USA

Peychl, J.
Max Planck Institute for Molecular Cell Biology and Genetics, Pfotenhauerstr. 108, 01329 Dresden, Germany

Sanderson, J.
Max Planck Institute for Molecular Cell Biology and Genetics, Pfotenhauerstr. 108, 01329 Dresden, Germany

Schäpe, A.
Definiens AG, Trappentreustr 1, 80339 Munich, Germany

Schmidt, G.
Definiens AG, Trappentreustr 1, 80339 Munich, Germany

Soll, D.R.
W.M. Keck Dynamic Image Analysis Facility, Department of Biological Sciences, The University of Iowa, Iowa City, IA 52242, USA

Swedlow, J.R.
Division of Gene Regulation and Expression, School of Life Sciences, University of Dundee, Dundee DD1 5EH, UK

van Cappellen, G.
Department of Reproduction and Development, Erasmus MC – University Medical Center Rotterdam, 3000 DR Rotterdam, The Netherlands

Voss, E.
W.M. Keck Dynamic Image Analysis Facility, Department of Biological Sciences, The University of Iowa, Iowa City, IA 52242, USA

Wachsmuth, M.
Cell Biophysics Group, Institut Pasteur Korea, 39-1 Hawolgok-dong, Seongbukgu, Seoul 136–791, Republic of Korea

Waters, A.
Department of Parasitology, Leiden University Medical Center, 2300 RC Leiden, The Netherlands

Weisshart, K.
Carl Zeiss MicroImaging GmbH, Carl-Zeiss-Promenade 10, 07745 Jena, Germany

Wessels, D.
W.M. Keck Dynamic Image Analysis Facility, Department of Biological Sciences, The University of Iowa, Iowa City, IA 52242, USA

Wölcke, J.
Novartis Institutes for BioMedical Research, 4002 Basel, Switzerland

Zerial, M.
Max Planck Institute of Molecular Cell Biology and Genetics, Pfotenhauerstrasse 108, 01307 Dresden, Germany

I
Considerations for Routine Imaging

1
Entering the Portal: Understanding the Digital Image Recorded Through a Microscope

Kristin L. Hazelwood, Scott G. Olenych, John D. Griffin, Judith A. Cathcart, and Michael W. Davidson

Abstract The primary considerations in imaging living cells in the microscope with a digital camera are detector sensitivity (signal-to-noise), the required speed of image acquisition, and specimen viability. The relatively high light intensities and long exposure times that are typically employed in recording images of fixed cells and tissues (where photobleaching is the major consideration) must be strictly avoided when working with living cells. In virtually all cases, live-cell microscopy represents a compromise between achieving the best possible image quality and preserving the health of the cells. Rather than unnecessarily oversampling time points and exposing the cells to excessive levels of illumination, the spatial and temporal resolutions set by the experiment should be limited to match the goals of the investigation. This chapter describes the fundamentals of digital image acquisition, spatial resolution, contrast, brightness, bit depth, dynamic range, and CCD architecture, as well as performance measures, image display and storage, and imaging modes in optical microscopy.

1.1 Introduction

For the most of the twentieth century, a photosensitive chemical emulsion spread on film was used to reproduce images from the optical microscope. It has only been in the past decade that improvements in electronic camera and computer technology have made digital imaging faster, cheaper, and far more accurate to use than conventional photography. A wide range of new and exciting techniques have subsequently been developed that enable researchers to probe deeper into tissues, observe extremely rapid biological processes in living cells, and obtain quantitative information about spatial and temporal events on a level approaching the single molecule.

The imaging device is one of the most critical components in optical microscopy because it determines at what level fine specimen detail may be detected, the relevant structures resolved, and/or the dynamics of a process visualized and recorded. The range of light-detection methods and the wide variety of imaging devices currently available to the microscopist make the equipment selection process difficult and often confusing. This discussion is intended to aid in understanding the basics of light detection, the fundamental properties of digital images, and the criteria relevant to selecting a suitable detector for specific applications.

1.2 Historical Perspective

Recording images with the microscope dates back to the earliest days of microscopy. The first single-lens instruments, developed by Dutch scientists Antoni van Leeuwenhoek and Jan Swammerdam in the late 1600s, were used by these pioneering investigators to produce highly detailed drawings of blood, microorganisms, and other minute specimens (Ruestow 1996). English scientist Robert Hooke engineered one of the first compound microscopes and used it to write *Micrographia*, his hallmark volume on microscopy and imaging published in 1665 (Jardine 2004). The microscopes developed during this period were incapable of projecting images, and observation was limited to close visualization of specimens through the eyepiece. True photographic images were first obtained with the microscope in 1835 when William Henry Fox Talbot applied a chemical emulsion process to capture *photomicrographs* at low magnification (Delly et al. 2007). Between 1830 and 1840 there was an explosive growth in the application of photographic emulsions to recording microscopic images. For the next 150 years, the art and science of capturing images through the microscope with photographic emulsions coevolved with advancements in film technology. During the late 1800s and early 1900s (Bradbury 1967), Carl Zeiss and Ernst Abbe perfected the manufacture of specialized optical glass and applied the new technology to many optical instruments, including compound microscopes.

The dynamic imaging of biological activity was introduced in 1909 by French doctorial student Jean Comandon (Gastou and Comandon 1909), who presented one of the earliest time-lapse videos of syphilis-producing spirochetes. Comandon's technique enabled movie production of the microscopic world. Between 1970 and 1980 researchers coupled tube-based video cameras with microscopes to produce time-lapse image sequences and real-time videos (Inoue and Spring 1997). In the 1990s the tube camera gave way to solid-state technology and the area-array charge-coupled device (CCD), heralding a new era in photomicrography (Inoue and Spring 1997; Murphy 2001). Current terminology referring to the capture of electronic images with the microscope is *digital* or *electronic imaging*.

1.3 Digital Image Acquisition: Analog to Digital Conversion

Regardless of whether light focused on a specimen ultimately impacts on the human retina, a film emulsion, a phosphorescent screen, or the photodiode array of a CCD, an *analog* image is produced (see Inoue and Spring 1997 for a comprehensive explanation). These images can contain a wide spectrum of intensities and colors. Images of this type are referred to as *continuous tone* because the various tonal shades and hues blend together without disruption, to generate a diffraction-limited reproduction of the original specimen. Continuous tone images accurately record image data by using a sequence of electrical signal fluctuations that vary continuously throughout the image.

An analog image must first be converted into a computer-readable or *digital* format before being processed or displayed by a computer. This applies to all images regardless of their origin and complexity. The analog image is digitized in the *analog to digital (A/D) converter* (Fig. 1.1). The continuous analog output of the camera is transformed into a sequence of discrete integers representing the binary code interpreted by computers. The analog image is divided into individual brightness values through two operational processes: *sampling* and *quantization* (Fig. 1.1b, c).

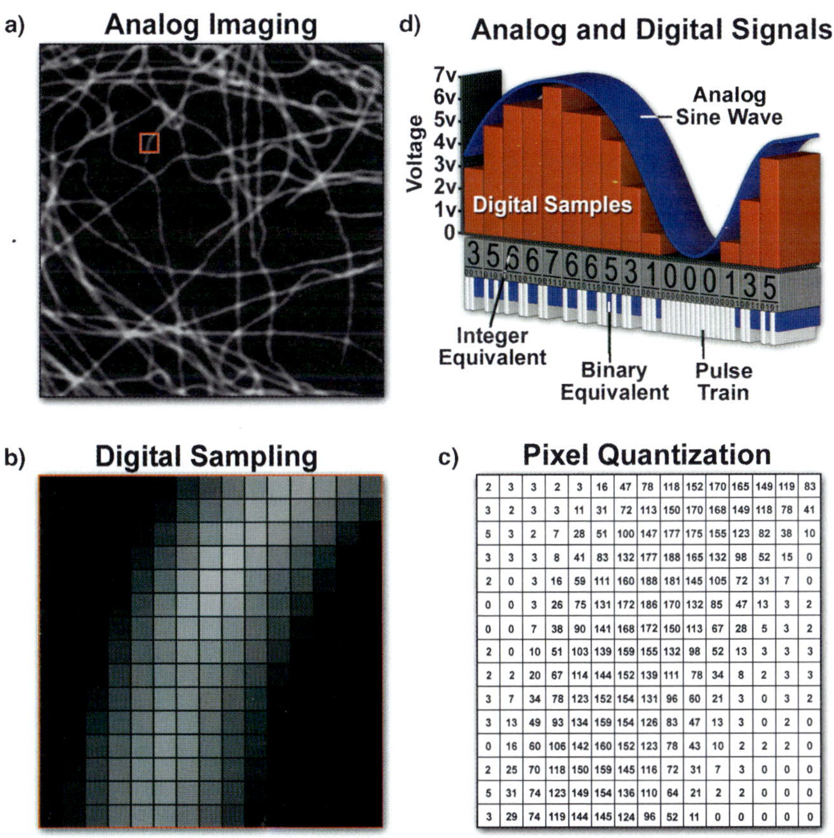

Fig. 1.1 Analog and digital Images. **a** The fluorescence image of human α-tubulin labeled with enhanced green fluorescent protein (EGFP). **b** Sampling of a small portion of **a** – the area with a red rectangle. **c** Quantization of pixel values. **d** The entire process

As we view them, images are generally square or rectangular in dimension; thus, each pixel is represented by a coordinate pair with specific x and y values, arranged in a typical Cartesian coordinate system (Fig. 1.1d). The x coordinate specifies the horizontal position or column location of the pixel, while the y coordinate indicates the row number or vertical position. Thus, a digital image is composed of a rectangular or square pixel array representing a series of intensity values that is ordered by an (x, y) coordinate system. In reality, the image exists only as a large serial array of data values that can be interpreted by a computer to produce a digital representation of the original scene.

The horizontal-to-vertical dimension ratio of a digital image is known as the *aspect ratio* and can be calculated by dividing the image width by the height. The aspect ratio defines the geometry of the image. By adhering to a standard aspect ratio for display of digital images, gross distortion of the image is avoided when the images are displayed on remote platforms. When a continuous tone image is sampled and quantized, the pixel dimensions of the resulting digital image acquire the aspect ratio of the original analog image. It is important that each pixel has a 1:1 aspect ratio (*square pixels*) to ensure compatibility with common digital image processing algorithms and to minimize distortion.

1.4 Spatial Resolution in Digital Images

The quality of a digital image, or *image resolution*, is determined by the total number of pixels and the range of brightness values available for each pixel. Image resolution is a measure of the degree to which the digital image represents the fine details of the analog image recorded by the microscope. The term *spatial resolution* is reserved to describe the number of pixels utilized in constructing and rendering a digital image (Inoue and Spring 1997; Murphy 2001). This quantity is dependent upon how finely the image is sampled during digitization, with higher spatial resolution images having a greater number of pixels within the same physical image dimensions. Thus, as the number of pixels acquired during sampling and quantization of a digital image increases, the spatial resolution of the image also increases.

The optimum sampling frequency, or number of pixels utilized to construct a digital image, is determined by matching the resolution of the imaging device and the computer system used to visualize the image. A sufficient number of pixels should be generated by sampling and quantization to dependably represent the original image. When analog images are inadequately sampled, a significant amount of detail can be lost or obscured, as illustrated in Fig. 1.2. The analog signal presented in Fig. 1.2a shows the continuous intensity distribution displayed by the original image, before sampling and digitization, when plotted as a function of sample position. When 32 digital samples are acquired (Fig. 1.2b), the resulting image retains a majority of the characteristic intensities and spatial frequencies present in the original analog image.

When the sampling frequency is reduced as in Fig. 2c and d, frequencies present in the original image are missed during A/D conversion and a phenomenon known as

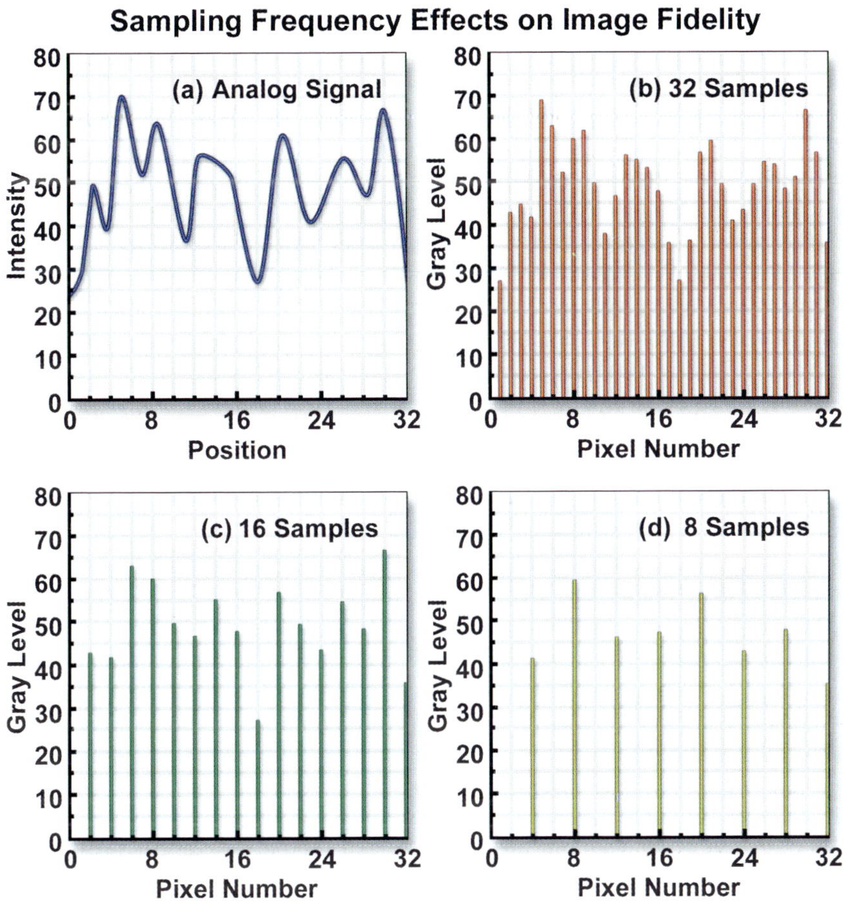

Fig. 1.2 The effects of sampling frequency on image fidelity. **a** Original analog signal; **b** 32 samples of **a**; **c** 16 samples of **a**; **d** eight samples of **a**

aliasing develops. Figure 1.2d illustrates the digital image with the lowest number of samples, where aliasing has produced a loss of high spatial frequency data while simultaneously introducing spurious lower frequency data that do not actually exist.

The *spatial resolution* of a digital image is related to the *spatial density* of the analog image and the *optical resolution* of the microscope or other imaging device. The number of pixels and the distance between pixels (*the sampling interval*) in a digital image are functions of the accuracy of the digitizing device. The optical resolution is a measure of the ability of the optical lens system (microscope and camera) to resolve the details present in the original scene. Optical resolution is affected by the quality of the optics, image sensor, and supporting electronics. *Spatial density* and the optical resolution determine the spatial resolution of the image (Inoue and Spring 1997). Spatial resolution of the image is limited solely by spatial density when the optical resolution of the imaging system is superior to the spatial density.

All of the details contained in a digital image are composed of brightness transitions that cycle between various levels of light and dark. The cycle rate between brightness transitions is known as the *spatial frequency* of the image, with higher rates corresponding to higher spatial frequencies (Inoue and Spring 1997). Varying levels of brightness in minute specimens observed through the microscope are common, with the background usually consisting of a uniform intensity and the specimen exhibiting a larger range of brightness levels.

The numerical value of each pixel in the digital image represents the intensity of the optical image averaged over the sampling interval; thus, background intensity will consist of a relatively uniform mixture of pixels, while the specimen will often contain pixels with values ranging from very dark to very light. Features seen in the microscope that are smaller than the digital sampling interval will not be represented accurately in the digital image. The *Nyquist criterion* requires a sampling interval equal to twice the highest spatial frequency of the specimen to accurately preserve the spatial resolution in the resulting digital image (Inoue and Spring 1997; Murphy 2001; Castleman 1993; Jonkman and Stelzer 2002; Pawley 2006a). If sampling occurs at an interval beneath that required by the Nyquist criterion, details with high spatial frequency will not be accurately represented in the final digital image. The Abbe limit of resolution for optical images is approximately 0.22 µm (using visible light), meaning that a digitizer must be capable of sampling at intervals that correspond in the specimen space to 0.11 µm or less. A digitizer that samples the specimen at 512 pixels per horizontal scan line would have to produce a maximum horizontal field of view of 56 µm (512 × 0.11 µm) in order to conform to the Nyquist criterion. An interval of 2.5–3 samples for the smallest resolvable feature is suggested to ensure adequate sampling for high-resolution imaging.

A serious sampling artifact known as s*patial* aliasing (*undersampling*) occurs when details present in the analog image or actual specimen are sampled at a rate less than twice their spatial frequency (Inoue and Spring 1997). When the pixels in the digitizer are spaced too far apart compared with the high-frequency detail present in the image, the highest-frequency information masquerades as low spatial frequency features that are not actually present in the digital image. Aliasing usually occurs as an abrupt transition when the sampling frequency drops below a critical level, which is about 25% below the Nyquist resolution limit. Specimens containing regularly spaced, repetitive patterns often exhibit moiré fringes that result from aliasing artifacts induced by sampling at less than 1.5 times the repetitive pattern frequency.

1.5 The Contrast Transfer Function

Contrast can be understood as a measure of changes in image signal intensity (ΔI) in relation to the average image intensity (I) as expressed by the following equation:

$$C = \Delta I / I.$$

Of primary consideration is the fact that an imaged object must differ in recorded intensity from that of its background in order to be perceived. Contrast and spatial resolution are closely related and both are requisite to producing a representative image of detail in a specimen (Pawley 2006a). The contrast transfer function (CTF) is analogous to the modulation transfer function (MTF), a measure of the microscope's ability to reproduce specimen contrast in the intermediate image plane at a specific resolution. The MTF is a function used in electrical engineering to relate the amount of modulation present in an output signal to the signal frequency. In optical digital imaging systems, contrast and spatial frequency are correlates of output modulation and signal frequency in the MTF. The CTF characterizes the information transmission capability of an optical system by graphing percentage contrast as a function of spatial frequency as shown in Fig. 1.3 (Pawley 2006b).

Spatial frequency can be defined as the number of times a periodic feature recurs in a given unit space or interval. The intensity recorded at zero spatial frequency in the CTF is a quantification of the average brightness of the image. Since contrast is diffraction-limited, spatial frequencies near zero will have high contrast (approximately 100%) and those with frequencies near the diffraction limit will have lower recorded contrast in the image. As the CTF graph in Fig. 1.3 illustrates, the Rayleigh criterion is not a fixed limit but rather the spatial frequency at which the contrast has dropped to about 25%. The CTF can therefore provide information about how well an imaging system can represent small features in a specimen (Pawley 2006a).

The CTF can be determined for any functional component of the imaging system and is a performance measure of the imaging system as a whole. System performance is evaluated as the product of the CTF curves determined for each component; therefore, it will be lower than that of any of the individual components. Small features that have limited contrast to begin with will become even less visible as the image passes through successive components of the system. The lowest

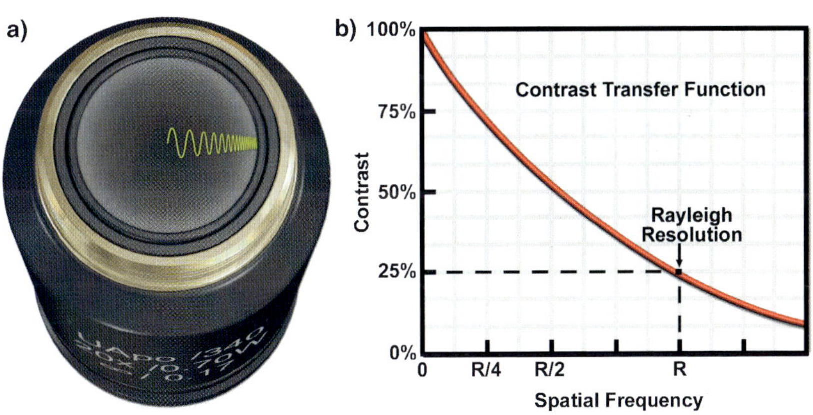

Fig. 1.3 The contrast transfer function and distribution of light waves at the objective rear focal planes. **a** Objective rear aperture demonstrating the diffraction of varying wavelengths. **b** Contrast transfer function indicating the Rayleigh criterion limit of optical resolution

CTFs are typically observed in the objective and CCD. Once the image has been digitally encoded, changes in magnification and concomitant adjustments of pixel geometry can result in improvement of the overall CTF.

1.6 Image Brightness and Bit Depth

The *brightness* of a digital image is a measure of relative intensity values across the pixel array after the image has been acquired with a digital camera or digitized by an A/D converter (Shotton 1993). Brightness should not be confused with *radiant intensity*, which refers to the magnitude or quantity of light energy actually reflected from or transmitted through the object being imaged. As concerns digital image processing, brightness is best described as the measured intensity of all the pixels comprising the digital image after it has been captured, digitized, and displayed. Pixel brightness is important to digital image processing because, other than color, it is the only variable that can be utilized by processing techniques to quantitatively adjust the image.

Regardless of the capture method, an image must be digitized to convert the specimen's continuous tone intensity into a digital brightness value. The accuracy of the digital value is directly proportional to the *bit depth* of the digitizing device (Inoue and Spring 1997; Pawley 2006a; Shotton 1993). If two bits are utilized, the image can only be represented by four brightness values or levels (2^2). Likewise, if three or four bits are processed, the corresponding images have eight (2^3) and 16 (2^4) brightness levels, respectively, as shown in Fig. 1.4.

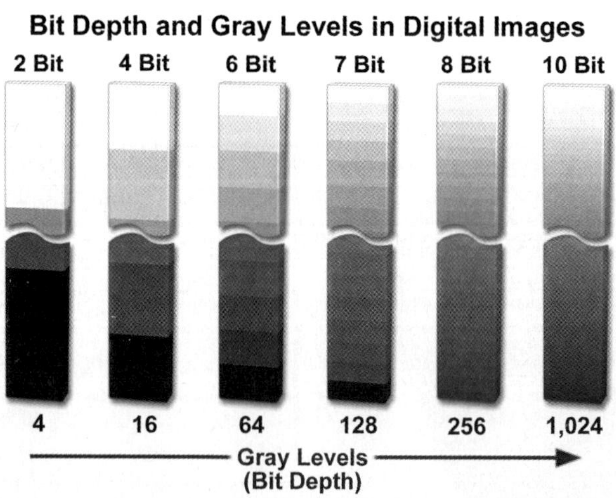

Fig. 1.4 Correlation between bit depth and the number of gray levels in digital images. If two bits are utilized, the image can only be represented by four brightness values or levels. Likewise, if three or four bits are processed, the corresponding images have eight and 16 brightness levels, respectively. In all of these cases, level 0 represents black, while the top level represents white, and each intermediate level is a different shade of gray

The *gray scale* or *brightness range* of a digital image consists of gradations of black, white, and gray brightness levels. The greater the bit depth, the more gray levels are available to represent the image, resulting in a greater signal *dynamic range*. For example, a 12-bit digitizer can display 4,096 gray levels (2^{12}), corresponding to a sensor dynamic range of 72 dB. When applied in this sense, dynamic range refers to the maximum signal level with respect to noise that the CCD sensor can transfer for image display. It can be defined in terms of pixel signal capacity and sensor noise characteristics. Similar terminology is used to describe the range of gray levels utilized in creating and displaying a digital image. This usage is relevant to the *intrascene dynamic range* (Inoue and Spring 1997).

The term *bit depth* refers to the binary range of possible gray scale values used by the A/D converter to translate analog image information into discrete digital values capable of being read and analyzed by a computer. For example, the most popular 8-bit digitizing converters have a binary range of 2^8 or 256 possible values and a 16-bit converter has 2^{16} or 65,536 possible values. The bit depth of the A/D converter determines the size of the gray scale increments, with higher bit depths corresponding to a greater range of useful image information available from the camera.

The number of gray scale levels that must be generated in order to achieve acceptable visual quality should be enough that the steps between individual gray scale values are not discernible to the human eye. The *just-noticeable difference* in intensity of a gray-level image for the average human eye is about 2% under ideal viewing conditions (Inoue and Spring 1997). At most, the human eye can distinguish about 50 discrete shades of gray within the intensity range of a video monitor (Inoue and Spring 1997; Murphy 2001), suggesting that the minimum bit depth of an image should be between 6 and 7 bits.

Digital images should have at least 8-bit to 10-bit resolution to avoid producing visually obvious gray-level steps in the enhanced image when contrast is increased during image processing. The number of pixels and gray levels necessary to adequately describe an image is dictated by the physical properties of the specimen. Low-contrast, high-resolution images often require a significant number of gray levels and pixels to produce satisfactory results, while other high-contrast and low-resolution images (such as a line grating) can be adequately represented with a significantly lower pixel density and gray-level range. Finally, there is a trade-off in computer performance between contrast, resolution, bit depth, and the speed of image-processing algorithms (Pawley 2006a).

1.7 Image Histograms

Grey-level or image histograms provide a variety of useful information about the intensity or brightness of a digital image (Russ 2006). In a typical histogram, the pixels are quantified for each grey level of an 8-bit image. The horizontal axis is scaled from 0 to 255 and the number of pixels representing each grey level is graphed on the vertical axis. Statistical manipulation of the histogram data allows the comparison of images in terms of their contrast and intensity. The relative number of pixels at each grey level can be used to indicate the extent to which the grey-level range is being utilized by a digital image. Pixel intensities are well distributed among grey levels in an image having normal contrast and indicate a large intrascene dynamic range. In low-contrast images only a small portion of available grey levels are represented and intrascene dynamic range is limited. When pixel intensities are distributed among high and low grey levels, leaving the intermediate levels unpopulated, there is an excess of black and white pixels and contrast is typically high.

1.8 Fundamental Properties of CCD Cameras

The fundamental processes involved in creating an image with a CCD camera include exposure of the photodiode array elements to incident light, conversion of accumulated photons to electrons, organization of the resulting electronic charge in potential wells and, finally, transfer of charge packets through the shift registers to the output amplifier (Janesick 2001; Holst 1998; Fig. 1.5). Charge output from the

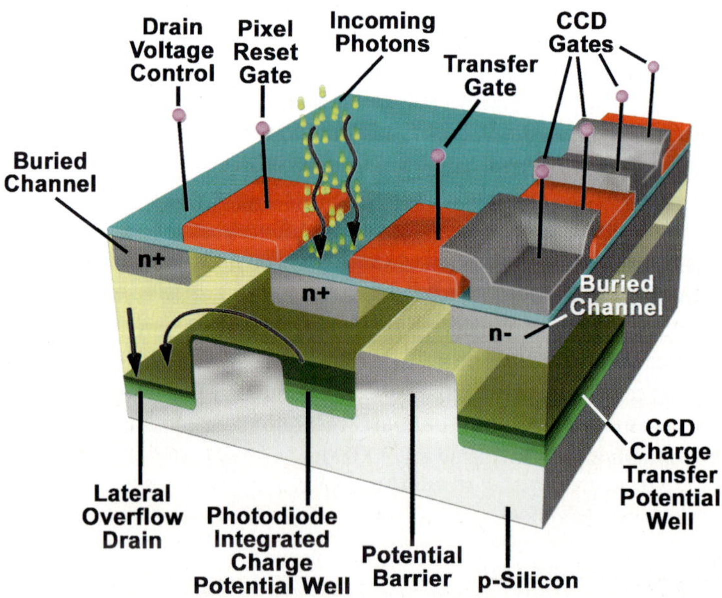

Fig. 1.5 The basic structure of a single metal oxide semiconductor element in a charge coupled device (*CCD*) array. The substrate is a p–n type silicon wafer insulated with a thin layer of silicon dioxide (approximately 100 nm) that is applied to the surface of the wafer. A grid pattern of electrically conductive, optically transparent, polysilicon squares or *gate electrodes* is used to control the collection and transfer of photoelectrons through the array elements

shift registers is converted to voltage and amplified prior to digitization in the A/D converter. Different structural arrangement of the photodiodes and photocapacitors result in a variety of CCD architectures. Some of the more commonly used configurations include frame transfer, full frame, and interline devices. Modifications to the basic architecture such as electron multiplication, backthinning/backillumination, and the use of microlenticular (lens) arrays have helped to increase the sensitivity and quantum efficiency of CCD cameras.

After being accumulated in a CCD during the exposure interval, photoelectrons accumulate when a positive voltage (0–10 V) is applied to an electrode. The applied voltage leads to a hole-depleted region beneath the electrode known as a *potential well*. The number of electrons that can accumulate in the potential well before their charge exceeds the applied electric field is known as the *full well capacity*. The full well capacity depends on pixel size. A typical full well capacity for CCDs used in fluorescence microscopy is between 20,000 and 40,000 photons (Berland et al. 1998). Excessive exposure to light can lead to saturation of the pixels where photoelectrons spill over into adjacent pixels and cause the image to smear or bloom.

The length of time electrons are allowed to accumulate in a potential well is a specified *integration* time controlled by a computer program. When a voltage is applied at a gate, electrons are attracted to the electrode and move to the oxide–silicon interface, where they collect in a 10-nm-thick region until the voltages at the electrodes are cycled or *clocked*.

Different bias voltages applied to the gate electrodes control whether a potential well or barrier will form beneath a particular gate. During charge transfer the charge packet held in the potential well is transferred from pixel to pixel in a cycling or clocking process often explained by analogy to a bucket brigade (Inoue and Spring 1997) as shown in Fig. 1.6. Depending on CCD type, various clocking circuit configurations may be used. Three-phase clocking schemes are commonly used in scientific cameras (Holst 1998; Berland et al. 1998).

The grid of electrodes forms a 2D, parallel register. When a programmed sequence of changing voltages is applied to the gate electrodes the electrons can be shifted across the parallel array. Each row in the parallel register is sequentially shifted into the serial register. The contents of the serial register are shifted one pixel at a time into the output amplifier, where a signal proportional to each charge packet is produced. When the serial register is emptied, the next row in the parallel register is shifted and the process continues until the parallel register has been emptied. This function of the CCD is known as *charge transfer* or *readout* and relies on the efficient transfer of charge from the photodiodes to the output amplifier. The rate at which image data are transferred depends on both the bandwidth of the output amplifier and the speed of the A/D converter.

CCD cameras use a variety of architectures to accomplish the tasks of collecting photons and moving the charge out of the registers and into the readout amplifier. The simplest CCD architecture is known as full frame (Fig. 1.7, architecture a). This configuration consists of a parallel photodiode shift register and a serial shift register (Spring 2000). Full-frame CCDs use the entire pixel array to simultaneously detect incoming photons during exposure periods and thus have a 100% fill

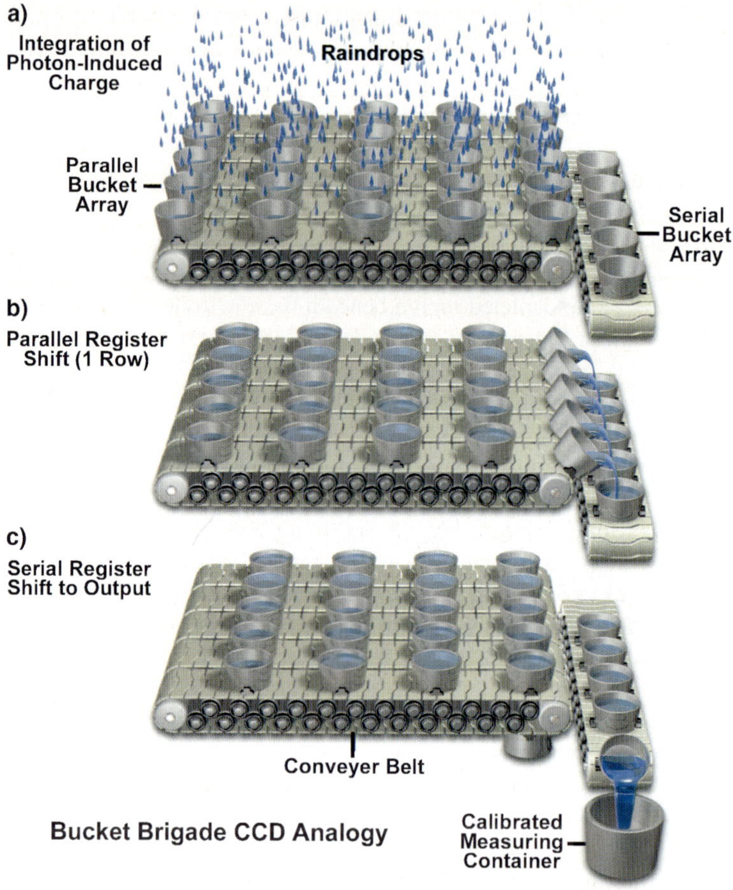

Fig. 1.6 Bucket brigade analogy for CCD technology. Raindrops are first collected in a parallel bucket array (**a**), and then transferred in parallel to the serial output register (**b**). The water accumulated in the serial register is output, one bucket at a time, to the output node (calibrated measuring container, **c**)

factor. Each row in the parallel register is shifted into the serial register. Pixels in the serial register are read out in discrete packets until all the information in the array has been transferred into the readout amplifier. The output amplifier then produces a signal proportional to that of each pixel in the array. Since the parallel array is used both to detect photons and to transfer the electronic data, a mechanical shutter or synchronized strobe light must be used to prevent constant illumination of the photodiodes. Full-frame CCDs typically produce high-resolution, high-density images but can be subject to significant readout noise.

Frame-transfer architecture (Fig. 1.7, architecture b) divides the array into a photoactive area and a light-shielded or masked array, where the electronic data are stored and transferred to the serial register (Holst 1998; Spring 2000). Transfer

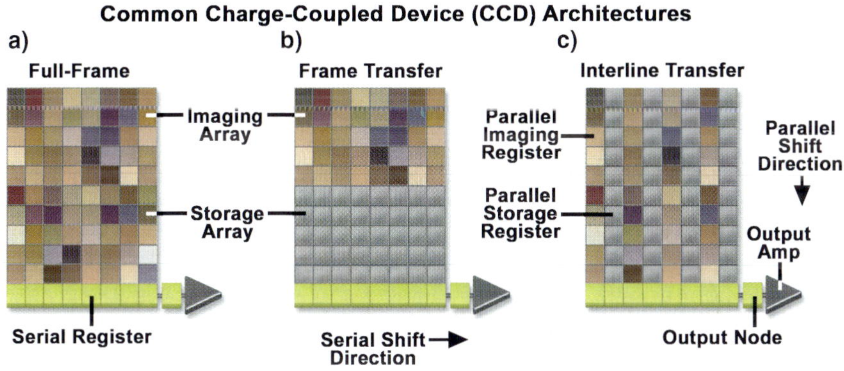

Fig. 1.7 Architectures of common CCDs. **a** full-frame CCD; **b** frame-transfer CCD; **c** interline-transfer CCD

from the active area to the storage array depends upon the array size, but can take less than 0.5 ms. Data captured in the active image area are shifted quickly to the storage register, where they are read out row by row into the serial register. This arrangement allows simultaneous readout of the initial frame and integration of the next frame. The main advantage of frame-transfer architecture is that it eliminates the need to shutter during the charge-transfer process and thus increases the frame rate of the CCD.

For every active row of pixels in an interline array (Fig. 1.7, architecture c) there is a corresponding masked transfer row. The exposed area collects image data and following integration each active pixel rapidly shifts its collected charge to the masked part of the pixel. This allows the camera to acquire the next frame while the data are shifted to charge-transfer channels. Dividing the array into alternating rows of active and masked pixels permits simultaneous integration of charge potential and readout of the image data. This arrangement eliminates the need for external shuttering and increases the device speed and frame rate. The incorporation of a microscopic lens partially compensates for the reduced light-gathering ability caused by pixel masking. Each lens directs a portion of the light that would otherwise be reflected by the aluminum mask to the active area of the pixel (Spring 2000).

Readout speed can be enhanced by defining one or more subarrays that represent areas of interest in the specimen. The reduction in pixel count results in faster readout of the data; however, increases in readout rate are accompanied by an increase in noise. In a clocking routine known as *binning,* charge is collected from a specified group of adjacent pixels and the combined signal is shifted into the serial register (Pawley 2006a; Spring 2000). The size of the binning array is usually selectable and can range from 2×2 pixels to most of the CCD array. The primary reasons for using binning are to improve the signal-to-noise ratio and dynamic range. These benefits come at the expense of spatial resolution. Therefore, binning is commonly used in applications where resolution of the image is less important than rapid throughput and signal improvement.

1.9 CCD Enhancing Technologies

In addition to microlens technology, a number of physical modifications have been made to CCDs to improve camera performance. Instruments used in contemporary biological research must be able to detect weak signals typical of low fluorophore concentrations and tiny specimen volumes, cope with low-excitation photon flux, and achieve the high speed and sensitivity required for imaging rapid cellular kinetics. The demands imposed on detectors can be considerable: ultralow detection limits, rapid data acquisition, and generation of a signal that is distinguishable from the noise produced by the device.

Most contemporary CCD enhancement is a result of *backthinning* and/or *gain register electron multiplication* (Coates et al. 2003). Photons are either absorbed or reflected from the overlying films on the pixels. Electrons created at the surface of the silicon by ultraviolet and blue wavelengths are often lost owing to recombination at the oxide–silicon interface, thus rendering traditional CCD chips less sensitive to high-frequency incident light. With an acid etching technique, the CCD silicon wafer can be uniformly thinned to about 10–15 μm. Incident light is directed onto the backside of the parallel register away from the gate structure. A potential accumulates on the surface and directs the generated charge to the potential wells. Backthinned CCDs exhibit photon sensitivity throughout a wide range of the electromagnetic spectrum, typically from ultraviolet to near-infrared wavelengths. Backthinning can be used with full-frame or frame-transfer architectures, in combination with solid-state electron-multiplication devices, to increase quantum efficiency to above 90% (Coates et al. 2003).

The electron-multiplying CCD (EMCCD) is a modification of the conventional CCD in which an electron-multiplying register is inserted between the serial register output and the charge amplifier (Denvir and Contry 2002). This *multiplication register* or *gain register* is designed with an extra grounded phase that creates a high-field region and a higher voltage (35–45 V) than the standard CCD horizontal register (5–15 V). Electrons passing through the high-field region are multiplied as a result of an approximately 1% probability that an electron will be produced as a result of collision. The multiplication register consists of four gates that use clocking circuits to apply potential differences (35–40 V) and generate secondary electrons by the process of *impact ionization*. Impact ionization occurs when an energetic charge carrier loses energy during the creation of other charge carriers. When this occurs in the presence of an applied electric field, an avalanche breakdown process produces a cascade of secondary electrons (*gain*) in the register. Despite the small (approximately 1%) probability of generating a secondary electron, the large number of pixels in the gain register can result in the production of electrons numbering in the hundreds or thousands.

Traditional slow-scan CCDs achieve high sensitivity and high speed but do so at the expense of readout rate. Readout speed is constrained in these cameras by the charge amplifier. In order to attain high speed, the bandwidth of the charge amplifier must be as wide as possible; however, as the bandwidth increases so too does the amplifier noise. The typically low bandwidths of slow-scan cameras mean they can only be read out at lower speeds (approximately 1 MHz). EMCCDs sidestep this constraint by amplifying the signal prior to the charge amplifier so that it is well above the read noise floor, thus providing both low detection limit and high speed. EMCCDs are thus able to produce low-light images rapidly, with good resolution, a large intensity range, and a wide dynamic range.

1.10 CCD Performance Measures

The term sensitivity, with respect to CCD performance, can be interpreted differently depending on the incident light level used in a particular application (Pawley 2006a). In imaging where signal levels are low, such as in fluorescence microscopy, sensitivity refers to the ability of the CCD to detect weak signals. In high light level applications (such as brightfield imaging of stained specimens) performance may be measured as the ability to determine small changes in the bright images. In either case, the *signal-to-noise ratio* is the measure of camera sensitivity. The signal-to-noise ratio as a rough measure of CCD device performance is the ratio of incident light signal to that of the combined noise of the camera. Signal (S) is determined as a product of input light level (I), quantum efficiency (QE) and the integration time (T) measured in seconds (Janesick 2001):

$$S = I \times \mathrm{QE} \times T.$$

Numerous types and sources of noise are generated throughout the digital imaging process. The amount and significance often depend on the application and type of CCD used to create the image. The primary sources of noise considered in determining the ratio are statistical noise (*shot noise*), thermal noise (*dark current*), and preamplification or *readout noise*, though other types of noise may be significant in some applications and types of camera. Total camera noise is usually calculated as the sum of readout noise, dark current, and statistical noise in quadrature as follows:

$$\Delta_{\mathrm{total}} = \sqrt{\delta_{\mathrm{readout}}^{2} + \delta_{\mathrm{dark}}^{2} + \delta_{\mathrm{shot}}^{2}}.$$

Preamplification or readout noise is produced by the readout electronics of the CCD. Readout noise is composed of two primary types or sources of noise, related to the operation of the solid-state electrical components of the CCD. *White noise* originates in the metal oxide semiconductor field effect transistor (MOSFET) of the output amplifier, where the MOSFET resistance generates thermal noise (Janesick 2001; Holst 1998; Pawley 2006c). *Flicker noise*, also known as *1/f noise* (Holst 1998), is also a product of the output amplifier that originates in the material interface between the silicon and silicon dioxide layers of the array elements.

Thermal noise or dark current is generated similarly, as a result of impurities in the silicon that allow energetic states within the silicon band gap. Thermal noise is generated within surface states, in the bulk silicon, and in the depletion region, though most is produced at surface states. Dark current is inherent to the operation of semiconductors as thermal energy allows electrons to undergo a stepped transition from the valence band to the conduction band, where they are added to the signal electrons and measured by the detector. Thermal noise is most often reduced by cooling the CCD. This can be accomplished using liquid nitrogen or a thermoelectric (*Peltier*) cooler (Spring 2000). The former method places the CCD in a nitrogen environment where the temperature is so low that significant thermal noise is eliminated. Thermoelectric cooling is commonly used to reduce the contribution of thermal noise to total camera noise. A Peltier-type cooler uses a semiconductor sandwiched between two metal plates. When a current is applied, the device acts like a heat pump and transfers heat from the CCD.

Amplification noise occurs in the gain registers of EMCCDs and is often represented by a quantity known as the *noise factor*. For low-light imaging systems the noise introduced by the multiplicative process or gain can be an important performance parameter (Robbins and Hadwen 2003). The electron-multiplication process amplifies weak signals above the noise floor, enabling detection of signals as low as those produced by single photon events, in some cases. In any process in which a signal is amplified, noise added to the signal is also amplified. For this reason it is important to cool EMCCDs to reduce dark current and its associated shot noise.

Whenever we undertake to quantify photons or photoelectric events, there is inherent uncertainty in the measurement that is due to the quantum nature of light. The absorption of photons is a quantum mechanical event and thus the number of photons absorbed varies according to a Poisson distribution. The accuracy of determinations of the number of photons absorbed by a particular pixel is fundamentally restrained by this inherent statistical error. This uncertainty is referred to as Poisson, statistical, or shot noise and is given by the square root of the signal or average number of photoelectrons detected. In a low-light fluorescence application the mean value of the brightest pixels might be as low as 16 photons. Owing to statistical uncertainty or Poisson noise, the actual number of photoelectrons collected in a potential well during an integration period could vary between 12 and 20 (16 ± 4). In mean values representing lower specimen signal levels, the uncertainty becomes more significant. For example, if the mean value is only four photoelectrons, the percentage of the signal representing statistical noise jumps to 50% (4 ± 2) (Pawley 2006b). Poisson or shot noise is an inherent physical limitation. Statistical noise decreases as signal increases and so can only be reduced by increasing the number of events counted. Although quantum efficiency is often considered separately from noise, a value indicating reduced numbers of quantum mechanical events implies an increase in statistical or Poisson noise.

Quantum efficiency is a measure of camera performance that determines the percentage of photons that are detected by a CCD (Spring 2000). It is a property of the photovoltaic response and is summarized by the following equation:

$$\mathrm{QE} = n_{\mathrm{e}} / n_{\mathrm{p}},$$

where the quantum efficiency (QE) is equal to the number of electron hole pairs generated as determined by the number of photoelectrons detected (n_{e}) divided by the average number of photons (n_p) incident on the pixel. Quantum efficiency will always be less than 1.

The number of photoelectrons generated is contingent upon the photovoltaic response of the silicon element to the incident photons and depends on a number of conditions. The amount of charge created during a photon–silicon interaction depends on a number of factors that include the absorption coefficient and diffusion length. The absorption coefficient of silicon varies as longer wavelengths penetrate further into the silicon substrate than do shorter wavelengths. Above a critical wavelength (above 1,100 nm) photons are not energetic enough to induce the photoelectric effect. Photons in the 450–700-nm range are absorbed in the location of the potential well and in the bulk silicon substrate. The quantum efficiency of photons absorbed in the depletion area approaches 100%, while those elsewhere in the substrate may cause release of electrons that move less efficiently.

The spectral sensitivity of a CCD depends on the quantum efficiency of the photoactive elements over the range of near-ultraviolet to near-infrared wavelengths, as illustrated in Fig. 1.8 (Janesick 2001; Holst 1998; Berland et al. 1998; Spring 2000). Modifications made to CCDs to increase performance have led to high quantum efficiencies in the blue–green portion of the spectrum. Backthinned CCDs can exhibit quantum efficiencies of greater than 90%, eliminating loss due to interaction with the charge-transfer channels.

A measure of CCD performance proposed by James Pawley is known as the *intensity spread function* (ISF) and measures the amount of error due to statistical noise in an intensity measurement (Pawley 2003; Pawley 2006b). The ISF relates the number measured by the A/D converter to the brightness of a single pixel. The ISF for a particular detector is determined first by making a series of measurements of a single pixel in which the source illumination is uniform and the integration

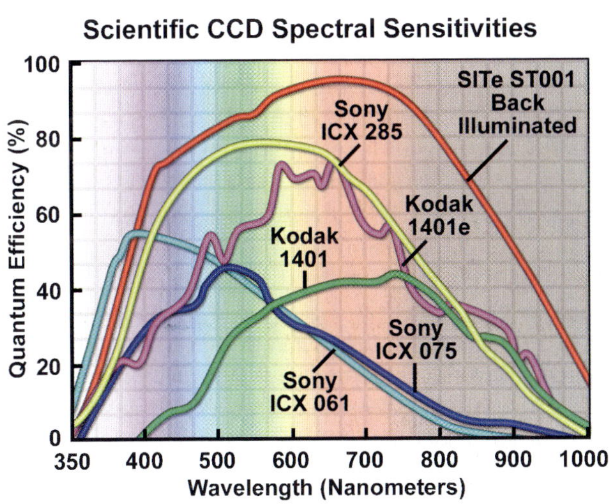

Fig. 1.8 CCD sensitivity across the near-ultraviolet, visible, and near-infrared spectral ranges of several common scientific image sensors

periods are identical. The data are then plotted as a histogram and the mean number of photons and the value at the *full width at half maximum* (FWHM) point (the standard deviation) are determined.

The ISF is equal to the mean divided by the FWHM calculated as the standard deviation. The value is expressed as photons, meaning it has been corrected for quantum efficiency and the known proportional relationship between photoelectrons and their representative numbers stored in memory. The quantity that is detected and digitized is proportional to the number of photoelectrons rather than the number of photons. The ISF is thus a measure of the amount of error in the output signal due to statistical noise that increases as the quantum efficiency (the ratio of photoelectrons to photons) decreases. The statistical error represents the minimum noise level attainable in an imaging system where readout and thermal noise have been adequately reduced.

The conversion of incident photons to an electronic output signal is a fundamental process in the CCD. The ideal relationship between the light input and the final digitized output is linear. As a performance measure, *linearity* describes how well the final digital image represents the actual features of the specimen. The specimen features are well represented when the detected intensity value of a pixel is linearly related to the stored numerical value and to the brightness of the pixel in the image display. Linearity measures the consistency with which the CCD responds to photonic input over its well depth. Most modern CCDs exhibit a high degree of linear conformity, but deviation can occur as pixels near their full well capacity. As pixels become saturated and begin to bloom or spill over into adjacent pixels or charge-transfer channels the signal is no longer affected by the addition of further photons and the system becomes nonlinear (Janesick 2001).

Quantitative evaluation of CCD linearity can be performed by generating sets of exposures with increasing exposure times using a uniform light source. The resulting data are plotted with the mean signal value as a function of exposure (integration) time. If the relationship is linear, a 1-s exposure that produces about 1,000 electrons predicts that a 10-s exposure will produce about 10,000 electrons. Deviations from linearity are frequently measured in fractions of a percent but no system is perfectly linear throughout its entire dynamic range. Deviation from linearity is particularly important in low-light, quantitative applications and for performing flat-field corrections (Murphy 2001). Linearity measurements differ among manufacturers and may be reported as a percentage of conformance to or deviation from the ideal linear condition.

In low-light imaging applications, the fluorescence signal is about one million times weaker than the excitation light. The signal is further limited in intensity by the need to minimize photobleaching and phototoxicity. When quantifying the small number of photons characteristic of biological fluorescent imaging, the process is photon-starved but also subject to the statistical uncertainty associated with enumerating quantum mechanical events. The measurement of linearity is further complicated by the fact that the amount of uncertainty increases with the square root of the intensity. This means that the statistical error is largest in the brightest regions of the image. Manipulating the data using a deconvolution algorithm is often the only way to address this problem in photon-limited imaging applications (Pawley 2006b).

1.11 Multidimensional Imaging

The term *multidimensional imaging* can be used to describe 3D imaging (volume), 4D imaging (volume plus time), or imaging in even more dimensions, each additional one representing different wavelengths. Modern bioscience applications increasingly require optical instruments and digital image processing systems capable of capturing quantitative, multidimensional information about dynamic, spatially complex specimens. Multidimensional, quantitative image analysis has become essential to a wide assortment of bioscience applications. The imaging of subresolution objects (Betzig et al. 2006; Roux et al. 2004), rapid kinetics (Lippincott-Schwartz et al. 2003), and dynamic biological processes (Day 2005; Zhang et al. 2002) present technical challenges for instrument manufacturers to produce ultrasensitive, extremely fast, and accurate image acquisition and processing devices.

The image produced by the microscope and projected onto the surface of the detector is a 2D representation of an object that exists in 3D space. As discussed previously, the image is divided into a 2D array of pixels, represented graphically by an x and a y axis. Each pixel is a typically square area determined by the lateral resolution and magnification of the microscope as well as the physical size of the detector array. Similar to the pixel in 2D imaging, a volume element or *voxel*, having dimensions defined by x, y, and z axes, is the basic unit or sampling volume in 3D imaging (Pawley 2006b; Roux et al. 2004). A voxel represents an optical section, imaged by the microscope, that comprises the area resolved in the x–y plane and a distance along the z axis defined by the depth of field, as illustrated in Fig. 1.9.

The depth of field is a measurement of object space parallel with the optical axis. It describes the numerical aperture (NA) dependent, axial resolution capability of the microscope objective and is defined as the distance between the nearest and farthest objects in simultaneous focus. The NA of a microscope objective is determined by multiplying the sine of half of the angular aperture by the refractive index of the imaging medium. Lateral resolution varies inversely with the first power of the NA, whereas axial resolution is inversely related to the square of the NA. The NA therefore affects axial resolution far more than lateral resolution. While spatial resolution depends only on the NA, voxel geometry depends on the spatial resolution as determined by the NA and magnification of the objective, as well as the physical size of the detector array. With the exception of multiphoton imaging, which uses femtoliter voxel volumes, widefield and confocal microscopy are limited to dimensions of about $0.2\,\mu m \times 0.2\,\mu m \times 0.4\,\mu m$ (Pawley 2006b; Roux et al. 2004) based on the highest NA objectives available.

Virus-sized objects that are smaller than the optical resolution limits can be detected but are poorly resolved. In thicker specimens, such as cells and tissues, it is possible to repeatedly sample at successively deeper layers so that each optical section contributes to a z series (or z stack). Microscopes that are equipped with computer-controlled step motors acquire an image then adjust the fine focus according to the sampling parameters, take another image, and continue until a large enough number of optical sections have been collected. The step size is

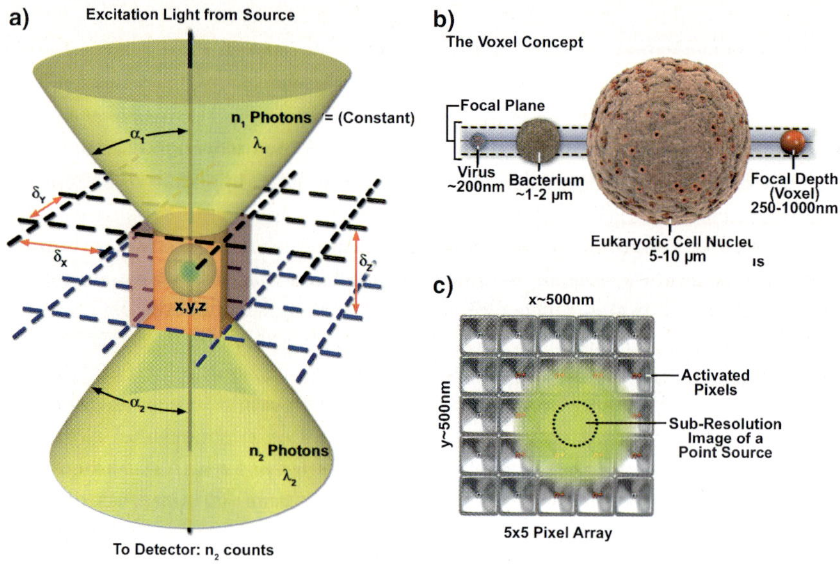

Fig. 1.9 The voxel concept. A subresolution fluorescent point object can be described in three dimensions with the coordinate system illustrated in **a**. The typical focal depth of an optical microscope is shown relative to the dimensions of a virus, a bacterium, and a mammalian cell nucleus (**b**). **c** A subresolution point image projected onto a 25-pixel array. Activated pixels (those receiving photons) span a much larger dimension than the original point source

adjustable and will depend, as for 2D imaging, on appropriate Nyquist sampling (Jonkman and Stelzer 2002; Pawley 2006b; Roux et al. 2004). The axial resolution limit is larger than the limit for lateral resolution. This means that the voxel may not be an equal-sided cube and will have a *z* dimension that can be several times greater than the *x* and *y* dimensions. For example, a specimen can be divided into 5-μm-thick optical sections and sampled at 20-μm intervals. If the *x* and *y* dimensions are both 0.5 μm, the resulting voxel will be 40 times longer than it is wide.

3D imaging can be performed with conventional widefield fluorescence microscopes equipped with a mechanism to acquire sequential optical sections. Objects in a focal plane are exposed to an illumination source and light emitted from the fluorophore is collected by the detector. The process is repeated at fine-focus intervals along the *z* axis, often hundreds of times, and a sequence of optical sections or a *z* series (also *z* stack) is generated. In widefield imaging of thick biological samples, blurred light and scatter can degrade the quality of the image in all three dimensions.

Confocal microscopy has several advantages that have made it a commonly used instrument in multidimensional, fluorescence microscopy (Pawley 2006d). In addition to slightly better lateral and axial resolution, a laser scanning confocal microscope has a controllable depth of field, eliminates unwanted wavelengths and out-of-focus light, and is able to finely sample thick specimens. A system of computer-controlled, galvanometer-driven dichroic mirrors direct an image of the pinhole aperture across the field of view, in a raster pattern similar to that used in a television. An exit pinhole is placed in a plane conjugate to the point on the object being scanned. Only light emitted from the point object is transmitted through the pinhole and reaches the detector element. Optical section thickness can be controlled by adjusting the diameter of the pinhole in front of the detector, a feature that enhances flexibility in imaging biological specimens (Pawley 2006b). Technological improvements such as computer-controlled and electronically controlled laser scanning and shuttering, as well as variations in the design of instruments (e.g., spinning disc, multiple pinhole, and slit scanning versions) have increased image acquisition speeds (see also Chap. 10 by Kaestner and Lipp). Faster acquisition and better control of the laser by shuttering the beam reduces the total exposure effects on light-sensitive fixed or live cells. This enables the use of intense, narrow-wavelength bands of laser light to penetrate deeper into thick specimens, making confocal microscopy suitable for many time-resolved, multidimensional imaging applications (Roux et al. 2004).

For multidimensional applications in which the specimen is very sensitive to visible wavelengths, the sample volume or fluorophore concentration is extremely small, or when the imaging is through thick tissue specimens, *laser scanning multiphoton microscopy* (LSMM; often simply referred to as multiphoton microscopy) is sometimes employed. While the scanning operation is similar to that of a confocal instrument, LSMM uses an infrared illumination source to excite a precise femtoliter sample volume (approximately 10^{-15} L). Photons are generated by an infrared laser and localized in a process known as photon crowding (Piston 1999). The simultaneous absorption of two low-energy photons is sufficient to excite the fluorophore and cause it to emit at its characteristic, Stokes-shifted wavelength. The longer-wavelength excitation light causes less photobleaching and phototoxicity and, as a result of reduced Rayleigh scattering, penetrates further into biological specimens. Owing to the small voxel size, light is emitted from only one diffraction-limited point at a time, enabling very fine and precise optical sectioning. Since there is no excitation of fluorophores above or below the focal plane, multiphoton imaging is less affected by interference and signal degradation. The absence of a pinhole aperture means that more of the emitted photons are detected, which, in the photon-starved applications typical of multidimensional imaging, may offset the higher cost of multiphoton imaging systems.

The *z* series is often used to represent the optical sections of a time-lapse sequence where the *z* axis represents time. This technique is frequently used in developmental biology to visualize physiological changes during embryo development. Live cell or dynamic process imaging often produces 4D data sets (Dailey et al. 2006). These time-resolved volumetric data are visualized using 4D viewing programs and can be reconstructed, processed, and displayed as a moving image or montage. Five or more dimensions can be imaged by acquiring the 3D or 4D sets at different wavelengths using different fluorophores. The multiwavelength optical sections can later be combined into a single image of discrete structures in the specimen that have been labeled with different fluorophores. Multidimensional imaging has the added advantage of being able to view the image in the *x*–*z* plane as a profile or vertical slice.

1.12 The Point-Spread Function

The ideal *point-spread function* (PSF) is the 3D diffraction pattern of light emitted from an infinitely small point source in the specimen and transmitted to the image plane through a high-NA objective (Inoue and Spring 1997). It is considered to be the fundamental unit of an image in theoretical models of image formation. When light is emitted from such a point object, a fraction of it is collected by the objective and focused at a corresponding point in the image plane. However, the objective lens does not focus the emitted light to an infinitely small point in the image plane. Rather, light waves converge and interfere at the focal point to produce a diffraction pattern of concentric rings of light surrounding a central, bright disk (termed an Airy disk), when viewed in the *x–y* plane. The radius of the disc is determined by the NA; thus, the resolving power of an objective lens can be evaluated by measuring the size of the Airy disc. The image of the diffraction pattern can be represented as an intensity distribution as shown in Fig. 1.10. The bright central portion of the Airy disc and concentric rings of light correspond to intensity peaks in the distribution.

In a perfect lens with no spherical aberration the diffraction pattern at the paraxial (perfect) focal point is both symmetrical and periodic in the lateral and axial planes. When viewed in either axial meridian (*x–z* or *y–z*) the diffraction image can have various shapes depending on the type of instrument used (i.e., widefield, confocal, or multiphoton) but is often hourglass- or football-shaped (Cannell et al. 2006). The PSF is generated from the *z* series of optical sections and can be used to evaluate the axial resolution. As with lateral resolution, the minimum distance

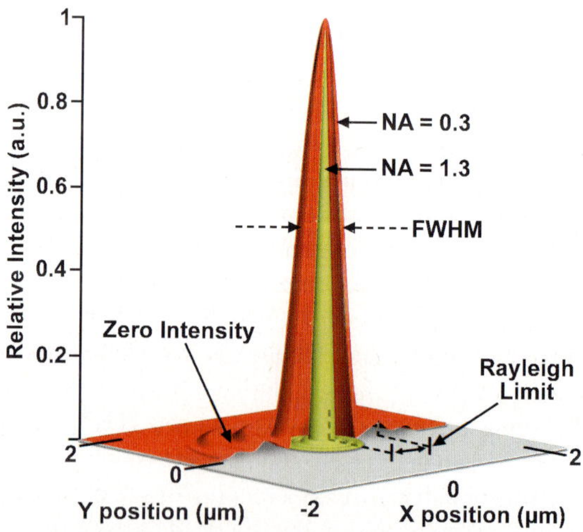

Fig. 1.10 The point-spread function. Relative intensity is plotted as a function of spatial position for point-spread function from objectives having a numerical aperture (*NA*) of 0.3 and 1.3. The full width at half maximum (*FWHM*) is indicated for the lower-NA objective along with the Rayleigh limit

the diffraction images of two points can approach each other and still be resolved is the axial resolution limit. The image data are represented as an axial intensity distribution in which the minimum resolvable distance is defined as the first minimum of the distribution curve (Pawley 2006b).

The PSF is often measured using a fluorescent bead embedded in a gel that approximates an infinitely small point object in a homogeneous medium. However, thick biological specimens are far from homogeneous. Differing refractive indices of cell materials, tissues, or structures in and around the focal plane can diffract light and result in a PSF that deviates from design specification, fluorescent bead determination, or the calculated theoretical PSF. A number of approaches to this problem have been suggested, including comparison of theoretical and empirical PSFs, embedding a fluorescent microsphere in the specimen, or measuring the PSF using a subresolution object native to the specimen (de Monvel et al. 2003).

The PSF is valuable not only for determining the resolution performance of different objectives and imaging systems, but also as a fundamental concept used in *deconvolution*. Deconvolution is a mathematical transformation of image data that reduces out-of-focus light or *blur*. Blurring is a significant source of image degradation in 3D widefield fluorescence microscopy. It is nonrandom and arises within the optical train and specimen, largely as a result of diffraction. A computational model of the blurring process, based on the *convolution* of a point object and its PSF, can be used to deconvolve or reassign out-of-focus light back to its point of origin. Deconvolution is used most often in 3D widefield imaging. However, images produced with confocal, spinning disc, and multiphoton microscopes can also be improved using image-restoration algorithms.

Image formation begins with the assumptions that the process is linear and shift-invariant. If the sum of the images of two discrete objects is identical to the image of the combined object, the condition of linearity is met, providing the detector is linear, and quenching and self-absorption by fluorophores are minimized. When the process is shift-invariant, the image of a point object will be the same everywhere in the field of view. Shift invariance is an ideal condition that no real imaging system meets. Nevertheless, the assumption is reasonable for high-quality research instruments (P.J. Shaw 2006).

Convolution mathematically describes the relationship between the specimen and its optical image. Each point object in the specimen is represented by a blurred image of the object (the PSF) in the image plane. An image consists of the sum of each PSF multiplied by a function representing the intensity of light emanating from its corresponding point object:

$$i(x) = \int_{-\infty}^{+\infty} o(x - x')\,\mathrm{PSF}(x')\,\mathrm{d}x'.$$

A *pixel blurring kernel* is used in convolution operations to enhance the contrast of edges and boundaries and the higher spatial frequencies in an image (Inoue and

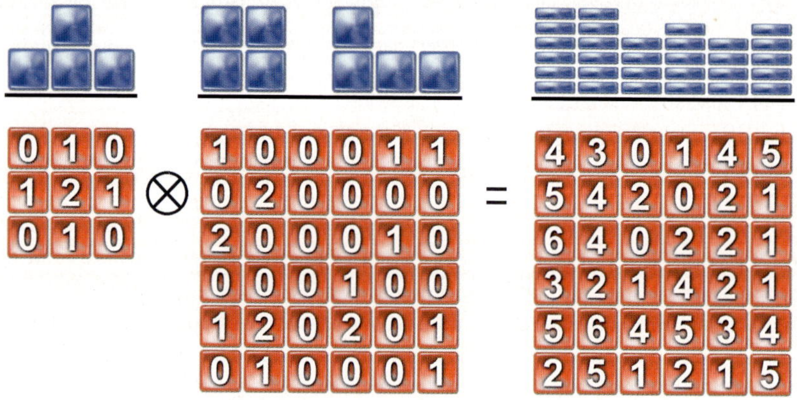

Fig. 1.11 Convolution operation. Illustration of a convolution operation with a 6×6 pixel array and a blurring kernel of 3×3 pixels. *Above the arrays* are profiles demonstrating the maximum projection of the 2D grids when viewed from above

Spring 1997; Russ 2006). Figure 1.11 illustrates the convolution operation using a 3×3 kernel to convolve a 6×6 pixel object.

An image *is* a convolution (⊗) of the object and the PSF and can be symbolically represented as follows:

$$i(r) = o(r) \otimes \mathrm{PSF}(r),$$

where the image, object, and PSF are denoted as functions of position (r) or an x, y, z, and t (time) coordinate. The Fourier transform shows the frequency and amplitude relationship between the object and the PSF, converting the space-variant function to a frequency-variant function. Because convolution in the spatial domain is equal to multiplication in the frequency domain, convolutions are more easily manipulated by taking their Fourier transform (F) (P.J. Shaw 2006):

$$F\left[i(x,y,z,t)\right] = F\left[o(x,y,z,t)\right] \times F\left[\mathrm{PSF}(x,y,z,t)\right].$$

In the spatial domain described by the PSF, a specimen is a collection of point objects and the image is a superposition or sum of point source images. The frequency domain is characterized by the *optical-transfer function* (OTF). The OTF is the Fourier transform of the PSF and describes how spatial frequency is affected by blurring. In the frequency domain the specimen is equivalent to the superposition of sine and cosine functions and the image consists of the sum of weighted sine and cosine functions. The Fourier transform further simplifies the representation of the convolved object and image such that the transform of the image is equal to the specimen multiplied by the OTF. The microscope passes low-frequency (large, smooth) components best, intermediate frequencies are attenuated, and high frequencies greater than $2\mathrm{NA}/\lambda$ are excluded. Deconvolution algorithms are therefore required to augment high spatial frequency components (P.J. Shaw 2006; Wallace et al. 2001).

Theoretically, it should be possible to reverse the convolution of object and PSF by taking the inverse of the Fourier-transformed functions. However, deconvolution increases noise which exists at all frequencies in the image. Beyond half the Nyquist sampling frequency no useful data are retained, but noise is nevertheless amplified by deconvolution. Contemporary image-restoration algorithms use additional assumptions about the object such as smoothness or nonnegative value and incorporate information about the noise process to avoid some of the noise-related limitations.

Deconvolution algorithms are of two basic types. Deblurring algorithms use the PSF to estimate blur then subtract it by applying the computational operation to each optical section in a *z* series. Algorithms of this type include nearest neighbor, multineighbor, no neighbor, and unsharp masking. The more commonly used nearest-neighbor algorithm estimates and subtracts blur from *z* sections above and below the section to be sharpened. While these run quickly and use less computer memory, they do not account for cross talk between distant optical sections. Deblurring algorithms may decrease the signal-to-noise ratio by adding noise from multiple planes. Images of objects whose PSFs overlap in the paraxial plane can often be sharpened by deconvolution; however, this is at the cost of displacement of the PSF. Deblurring algorithms introduce artifacts or changes in the relative intensities of pixels and thus cannot be used for morphometric measurements, quantitative intensity determinations, or intensity ratio calculations (Wallace et al. 2001).

Image-restoration algorithms use a variety of methods to reassign out-of-focus light to its proper position in the image. These include inverse filter types such as Wiener deconvolution or linear least squares, constrained iterative methods such as Jansson van Cittert, statistical image restoration, and blind deconvolution (Jansson 1997). Constrained deconvolution imposes limitations by excluding nonnegative pixels and placing finite limits on size or fluorescent emission, for example. An estimation of the specimen is made and an image is calculated and compared with the recorded image. If the estimation is correct, constraints are enforced and unwanted features are excluded. This process is convenient to iterative methods that repeat the constraint algorithm many times. The Jansson van Cittert algorithm predicts an image, applies constraints, and calculates a weighted error that is used to produce a new image estimate for multiple iterations. This algorithm has been effective in reducing high-frequency noise.

Blind deconvolution does not use a calculated or measured PSF, but rather calculates the most probable combination of object and PSF for a given data set. This method is also iterative and has been successfully applied to confocal images. Actual PSFs are degraded by the varying refractive indices of heterogeneous specimens. In laser scanning confocal microscopy (LSCM) where light levels are typically low, this effect is compounded. Blind deconvolution reconstructs both the PSF and the deconvolved image data. Compared with deblurring algorithms, image-restoration methods are faster, frequently result in better image quality, and are amenable to quantitative analysis (Holmes et al. 2006).

Deconvolution performs its operations using floating-point numbers and consequently uses large amounts of computing power. Four bytes per pixel are required, which translates to 64 MB for a $512 \times 512 \times 64$ image stack. Deconvolution is also CPU-intensive and large data sets with numerous iterations may take several hours to produce a fully restored image, depending on processor speed. Choosing an appropriate deconvolution algorithm involves determining a delicate balance of resolution, processing speed, and noise that is correct for a particular application (Holmes et al. 2006; Jansson 1997; von Tiedemann et al. 2006; Wallace et al. 2001).

1.13 Digital Image Display and Storage

The display component of an imaging system reverses the digitizing process accomplished in the A/D converter. The array of numbers representing image signal intensities must be converted back into an analog signal (voltage) in order to be viewed on a computer monitor (Inoue and Spring 1997; Shotton 1993). A problem arises when the function sinx/x representing the waveform of the digital information must be made to fit the simpler Gaussian curve of the monitor scanning spot. To perform this operation without losing spatial information, the intensity values of each pixel must undergo interpolation, a type of mathematical curve-fitting. The deficiencies related to the interpolation of signals can be partially compensated for by using a high-resolution monitor that has a bandwidth greater than 20 MHz, as do most modern computer monitors. Increasing the number of pixels used to represent the image by sampling in excess of the Nyquist limit (oversampling) increases the pixel data available for image processing and display.

A number of different technologies are available for displaying digital images though microscopic imaging applications most often use monitors based on either cathode ray tube (CRT) or liquid crystal display (LCD) technology. These display technologies are distinguished by the type of signals each receives from a computer. LCD monitors accept digital signals which consist of rapid electrical pulses that are interpreted as a series of binary digits (0 or 1). CRT displays accept analog signals and thus require a digital to analog converter (DAC) that precedes the monitor in the imaging process train.

Digital images can be stored in a variety of file formats that have been developed to meet different requirements. The format used depends on the type of image and how it will be presented. Quality, high-resolution images require large file sizes. File sizes can be reduced by a number of different compression algorithms but image data may be lost depending on the type. Lossless compressions (such as Tagged Image File Format, TIFF) encode information more efficiently by identifying patterns and replacing them with short codes. These algorithms can reduce an original image by about 50–75%. This type of file compression can facilitate transfer and sharing of images and allows decompression and restoration to the original image parameters. Lossy compression algorithms, such as that used to define pre-2000 JPEG image files, are capable of reducing images to less than 1% of their original size. The JPEG 2000 format uses both types of compression. The large reduction is accomplished by a type of undersampling in which imperceptible gray-level steps are eliminated. Thus, the choice is often a compromise between image quality and manageability.

Bit-mapped or raster-based images are produced by digital cameras, screen, and print output devices that transfer pixel information serially. A 24-bit color (RGB) image uses 8 bits per color channel, resulting in 256 values for each color for a total of 16.7 million colors. A high-resolution array of 1,280 × 1,024 pixels representing a true-color 24-bit image would require more than 3.8 MB of storage space. Commonly used raster-based file types include GIF, TIFF, and JPEG. Vector-based images are defined mathematically and used primarily for storage of images created by drawing and animation software. Vector imaging typically requires less storage space and is amenable to transformation and resizing. Metafile formats, such as PDF, can incorporate files created by both raster- and vector-based images. This file format is useful when images must be consistently displayed in a variety of applications or transferred between different operating systems.

As the dimensional complexity of images increases, image file sizes can become very large. For a single-color, 2,048 × 2,048 image, file size is typically about 8 MB. A multicolor image of the same resolution can reach 32 MB. For images with three spatial dimensions and multiple colors a smallish image might require 120 MB of storage. In live-cell imaging where time-resolved, multidimensional images are collected, image files can become extremely large. For example, an experiment that uses ten stage positions, imaged over 24 h with three to five colors at one frame per minute, a 1,024 × 1,024 frame size, and 12-bit image could amount to 86 GB/day. High-speed confocal imaging with special storage arrays can produce up to 100 GB/h. Image files of this size and complexity must be organized and indexed and often require massive directories with hundreds of thousands of images saved in a single folder as they are streamed from the digital camera. Modern hard drives are capable of storing at least 500 GB. The number of images that can be stored depends on the size of the image file. About 250,000 2–3 MB images can be stored on most modern hard drives. External storage and backup can be performed using CDs that hold about 650 MB or DVDs that have 4.7-GB capacities. Image analysis typically takes longer than collection and is presently limited by computer memory and drive speed. Storage, organization, indexing, analysis, and presentation will be improved as 64-bit multiprocessors with large memory cores become available.

1.14 Imaging Modes in Optical Microscopy

The imaging of living cells and organisms has traditionally been based on long-term time-lapse experiments designed to observe cell movement and dynamic events. Techniques have typically included brightfield, polarized light microscopy, differential interference contrast (DIC), Hoffman modulation contrast (HMC), phase contrast, darkfield, and widefield fluorescence (Davidson and Abramowitz 2002). In the past decade, a number of new imaging technologies have been developed that have enabled time-lapse imaging to be integrated with techniques that monitor, quantify, and perturb dynamic processes in living cells and organisms. LSCM, spinning disc confocal microscopy, LSMM, and total internal reflection microscopy (TIRFM) have generated a wide variety of techniques that have facilitated greater insights into dynamic biological processes (reviewed in Pawley 2006d).

Until recently, live-cell imaging has involved adherent mammalian cells, positioned a short distance (approximately 10 µm or less) from the cover slip–medium interface. Specimens in a growing number of contemporary investigations are often 10–200-µm thick. There are a number of problems associated with imaging beyond a depth of 20–30 µm within a living specimen. Primary among the difficulties are blurring caused by out-of-focus light, movement within the cytoplasm that limits exposure time, and the photosensitivity of fluorophores and living cells that makes them vulnerable to photobleaching and phototoxic effects. The imaging of living cells, tissues, and organisms usually involves a compromise between image resolution and maintaining conditions requisite to the survival and normal biological functioning of the specimen (Goldman and Spector 2005).

Traditional approaches to live-cell imaging are often based on short-term or long term time-lapse investigations designed to monitor cellular motility and dynamic events using common contrast enhancement techniques, including brightfield, DIC, HMC, phase contrast, and widefield fluorescence. However, modern techniques and newly introduced methods are extending these observations well beyond simply creating cinematic sequences of cell structure and function, thus enabling time-lapse imaging to be integrated with specialized modes for monitoring, measuring, and perturbing dynamic activities of tissues, cells, and subcellular structures.

A majority of live-cell imaging investigations are conducted with adherent mammalian cells, which are positioned within 10 µm of the cover slip–medium interface. Increasingly, however, investigators are turning their attention to thicker animal and plant tissue specimens that can range in thickness from 10 to 200 µm. In this case, out-of-focus information blurs the image and the constant churning of the cytoplasm creates limitations on exposure times. Both brightfield and fluorescence methods used in imaging thicker animal tissues and plants must take into account the sensitivity of these specimens to light exposure and the problems associated with resolving features that reside more than 20–30 µm within the specimen.

Brightfield techniques are often less harmful to living cells, but methods for observing specific proteins using transillumination have not been widely developed. Generating a high-contrast chromatic (color) or intensity difference in a brightfield image is more difficult than identifying a luminous intensity change (in effect, due to fluorescence) against a dark or black background. Therefore, brightfield techniques are used for following organelles or cellwide behavior, while fluorescence methods, including confocal techniques, are generally used for following specific molecules.

Presented in Fig. 1.12 is a schematic illustration of popular imaging modes in widefield and scanning modes of fluorescence microscopy (Pawley 2006d). Widefield, laser scanning, spinning disc, and multiphoton techniques employ vastly different illumination and detection strategies to form an image. The diagram illustrates an adherent mammalian cell on a cover slip being illuminated with total internal reflection, laser scanning, and spinning disc confocal, in addition to traditional

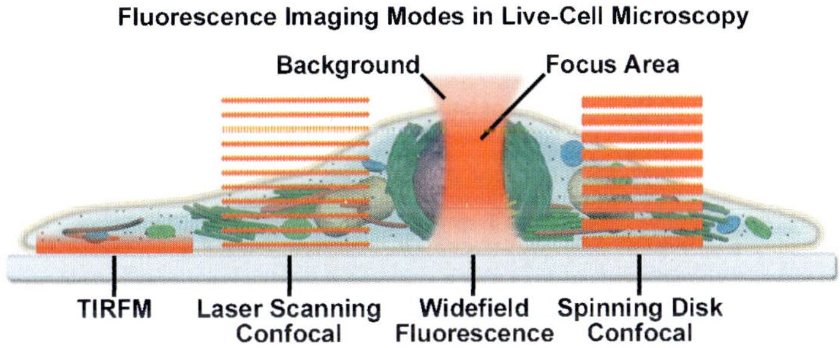

Fig. 1.12 Fluorescence imaging modes in live-cell microscopy (see text for details). *TIRFM* total internal reflection microscopy

widefield fluorescence. The excitation patterns for each technique are indicated in red overlays. In widefield, the specimen is illuminated throughout the field as well as above and below the focal plane. Each point source is spread into a shape resembling a double-inverted cone (the PSF). Only the central portion of this shape resides in the focal plane, with the remainder contributing to out-of-focus blur, which degrades the image.

In contrast the laser scanning, multiphoton, and spinning disc confocal microscopes scan the specimen with a tightly focused laser or arc-discharge lamp (spinning disc). The pattern of excitation is a PSF, but a conjugate pinhole in the optical path of the confocal microscopes prevents fluorescence originating away from the focal plane from impacting the photomultiplier or digital camera detector. The laser scanning confocal microscope has a single pinhole and a single focused laser spot that is scanned across the specimen. In the spinning disc microscope, an array of pinhole or slit apertures, in some cases fitted with microlenses, is placed on a spinning disc such that the apertures rapidly sweep over the specimen and create an image recorded with an area array detector (digital camera). In the multiphoton microscope, the region at which photon flux is high enough to excite fluorophores with more than one photon resides at the in-focus position of the PSF (Piston 1999); thus, fluorophore excitation only occurs in the focal plane. Because all fluorescence emanates from in-focus fluorophores, no pinhole is required and the emitted fluorescence generates a sharp, in-focus image.

One of the primary and favorite techniques used in all forms of optical microscopy for the past three centuries, brightfield illumination, relies upon changes in light absorption, refractive index, or color for generating contrast (Davidson and Abramowitz 2002). As light passes through the specimen, regions that alter the direction, speed, and/or spectrum of the wavefronts generate optical disparities (contrast) when the rays are gathered and focused by the objective. Resolution in a brightfield system depends on both the objective and the condenser NAs, and an immersion medium is often required on both sides of the specimen (for NA combinations exceeding a value of 1.0). Digital cameras provide the wide dynamic range and spatial resolution required to capture the information present in

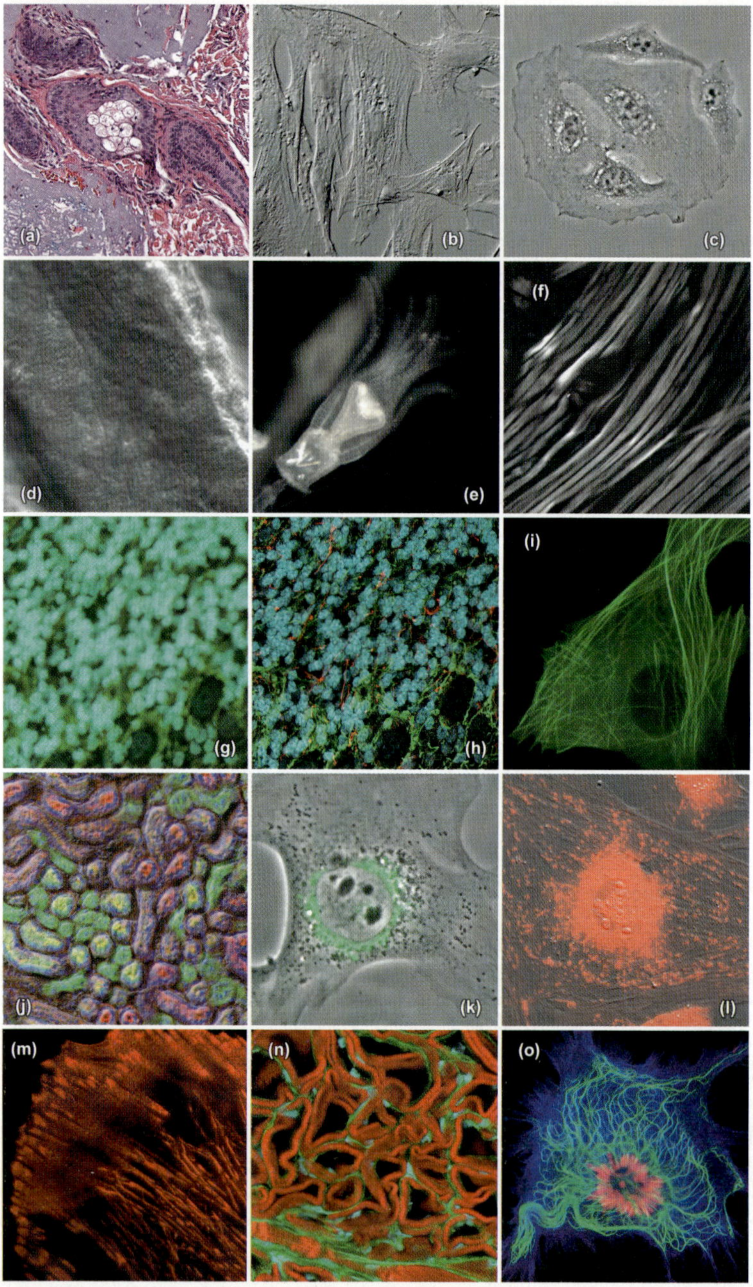

Fig. 1.13 Contrast-enhancing imaging modes in brightfield and fluorescence microscopy. **a** Brightfield; human basal cell carcinoma stained with eosin and hematoxylin. **b** Differential interference contrast (DIC); living Indian Muntjac fibroblast cells. **c** Phase contrast; HeLa cells in plastic culture vessel. **d** Hoffman modulation contrast (HMC); mouse heart tissue in saline.

a brightfield image. In addition, background-subtraction algorithms, using averaged frames taken with no specimen in the optical path, increase contrast dramatically.

Simple brightfield imaging, with the microscope properly adjusted for Köhler illumination, provides a limited degree of information about the cell outline, nuclear position, and the location of larger vesicles in unstained specimens. Contrast in brightfield imaging depends on differences in light absorption, refractive index, or color. Optical disparities (contrast) are developed as light passes through the specimen altering the direction, speed, or spectral characteristics of the imaging wavefront. The technique is more useful with specimens stained with visible light absorbing dyes (such as eosin and hematoxylin; Fig. 1.13a). However, the general lack of contrast in brightfield mode when examining unstained specimens renders this technique relatively useless for serious investigations of living-cell structure.

Methods that enhance contrast include DIC, polarized light, phase contrast, HMC, and darkfield microscopy (examples are illustrated in Fig. 1.13). Several of these techniques are limited by light originating in regions removed from the focal plane when imaging thicker plant and animal tissues, while polarized light requires birefringence (usually not present to a significant degree in animal cells) to generate contrast.

DIC microscopy (Fig. 1.13b) requires plane-polarized light and additional light-shearing (Nomarski) prisms to exaggerate minute differences in specimen thickness gradients and refractive index (Davidson and Abramowitz 2002). Lipid bilayers, for example, produce excellent contrast in DIC because of the difference in refractive index between aqueous and lipid phases of the cell. In addition, cell boundaries in relatively flat adherent mammalian and plant cells, including the plasma membrane, nucleus, vacuoles, mitochondria, and stress fibers, which usually generate significant gradients, are readily imaged with DIC. In plant tissues, the birefringent cell wall reduces contrast in DIC to a limited degree, but a properly aligned system should permit visualization of nuclear and vacuolar membranes, some mitochondria, chloroplasts, and condensed chromosomes in epidermal cells. DIC is an important technique for imaging thick plant and animal tissues because, in addition to the increased contrast, DIC exhibits decreased depth of focus at wide apertures, creating a thin optical section of the thick specimen. This effect is also advantageous for imaging adherent cells to minimize blur arising from floating debris in the culture medium.

Polarized light microscopy (Fig. 13f) is conducted by viewing the specimen between crossed polarizing elements (Davidson and Abramowitz 2002; Murphy 2001). Assemblies within the cell having birefringent properties, such as the plant

Fig. 1.13 (continued) **e** Darkfield; Obelia hydroid in culture. **f** Polarized light; rabbit skeletal muscle. **g** Widefield fluorescence; rat brain hippocampus. **h** Laser scanning confocal; same area of rat brain as for **g**. **i** Spinning disc confocal; microtubules in living cell. **j** DIC–fluorescence; mouse kidney tissue with immunofluorescence. **k** Phase contrast–fluorescence; Golgi apparatus in epithelial cell. **l** HMC–fluorescence; mitochondria in fibroblast cell. **m** TIRFM; α-actinin cytoskeletal network near the cover slip. **n** Multiphoton; rabbit skeletal muscle with immunofluorescence. **o** Widefield–deconvolution; mitosis in epithelial cell with immunofluorescence cell wall, starch granules, and the mitotic spindle, as well as muscle tissue, rotate the plane of light polarization, appearing bright on a dark background. The rabbit muscle tissue illustrated in Fig. 13f is an example of polarized light microscopy applied to living-tissue observation. Note that this technique is limited by the rare occurrence of birefringence in living cells and tissues, and has yet to be fully explored. As mentioned above, DIC operates by placing a matched pair of opposing Nomarski prisms between crossed polarizers, so any microscope equipped for DIC observation can also be employed to examine specimens in plane-polarized light simply by removing the prisms from the optical pathway.

The widely popular phase-contrast technique (as illustrated in Fig. 1.13c) employs an optical mechanism to translate minute variations in phase into corresponding changes in amplitude (Murphy 2001), which can be visualized as differences in image contrast. The microscope must be equipped with a specialized condenser containing a series of annuli matched to a set of objectives containing phase rings in the rear focal plane (phase-contrast objectives can also be used with fluorescence, but with a slight reduction in transmission). Phase contrast is an excellent method to increase contrast when viewing or imaging living cells in culture, but typically results in excessive halos surrounding the outlines of edge features. These halos are optical artifacts that often reduce the visibility of boundary details. The technique is not useful for thick specimens (such as plant and animal tissue sections) because shifts in phase occur in regions removed from the focal plane that distort image detail. Furthermore, floating debris and other out-of-focus phase objects interfere with imaging adherent cells on cover slips.

Often metaphorically referred to as "poor man's DIC," HMC is an oblique illumination technique that enhances contrast in living cells and tissues by detection of optical phase gradients (Fig. 13d). The basic microscope configuration includes an optical amplitude spatial filter, termed a modulator, which is inserted into the rear focal plane of the objective (Davidson and Abramowitz 2002; Murphy 2001). The intensity of light passing through the modulator varies above and below an average value, which by definition, is then said to be modulated. Coupled to the objective modulator is an off-axis slit aperture that is placed in the condenser front focal plane to direct oblique illumination towards the specimen. Unlike the phase plate in phase-contrast microscopy, the Hoffman modulator is designed not to alter the phase of light passing through; rather it influences the principal zeroth-order maxima to produce contrast. HMC is not hampered by the use of birefringent materials (such as plastic Petri dishes) in the optical pathway, so the technique is more useful for examining specimens in containers constructed with polymeric materials. On the downside, HMC produces a number of optical artifacts that render the technique somewhat less useful than phase contrast or DIC for live-cell imaging on glass cover slips.

The method surrounding darkfield microscopy, although widely used for imaging transparent specimens throughout the nineteenth and twentieth centuries, is limited in use to physically isolated cells and organisms (as presented in Fig. 1.13e). In this technique, the condenser directs a cone of light onto the specimen at high azimuths so first-order wavefronts do not directly enter the objective front lens element. Light passing through the specimen is diffracted, reflected, and/or refracted by optical discontinuities (such as the cell membrane, nucleus, and internal organelles), enabling these faint rays to enter the objective (Davidson and Abramowitz 2002). The specimen can then be visualized as a bright object on an otherwise black background. Unfortunately, light scattered by objects removed from the focal plane also contribute to the image, thus reducing contrast and obscuring specimen detail. This artifact is compounded by the fact that dust and debris in the imaging chamber also contribute significantly to the resulting image. Furthermore, thin adherent cells often suffer from very faint signal, whereas thick plant and animal tissues redirect too much light into the objective path, reducing the effectiveness of the technique.

Widefield and point or slit scanning fluorescence imaging modes use divergent strategies to excite samples and detect the fluorescence signals as reviewed in Fig. 1.12 and Pawley (2006d). Figure 1.12 illustrates the different excitation patterns used in TIRFM, LSCM, LSMM and widefield fluorescence microscopy. In widefield fluorescence microscopy the sample is illuminated throughout the entire field, including the regions above and below the focal plane. The PSF in widefield fluorescence microscopy resembles a double-inverted cone with its central portion in the focal plane. Light originating in areas adjacent to the focal plane contributes to blurring and image degradation (Fig. 13g). While deconvolution can be used to reduce blur (Fig. 1.13o), computational methods work better on fixed specimens than on live cell cultures owing to the requirement for larger signal (longer exposure) and a homogeneous sample medium.

The advent of confocal (Fig. 1.13h), spinning disc (Fig. 1.13i), and multiphoton (Fig. 1.13n) microscopy enabled thin and precise optical sectioning to greater depths within living samples. These imaging modes use a precisely focused laser or arc lamp (in the case of spinning disk microscope) to scan the specimen in a raster pattern, and are often combined with conventional transmitted brightfield techniques, such as DIC, phase contrast, and HMC (Fig. 1.13j–l). LSCM uses a single pinhole to produce an illumination spot that is scanned across the specimen. The use of conjugate pinholes in LSCM prevents out-of-focus light from reaching the detector. Spinning disc microscopy uses an array of pinhole or slit apertures and is able to scan rapidly across a specimen, though it produces thicker optical sections than the single, stationary pinhole used in LSCM. Spinning disc modes are less effective at excluding out-of-focus information than LSCM but scan more rapidly without compromising photon throughput. Both confocal and spinning disc modes reduce blur and improve axial resolution. Confocal fluorescence microscopy is frequently limited by the low number of photons collected in the brightest pixels in the image. Multiphoton microscopy uses two or more lower-energy photons (infrared) to excite a femtoliter sample volume, exciting only the fluorophores at the in-focus position of the PSF. Multiphoton imaging therefore does not require pinholes to exclude out-of-focus light and collects a greater portion of the emitted fluorescence (Piston 1999).

An emerging technique known as total internal reflection fluorescence microscopy (TIRFM; discussed above and see Fig. 1.13m) employs a laser source that enters the cover slip at a shallow angle and reflects off the surface of the glass without entering the specimen (Axelrod 2003). Differences in refractive index (n_1/n_2) between the glass and the interior of a cell determine how light is refracted or reflected at the interface as a function of the incident angle. At the critical angle,

$$\theta_{\text{critical}} = \sin^{-1}\left(n_1/n_2\right),$$

a majority of the incident light is completely (totally) reflected from the glass–medium interface. The reflection within the cover slip leads to an evanescent surface wave (electromagnetic field) that has a frequency equal to the incident energy and is able to excite flurophores within 50–100 nm of the surface of the cover slip. TIRFM works well for single-molecule determinations and adherent mammalian cells because of the extreme limitation on the depth of excitation. Thick specimens are not well imaged because of the limited band of excitation. TIRFM has wide application in imaging surface and interface fluorescence. For example, TIRFM can be used to visualize cell–substrate interface regions, track granules during the secretory process in a living cell, determine micromorphological structures and the dynamics of live cells, produce fluorescence movies of cells developing in culture, compare ionic transients near membranes, and measure kinetic binding rates of proteins and surface receptors (Toomre and Manstein 2001).

The properties of fluorescent molecules allow quantification and characterization of biological activity within living cells and tissues. The capture (absorption) and release (emission) of a photon by a fluorophore is a probabilistic event (Lackowicz 1999). The probability of absorption (extinction coefficient) occurs within a narrow bandwidth of excitation energy and emission is limited to even longer wavelengths. The difference in excitation and emission wavelength is known as Stokes shift. Fluorescent molecules exhibit a phenomenon called photobleaching in which the ability of the molecule to fluoresce is permanently lost as a result of photon-induced chemical changes and alteration of covalent bonds. Some fluorophores bleach easily and others can continue to fluoresce for thousands or millions of cycles before they become bleached. Though the interval between absorption and emission is random, fluorescence is an exponential decay process and fluorophores have characteristic half-lives. Fluorescence is a dipolar event. When a fluorophore is excited with plane-polarized light, emission is polarized to a degree determined by the rotation of the molecule during the interval between absorption and emission. The properties of fluorophores depend on their local environment and small changes in ion concentration, the presence of electron acceptors and donors, as well as solvent viscosity, which can affect both the intensity and the longevity of fluorescent probes.

Ratio imaging takes advantage of the sensitivity of fluorophores in order to quantitatively determine molecular changes within the cell environment. Ratio dyes are often used to indicate calcium ion (Ca^{2+}) concentration, pH, and other changes in the cellular environment. These dyes change their absorption and fluorescence characteristics in response to changes in the specimen environment. The fluorescence properties of Fura 2, for example, change in response to the concentration of free calcium, while the SNARF 1 dye fluoresces differently depending on pH (S.L. Shaw 2006). Both excitation and emission dyes are available and can be used to determine differences in fluorescence excitation and emission. Ratio imaging can distinguish between intensity differences due to probe properties and those resulting from probe distribution. The ratio dye can be excited at two different wavelengths, one of which must be sensitive to the environment change being measured. As calcium binds to the dye molecule the primary excitation peak can shift by more than 30 nm, making the dye intensity appear to decrease with increasing Ca^{2+} concentration. If the fluorescent probe is then excited at the shifted wavelength, the intensity appears to increase with increasing Ca^{2+} concentration. Intensity changes are normalized to the amount of dye in a particular position in the cell by dividing one image by the other. The change in intensity can then be attributed to the dye property rather than its distribution or the ratio can be calibrated to determine intracellular Ca^{2+} concentration (Haugland 2005).

Ratio imaging can be performed using widefield, confocal, or multiphoton microscopy. Labeling cells for a ratio method is usually accomplished either by microinjection of ratio dyes or by acetoxymethyl ester loading (a technique using membrane-permeable dyes), a less invasive technique. Living cells are often damaged by microinjection or sequester dye in unwanted locations within the cell. In acetoxymethyl ester loading, a membrane-permeable (nonpolar) ester, Ca^{2+}-insensitive version of the dye enters the cell, where it is hydrolyzed by intracellular esterases. The resulting polyanionic molecule is polar and thus sensitive to calcium ions.

In photouncaging, fluorescent molecules are designed to be inactive until exposed to high-energy wavelengths (approximately 350 nm), at which time bonds joining the caging group with the fluorescent portion of the molecule are cleaved and produce an active fluorescent molecule. Similarly, the use of genetically encoded, photoactivated probes provides substantially increased fluorescence at particular wavelengths. For example, the caged fluorescein is excited at 488 nm and emits at 517 nm. Photouncaging and photoactivation can be used with time-lapse microscopy to study the dynamics of molecular populations within live cells (Lippincott-Schwartz et al. 2003). Recently introduced optical highlighter (Chudakov et al. 2005) fluorescent proteins offer new avenues to research in photoconvertible fluorescent probes.

Fluorescence resonance energy transfer (FRET) is an interaction between the excited states of a donor and acceptor dye molecule that depends on their close proximity (approximately 30–60 Å). When donor and acceptor are within 100 Å of each other, and the emission spectrum of the donor overlaps the absorption spectrum of the acceptor, provided the dipole orientations of the two molecules are parallel, energy is transferred from the donor to the acceptor without the emission and reabsorption of a photon (Periasamy and Day 2005; see also Chap. 6 by Hoppe). While the donor molecule still absorbs the excitation energy, it transfers this energy without fluorescence to the acceptor dye, which then fluoresces. The efficiency of FRET is determined by the inverse sixth power of the intermolecular separation and is often defined in terms of the Förster radius. The Förster radius (R_0) is the distance at which 50% of the excited donor molecules are deactivated owing to FRET and is given by the equation

$$R_0 = [8.8 \times 10^{23} \kappa^2 n^{-4} QY_D J(\lambda)]^{1/6} \text{ Å},$$

where κ^2 is the dipole orientation factor, QY_D is the quantum yield of the donor in the absence of the acceptor molecule, n is the refractive index of the medium, and $J(\lambda)$ is the spectral overlap integral of the two dyes. Different donor and acceptor molecules have different Förster radii and R_0 for a given dye depends on its spectral properties (Periasamy and Day 2005). FRET can also be measured simply as a ratio of donor to acceptor molecules (F_D/F_A) (Periasamy and Day 2005).

FRET is an important technique for imaging biological phenomena that can be characterized by changes in molecular proximity. For example, FRET can be used to assess when and where proteins interact within a cell or can document large conformational changes in single proteins. Additionally, FRET biosensors based on fluorescent proteins are emerging as powerful indicators of intracellular dynamics (Chudakov et al. 2005; Zhang et al. 2002). Typical intermolecular distances between donor and acceptor are within the range of dimensions found in biological macromolecules. Other mechanisms to measure FRET include acceptor photobleaching, lifetime imaging, and spectral resolution. FRET can be combined with ratio imaging methods but requires rigorous controls for measurement (Chudakov et al. 2005; S.L. Shaw 2006).

Fluorescence recovery after photobleaching (FRAP) is a commonly used method for measuring dynamics in proteins within a defined region of a cell (Lippincott-Schwartz et al. 2003). When exposed to intense blue light, fluorescent probes photobleach or lose their ability to fluoresce. While this normally results in image degradation, the photobleaching phenomenon can be used to determine diffusion rates or perform kinetic analyses. Fluorophores are attached to the molecule of interest (protein, lipid, carbohydrate, etc.) and a defined area of the specimen is deliberately photobleached. Images captured at intervals following the bleaching process show recovery as unbleached molecules diffuse into the bleached area. In a similar process known as fluorescence loss in photobleaching (FLIP), intracellular connectivity is investigated by bleaching fluorophores in a small region of the cell while simultaneous intensity measurements are made in related regions. FLIP can be used to evaluate the continuity of membrane-enclosed structures such as the endoplasmic reticulum or Golgi apparatus as well as to define the diffusion properties of molecules within these cellular components (Dailey et al. 2006; Lippincott-Schwartz et al. 2003; S.L. Shaw 2006).

Fluorescence lifetime imaging (FLIM) measures the kinetics of exponential fluorescence decay in a dye molecule (Bastiaens and Squire 1999). The duration of the excited state in fluorophores ranges between 1 and 20 ns and each dye has a characteristic lifetime. The intensity value in each pixel is determined by time and thus contrast is generated by imaging multiple fluorophores with differing decay rates. FLIM is often used during FRET analysis since the donor fluorophore lifetime is shortened by FRET. The fact that fluorescence lifetime is independent of fluorophore concentration and excitation wavelength makes it useful for enhancing measurement during FRET experiments. Because FLIM measures the duration of fluorescence rather than its intensity, the effect of photon scattering in thick specimens is reduced, as is the need to precisely know concentrations. For this reason FLIM is often used in biomedical tissue imaging to examine greater specimen depths.

Emission spectra often overlap in specimens having multiple fluorescent labels or exhibiting significant autofluorescence, making it difficult to assign fluorescence to a discrete and unambiguous origin. In multispectral imaging, overlapping of the channels is referred to as *bleedthrough* and can be easily misinterpreted as colocalization (Zimmermann 2005). Fluorescent proteins such as cyan fluorescent protein (CFP), green fluorescent protein (GFP), yellow fluorescent protein (YFP), and *Discosoma* sp. red fluorescent protein (DsRed) have transfection properties that make them useful in many multichannel experiments, but they also have broad excitation and emission spectra and bleedthrough is a frequent complication. Bleedthrough can be minimized by a computational process similar to deconvolution. Known as *linear unmixing* or *spectral reassignment*, this process analyzes the spectra of each fluorescent molecule as a PSF on a pixel-by-pixel basis in order to separate the dye signals and reassign them to their correct location in the image array. These image-processing algorithms are able to separate multiple overlapping spectra but like deconvolution, accurate separation necessitates collecting more photons at each pixel.

With use of a technique known as fluorescence correlation spectroscopy (FCS), the variations in fluorophore intensity can be measured with an appropriate spectroscopic detector in stationary femtoliter-volume samples (Kim and Schwille 2003; see also Chap. 7 by Wachsmuth and Weisshart). Fluctuations represent changes in the quantum yield of fluorescent molecules and can be statistically analyzed to determine equilibrium concentrations, diffusion rates, and functional interaction of fluorescently labeled molecules. The FCS technique is capable of quantifying such interactions and processes at the single-molecule level with light levels that are orders of magnitude lower than for FRAP.

Fluorescence speckle microscopy (FSM) is a technique used with widefield or confocal microscopy that employs a low concentration of fluorophores to reduce out-of-focus fluorescence and enhance contrast and resolution of structures and processes in thick portions of a live specimen (Danuser and Waterman-Storer 2006). Unlike FCS, where the primary focus is on quantitative temporal features, FSM labels a small part of the structure of interest and is concerned with determining spatial patterns. FSM is often used in imaging cytoskeletal structures such as actin and microtubules in cell-motility determinations (see also Chap. 9 by Jaqaman et al.).

1.15 Summary

Many of the techniques and imaging modes described in this chapter can be used in combination to enhance visibility of structures and processes and to provide greater information about the dynamics of living cells and tissues. DIC microscopy, for example, is frequently used with LSCM to observe the entire cell while fluorescence information relating to uptake and distribution of fluorescent probes is imaged with the single confocal beam.

Live-cell imaging requires consideration of a number of factors that depend not only on the technique or imaging mode used but also rely on appropriate labeling in order to visualize the structure or process of interest. Specimens must be prepared and handled in ways that maintain conditions supportive of normal cell or tissue health. Spatial and temporal resolution must be achieved without damaging the cell or organism being imaged, or compromising the image data obtained. Most organisms, and thus living cell cultures and biological processes, are sensitive to changes in temperature and pH. Heated stages, objective lens heaters, and other mechanisms for controlling temperature are usually required for imaging live cells. Metabolism of the specimen itself may induce significant changes in the pH of the medium over time. Some type of pH monitoring, buffered media, and or perfusion chamber is used to keep the pH within an acceptable range. Most living organisms require the presence of sufficient oxygen and removal of respired carbon dioxide, which can be problematic in closed chambers. Humidity is often controlled to prevent evaporation and subsequent increases in salinity and pH. Perfusion chambers, humidifiers, and other atmospheric controls must be used to keep living cells viable.

Signal strength is usually critical for fluorescence imaging methods as probes are sometimes weakly fluorescent or at such low concentrations that the images produced have low signal-to-noise ratios. Possible solutions include increasing integration time or the size of the confocal pinhole, although increasing the signal may result in photobleaching or phototoxicity. Alternatively, noise can be reduced wherever possible and line or frame averaging used to increase the signal-to-noise ratio.

Bleedthrough and cross talk are often an issue in specimens labeled with multiple fluorescent proteins. Improvement can be made by imaging different channels sequentially rather than simultaneously. Spectral imaging techniques or linear unmixing algorithms, interference filters, and dichroics can be used to separate overlapping fluorophore spectra. Unintentional photobleaching is a risk attendant with frequent or repeated illumination and some fluorescent probes bleach more easily and quickly than others. Photobleaching can be minimized by reducing incident light, using fade-resistant dye, reducing integration time, reducing the frequency of image capture, using a beam shuttering mechanism, and scanning only when collecting image data.

Many experimental determinations require high spatial resolution in all three dimensions. Spatial resolution can be enhanced by using high-NA objectives, reducing the size of the confocal pinhole aperture, increasing sampling frequency according to the Nyquist criterion, decreasing the step size used to form the z series, using water immersion objectives to reduce spherical aberrations, and by using deconvolution algorithms to reduce blurring. Biological processes are often rapid compared with the rate of image acquisition, especially in some scanning confocal systems. Temporal resolution can be improved by reducing the field of view and pixel integration time or increasing the scan speed as well as reducing the sampling frequency. Live specimens or features within living cells may move in or out of the focal plane during imaging, requiring either manual or autofocus adjustments or collection of z stacks followed by image reconstruction. The emergence of diffraction-breaking optical techniques (Hell 2003) opens the door to even higher resolutions in all forms of fluorescence microscopy and live-cell imaging. Among the most important advances are stimulated emission depletion (STED) (Hell and Wichmann 1994), spot-scanning 4Pi confocal (Hell and Stelzer 1992), widefield I^5M (Gustafsson et al. 1995), photoactivated localization microscopy (PALM) (Betzig et al. 2006), and stochastic optical reconstruction microscopy (STORM) (Rust et al. 2006). All of these techniques rely on the properties of fluorescent molecules and promise to deliver spatial resolutions that vastly exceed that of conventional optical microscopes.

The quality of any final image, analog or digital, depends fundamentally on the properties and precise configuration of the optical components of the imaging system. Correct sampling of the digital data is also critical to the fidelity of the final image. For this reason it is important to understand the relationships between spatial resolution and contrast as well as their theoretical and practical limitations. Recognition of the inherent uncertainties involved in manipulating and counting photoelectrons is important to quantitative imaging, especially as applied to photon-limited applications. In conclusion, with an understanding and appreciation of the potentials and limitations of digital imaging and the special considerations related to living cells, the microscopist can produce high-quality, quantitative, color images in multiple dimensions that enhance investigations in optical microscopy.

1.16 Internet Resources

The Web sites listed below are continuously updated and provide a wealth of information on all phases of optical microscopy and digital imaging:

- Molecular Expressions: Images from the Microscope (http://microscopy.fsu.edu)
- Nikon MicroscopyU (http://www.microscopyu.com)
- Olympus Microscopy Resource Center (http://www.olympusmicro.com)
- Olympus FluoView Resource Center (http://www.olympusconfocal.com)

References

Axelrod D (2003) Total internal reflection fluorescence microscopy in cell biology. Methods Enzymol 361:1–33

Bastiaens PIH, Squire A (1999) Fluorescence lifetime imaging microscopy: spatial resolution of biochemical processes in a cell. Trends Cell Biol 9:48–52

Betzig E, Patterson GH, Sougrat R, Lindwasser OW, Olenych S, Bonifacino JS, Davidson MW, Lippincott-Schwartz J, Hess HF (2006) Imaging intracellular fluorescent proteins at nanometer resolution. Science 313:1642–1645

Berland K, Jacobson K, French T (1998) Electronic cameras for low-light microscopy. Methods Cell Biol 56:19–44

Bradbury S (1967) The evolution of the microscope. Pergamon, New York

Cannell MB, McMorlad A, Soeller C (2006) Image enhancement by deconvolution. In: Pawley JB (ed) Handbook of biological confocal microscopy, 3rd edn. Springer, New York, pp 488–500

Castleman KR (1993) Resolution and sampling requirements for digital image processing, analysis, and display. In: Shotton D (ed) Electronic light microscopy: techniques in modern biomedical microscopy. Wiley-Liss, New York, pp 71–93

Chudakov DM, Lukyanov S, Lukyanov KA (2005) Fluorescent proteins as a toolkit for *in vivo* imaging. Trends Biotechnol 23:605–613

Coates C, Denvir D, Conroy E, McHale N, Thornbury K, Hollywood M (2003) Back illuminated electron multiplying technology: the world's most sensitive CCD for ultra low light microscopy. J Biomed Opt 9:1244–2004

Dailey ME, Manders E, Soll DR, Terasaki M (2006) Confocal microscopy of living cells. In: Pawley JB (ed) Handbook of biological confocal microscopy, 3rd edn. Springer, New York, pp 381–403

Danuser G, Waterman-Storer CM (2006) Quantitative fluorescent speckle microscopy of cytoskeletal dynamics. Annu Rev Biophys Biomol Struct 35:361–387

Davidson MW, Abramowitz M (2002) Optical microscopy. In: Hornak JP (ed) Encyclopedia of imaging science and technology. Wiley, New York, pp 1106–1140

Day RN (2005) Imaging protein behavior inside the living cell. Mol Cell Endocrinol 230:1–6

Delly JG, Olenych S, Claxton N, Davidson MW (2007) Digital photomicrography. In: The focal encyclopedia of photography. Focal, New York, pp 592–601

de Monvel JB, Scarfone E, Le Calvez S, Ulfendahl M (2003) Image adaptive deconvolution for three dimensional deep biological imaging. Biophys J 85:3991–4001

Denvir DJ, Contry E (2002) Electron multiplying CCDs. Proc SPIE 4877:55–68

Gastou P, Comandon J (1909) L'ultramicroscope et son role essential dans le diagnostic de la syphilis. J Med Fr 4

Goldman RD, Spector DL (2005) Live cell imaging: a laboratory manual. Cold Spring Harbor Laboratory Press, Cold Spring Harbor

Gustafsson MGL, Agard DA, Sedat JW (1995) Sevenfold improvement of axial resolution in 3D widefield microscopy using two objective lenses. Proc Soc Photo-Opt Instrum Eng 2412:147–156

Haugland RP (2005) A guide to fluorescent probes and labeling technologies. Invitrogen/Molecular Probes, Eugene

Hell SW (2003) Toward fluorescence nanoscopy. Nat Biotechnol 21:1347–1355

Hell SW, Stelzer EHK (1992) Properties of a 4Pi-confocal fluorescence microscope. J Opt Soc Am A 9:2159–2166

Hell SW Wichmann J (1994) Breaking the diffraction resolution limit by stimulated emission: stimulated emission depletion microscopy. Opt Lett 19:780–782

Holmes TJ, Biggs D, Abu-Tarif A (2006) Blind deconvolution. In: Pawley JB (ed) Handbook of biological confocal microscopy, 3rd edn. Springer, New York, pp 468–487

Holst GC (1998) CCD arrays, cameras, and displays. SPIE, Bellingham

Inoue S, Spring KG (1997) Video microscopy: the fundamentals. Plenum, New York

Janesick JR (2001) Scientific charge-coupled devices. SPIE, Bellingham

Jansson PA (1997) Deconvolution of images and spectra, 2nd edn. Academic, New York

Jardine L (2004) The curious life of Robert Hooke. HarperCollins, New York

Jonkman JEN, Stelzer EHK (2002) Resolution and contrast in confocal and two-photon microscopy. In: Diaspro A (ed) Confocal and two-photon microscopy: foundations, applications, and advances. Wiley-Liss, New York, pp 101–125

Kim SA, Schwille P (2003) Intracellular applications of fluorescence correlation spectroscopy: prospects for neuroscience. Curr Opin Neurobiol 13:583–590

Lackowicz JR (1999) Principles of fluorescence spectroscopy, 2nd edn. Kluwer/Plenum, New York

Lippincott-Schwartz J, Altan-Bonnet N, Patterson GH (2003) Photobleaching and photoactivation: following protein dynamics in living cells. Nat Cell Biol S7–S14

Murphy DB (2001) Fundamentals of light microscopy and digital imaging. Wiley-Liss, New York

Pawley J (2003) The intensity spread function (ISF): a new metric of photodetector performance. http://www.focusonmicroscopy.org/2003/abstracts/107-Pawley.pdf
Pawley J (2006a) Points, pixels, and gray Levels: digitizing image data. In: Pawley JB (ed) Handbook of biological confocal microscopy, 3rd edn. Springer, New York, pp 59–79
Pawley J (2006b) Fundamental limits in confocal microscopy. In: Pawley JB (ed) Handbook of biological confocal microscopy, 3rd edn. Springer, New York, pp 20–42
Pawley J (2006c) More than you ever really wanted to know about CCDs. In: Pawley JB (ed) Handbook of biological confocal microscopy, 3rd edn. Springer, New York, pp 919–932
Pawley JB (2006d) Handbook of biological confocal microscopy, 3rd edn. Springer, New York
Periasamy A, Day RN (2005) Molecular imaging: FRET microscopy and spectroscopy. Oxford University Press, New York
Piston DW (1999) Imaging living cells and tissues by two-photon excitation microscopy. Trends Cell Biol 9:66–69
Robbins M, Hadwen B (2003) The noise performance of electron multiplying charge coupled devices. IEEE Trans Electron Devices 50:1227–1232
Roux P, Münter S, Frischknecht F, Herbomel P, Shorte SL (2004) Focusing light on infection in four dimensions. Cell Microbiol 6:333–343
Ruestow EG (1996) The microscope in the Dutch Republic. Cambridge University Press, New York
Russ JC (2006) The image processing handbook, 5th edn. CRC, Boca Raton
Rust MJ, Bates M, Zhuang X (2006) Sub-diffraction-limit imaging by stochastic optical reconstruction microscopy (STORM). Nat Methods 3:793–795
Shaw PJ (2006) Comparison of widefield/deconvolution and confocal microscopy for three-dimensional imaging. In: Pawley JB (ed) Handbook of biological confocal microscopy, 3rd edn. Springer, New York, pp 453–467
Shaw SL (2006) Imaging the live plant cell. Plant J 45:573–598
Shotton D (1993) An introduction to digital image processing and image display in electronic light microscopy. In: Shotton D (ed) Electronic light microscopy: techniques in modern biomedical microscopy. Wiley-Liss, New York, pp 39–70
Spring K (2000) Scientific imaging with digital cameras. BioTechniques 29:70–76
Toomre D, Manstein DJ (2001) Lighting up the cell surface with evanescent wave microscopy. Trends Cell Biol 11:298–303
von Tiedemann M, Fridberger A, Ulfendahl M, de Monvel JB (2006) Image adaptive point-spread function estimation and deconvolution for in vivo confocal microscopy. Microsc Res Tech 69:10–20
Wallace W, Schaefer LH, Swedlow JR (2001) A workingperson's guide to deconvolution in light microscopy. BioTechniques 31:1076–1097
Zhang J, Campbell RE, Ting AY, Tsien RY (2002) Creating new fluorescent probes for cell biology. Nat Rev Mol Cell Biol 3:906–918
Zimmermann T (2005) Spectral imaging and linear unmixing in light microscopy. Adv Biochem Eng Biotechnol 95:245–265

2
Quantitative Biological Image Analysis

Erik Meijering and Gert van Cappellen

Abstract Progress in biology is increasingly relying on images. As image data sets become larger and larger, and potentially contain more and more biologically relevant information, there is a growing need to replace subjective visual inspection and manual measurement by quantitative computerized image processing and analysis. Apart from reducing manual labor, computerized methods offer the possibility to increase the sensitivity, accuracy, objectivity, and reproducibility of data analysis. This chapter discusses the basic principles underlying automated image processing and analysis tools, with the aim of preparing the reader to get started and to avoid potential pitfalls in using these tools. After defining the necessary terminology and putting image processing and analysis into historical and future perspective, it subsequently explains important preprocessing operations, gives an introduction to more advanced processing methods for specific biological image analysis tasks, discusses the main methods for visualization of higher-dimensional image data, and addresses issues related to the use and development of software tools.

2.1 Introduction

Images play an increasingly important role in many fields of science and its countless applications. Biology is without doubt one of the best examples of fields that have come to depend heavily upon images for their progress. As a consequence of the ever-increasing resolving power and efficiency of microscopic image acquisition hardware and the rapidly decreasing cost of mass storage and communication media, biological image data sets are growing exponentially in size and are carrying more and more information. Extracting this information by visual inspection and manual measurement is labor-intensive, and the results are potentially inaccurate and poorly reproducible. Hence, there is a growing need for computerized image processing and analysis, not only to cope with the rising rate at which images are acquired, but also to reach a higher level of sensitivity, accuracy, and objectivity than can be attained by human observers (Murphy et al. 2005). It seems inevitable,

S.L. Shorte and F. Frischknecht (eds.), *Imaging Cellular and Molecular Biological Functions.*

therefore, that biologists will increasingly resort to automated image processing and analysis technology in exploiting their precious data.

In order to benefit from any technology it is of paramount importance to have at least a basic understanding of its underlying principles. This universal rule applies undiminished to computerized image processing and analysis: biologically highly relevant information may easily go unnoticed or get destroyed (or may even be created ex nihilo!) by improper use of such technology. The present chapter, which updates earlier (partial) reviews in the field (Chen et al. 1995; Glasbey and Horgan 1995; Sabri et al. 1997; Eils and Athale 2003; Gerlich et al. 2003), was written with the aim of providing the biologist with the necessary know-how to get started and to avoid potential pitfalls in using image processing and analysis tools. We begin by defining the necessary terminology and putting image processing and analysis into historical and future perspective. In the subsequent two sections we explain important image preprocessing operations and give an introduction to advanced image-processing methods for biological image analysis. Next we discuss the main methods for visualization of higher-dimensional image data. In the last section we address several issues related to the use and development of software tools. Throughout the chapter, ample reference is made to the (mostly recent) literature for those interested in more in-depth information.

2.2 Definitions and Perspectives

Because of the rapid rise of imaging technology in the sciences as well as in everyday life, several terms have become very fashionable even among a large percentage of the general public but whose precise meanings appear to vary. Before we go into details it is necessary to define these terms to avoid confusion. The word “image” itself, for starters, already has at least five different meanings. In the most general sense of the word, an image is a representation of something else. Depending on the type of representation, images can be divided into several classes (Castleman 1996). These include images perceivable by the human eye, such as pictures (photographs, paintings, drawings), or those formed by lenses or holograms (optical images), as well as nonvisible images, such as continuous or discrete mathematical functions or distributions of measurable physical properties. In the remainder of this chapter, when we speak of an image, we mean a *digital* image, defined as a representation obtained by taking finitely many samples expressed as numbers that can take on only finitely many values. In the present context of in vivo biological imaging, the objects we make representations of are living cells and molecules, and the images are usually acquired by taking samples of (fluorescent) light at given intervals in space and time and wavelength.

Mathematically speaking, images are n-dimensional matrices, or discrete functions (with n typically 1–5), where each dimension corresponds to a parameter, a degree of freedom, or a coordinate needed to uniquely locate a sample value (Fig. 2.1).

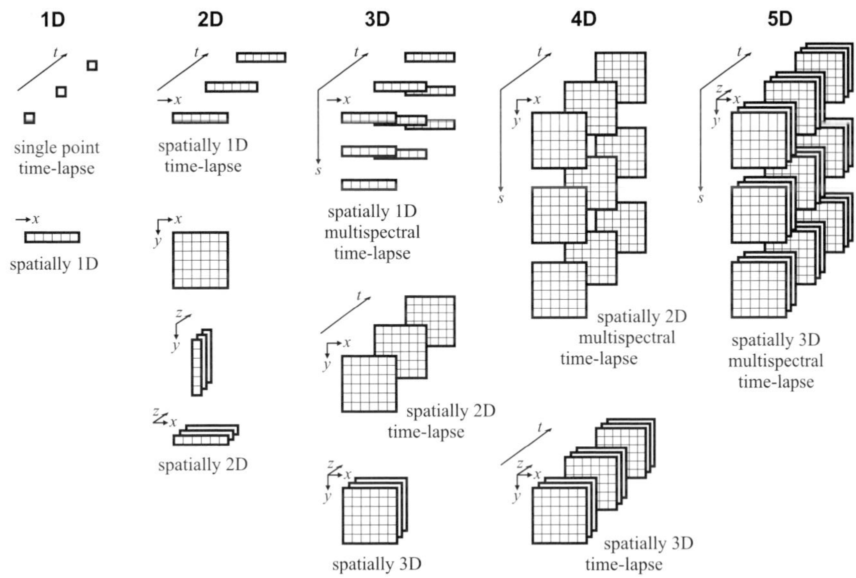

Fig. 2.1 Images viewed as *n*-dimensional matrices. The overview is not meant to be exhaustive but reflects some of the more frequently used modes of image acquisition in biological and medical imaging, where the number of dimensions is typically 1–5, with each dimension corresponding to an independent physical parameter: three (usually denoted x, y, and z) to space, one (usually denoted t) to time, and one to wavelength, or color, or more generally to any spectral parameter (we call this dimension s here). In other words, images are discrete functions, $I(x,y,z,t,s)$, with each set of coordinates yielding the value of a unique sample (indicated by the *small squares*, the number of which is obviously arbitrary here). Note that the dimensionality of an image (indicated in the *top row*) is given by the number of coordinates that are varied during acquisition. To avoid confusion in characterizing an image, it is advisable to add adjectives indicating which dimensions were scanned, rather than mentioning just dimensionality. For example, a 4D image may be either a spatially 2D multispectral time-lapse image or a spatially 3D time-lapse image

How, exactly, these matrices are obtained and how they relate to the physical world is described in Chap. 1 by Hazelwood et al. (see also Pawley 2006). Each sample corresponds to what we call an image element. If the image is spatially two-dimensional (2D), the elements are usually called pixels ("picture elements," even though the image need not necessarily be a picture). In the case of spatially three-dimensional (3D) images, they are called voxels ("volume elements"). However, since data sets in optical microscopy usually consist of series of 2D images (time frames or optical sections) rather than truly volumetric images, we refer to an image element of any dimensionality as a "pixel" in this chapter.

Image processing is defined as the act of subjecting an image to a series of operations that alter its form or its value. The result of these operations is again an image. This is distinct from image analysis, which is defined as the act of measuring (biologically) meaningful object features in an image. Measurement results can be either qualitative (categorical data) or quantitative (numerical data) and both types of results can be either subjective (dependent on the personal feelings and prejudices of the subject doing the measurements) or objective (solely dependent on the object itself and the measurement method). In many fields of research there is a tendency towards quantification and objectification, feeding the need for fully automated image analysis methods. Ultimately, image analysis results should lead to understanding the nature and interrelations of the objects being imaged. This requires not only measurement data, but also reasoning about the data and making inferences, which involves some form of intelligence and cognitive processing. Computerizing these aspects of human vision is the long-term goal of computer vision. Finally we mention computer graphics and visualization. These terms are strongly related (Schroeder et al. 2002), but strictly speaking the former refers to the process of *generating* images for display of given data using a computer, while the latter is more concerned with *transforming* data to enable rendering and exploring it. An illustration of all these terms (Fig. 2.2) may help their meaning to be memorized. In this chapter we focus mainly on image processing and image analysis and also briefly touch upon visualization.

The idea of processing images by computer was conceived in the late 1950s, and over the decades to follow was further developed and applied to such diverse fields as astronomy and space exploration, remote sensing for earth resources research, and diagnostic radiology, to mention but a few. In our present-day life, image processing and analysis technology is employed in surveillance, forensics, military defense, vehicle guidance, document processing, weather prediction, quality inspection in automated manufacturing processes, etc. Given this enormous success, one might think that computers will soon be ready to take over most human vision tasks, also in biological investigation. This is still far from becoming a reality, however. After 50 years of research, our knowledge of the human visual system and how to excel it is still very fragmentary and mostly confined to the early stages, that is to image processing and image analysis. It seems reasonable to predict that another 50 years of multidisciplinary efforts involving vision research, psychology, mathematics, physics, computer science, and artificial intelligence will be required before we can begin to build highly sophisticated computer vision systems that outperform human observers in all respects. In the meantime, however, currently available methods may already be of great help in reducing manual labor and increasing accuracy, objectivity, and reproducibility.

2.3 Image Preprocessing

A number of fundamental image processing operations have been developed over the past decades that appear time and again as part of more involved image processing and analysis procedures. Here we discuss four classes of operations that are most commonly used in image preprocessing: intensity transformation, linear and nonlinear image filtering, geometrical transformation, and image restoration operations. For ease of illustration, examples are given for spatially 2D images, but

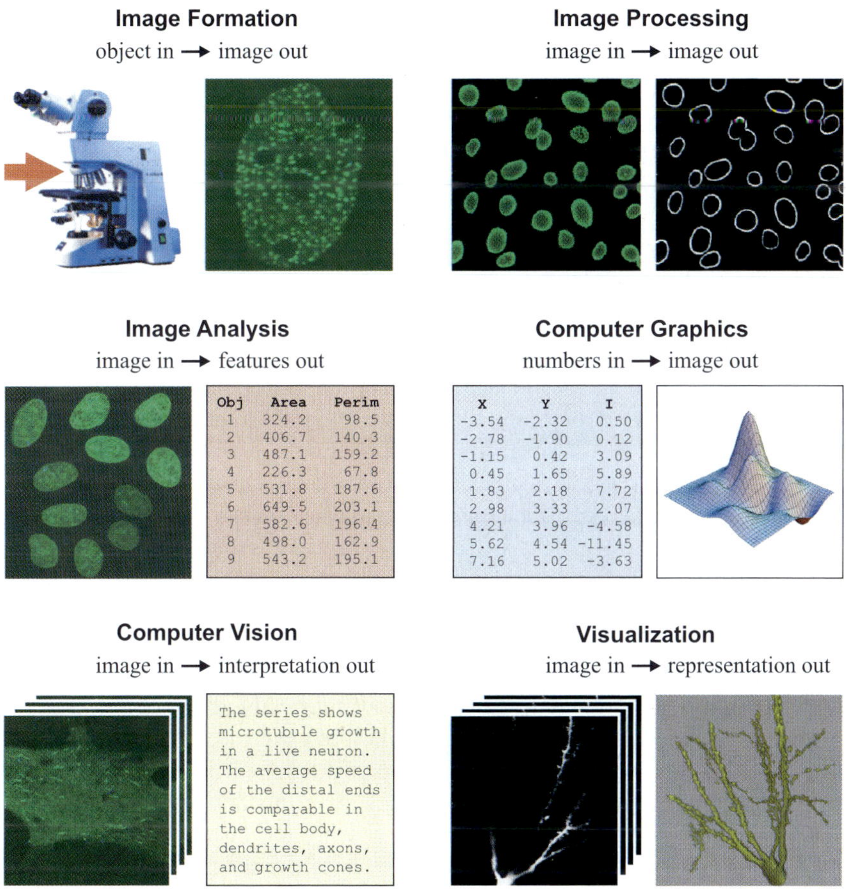

Fig. 2.2 Illustration of the meaning of commonly used terms. The process of digital image formation in microscopy is described in Chap. 1 by Hazelwood et al. Image processing takes an image as input and produces a modified version of it (in the case shown, the object contours are enhanced using an operation known as edge detection, described in more detail in the text). Image analysis concerns the extraction of object features from an image. In some sense, computer graphics is the inverse of image analysis: it produces an image from given primitives, which could be numbers (the case shown), or parameterized shapes, or mathematical functions. Computer vision aims at producing a high-level interpretation of what is contained in an image. This is also known as image understanding. Finally, the aim of visualization is to transform higher-dimensional image data into a more primitive representation to facilitate exploring the data

they easily extend to higher-dimensional images. Also, the examples are confined to intensity (gray-scale) images only. In the case of multispectral images, some operations may need to be applied separately to each channel, possibly with different parameter settings. A more elaborate treatment of the mentioned (and other) basic image processing operations can be found in the cited works as well as in a great variety of textbooks (Jain 1989; Baxes 1994; Castleman 1996; Sonka et al. 1999; Russ 2002; Gonzalez and Woods 2002; Jähne 2004).

2.3.1 Image Intensity Transformation

Among the simplest image processing operations are those that pass along each image pixel and produce an output value that depends only on the corresponding input value and some mapping function. These are also called point operations. If the mapping function is the same for each pixel, we speak of a global intensity transformation. An infinity of mapping functions can be devised, but most often a (piecewise) linear function is used, which allows easy (interactive) adjustment of image brightness and contrast. Two extremes of this operation are intensity inversion and intensity thresholding. The latter is one of the easiest (and most error-prone!) approaches to divide an image into meaningful objects and background, a task referred to as image segmentation. Logarithmic mapping functions are also sometimes used to better match the light sensitivity of the human eye when displaying images. Another type of intensity transformation is pseudocoloring. Since the human eye is more sensitive to changes in color than to changes in intensity, more detail may be perceived when mapping intensities to colors.

Mapping functions usually have one or more parameters that need to be specified. A useful tool for establishing suitable values for these is the intensity histogram, which lists the frequency (number of occurrences) of each intensity value in the image (Fig. 2.3). For example, if the histogram indicates that intensities occur mainly within a limited range of values, the contrast may be improved considerably by mapping this input range to the full output range (this operation is therefore called contrast stretching). Instead of being derived by the user, mapping functions may also be computed automatically from the histogram. This is done, for example, in histogram equalization, where the mapping function is derived from the cumulative histogram of the input image, causing the histogram of the output image to be more uniformly distributed. In cases where the intensity histogram is multimodal, this operation may be more effective in improving image contrast between different types of adjacent tissues than simple contrast stretching. Another example is the automatic determination of a global threshold value as the minimum between the two major modes of the histogram (Glasbey 1993).

2.3.2 Local Image Filtering

Instead of considering just the corresponding input pixel when computing a value for each output pixel (as in intensity transformation), one could also take into account the values of adjacent input pixels. Image processing operations that are based on this principle are called neighborhood operations, or alternatively image filtering operations, as they are usually designed to filter out (enhance or reduce) specific image information. They can be classified into linear and nonlinear. Linear filtering operations compute the output pixel value as a linear combination (weighing and summation) of the values of the corresponding input pixel and its neighbors. This process can be described mathematically as a convolution operation, and the mask (or filter)

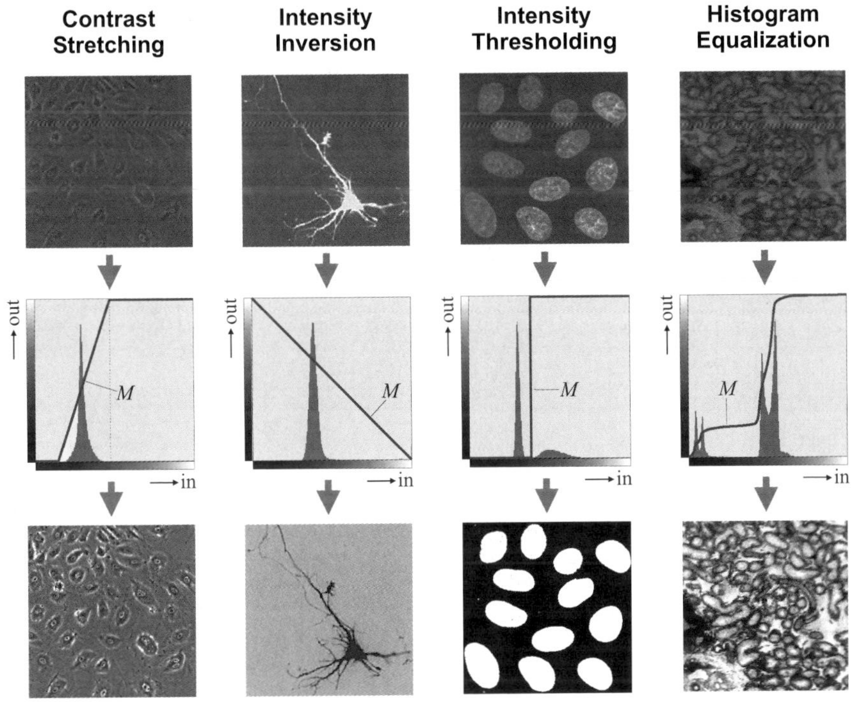

Fig. 2.3 Examples of intensity transformations based on a global mapping function: contrast stretching, intensity inversion, intensity thresholding, and histogram equalization. The *top row* shows the images used as input. The *second row* shows for each image the mapping function used (denoted *M*), with the histogram of the input image shown in the background (*gray area*). The *bottom row* shows for each input image the corresponding output image resulting from applying the mapping function: $O(x,y)=M[I(x,y)]$. It is clear from the mapping functions that contrast stretching and histogram equalization both distribute the most frequently occurring intensities over a wider range of values, thereby increasing image contrast. The former is suitable in the case of unimodal histograms, whereas the latter is better suited for images having multimodal histograms

specifying the weight factor for each neighboring pixel value is accordingly called a convolution kernel. Examples of kernels include averaging filters, sharpening filters, and Gaussian smoothing and derivative filters of varying sizes (Fig. 2.4). The last of these can be used, for example, to detect object edges by a procedure known as edge detection (Canny 1986). Convolution of an image with a kernel is equivalent to multiplication of their respective Fourier transformations, followed by inverse transformation of the result (Bracewell 2000). Certain filtering operations, for example, to remove specific intensity oscillations, are better done in the Fourier domain, as the corresponding convolution kernel would be very large, requiring excessive computation times.

Nonlinear filtering operations are those that combine neighboring input pixel values in a nonlinear fashion in producing an output pixel value. They cannot be described as a convolution process. Examples of these include median filtering (which for each output pixel computes the value as the median of the corresponding

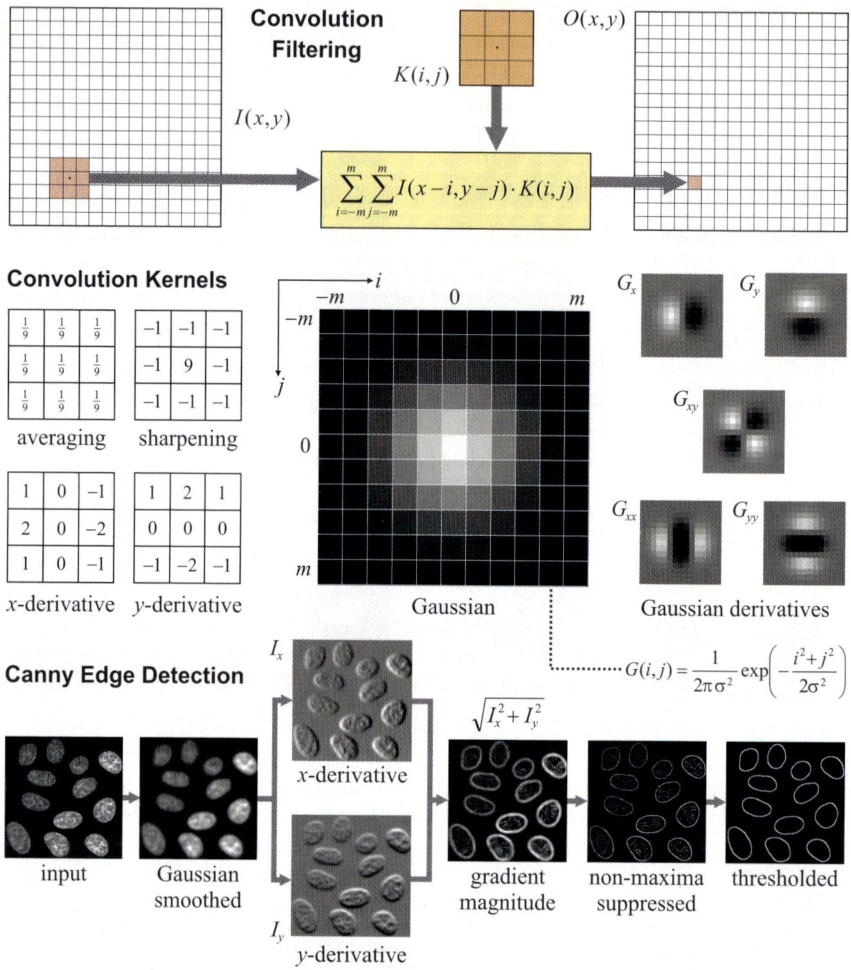

Fig. 2.4 Principles and examples of convolution filtering. The value of an output pixel is computed as a linear combination (weighing and summation) of the value of the corresponding input pixel and of its neighbors. The weight factor assigned to each input pixel is given by the convolution kernel (denoted K). In principle, kernels can be of any size. Examples of commonly used kernels of size 3×3 pixels include the averaging filter, the sharpening filter, and the Sobel x- or y-derivative filters. The Gaussian filter is often used as a smoothing filter. It has a free parameter (standard deviation σ) that determines the size of the kernel (usually cut off at $m=3\sigma$) and therefore the degree of smoothing. The derivatives of this kernel are often used to compute image derivatives at different scales, as for example, in edge detection using the approach of Canny (1986). The scale parameter, σ, should be chosen such that the resulting kernel matches the structures to be filtered

input values in a neighborhood of given size) and min-filtering or max-filtering (where the output value is computed as, respectively, the minimum or the maximum value in a neighborhood around the corresponding input pixel). Another class of nonlinear filtering operations comes from the field of mathematical morphology

(Serra 1982) and deals with the processing of object shape. Of particular interest to image analysis is binary morphology, which applies to two-valued (binary) images and is often applied as a postprocessing step to clean up imperfect segmentations. Morphological filtering is described in terms of the interaction of an image and a structuring element (a small mask reminiscent of a convolution kernel in the case of linear filtering). Basic morphological operations include erosion, dilation, opening, and closing (Fig. 2.5). By combining these, we can design many interesting filters to prepare for (or even perform) image analysis. For example, subtracting the results of dilation and erosion yields object edges. Or by analyzing the results of a family of openings, using increasingly larger structuring elements, we may perform granulometry of objects. Another operation that is frequently used in biological shape analysis (He et al. 2003; Houle et al. 2003; Evers et al. 2005) is skeletonization, which yields the basic shape of segmented objects.

2.3.3 Geometrical Image Transformation

In many situations it may occur that the images acquired by the microscope are spatially distorted or lack spatial correspondence. In colocalization experiments, for example, images of the same specimen imaged at different wavelengths may show mismatches due to chromatic aberration. Nonlinear magnification from the center to the edge of the field of view may result in deformations known as barrel distortion or pincushion distortion. In live-cell experiments, one may be interested in studying specific intracellular components over time, which appear in different places in each image owing to the motion of the cell itself. Such studies require image alignment, also referred to as image registration in the literature (Maintz and Viergever 1998; Pluim et al. 2003; Sorzano et al. 2005). Other studies, for example, karyotype analyses, require the contents of images to be reformatted to some predefined configuration. This is also known as image reformatting.

In all such cases, the images (or parts thereof) need to undergo spatial or geometrical transformation prior to further processing or analysis. There are two aspects to this type of operation: coordinate transformation and image resampling. The former concerns the mapping of input pixel positions to output pixel positions (and vice versa). Depending on the complexity of the problem, one commonly uses a rigid transformation, an affine transformation, or a curved transformation (Fig. 2.6). Image resampling concerns the issue of computing output pixel values based on the input pixel values and the coordinate transformation. This is also known as image interpolation, for which many methods exist. It is important to realize that every time an image is resampled, some information is lost. Studies in medical imaging (Thévenaz et al. 2000; Meijering et al. 2001) have indicated that higher-order spline interpolation methods (for example, cubic splines) are much less harmful in this regard than some standard approaches, such as nearest-neighbor interpolation and linear interpolation, although the increased computational load may be prohibitive in some applications.

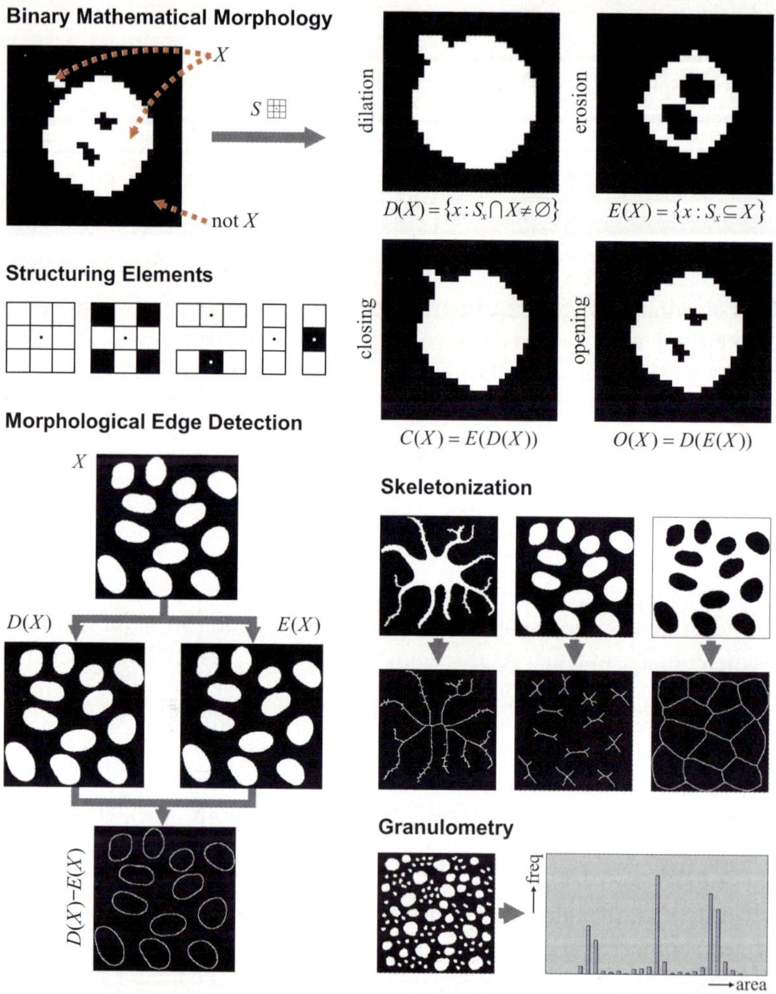

Fig. 2.5 Principles and examples of binary morphological filtering. An object in the image is described as the set (denoted X) of all coordinates of pixels belonging to that object. Morphological filters process this set using a second set, known as the structuring element (denoted S). Here the discussion is limited to structuring elements that are symmetrical with respect to their center element, $s = (0,0)$, indicated by the *dot*. In that case, the dilation of X is defined as the set of all coordinates x for which the cross section of S placed at x (denoted S_x) with X is not empty, and the erosion of X as the set of all x for which S_x is a subset of X. A dilation followed by an erosion (or vice versa) is called a closing (versus opening). All these operations are named after the effects they produce, as illustrated. Many interesting morphological filters can be constructed by taking differences of two or more operations, such as in morphological edge detection. Other applications include skeletonization, which consists of a sequence of thinning operations producing the basic shape of objects, and granulometry, which uses a family of opening operations with increasingly larger structuring elements to compute the size distribution of objects in an image

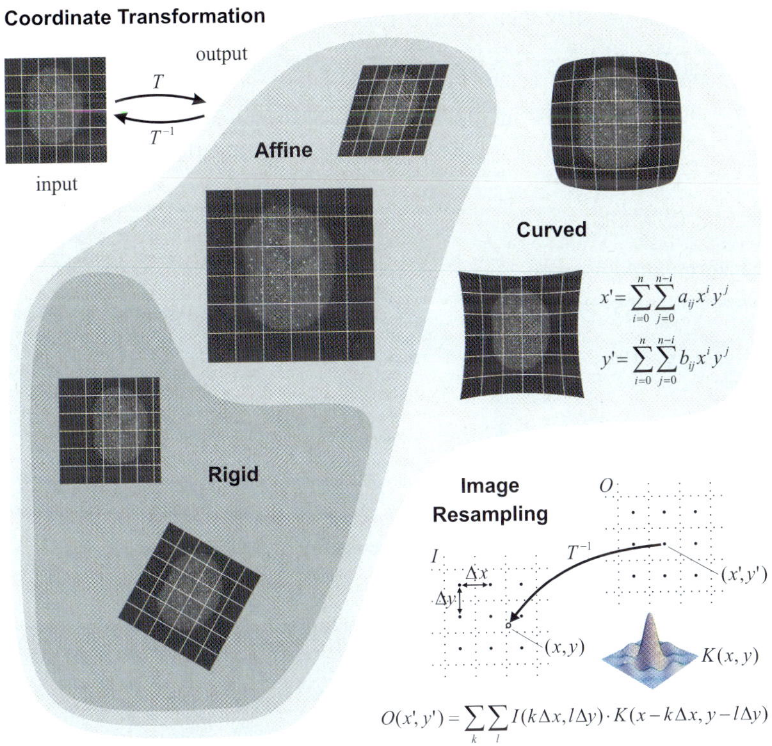

Fig. 2.6 Geometrical transformation of images by coordinate transformation and image resampling. The former is concerned with how input pixel positions are mapped to output pixel positions. Many types of transformations (denoted T) exist. The most frequently used types are (in increasing order of complexity) rigid transformations (translations and rotations), affine transformations (rigid transformations plus scalings and skewings), and curved transformations (affine transformations plus certain nonlinear or elastic deformations). All of these are defined (or can be approximated) by polynomial functions (with degree n depending on the complexity of the transformation). Image resampling concerns the computation of the pixel values of the output image (denoted O) from the pixel values of the input image (denoted I). This is done by using the inverse transformation (denoted T^{-1}) to map output grid positions (x',y') to input positions (x,y). The value at this point is then computed by interpolation from the values at neighboring grid positions, using some weighing function, also known as the interpolation kernel (denoted K)

2.3.4 *Image Restoration*

There are many factors in the acquisition process that cause a degradation of image quality in one way or another, resulting in a corrupted view of reality. Chromatic and other aberrations in the imaging optics may result in spatial distortions (already mentioned). These may be corrected by image registration methods. Certain illumination modes result in (additive) intensity gradients or shadows, which may be corrected by subtracting an image showing only these phenomena, not the specimen. This is known as background subtraction. If it is not possible to capture a background image, it may in some cases be obtained from the image to be corrected (Fig. 2.7). Another major source of intensity corruption is noise, due to the quantum nature of light (signal-dependent noise, following Poisson statistics) and imperfect electronics (mostly signal-independent,

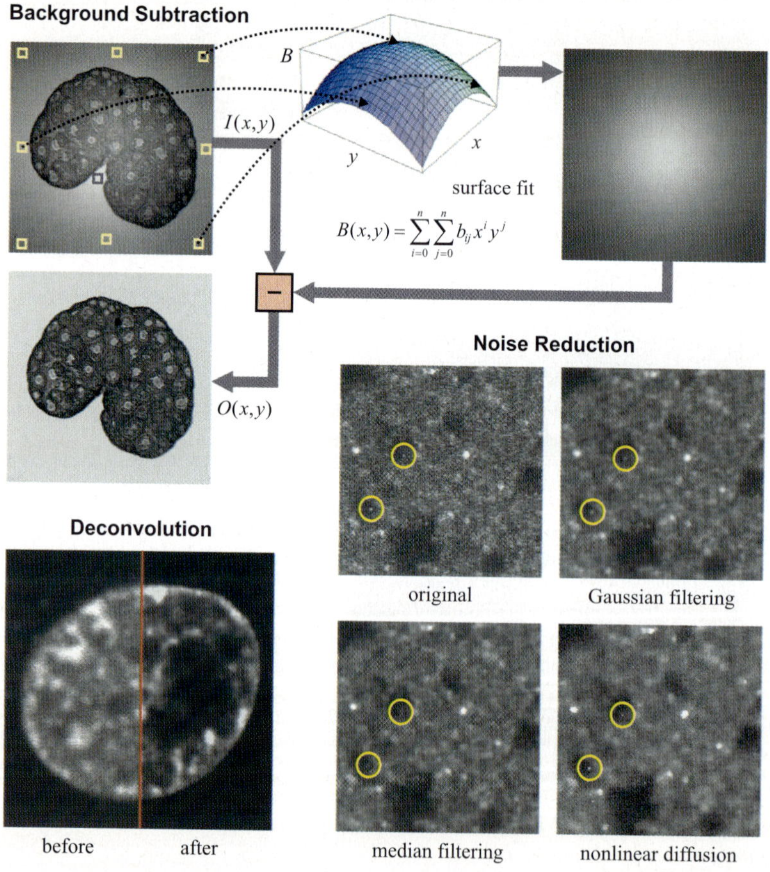

Fig. 2.7 Examples of the effects of image restoration operations: background subtraction, noise reduction, and deconvolution. Intensity gradients may be removed by subtracting a background image. In some cases, this background image may be obtained from the raw image itself by mathematically fitting a polynomial surface function through the intensities at selected points (indicated by the *squares*) corresponding to the background. Several filtering methods exist to reduce noise. Gaussian filtering blurs not only noise but all image structures. Median filtering is somewhat better at retaining object edges but has the tendency to eliminate very small objects (compare the *circles* in each image). Needless to say, the magnitude of these effects depends on the filter size. Nonlinear diffusion filtering was designed specifically to preserve object edges while reducing noise. Finally, deconvolution methods aim to undo the blurring effects of the microscope optics and to restore small details. More sophisticated methods are also capable of reducing noise

Gaussian noise). One way to reduce noise is local averaging of image intensities using a uniform or Gaussian convolution filter. While improving the overall signal-to-noise ratio (SNR), this has the disadvantage that structures other than noise are also blurred. Median filtering is an effective way to remove shot noise (as caused, for example, by bright or dark pixels). It should be used with great care, however, when small objects are studied (such as in particle tracking), as these may also be (partially) filtered out. A more sophisticated technique is nonlinear diffusion filtering (Perona and Malik 1990), which smoothes noise while preserving sharpness at object edges, by taking into account local image properties (notably the gradient magnitude).

Especially widefield microscopy images may suffer from excessive blurring due to out-of-focus light. But even in confocal microscopy, where most of these effects are suppressed, images are blurred owing to diffraction effects (Born and Wolf 1980; Gu 2000). To good accuracy, these effects may be modeled mathematically as a convolution of the true optical image with the 3D point-spread function (PSF) of the microscope optics. Methods that try to undo this operation, in other words that try at every point in the image to reassign light to the proper in-focus location, are therefore called deconvolution methods (Van der Voort and Strasters 1995; Pawley 2006; Jansson 1997). Simple examples include nearest-neighbor or multineighbor deblurring and Fourier-based inverse-filtering methods. These are computationally fast but have the tendency to amplify noise. More sophisticated methods, which also *reduce* noise, are based on iterative regularization and other (constrained or statistical) iterative algorithms. The underlying principles of deconvolution are described in more detail elsewhere (Pawley 2006). In principle, deconvolution preserves total signal intensity while improving contrast by restoring signal position; therefore, it is often desirable prior to quantitative image analysis.

2.4 Advanced Processing for Image Analysis

The image preprocessing operations described in the previous section are important in enhancing or correcting image data, but by themselves do not answer any specific biological questions. Addressing such questions requires much more involved image processing and analysis algorithms, consisting of a series of operations working closely together in "interrogating" the data and extracting biologically meaningful information. Because of the complexity of biological phenomena and the variability or even ambiguity of biological image data, many analysis tasks are difficult to automate fully and require expert-user input or interaction. In contrast with most image preprocessing operations, image analysis methods are therefore often semiautomatic. Here we briefly describe state-of-the-art methods for biological image analysis problems that are of particular relevance in the context of this book: colocalization analysis, neuron tracing and quantification, and the detection or segmentation, tracking, and motion analysis of particles and cells. Several technical challenges in these areas are still vigorously researched.

2.4.1 Colocalization Analysis

An interesting question in many biological studies is to what degree two or more molecular species (typically proteins) are active in the same specimen (see also Chap. 5 by Oheim and Li). This co-occurrence phenomenon can be imaged by using a different fluorescent label for each species, combined with multicolor optical microscopy imaging. A more specific question is whether or not proteins reside in the same (or proximate) physical locations in the specimen. This is the problem of colocalization. For such experiments it is of paramount importance that the emission spectra (rather than just the peak wavelengths) of the fluorophores are sufficiently well separated and that the correct filter sets are used during acquisition to reduce artifacts due to spectral bleed-through or fluorescence resonance energy transfer (FRET) as much as possible. Quantitative colocalization is perhaps the most extreme example of image analysis: it takes two images (typically containing millions of pixels) and produces only a few numbers: the colocalization measures (Fig. 2.8; see also Chap. 5 by Oheim and Li). Pearson's correlation coefficient is often used for this purpose but may produce negative values, which is counterintuitive for a measure expressing the degree of overlap. A more intuitive measure, ranging from 0 (no colocalization) to 1 (full colocalization), is the so-called overlap coefficient, but it is appropriate only when the number of fluorescent targets is more or less equal in each channel. If this is not the case, multiple coefficients (two in the case of dual-color imaging) are required to quantify the degree of colocalization in a meaningful way (Manders et al. 1993). These, however, tend to be rather sensitive to background offsets and noise, and require careful image restoration (Landmann and Marbet 2004). The most important step in colocalization analysis is the separation of signal and background, which is often done by intensity thresholding at visually determined levels (Peñarrubia et al. 2005). The objectivity and reproducibility of this step may be improved considerably by applying statistical significance tests and automated threshold search algorithms (Costes et al. 2004). Clearly, the resolution of colocalization is limited to the optical resolution of the microscope (about 200 nm laterally and about 600 nm axially), which is insufficient to determine whether two fluorescent molecules are really attached to the same target or reside within the same organelle. If colocalization or molecular interaction needs to be quantitatively studied at much higher resolutions (less than 10 nm), FRET imaging and analysis is more appropriate, which is discussed in more detail in Chap. 6 by Hoppe (see also Berney and Danuser 2003).

2.4.2 Neuron Tracing and Quantification

Another biological image analysis problem, which occurs, for example, when studying the molecular mechanisms involved in neurite outgrowth and differentiation, is the length measurement of elongated image structures. For practical reasons, many neuronal morphology studies were and still are performed using 2D imaging. This often results in ambiguous images: at many places it is unclear whether neurites are

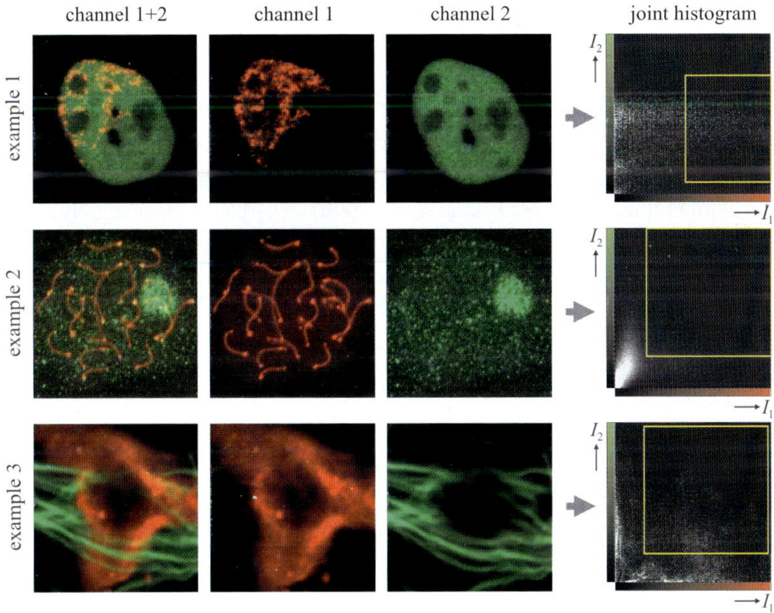

Colocalization Measures

Pearson's correlation coefficient: $$r_P = \frac{\sum (I_1 - \bar{I}_1)(I_2 - \bar{I}_2)}{\sqrt{\sum (I_1 - \bar{I}_1)^2 \sum (I_2 - \bar{I}_2)^2}}$$

example	r_P	r	k_1	k_2	m_1	m_2
1	0.49	0.58	0.79	0.43	0.99	0.15
2	0.39	0.71	0.81	0.62	0.35	0.24
3	-0.08	0.37	0.24	0.57	0.23	0.38

overlap coefficient: $r = \sqrt{k_1 k_2}$ with

$$k_1 = \frac{\sum I_1 I_2}{\sum I_1^2} \quad \text{and} \quad k_2 = \frac{\sum I_1 I_2}{\sum I_2^2}$$

Manders' colocalization coefficients:

$$m_1 = \frac{\sum I_{1\,\mathrm{coloc}\,2}}{\sum I_1} \quad \text{and} \quad m_2 = \frac{\sum I_{2\,\mathrm{coloc}\,1}}{\sum I_2}$$

Fig. 2.8 Commonly used measures for quantitative colocalization analysis. The aim of all these measures is to express in numbers the degree of overlap between two fluorophores (captured in well-separated channels), indicating the presence of the corresponding labeled molecules in the same or proximate physical locations (up to the optical resolution of the microscope). A visual impression of the co-occurrence of fluorophore intensities (I_1 and I_2) is given by the joint histogram (also referred to as the scatter plot or fluorogram). Some colocalization measures are computed over the entire images, while some are restricted to certain intensity ranges (indicated by the *squares in the joint histograms*). Among the first are Pearson's correlation coefficient (denoted r_P) and the so-called overlap coefficient (denoted r and computed from the subcoefficients k_1 and k_2). Both coefficients are insensitive to intensity scalings (due to photobleaching or a difference in signal amplification), while the former is also insensitive to intensity offsets (different background levels). The value of r_P may range from −1 to 1 and is therefore at odds with intuition. Its squared value is perhaps more valuable as it expresses the quality of a least-squares fitting of a line through the points in the scatter plot. The other measures range from 0 to 1. The value of r is meaningful only when the amount of fluorescence is approximately equal in both channels, that is, when k_1 and k_2 have similar values. Manders's colocalization coefficients (denoted m_1 and m_2) are intuitively most clear but require careful separation of signal and background in both channels: the denominators are computed over the entire images, but the numerators sum only those intensities in one channel for which the corresponding intensity in the other channel is within a predefined range (the *left and right and the top and bottom lines of the square region indicated in the joint histogram*, for I_1 and I_2 respectively) branching or crossing. Tracing such structures and building neuritic trees for morphological analysis requires the input of human experts to resolve ambiguities. This resorting to human input is not unique to neuron tracing but is inevitable in many other complicated image analysis tasks and has led to the development of a variety of interactive segmentation methods. An example is live-wire segmentation, which was originally designed to perform computer-supported delineation of object edges (Barrett and Mortensen 1997; Falcão et al. 1998). It is based on a search algorithm that finds a path from a single user-selected pixel to all other pixels in the image by minimizing the cumulative value of a predefined cost function computed from local image features (such as gradient magnitude) along the path. The user can then interactively select the path that according to his/her own judgment best follows the structure of interest and fix the tracing up to some point, from where the process is iterated until the entire structure is traced. This technique has been adapted to enable tracing of neuritelike image structures in two dimensions (Meijering et al. 2004) and similar methods have been applied to neuron tracing in three dimensions (Fig. 2.9). Fully automated methods for 3D neuron tracing have also been published recently (He et al. 2003; Schmitt et al. 2004; Evers et al. 2005). In the case of poor image quality, however, these may require manual postprocessing of the results.

2.4.3 Particle Detection and Tracking

One of the major challenges of biomedical research in the postgenomic era is the unraveling of not just the spatial, but also the *spatiotemporal* relationships of complex biomolecular systems (Tsien 2003). Naturally this involves the acquisition of time-lapse image series and the tracking of objects over time (see also Chap. 9 by Jaqaman et al. and Chap. 13 by Soll et al.). From an image analysis point of view, a distinction can be made between tracking of single molecules (or complexes) and tracking of entire cells (Sect. 2.4.4). A number of tools are available for studying the dynamics of proteins based on fluorescent labeling and time-lapse imaging, such as fluorescence recovery after photobleaching (FRAP) and fluorescence loss in photobleaching (FLIP), but these yield only ensemble-average measurements of properties. More detailed studies into the different modes of motion of subpopulations require single-particle tracking (Qian et al. 1991; Saxton and Jacobson 1997), which aims at motion analysis of individual proteins or microspheres. Computerized image analysis methods for this purpose have been developed since the early 1990s and are constantly being improved (Bacher et al. 2004; Dorn et al. 2005) to deal with increasingly sophisticated biological experimentation.

Generally, particle tracking methods consist of two stages (Meijering et al. 2006): (1) the detection of individual particles per time frame and (2) the linking of particles detected in successive frames (Fig. 2.10). Regarding the former, it has been shown theoretically as well as empirically (Cheezum et al. 2001; Thomann et al. 2002; Ober et al. 2004; Ram et al. 2006) that the localization error can be at least 1 order of magnitude lower than the extension of the microscope PSF, and that the SNR is among the main

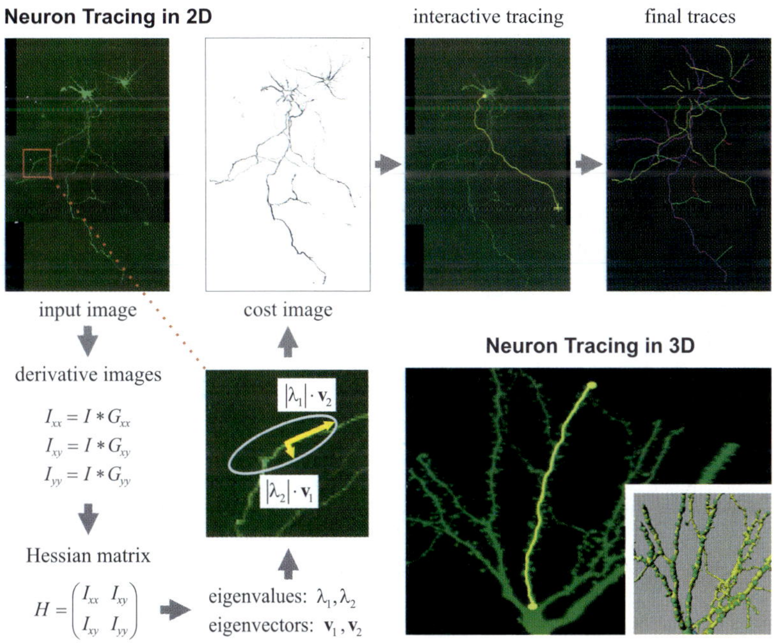

Fig. 2.9 Tracing of neurite outgrowth using interactive segmentation methods. To reduce background intensity gradients (shading effects) or discontinuities (due to the stitching of scans with different background levels), the image features exploited here are the second-order derivatives, obtained by convolution with the second-order Gaussian derivative kernels (Fig. 2.4) at a proper scale (to suppress noise). These constitute a so-called Hessian matrix at every pixel in the image. Its eigenvalues and eigenvectors are used to construct an ellipse (as indicated), whose size is representative of the local neurite contrast and whose orientation corresponds to the local neurite orientation. In turn, these properties are used to compute a cost image (with *dark values* indicating a lower cost and *bright values* a higher cost) and vector field (not shown), which together guide a search algorithm that finds the paths of minimum cumulative cost between a start point and all other points in the image. With use of graphics routines, the path to the current cursor position (indicated by the *cross*) is shown at interactive speed while the user selects the optimal path on the basis of visual judgment. Once tracing is finished, neurite lengths and statistics can be computed automatically. This is the underlying principle of the NeuronJ tracing tool, freely available as a plug-in to the ImageJ program (discussed in Sect. 2.6). The Filament Tracer tool, commercially available as part of the Imaris software package (Bitplane), uses similar principles for tracing in 3D images, based on volume visualization

factors limiting the localization accuracy. Currently, one of the best approaches to particle detection is by least-squares fitting of a Gaussian (mixture) model to the image data. In practice, the real difficulty in particle tracking is the data-association problem: determining which particles as detected in one frame correspond to which particles in the next is not trivial, as the number of (real or detected) particles may not be constant over time (particles may enter or exit the field of view, they may assemble or disassemble, or limitations in the detection stage may cause varying degrees of underdetection or overdetection). Therefore, most current particle tracking tools should be used with care (Carter et al. 2005) and may still require manual checking and correction of the results. Several examples of particle tracking applications are discussed elsewhere in this book.

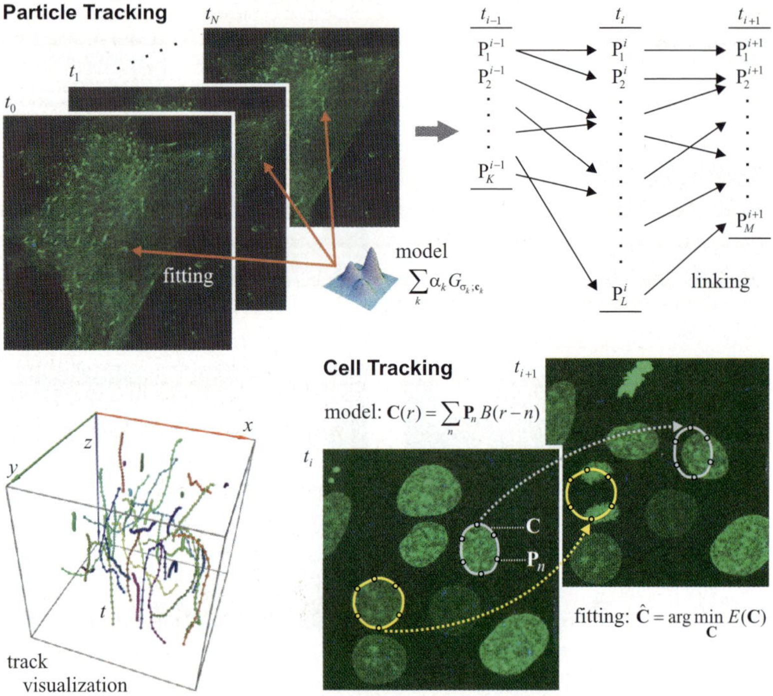

Fig. 2.10 Challenges in particle and cell tracking. Regarding particle tracking, currently one of the best approaches to detection of fluorescent tags is by least-squares fitting of a model of the intensity distribution to the image data. Because the tags are subresolution particles, they appear as diffraction-limited spots in the images and therefore can be modeled well by a mixture of Gaussian functions, each with its own amplitude scaling factor, standard deviation, and center position. Usually the detection is done separately for each time step, resulting in a list of potential particle positions and corresponding features, to be linked between time steps. The linking is hampered by the fact that the number of particles detected may be different for each time step. In cell tracking, a contour model (surface model in the case of 3D time-lapse experiments) is often used for segmentation. Commonly used models consist of control points, which are interpolated using smooth basis functions (typically B-splines) to form continuous, closed curves. The model must be flexible enough to handle geometrical as well as topological shape changes (cell division). The fitting is done by (constrained) movement of the control points to minimize some predefined energy functional computed from image-dependent information (intensity distributions inside and outside the curve) as well as image-independent information (a priori knowledge about cell shape and dynamics). Finally, trajectories can be visualized by representing them as tubes (segments) and spheres (time points) and using surface rendering

2.4.4 *Cell Segmentation and Tracking*

Motion estimation of cells is another frequently occurring problem in biological research (see also Chap. 12 by Amino et al. and Chap. 13 by Soll et al.). In particle tracking studies, for example, cell movement may muddle the motion analysis of intracellular components and needs to be corrected for. In some cases this is may be accomplished by applying (nonrigid) image registration methods (Eils and

Athale 2003; Gerlich et al. 2003; Rieger et al. 2004; Sorzano et al. 2005). However, cell migrations and deformations are also interesting in their own right owing to their role in a number of biological processes, including immune response, wound healing, embryonic development, and cancer metastasis (Chicurel 2002). Understanding these processes is of major importance in combating various types of human disease. Typical 3D time-lapse data sets acquired for studies in this area consist of thousands of images and are almost impossible to analyze manually, both from a cost-efficiency perspective and because visual inspection lacks the sensitivity, accuracy, and reproducibility needed to detect subtle but potentially important phenomena. Therefore, computerized, quantitative cell tracking and motion analysis is a requisite (Dufour et al. 2005; Zimmer et al. 2006).

In contrast to single molecules or molecular complexes, which are subresolution objects appearing as PSF-shaped spots in the images, cells are relatively (with respect to pixel size) large objects having a distinct shape. Detecting (or segmenting) entire cells and tracking position and shape changes requires quite different image processing methods. Owing to noise and photobleaching effects, simple methods based on intensity thresholding are generally inadequate. To deal with these artifacts and with obscure boundaries in the case of touching cells, recent research has focused on the use of model-based segmentation methods (Kass et al. 1988; McInerney and Terzopoulos 1996), which allow the incorporation of prior knowledge about object shape. Examples of such methods are active contours (also called snakes) and active surfaces, which have been applied to a number of cell tracking problems (Ray et al. 2002; Debeir et al. 2004; Dufour et al. 2005). They involve mathematical, prototypical shape descriptions having a limited number of degrees of freedom, which enable shape-constrained fitting to the image data based on data-dependent information (image properties, in particular intensity gradient information) *and* data-independent information (prior knowledge about the shape). Tracking is achieved by using the contour or surface obtained for one image as initialization for the next and repeating the fitting procedure (Fig. 2.10).

2.5 Higher-Dimensional Data Visualization

Advances in imaging technology are rapidly turning higher-dimensional image data acquisition into the rule rather than the exception. Consequently, there is an increasing need for sophisticated visualization technology to enable efficient presentation and exploration of this data and associated image analysis results. Early systems supported browsing the data in a frame-by-frame fashion (Thomas 1996), which provided only limited insight into the interrelations of objects in the images. Since visualization means generating representations of higher-dimensional data, this necessarily implies reducing dimensionality and possibly even reducing information in some sensible way. How to do this in an optimal way depends on the application, the dimensionality of the data, and the physical nature of its respective dimensions (Fig. 2.1). In any case, visualization methods usually consist of highly sophisticated information processing steps that may have a strong influence on the final result, making them very susceptible to misuse. Here we briefly explain the two main modes of visualization and point at critical steps in the process. More information can be found in textbooks on visualization and computer graphics (Schroeder et al. 2002; Foley et al. 1997).

2.5.1 Volume Rendering

Visualization methods that produce a viewable image of higher-dimensional image data without requiring an explicit geometrical representation of that data are called volume rendering methods. A commonly used, flexible and easy-to-understand volume rendering method is ray casting, or ray tracing. With this method, the value of each pixel in the view image is determined by "casting a ray" into the image data and evaluating the data encountered along that ray using a predefined ray function (Fig. 2.11). The direction of the rays is determined by the viewing angles and the mode of projection, which can be orthographic (the rays run parallel to each other) or perspective (the rays have a common focal point). Analogous to the real-world situation, these are called camera properties. The rays pass through the data with a certain step size, which should be smaller than the pixel size to avoid skipping important details. Since, as a consequence, the ray sample points will generally not coincide with grid positions, this requires data interpolation. The ray function determines what information is interpolated and evaluated and how the sample values are composed into a single output value. For example, if the function considers image intensity only and stores the maximum value found along the ray, we obtain a maximum intensity projection (MIP). Alternatively, it may sum all values and divide by the number to yield an average intensity projection. These methods are useful to obtain first impressions, even in the case of very noisy data, but the visualizations are often ambiguous owing to overprojections. More complex schemes may consider gradient magnitude, color, or distance information. They may also include lighting effects to produce nicely shaded results. Each method yields a different view of the image data and may give rise to a slightly different interpretation. It is therefore often beneficial to use multiple methods.

2.5.2 Surface Rendering

In contrast to volume rendering methods, which in principle take into account all data along rays and therefore enable the visualization of object interiors, surface rendering methods visualize only object surfaces. Generally, this requires a mathematical description of the surfaces in terms of primitive geometrical entities: points, lines, triangles, polygons, or polynomial curves and surface patches, in particular splines. Such descriptions are derived from a segmentation of the image data into meaningful parts (objects versus background). This constitutes the most critical aspect of surface rendering: the value of the visualization depends almost entirely

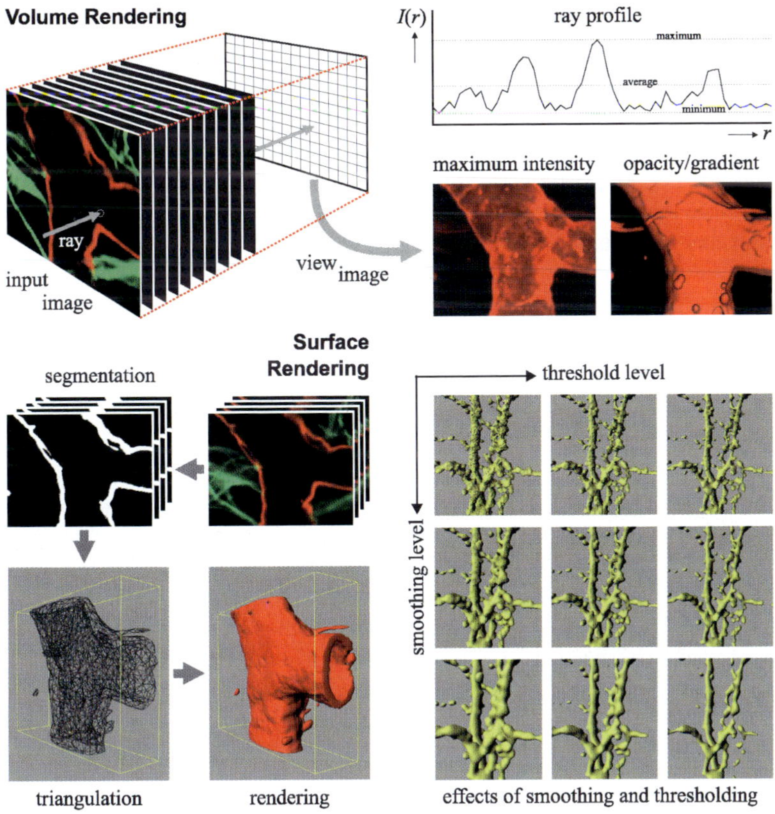

Fig. 2.11 Visualization of volumetric image data using volume rendering and surface rendering methods. Volume rendering methods do not require an explicit geometrical representation of the objects of interest present in the data. A commonly used volume rendering method is ray casting: for each pixel in the view image, a ray is cast into the data, and the intensity profile along the ray is fed to a ray function, which determines the output value, such as the maximum, average, or minimum intensity, or accumulated "opacity" (derived from intensity or gradient magnitude information). By contrast, surface rendering methods require a segmentation of the objects (usually obtained by thresholding), from which a surface representation (triangulation) is derived, allowing for very fast rendering by graphics hardware. To reduce the effects of noise, Gaussian smoothing is often applied as a preprocessing step prior to segmentation. As shown, both operations have a substantial influence on the final result: by slightly changing the degree of smoothing or the threshold level, objects may appear (dis)connected while in fact they are not. Therefore, it is recommended to establish optimal parameter values for both steps while inspecting the effects on the original image data rather than looking directly at the renderings

on the correctness of the segmentation (Fig. 2.11). Once a correct segmentation is available, however, a representation of the object surfaces in terms of primitives, in particular a surface triangulation, is easily obtained by applying the so-called marching cubes algorithm (Lorensen and Cline 1987). Having arrived at this point, the visualization task has reduced to a pure computer graphics problem: generating an image from numbers representing primitive geometrical shapes. This could be done, again, by ray tracing: for each pixel in the view image, a ray is defined and its intersections with the surfaces are computed, at which points the effect of the light source(s) on the surfaces (based on their orientation, opacity, color, texture) are determined to yield an output pixel value. This is called an image-order rendering approach (from pixels to surfaces). Most modern computer graphics hardware, however, uses an object-order rendering approach (from surfaces, or primitives, to pixels). Note that using such methods, we can visualize not just segmented image data, but any information that can be converted somehow to graphics primitives. Examples of this are tracing and tracking results, which can be represented by tubes and spheres (Figs. 2.9, 2.10).

2.6 Software Tools and Development

It will be clear from the (rather compressed) overview in the previous sections that after decades of research a host of methods for image processing, analysis, and visualization have been developed, but also that there exists no such thing as a universal method capable of solving all problems. Although it is certainly possible to categorize problems, in a sense each biological study is unique: being based on specific premises and hypotheses to be tested, giving rise to unique image data to be analyzed, and requiring dedicated image analysis methods in order to take full advantage of this data. As a consequence, there is also a great variety of software tools. Roughly, they can be divided into four categories, spanning the entire spectrum from least to most dedicated. At one end are tools that are mainly meant for image acquisition but that also provide basic image processing, measurement, visualization, and documentation facilities. Examples include some tools provided by microscope manufacturers, such as the AIM tool (LSM Image Browser, Carl Zeiss), QWin (Leica Microsystems), and analySIS (Olympus and Soft Imaging System). Next are tools that in addition to offering basic facilities were designed to also address a range of more complicated biological image analysis problems. Often these tools consist of a core platform with the possibility to add modules developed for dedicated applications, such as deconvolution, colocalization, filament tracing, image registration, or particle tracking. Examples of these include Imaris (Bitplane), AxioVision (Carl Zeiss), Image-Pro Plus (MediaCybernetics), MetaMorph (Molecular Devices Corporation), and ImageJ (National Institutes of Health). At the other end of the spectrum are tools that are much more dedicated to specific tasks, such as Huygens (Scientific Volume Imaging) or AutoDeblur (AutoQuant Imaging) for deconvolution, Amira (Mercury Computer Systems) for visualization, Volocity (Improvision) for tracking and motion analysis, and Neurolucida (MicroBrightField) for neuron tracing.

As a fourth category we mention software packages offering researchers much greater flexibility in developing their own, dedicated image analysis algorithms. An example of this is MATLAB (The MathWorks), which offers an interactive developing environment and a high-level programming language for which extensive image processing toolboxes are available, such as DIPimage (Quantitative Imaging Group, Delft University of Technology). It is used by engineers and scientists in many fields for rapid prototyping and validation of new algorithms but has not (yet) gained wide acceptance in biology. An example of an interesting MATLAB-based software tool for high-content, high-throughput image-based cell screening is CellProfiler (Carpenter et al. 2006). A software tool that is rapidly gaining popularity is ImageJ (National Institutes of Health), already mentioned. It is a public-domain tool and developing environment based on the Java programming language (Sun Microsystems): it can be used without the need for a license, it runs on any computer platform (Windows, Macintosh, Linux, and a variety of UNIX variants), and its source code is openly available. The core distribution of the program supports most of the common image file formats and offers a host of facilities for manipulation and analysis of image data (up to 5D), including all basic image processing methods described in this chapter. Probably the strongest feature of the program is its extensibility: existing operations can be combined into more complex algorithms by means of macros, and new functionality can easily be added by writing plug-ins (Abràmoff et al. 2004). Hundreds of plug-ins are already available, considerably increasing its image file support and image processing and analysis capabilities, ranging from very basic but highly useful pixel manipulations to much more involved algorithms for image segmentation, registration, or transformation, visualization (Abràmoff and Viergever 2002; Rueden et al. 2004), deconvolution, extended depth of field (Forster et al. 2004), neuron tracing (Meijering et al. 2004), FRET analysis (Feige et al. 2005), particle tracking (Sbalzarini and Koumoutsakos 2005; Sage et al. 2005), colocalization, texture analysis, cell counting, granulometry, and more.

Finally we wish to make a few remarks regarding the use and development of software tools for biological image analysis. In contrast to diagnostic patient studies in clinical medical imaging practice, biological investigation is rather experimental by nature, allowing researchers to design their own experiments, including the imaging modalities to be used and how to process and analyze the resulting data. While freedom is a great virtue in science, it may also give rise to chaos. All too often, scientific publications report the use of image analysis tools without specifying which algorithms were involved and how parameters were set, making it very difficult for others to reproduce or compare results. What is worse, many software tools available on the market or in the public domain have not been thoroughly scientifically validated, at least not in the open literature, making it impossible for reviewers to verify the validity of using them under certain conditions. It requires no explanation that this situation needs to improve. Another consequence of freedom, on the engineering side of biological imaging, is the (near) total lack of standardization in image data management and information exchange. Microscope manufacturers, software companies, and sometimes even research laboratories have their own image file formats, which generally are rather rigid. As the development of new imaging technologies and analytic tools accelerates, there is an increasing need for an adaptable data model for multidimensional images, experimental metadata, and analytical results, to increase the compatibility of software tools and facilitate the sharing and exchange of information between laboratories. First steps in this direction have already been taken by the microscopy community by developing and implementing the Open Microscopy Environment (OME), whose data model and file format, based on the Extensible Markup Language (XML), is gaining acceptance (Swedlow et al. 2003; Goldberg et al. 2005). More information on the OME initiative is provided in Chap. 3 by Swedlow.

Acknowledgements The authors are grateful to Niels Galjart, Jeroen Essers, Carla da Silva Almeida, Adriaan Houtsmuller, Remco van Horssen, and Timo ten Hagen (Erasmus MC, The Netherlands), J.-C. Floyd Sarria and Harald Hirling (Swiss Federal Institute of Technology Lausanne, Switzerland), Anne McKinney (McGill University, Canada), and Elisabeth Rungger-Brändle (University of Geneva, Switzerland) for providing image data for illustrational purposes. E.M. was supported financially by the Netherlands Organization for Scientific Research (NWO) through VIDI grant 639.022.401.

References

Abràmoff MD, Viergever MA (2002) Computation and visualization of three-dimensional soft tissue motion in the orbit. IEEE Trans Med Imaging 21:296–304

Abràmoff MD, Magalhães PJ, Ram SJ (2004) Image processing with ImageJ. Biophotonics Int 11:36–42

Bacher CP, Reichenzeller M, Athale C, Herrmann H, Eils R (2004) 4-D single particle tracking of synthetic and proteinaceous microspheres reveals preferential movement of nuclear particles along chromatin-poor tracks. BMC Cell Biol 5:1–14

Barrett WA, Mortensen EN (1997) Interactive live-wire boundary extraction. Med Image Anal 1:331–341

Baxes GA (1994) Digital image processing: principles and applications. Wiley, New York.

Berney C, Danuser G (2003) FRET or no FRET: a quantitative comparison. Biophys J 84: 3992–4010

Born M, Wolf E (1980) Principles of optics: electromagnetic theory of propagation, interference and diffraction of light, 6th edn. Pergamon, Oxford

Bracewell RN (2000) The Fourier transform and its applications, 3rd edn. McGraw-Hill, New York

Canny JF (1986) A computational approach to edge detection. IEEE Trans Pattern Anal Mach Intell 8:679–698

Carpenter AE, Jones TR, Lamprecht MR, Clarke C, Kang IH, Friman O, Guertin DA, Chang JH, Lindquist RA, Moffat J, Golland P, Sabatini DM (2006) CellProfiler: image analysis software for identifying and quantifying cell phenotypes. Genome Biol 7:R100.

Carter BC, Shubeita GT, Gross SP (2005) Tracking single particles: a user-friendly quantitative evaluation. Phys Biol 2:60–72

Castleman KR (1996) Digital image processing. Prentice Hall, Englewood Cliffs

Cheezum MK, Walker WF, Guilford WH (2001) Quantitative comparison of algorithms for tracking single fluorescent particles. Biophys J 81:2378–2388

Chen H, Swedlow JR, Grote M, Sedat JW, Agard DA (1995) The collection, processing, and display of digital three-dimensional images of biological specimens. In: Pawley JB (ed) Handbook of biological confocal microscopy, 2nd edn. Plenum, London, pp 197–210

Chicurel M (2002) Cell migration research is on the move. Science 295:606–609

Costes SV, Daelemans D, Cho EH, Dobbin Z, Pavlakis G, Lockett S (2004) Automatic and quantitative measurement of protein-protein colocalization in live cells. Biophys J 86:3993–4003

Debeir O, Camby I, Kiss R, Van Ham P, Decaestecker C (2004) A model-based approach for automated in vitro cell tracking and chemotaxis analyses. Cytometry Part A 60:29–40

Dorn JF, Jaqaman K, Rines DR, Jelson GS, Sorger PK, Danuser G (2005) Yeast kinetochore microtubule dynamics analyzed by high-resolution three-dimensional microscopy. Biophys J 89:2835–2854

Dufour A, Shinin V, Tajbakhsh S, Guillen-Aghion N, Olivo-Marin JC, Zimmer C (2005) Segmenting and tracking fluorescent cells in dynamic 3-D microscopy with coupled active surfaces. IEEE Trans Image Process 14:1396–1410

Eils R, Athale C (2003) Computational imaging in cell biology. J Cell Biol 161:477–481

Evers JF, Schmitt S, Sibila M, Duch C (2005) Progress in functional neuroanatomy: precise automatic geometric reconstruction of neuronal morphology from confocal image stacks. J Neurophysiol 93:2331–2342

Falcão AX, Udupa JK, Samarasekera S, Sharma S, Hirsch BE, de A. Lotufo R (1998) User-steered image segmentation paradigms: live wire and live lane. Graphical Models Image Process 60:233–260

Feige JN, Sage D, Wahli W, Desvergne B, Gelman L (2005) PixFRET, an ImageJ plug-in for FRET calculation that can accommodate variations in spectral bleed-throughs. Microsc Res Tech 68:51–58

Foley JD, van Dam A, Feiner SK, Hughes JF (1997) Computer graphics: principles and practice, 2nd edn in C. Addison-Wesley, Reading

Forster B, Van De Ville D, Berent J, Sage D, Unser M (2004) Complex wavelets for extended depth-of-field: a new method for the fusion of multichannel microscopy images. Microsc Res Tech 65:33–42

Gerlich D, Mattes J, Eils R (2003) Quantitative motion analysis and visualization of cellular structures. Methods 29:3–13

Glasbey CA (1993) An analysis of histogram-based thresholding algorithms. Graphical Models Image Process 55:532–537

Glasbey CA, Horgan GW (1995) Image analysis for the biological sciences. Wiley, New York

Goldberg IG, Allan C, Burel J-M, Creager D, Falconi A, Hochheiser H, Johnston J, Mellen J, Sorger PK, Swedlow JR (2005) The Open Microscopy Environment (OME) data model and XML file: open tools for informatics and quantitative analysis in biological imaging. Genome Biol 6:R47

Gonzalez RC, Woods RE (2002) Digital image processing, 2nd edn. Prentice Hall, Upper Saddle River

Gu M (2000) Advanced optical imaging theory. Springer, Berlin

He W, Hamilton TA, Cohen AR, Holmes TJ, Pace C, Szarowski DH, Turner JN, Roysam B (2003) Automated three-dimensional tracing of neurons in confocal and brightfield images. Microsc Microanal 9:296–310

Houle D, Mezey J, Galpern P, Carter A (2003) Automated measurement of *Drosophila* wings. BMC Evol Biol 3:1–13

Jähne B (2004) Practical handbook on image processing for scientific applications, 2nd edn. CRC, Boca Raton

Jain AK (1989) Fundamentals of digital image processing. Prentice-Hall, Englewood Cliffs

Jansson PA (ed) (1997) Deconvolution of images and spectra. Academic, San Diego

Kass M, Witkin A, Terzopoulos D (1988) Snakes: active contour models. Int J Comput Vis 1:321–331

Landmann L, Marbet P (2004) Colocalization analysis yields superior results after image restoration. Microsc Res Tech 64:103–112

Lorensen WE, Cline HE (1987) Marching cubes: a high resolution 3D surface construction algorithm. Comput Graphics 21:163–169

Maintz JBA, Viergever MA (1998) A survey of medical image registration. Med Image Anal 2:1–36

Manders EMM, Verbeek FJ, Aten JA (1993) Measurement of colocalization of objects in dual-colour confocal images. J Microsc 169:375–382

McInerney T, Terzopoulos D (1996) Deformable models in medical image analysis: a survey. Med Image Anal 1:91–108

Meijering EHW, Niessen WJ, Viergever MA (2001) Quantitative evaluation of convolution-based methods for medical image interpolation. Med Image Anal 5:111–126

Meijering E, Jacob M, Sarria J-CF, Steiner P, Hirling H, Unser M (2004) Design and validation of a tool for neurite tracing and analysis in fluorescence microscopy images. Cytometry Part A 58:167–176

Meijering E, Smal I, Danuser G (2006) Tracking in molecular bioimaging. IEEE Signal Process Mag 23:46–53

Murphy RF, Meijering E, Danuser G (2005) Special issue on molecular and cellular bioimaging. IEEE Trans Image Proc 14:1233–1236

Ober RJ, Ram S, Ward ES (2004) Localization accuracy in single-molecule microscopy. Biophys J 86:1185–1200

Pawley JB (ed) (2006) Handbook of biological confocal microscopy, 3rd edn. Springer, New York

Peñarrubia PG, Ruiz XF, Gálvez J (2005) Quantitative analysis of the factors that affect the determination of colocalization coefficients in dual-color confocal images. IEEE Trans Image Proc 14:1151–1158

Perona P, Malik J (1990) Scale-space and edge detection using anisotropic diffusion. IEEE Trans Pattern Anal Mach Intell 12:629–639

Pluim JPW, Maintz JBA, Viergever MA (2003) Mutual-information-based registration of medical images: A survey. IEEE Trans Med Imaging 22:986–1004

Qian H, Sheetz MP, Elson EL (1991) Single particle tracking: analysis of diffusion and flow in two-dimensional systems. Biophys J 60:910–921

Ram S, Ward ES, Ober RJ (2006) Beyond Rayleigh's criterion: a resolution measure with application to single-molecule microscopy. Proc Natl Acad Sci USA 103:4457–4462

Ray N, Acton ST, Ley K (2002) Tracking leukocytes in vivo with shape and size contrained active contours. IEEE Trans Med Imaging 21:1222–1235

Rieger B, Molenaar C, Dirks RW, van Vliet LJ (2004) Alignment of the cell nucleus from labeled proteins only for 4D in vivo imaging. Microsc Res Tech 64:142–150

Rueden C, Eliceiri KW, White JG (2004) VisBio: a computational tool for visualization of multidimensional biological image data. Traffic 5:411–417

Russ JC (2002) The image processing handbook, 4th edn. CRC, Boca Raton

Sabri S, Richelme F, Pierres A, Benoliel A-M, Bongrand P (1997) Interest of image processing in cell biology and immunology. J Immunol Methods 208:1–27

Sage D, Neumann FR, Hediger F, Gasser SM, Unser M (2005) Automatic tracking of individual fluorescence particles: application to the study of chromosome dynamics. IEEE Trans Image Proc 14:1372–1383

Saxton MJ, Jacobson K (1997) Single-particle tracking: applications to membrane dynamics. Annu Rev Biophys Biomol Struct 26:373–399

Sbalzarini IF, Koumoutsakos P (2005) Feature point tracking and trajectory analysis for video imaging in cell biology. J Struct Biol 151:182–195

Schmitt S, Evers JF, Duch C, Scholz M, Obermayer K (2004) New methods for the computer-assisted 3-D reconstruction of neurons from confocal image stacks. Neuroimage 23:1283–1298

Schroeder W, Martin K, Lorensen B (2002) The visualization toolkit: an object-oriented approach to 3D graphics, 3rd edn. Kitware, New York

Serra J (1982) Image analysis and mathematical morphology. Academic, London

Sonka M, Hlavac V, Boyle R (1999). Image processing, analysis, and machine vision, 2nd edn. PWS, Pacific Grove

Sorzano CÓS, Thévenaz P, Unser M (2005) Elastic registration of biological images using vector-spline regularization. IEEE Trans Biomed Eng 52:652–663

Swedlow JR, Goldberg I, Brauner E, Sorger PK (2003) Informatics and quantitative analysis in biological imaging. Science 300:100–102

Thévenaz P, Blu T, Unser M (2000) Interpolation revisited. IEEE Trans Med Imaging 19:739–758

Thomann D, Rines DR, Sorger PK, Danuser G (2002) Automatic fluorescent tag detection in 3D with super-resolution: application to the analysis of chromosome movement. J Microsc 208:49–64

Thomas C, DeVries P, Hardin J, White J (1996) Four-dimensional imaging: computer visualization of 3D movements in living specimens. Science 273:603–607

Tsien RY (2003) Imagining imaging's future. Nat Rev Mol Cell Biol 4:S16–S21

Van der Voort HTM, Strasters KC (1995) Restoration of confocal images for quantitative image analysis. J Microsc 178:165–181

Zimmer C, Zhang B, Dufour A, Thébaud A, Berlemont S, Meas-Yedid V, Olivo-Marin JC (2006) On the digital trail of mobile cells. IEEE Signal Process Mag 23:54–62

3
The Open Microscopy Environment: A Collaborative Data Modeling and Software Development Project for Biological Image Informatics

Jason R. Swedlow

Abstract The transition of a microscope's output from an "image," recorded on paper or film, to digitally recorded "data" has created new demands for storage, analysis and visualization that are not adequately met in current software packages. The Open Microscopy Environment (OME) Consortium is dedicated to developing open available tools to meet this challenge. We have developed and released the OME data model that provides a thorough description of the image data acquisition, structure and analysis results. An XML representation of the OME data model provides convenient standardized file formats known as OME-XML and OME-TIFF. In addition, OME has built two software tools, the OME and OME Remote Objects (OMERO) servers that enable visualization, management and analysis of multidimensional image data in structures that enable remote access. The OME server provides a flexible data model and an interface into complex analysis workflows. The OMERO server and clients provide image data visualization and management. A major goal for the next year is the provision of well-developed libraries and documentation to support the OME file formats, and enhanced functionality in our OME and OMERO applications to provide complete solutions for imaging in cell biology.

3.1 Introduction

The transition of a microscope's output from an "image," recorded on paper or film, to digitally recorded "data" has created new demands for storage, analysis and visualization that are not adequately met in current software packages. The absence of suitable software for image management currently hinders many projects from exploiting the full potential of digital microscopy to solve biological problems, such as live-cell dynamics, photobleaching and fluorescence resonance energy transfer studies on cells expressing fluorescent protein fusions (Eils and Athale 2003; Lippincott-Schwartz et al. 2001; Phair and Misteli 2001; Wouters et al. 2001). In addition, cell-based high-content assays are under development in many

S.L. Shorte and F. Frischknecht (eds.), *Imaging Cellular and Molecular Biological Functions.*

academic and commercial laboratories, but tools for managing these types of data and integrating all experimental information and data analysis are lacking (Conrad et al. 2004; Kiger et al. 2003; Simpson et al. 2000; Yarrow et al. 2003). Overcoming these difficulties will therefore have a valuable impact on many areas of cell biology and drug discovery, and ultimately human health.

The Open Microscopy Environment (OME) project was initiated to build tools to address this problem. Specifically the goals of OME are:

- To enable the integration of image storage, visualization, annotation, management and analysis.
- To provide tools that solve the access difficulties caused by large numbers of proprietary file formats.
- To provide open, freely accessible and usable software to support the biological microscope imaging community.

The challenges for providing such software tools in a form that is robust, powerful and usable are immense. Reasonable progress has been made, and OME tools are increasingly in use throughout the world. In this chapter, I summarize the activities of the OME project up to the current date (end of 2006) and look forward to some of the challenges facing us in 2007 and 2008.

3.1.1 What Is OME?

OME is a consortium of groups working to produce, release and support software for biological microscope image informatics (see Open Microscopy Environment 2007a for more information). The main development groups are currently based at the University of Dundee (Swedlow laboratory), National Institute on Aging, National Institutes of Health (Goldberg laboratory), Laboratory for Optical and Computational Instrumentation, University of Wisconsin, Madison (White and Eliceiri laboratories), and Harvard Medical School (Sorger laboratory). The OME project welcomes the participation of other groups. For example, groups at Vanderbilt University and the University of California, Santa Barbara are currently using the OME and developing their own tools around it. We collect examples of this kind of work at Open Microscopy Environment (2007b).

The OME project releases open-source software, licensed under the GNU general public license or the GNU lesser general public license (Free Software Foundation 2007a, b).

3.1.2 Why OME – What Is the Problem?

Modern microscope imaging systems record images through space (by changing focus), time (by taking images at defined time points) and channel (by adjusting the wavelength of light the detector measures). Such "5D" imaging is the basis of most modern imaging experiments, whether using traditional microscopes or more modern high-content imaging systems. Systems to acquire these 5D images either are commonly available from a large number of commercial suppliers or can be built from "scratch" using individual components.

5D images are often transformed to improve contrast and extend resolution. This transformation can be as simple as spatial filtering (Castleman 1979) and as sophisticated as image restoration by iterative deconvolution (Swedlow et al. 1997). Quantitative measurement of key parameters (e.g., object size, signal content or shape) along with visualization of the spatial, temporal or spectral dimensions of the image are then used to generate a "result" – a statement of what the image means. In the many excellent commercial systems currently available, the acquisition, transformation, viewing and analysis functions are often integrated into a single software suite (Fig. 3.1). However, invariably, results from image visualization or analysis are exported to other software (e.g., Adobe Photoshop or Microscoft Excel) for layout, graphing or further analysis. While these are

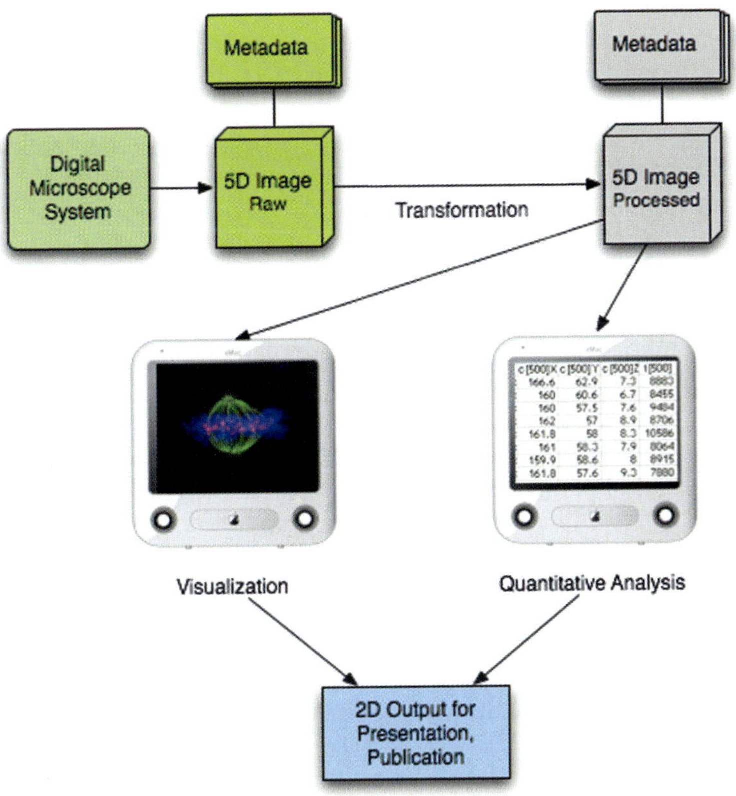

Fig. 3.1 The standard paradigm for image data acquisition, processing and analysis

excellent applications, all metadata and links to the original experiment are lost during this export.

This approach has been used since the dawn of digital microscope imaging in the 1980s. It has proven hugely successful, but with the growth of image-based assays, especially using fluorescent proteins (Giepmans et al. 2006) and the increasingly routine collection of large 5D images, the standard paradigm is too unwieldy because available software does not provide facilities to support the integration of a number of disparate data types associated with a single 5D image:

- The binary data themselves (the actual pixel data, in OME referred to as "pixels").
- The acquisition metadata, the information about the imaging experiment, including instrument settings, timestamps, configuration, etc.
- Results from processing and analyses and analysis of the required image.
- Annotations of other additional data that are attached to specific images or their subregions.

These limitations are severely compounded by the explosion of proprietary file formats, each with different metadata and binary data structures. In the past, the integration and management of these data was done by hand. As the size and complexity of image data has grown, manual data management is no longer feasible. Clearly, a sophisticated software tool that provides access to all these data types, regardless of acquisition file format, is required for large-scale imaging. This tool must provide access to data regardless of the file format used for collection and enable sophisticated image visualization, analysis, querying and annotation. Providing the necessary specifications, documents and software for this application is the goal of the OME project.

3.2 OME Specifications and File Formats

The OME project has produced a number of tools and resources to support the requirements laid out above. The OME project does release specifications, but always includes releases of software that use those specifications. This allows us to ensure the specifications actually work and also provide reference implementations for other interested developers. The following sections detail the different software and specifications of the OME project.

3.2.1 OME Data Model

The OME project bases all of its work on the OME data model, a description of the relationships between various image data and metadata elements (Swedlow et al. 2003). This model forms the foundation for the OME file formats and software the project releases. The full model is available at Open Microscopy Environment (2007c) (requires a browser that can read Adobe SVG). Figure 3.2 shows a portion of the OME data model that describes the image data and metadata (metadata describing

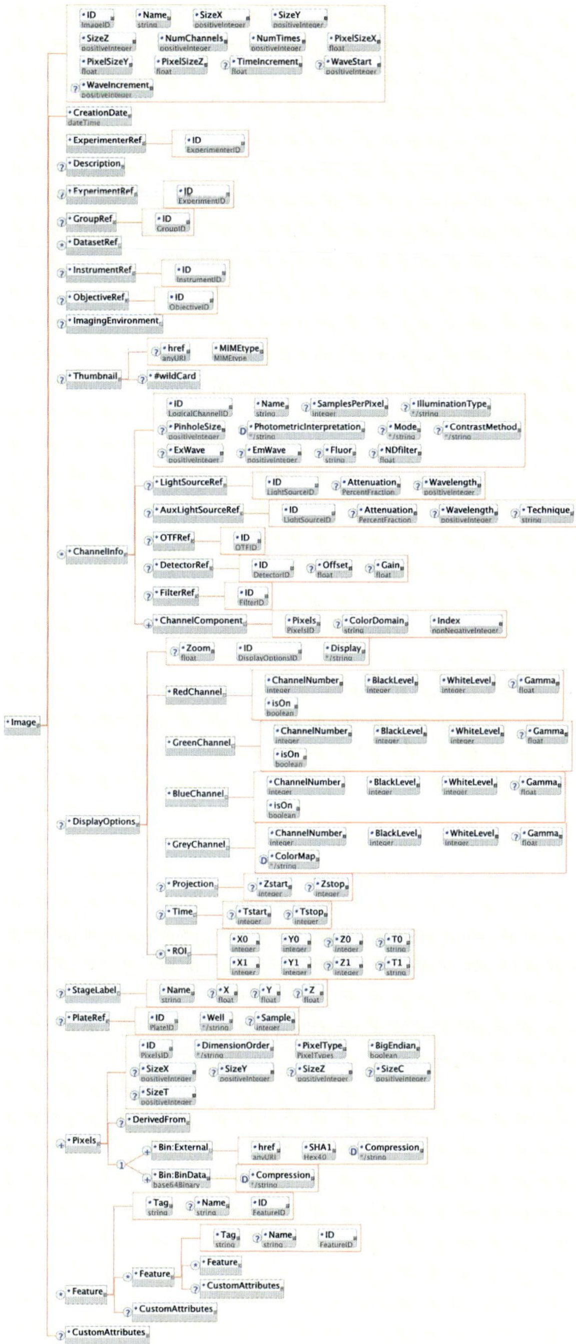

Fig. 3.2 The image element from the Open Microscopy Environment (OME) data model. The relationships between the various data elements are shown. For more information, see Open Microscopy Environment (2007m)

the acquisition instrument are stored under the "Instrument" element – see Open Microscopy Environment 2007c for more information). The model considers an "image" to be a 5D data structure, and thus supports 3D space, time and channel. This model contains descriptors for image dimensions and some of the experimental metadata and definitions and display settings for each of the channels associated with the image. In addition, the model supports results from image analysis (for example, a segmentation algorithm that defines objects within the image) as "Features." Finally, the model includes a facility for updating the metadata associated with an Image, to support any requirements of a specific application, within the "CustomAttributes" field. Any changes in this field become nonstandardized, but can be incorporated into the model in later releases if they appear to be generally useful.

Most recently, we have begun a process of defining updates to the model, to fix a number of omissions in the first version and to support new types of data. These include explicit support for high-content screening data and better support for some new detectors. In general, we collect suggestions for changes to the OME data model at Open Microscopy Environment (2007d) after an evaluation and discussion process on other pages. Note that updates on the OME-XML Evolution page will be released once they have been implemented in OME-XML and OME-TIFF files and in OME software.

3.2.2 OME-XML, OME-TIFF and Bio-Formats

The complexity of imaging experiments mandates the storage of not only images, but also the associated metadata that describe acquisition instrument settings, the user, etc. There is general agreement on the importance of metadata through biological research, and a variety of strategies have been tried to standardize data-storage formats (Swedlow and Goldberg 2006). Most proprietary software faithfully stores these metadata in a proprietary format, making it accessible only with a particular proprietary software package. The only method for migrating data from one proprietary software tool to another is to export image data in a neutral format (usually TIFF), which almost always results in the loss of image metadata.

To solve this problem, we have expressed the OME data model as an Extensible Markup Language (XML; XML.org 2007) file, resulting in the creation of a human- and machine-readable data file that uses a standardized format (Goldberg et al. 2005). All metadata in the OME-XML file are defined by tags that derive from the OME data model. The file explicitly supports 5D data and stores binary image data as base64 text. This strategy is inherently less space efficient than a binary format, but compresses very well, so it does not usually cause a significant storage burden.

OME-XML is thus an effective means for nonproprietary data storage and appears to be a useful tool for data migration – transferring image data between software tools or even collaborators. However, once binary data are stored as base64, especially in compressed form, access to any individual 2D image frame is much slower than with binary data, so reading OME-XML with many frames is relatively slow. To circumvent this problem, OME partners at the Laboratory for Optical and Computational Instrumentation (Laboratory for Optical Computation and Instrumentation 2007a) have developed an alternative use of OME-XML, known as OME-TIFF (Laboratory for Optical Computation and Instrumentation 2007b). This format takes the defined structure of OME-XML and combines it with the de facto standard format for binary images, the TIFF file. An OME-TIFF file can contain single or multiple planes, and thus can support 5D images stored as single files. Since TIFF and XML libraries are commonly available and reading and writing TIFF files is relatively very fast, it seems likely that OME-TIFF can provide an efficient, open format for sharing data between laboratories.

Providing specifications for these formats is useful, but for general acceptance, they must be supported by openly available software. To support OME-TIFF, the OME project has been developing an open library for file format conversion called Bio-Formats (Laboratory for Optical Computation and Instrumentation 2007c). This Java library provides tools for reading over 40 proprietary file formats and if necessary, converting them into OME-TIFF. A plug-in for ImageJ (2007) is available and the library is also in use in the OMERO.Importer (see later). Bio-Formats represents a huge curatorial undertaking, as the metadata in each proprietary file format are identified and converted into the correct data element in OME-XML. Our hope is that as Bio-Formats matures, it will become the standard library for image data translation. In addition to Bio-Formats, OME will begin releasing OME-XML readers, writers and validators in 2007 to help expand the toolset that supports OME-XML.

3.3 OME Data Management and Analysis Software

3.3.1 OME Server and Web User Interface

The first major software application released by the OME project was the OME server. This server and its user interface system are intended to manage the image data from a single laboratory or imaging facility. The design and requirements for this system were laid out in our first paper, which argued for an integrated solution for image informatics (Swedlow et al. 2003). The OME server is based on a Web browser interface and was first released in 2000, and the OME project has released a series of updates to the OME server, most recently in December 2006 with OME2.6.0. At Dundee, our production OME server manages about 1 TB of image data, comprising about one million individual image frames.

Functionally, the OME server imports proprietary files using a series of file translators that mediate the import of data from a number of different proprietary file formats. This strategy allows the data from a number of different acquisition systems to be integrated and accessed on the same system. All acquisition metadata, annotations, textual descriptions, hierarchy assignments and analytic results are stored in a relational database in a data model designed to link all of these different, but related data types. The OME server provides facilities for visualizing and analyzing image data and metadata using its own facilities or links into external analytic tools like MATLAB, and for collecting these data into user-defined hierarchies (Sect. 3.3.1.1).

3.3.1.1 Data Hierarchies

The OME server uses two kinds of hierarchies to help users organize their data. The first, called project/dataset/image (PDI), groups images (each image is a 5D data structure) into "datasets," and datasets into "projects," these groupings, and the names of these structures are defined by the user. An example of this kind of hierarchy is shown in Fig. 3.3. Note that the relationships in PDI are "many-to-many": a single image can belong to multiple datasets, and a single dataset can belong to multiple projects. The PDI hierarchy provides a mechanism for organizing and managing large sets of image data and for collaborating and sharing defined sets of data.

The second data hierarchy, CG/C, is more flexible than PDI. This consists of a "CategoryGroup," which is a collection of categories. Users define their own CG/C hierarchies, but in the OME server the decision was made to make these "global," or allow all users access to all CG/Cs defined on the system. The idea of this facility was to provide a flexible tool for classifying data using user-defined phenotypes or classes. For example, one possible CategoryGroup might be "cell cycle position" and the categories included might be "prophase," "metaphase," "anaphase," etc. These represent visual phenotypes assigned by a user. Another example of a useful CategoryGroup would be "use for paper," with its component categories being "figure 1," "figure 2," etc. As with the PDI hierarchy, an image can belong to multiple different CG/C hierarchies. Currently, in the OME server an image can only belong to a single category within each CategoryGroup. We are still evaluating

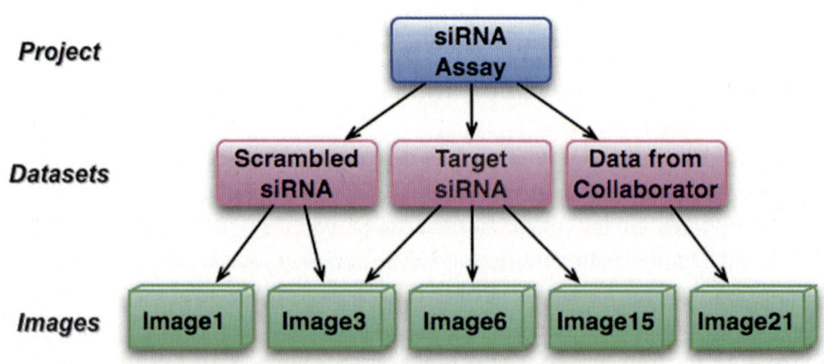

Fig. 3.3 The project/dataset/image (PDI) hierarchy in OME. Each of these are user-defined containers that are used for organizing large sets of images and accessory metadata and analytics on infectious material

how users react to this facility, but the flexibility will hopefully provide a useful mechanism of organizing image data.

3.3.1.2 Semantic Typing in the OME Server

The OME server uses "hard" semantic typing to unambiguously define all elements of data and their relationships (Goldberg et al. 2005). In brief, all data are stored as defined "semantic types" (STs), so that not only the value, but also the meaning of an individual data element is known. Using STs has a number of advantages:

- It uniquely defines data, removing any ambiguity.
- It allows new data elements to be added, again without ambiguity; this facility is at the heart of the schema updates that are possible with the OME server (see later).
- It allows some level of reasoning, especially for deciding if the computational prerequisites have been satisfied (this is used extensively in the analysis engine (see later).

While STs are quite powerful, their use requires substantial knowledge of the underlying OME data model and they are therefore not often appropriate for normal users. Nonetheless, they provide a useful tool for managing complex data types.

3.3.1.3 Image Analysis in the OME Server

The OME server includes an analysis engine that provides a framework for analyzing images contained in an OME server instance. The OME server does not include a fully developed image analysis suite. Rather, it provides a facility for running image analysis and processing on large collections of images.

Currently, analysis is run on datasets in an OME server, with each image in a dataset processed sequentially (a distributed analysis system is being implemented and tested). The OME analysis engine supports the idea of module executions (MEXs), or executions of single atomic image analysis algorithms (e.g., an image segmentation algorithm) and chain executions (CHEXs), or linked runs of multiple MEXs or image analysis steps (e.g., an image statistic calculator, an image segmentation routine and an object tracker). A critical step in defining a chain is satisfying the computational requirements of each module in the chain. This is achieved by declaring the "FormalInputs" and "FormalOutputs" of each module (as always, declared as STs, and stored in the OME database). With this information, it is possible to determine if a chain is logically correct and ready for execution – if appropriate values are stored or will be generated by a previous module, then the next module in the chain can run. Technical details on how to define a new chain are at Open Microscopy Environment (2007e).

The OME server includes some simple image statistics methods, a segmentation algorithm (FindSpots; Goldberg et al. 2005), and an object tracker (TrackSpots). These tools are also installed during a standard installation and can be used for relatively simple object identification and tracking (see Schiffmann et al. 2007 for a tutorial and Schiffmann et al. 2006 for an example of use). Both FindSpots and TrackSpots are C programs, written before the OME server was developed and thus serve as an example of how legacy programs can be integrated into an OME server. However, it is more likely that new programs will be integrated, most often through defined scripting environments. For this reason, the OME server contains a MATLAB handler that allows an interface between the analysis engine and this commercial scripting tool. This allows users to define their own analysis algorithms and apply them to data on an OME server. The OME server MATLAB handler is documented at Open Microscopy Environment (2007f).

3.3.1.4 The OME Server and Dynamic Data Models

The creation and use of customized chains using the OME server is based on known STs within the OME server that describe all the required inputs and outputs for any single image analysis module. However, the flexibility provided by the MATLAB interface and the capability of the OME server analysis engine to support potentially any image analysis module demands that the STs supported in the database can be updated to meet the needs of new modules, and experiments. For this reason, the OME server supports on-the-fly updates to its data model, by importing new ST definitions using XML. Once an ST has been defined, data relevant to that type can be written in OME-XML using the CustomAttributes element in OME-XML (see Open Microscopy Environment 2007g for more info). This mechanism then allows new types of results to be stored on an OME server, as required by new experimental methods or analytic methods. This is perhaps one of the most important capabilities of the OME server, and the one that distinguishes it from most other data server systems.

3.3.1.5 OME Server Architecture

A diagram of the technical design of the OME server is shown in Fig. 3.4. Fundamentally, the OME server is a middleware and server application written in Perl that links a relational database for storage of image metadata and analytic results and a large repository for storage of binary image data (the pixels) with a remote user interface, usually a Web browser. The OME data server (OMEDS) and the OME image server (OMEIS) provide all the interfaces into the relational database (running under the open-source relational database management system PostgreSQL) and the binary image repository, respectively. The analysis engine is included within the data services in Fig. 3.4.

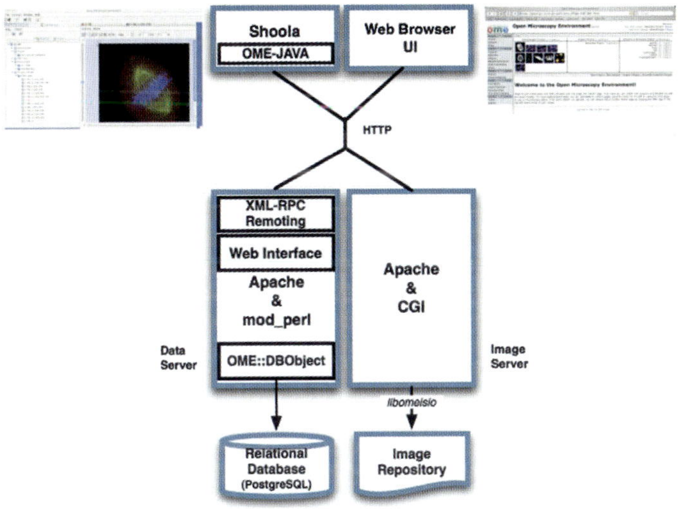

Fig. 3.4 OME server architecture. The Perl-based OME server and user interfaces are built on a number of separate software modules; DB:Object, OME-JAVA, and the remoting architecture is modules built by the OME project. Screenshots of the two OME user interfaces, the Web-browser-based interface and Shoola 2.4.0, a remote Java-based application, are shown

3.3.1.6 OME Server User Interfaces

The OME server Web user interface contains a number of facilities for visualizing, searching, analyzing, and managing image data. A full description of the facilities of this system is at Open Microscopy Environment 2007h. In short, this tool uses a standard Web browser as an interface into an OME server. Facilities are provided for viewing and searching the PDI hierarchy, CG/C hierarchies and image metadata and analytic results. Figure 3.5 shows a screenshot of a Web user interface view of data on an OME server. Image annotations, descriptions and thumbnails, are all indicated, and links for access to previous annotations and runs of analysis chains ("Find and Track Spots") are shown. Following these analysis chain links provides access to actual results – centroids, integrated signal, etc. However, having the data in a Web form is not very useful, especially for further analysis and graphing. For this purpose, a Microsoft Excel file (OME-Excel) is available (download from Open Microscopy Environment 2007i) that can query an OME database and download analytic results from a specific dataset, analysis run, etc. A demonstration of this tool in use is available in Schiffmann et al. (2007). In our own experience, combining the Web user interface with OME-Excel supports a complete workflow: large sets of image data can be imported onto an OME server, organized into a single dataset, processed as a unit and then downloaded for graphing in Excel, all in a completely automated fashion.

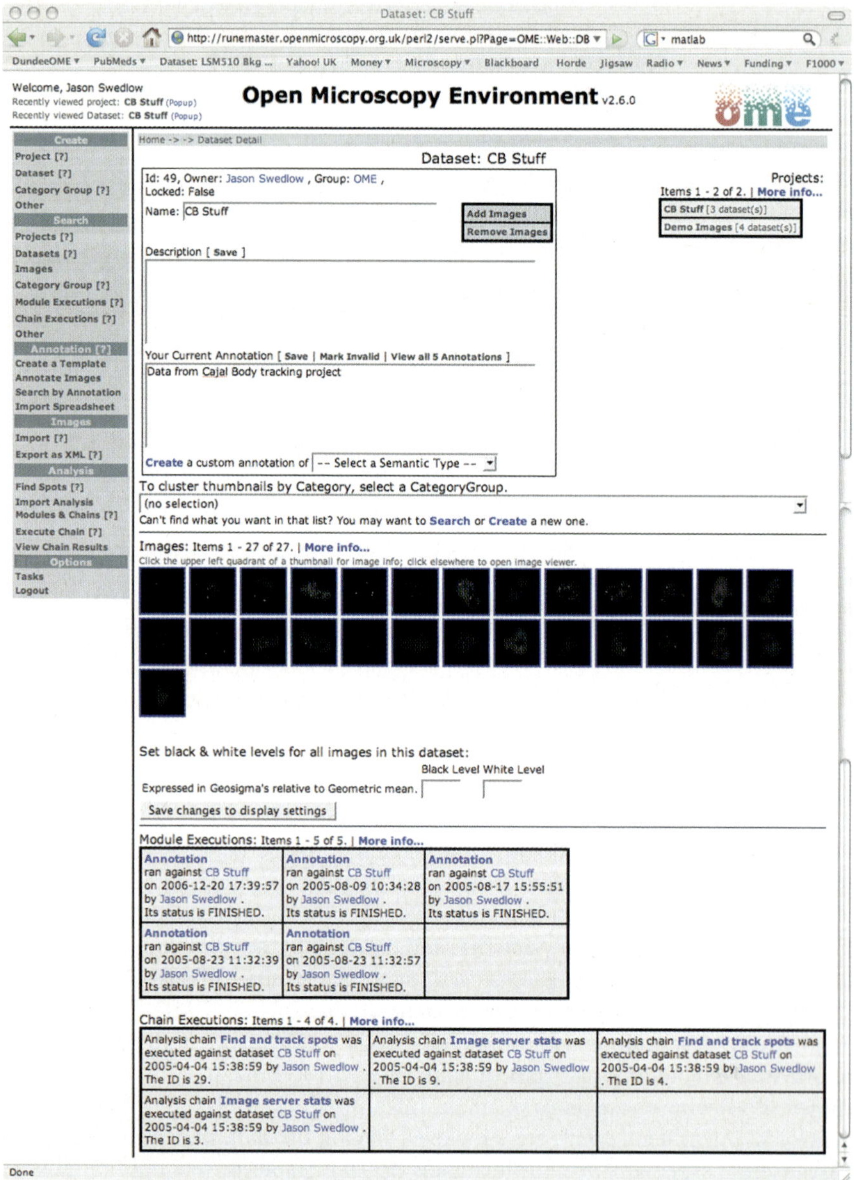

Fig. 3.5 A screenshot of the OME Web user interface, showing a view of a dataset, its associated images, annotations and analyses

3.3.1.7 Limitations to the OME Server and Web User Interface

The OME server (currently released in version 2.6.0) represents the culmination of a large amount of work by the OME team to develop an image informatics tool that could be used by biologists. In general, the system works as advertised and is in use, so far in a relatively small number of laboratories, for large-scale image analysis. Despite the powerful functionality described above, it suffers from a number of limitations:

- Many of the developers and users that have tried to use the OME server find it too complex. This is a common problem with scientific software – too much of the underlying structure is exposed to the user, and access for new developers is too hard.
- The OME server contains a significant number of software components that were written by the OME team, notably DB:Object and OME-JAVA (Fig. 3.4). These have very significant dependencies that have to be maintained as operating systems and other libraries (especially Perl) are updated. This is a significant burden for the OME development team.
- While a Web browser is an effective environment for image data management analysis, it cannot provide as rich a user experience as a standalone application. The Dundee OME group has been developing a remote application for the OME server since 2003 (Shoola; Open Microscopy Environment 2007j), including development of our own Java remote interface (OME-JAVA; Fig. 3.4). However the protocols for data transfer between client and server, because of the underlying structure of the Perl-based OME server are limited to XML-RPC, which is relatively slow for large data graphs, which are necessary for the types of data often stored on an OME server (hierarchies, metadata, etc). In general, these interfaces were much too slow (by about 100–1,000-fold) to be useful.
- Many users requested an easier, simpler system, often requesting "OME Lite." Specifically, many users asked that the flexibility in the server's data model be sacrificed for better support of file formats, better visualization and performance, less emphasis on analysis and easier installation.

For these reasons, the Dundee development team embarked on the development of an alternative OME server that includes much of the functionality of the original server, but in a simpler system. This project, OME Remote Objects (OMERO) was just being released as this chapter was being written, and is the subject of the next section. However, it is important to emphasize that the OME server and Web user interface are quite powerful tools that are in use, have produced published results (Platani et al. 2002; Schiffmann et al. 2006) and are actively developed. The two technologies are complementary and solve different parts of the image informatics problem. The development of both server systems continues and we are currently examining methods of making them interact to leverage the advantages of both systems.

3.3.2 *OMERO Server, Client and Importer*

The OMERO project is focused on providing high-performance and powerful remote-access tools for large image datasets. The OMERO project involves a port of much of the functionality in the OME server to a Java Enterprise application, running inside a JBOSS (JBoss.org 2007) application server. The resulting application, known as OMERO server, is designed for flexibility – as detailed below, it can run on a number of different relational database systems and is designed to support a number of different remote client environments (Java, C++, .NET, Python, etc.). Currently, two Java-based clients, OMERO.insight and OMERO.importer, have been developed that allow interaction with the OMERO server. The following sections detail the design and capabilities of the system and demonstrate its current uses and future goals. All OMERO resources are available at Open Microscopy Environment (2007k).

The OMERO server is based on the OME data model, so all data types describing image data acquisition and data hierarchies are supported. These facilities make the OMERO server useful for image data management – visualization, annotation and organization of large data sets. In its first incarnation, the OMERO server supports the PDI and CG/C hierarchies described already, and provides basic visualization features for multidimensional images via its rendering engine (Sect. 3.3.2.1). Image and hierarchy annotations are supported, and all metadata are searchable. Most importantly, all of these functions are provided in a server that supports a remote client, so images and metadata are accessible via network access yielding a high-performance client environment. Currently, the OMERO server's functionality is limited, but it provides a strong foundation for adding new image browsing and analysis functionality.

The communication between the OMERO server and its underlying database uses a software library called Hibernate (2007). This object-relational mapping (ORM) tool provides the necessary infrastructure for the OMERO server to communicate with a number of database systems. The OMERO server currently is deployed using the PostgreSQL (2007) open-source relational database management system (), but using the facilities in Hibernate has been easily migrated to use Oracle (2007) and MySQL (MySQL.com 2007). Rather than using DB:Object, the custom-built ORM system included in the OME server (Fig. 3.4), Hibernate provides an actively developed library that supports most of the common relational database systems. This provides significant flexibility, with a preexisting tool, and thus represents a substantial simplification of the underlying server base code.

3.3.2.1 OMERO Rendering Engine

A critical requirement for any image software tool is the ability to actually view images. In biological microscopy, even though each pixel value is a quantitative measure of the photon flux at each volume of the sample, it is rarely necessary to present an image to the user that exactly reports each pixel value. Instead, images are "rendered" (sometimes referred to as "scaled") to present a version of the image that is suitable for display on a monitor, and perhaps more importantly that is sufficient to convey the measurement contained in the image. This is not a trivial challenge – almost all modern microscope imaging includes multiple spatial, temporal or spectral dimensions (Andrews et al. 2002) and these must be rapidly presented to users. A particular challenge occurs in high-content imaging, where screens include many multiwell plates, where each well holds a different experimental treatment (e.g., small molecule or small interfering RNA) – how should each well be displayed? In some cases, a full-sized image is required; in others, small thumbnails are sufficient; in others, only a single color that represents a calculated value related to the image is necessary (a "heat map").

To support all of these requirements, the OMERO server includes a rendering engine, a facility for converting pixel data into images that are ready to be painted onto the screen of the client's monitor. For multiple channel images, the a rendering engine contains a multithreading facility to make use of multiple processors in a single machine. In test cases, the rendering engine can deliver rendered 512×512 pixel images to a client, over a gigabit network, in less than 100 ms. In practice, this makes the delivery of images to a client limited by network bandwidth and is sufficient to support rapid delivery of large numbers of thumbnails, scrolling through optical section or time-lapse data or rapid changing of image display parameters. For access to data at truly remote sites, the a rendering engine contains a compression library, so that compressed versions of images that require substantially less bandwidth can be sent to the client. This facility provides fast and efficient image access even at substantial distances.

3.3.2.2 OMERO Server Architecture

The design of the OMERO server is shown in Fig. 3.6. The general layout is very similar to that used in the OME server (Fig. 3.4), with metadata services provided by a data server and image data access provided by an image server. One important design difference is that the OMERO image service (including the rendering engine) is contained in the same process as the OMERO data service, allowing the rendering engine to query the database for any image properties or metrics necessary for proper image rendering. This approach enables future image rendering imaging schemes to be added easily without significant computational burden.

Currently, the only OMERO remote clients are built in Java, allowing the use of Java Remote Method Invocation (Java RMI; Sun Microsystems 2007) as a remoting method. This is sufficient for our current needs, but in the future, to support a broader range of clients, a more flexible remoting strategy, like ICE (ZeroC 2007) will be used. OMERO has been designed with this flexibility in mind, with the aim of supporting many different client environments.

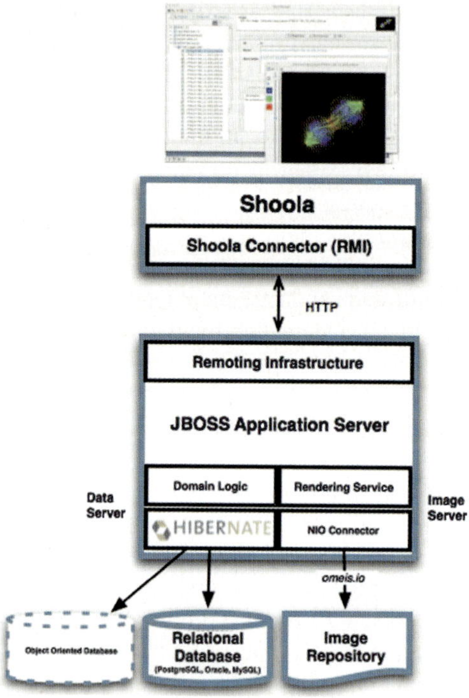

Fig. 3.6 OME Remote Objects (OMERO) server architecture. The Java-based OMERO server is deployed in a JBOSS application server (JBoss.org 2007) and uses Hibernate (2007) for object-relational mapping. These choices provide substantial flexibility – Hibernate can support multiple relational database systems, including PostgreSQL (shipped with the OMERO server), Oracle (2007) and MySQL (MySQL.com 2007). The Java Enterprise framework can support a number of remote application environments

3.3.2.3 OMERO.Importer

The OMERO system implements the same concept of importing data as in the OME server – image data files stored in proprietary image formats are imported into the server, converting all available image metadata into entries into the database. However, there are two differences in OMERO's approach to import. At least initially, import in OMERO occurs via a remote client application and thus enables import from a file system on a remote workstation (e.g., the workstation used to acquire data on a microscope). A user indicates the files to be uploaded and the project and dataset that they will be added to, and then starts the import process (Fig. 3.7).

OMERO.Importer uses the Bio-Formats library (Laboratory for Optical Computation and Instrumentation 2007c) as a source of converters for proprietary files. The Bio-Formats library reads over 40 proprietary file formats, although the wide range of formats and metadata means that ensuring full support for all metadata in each file format is a very involved task that requires constant updating

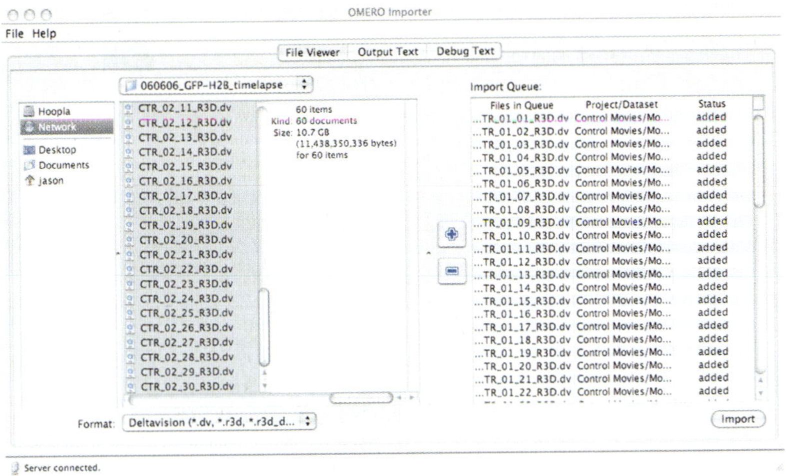

Fig. 3.7 OMERO.Importer A screenshot of the OMERO.Importer tool. The application is reading proprietary DeltaVision files and importing them into an OMERO server

as file formats evolve. Currently (December 2006), OMERO.Importer only supports DeltaVision, MetaMorph and TIFF files, as these are fully supported by Bio-Formats. A major goal for early 2007 is the substantial extension of the range of formats supported by OMERO.Importer.

3.3.2.4 OMERO.insight

The first client built for the OMERO server, OMERO.insight, is a Java-based application that provides access to data in an OMERO server. Initially, this tool supports data management and visualization function only, with image analysis support coming in a future version. Once data have been imported into an OMERO server using OMERO.Importer, OMERO.insight supports the organization of image data into the PDI and CG/C hierarchies, image and hierarchy annotation and visualization, and 5D image viewing (a screenshot of this functionality is shown in Fig. 3.8). All image display settings can be saved to an OMERO server. A key theme of OMERO.Client design is the use of visual hints to help a user to see what is inside different data containers. For instance, each project and dataset name in a classical hierarchy tree is displayed alongside the number of data elements inside it. Annotated or classified images (images assigned to a CG/C hierarchy) are indicated with small icons in the file tree. Finally, a thumbnail of the image, using the latest saved image settings, is shown in the top-right-hand corner of the DataManager. The DataManager (in the current beta version) supports text-based searches of image descriptions and titles, and will support annotation searching in an upcoming version.

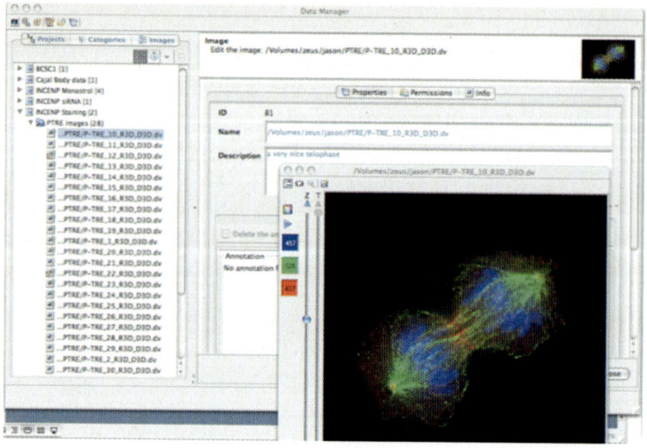

Fig. 3.8 OMERO.insight. A screenshot of the OMERO.insight data visualization tool showing the DataManager, a traditional tree-based data viewing tool and the Image Viewer, a tool for viewing 5D images

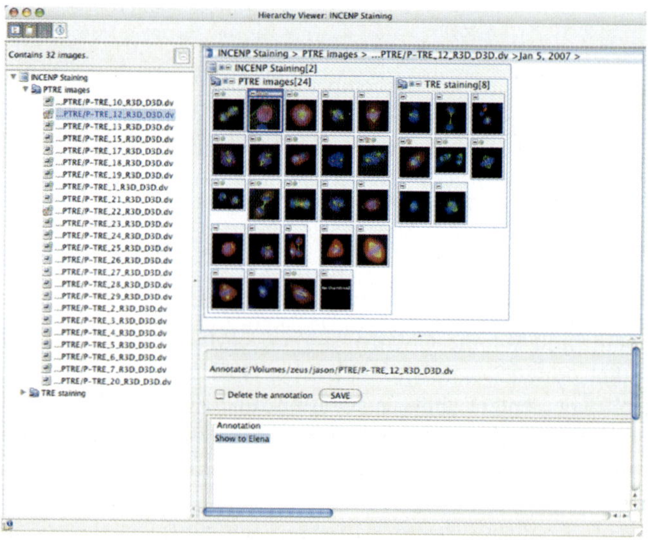

Fig. 3.9 The OMERO.insight HiViewer. The screenshot shows a graphical view of the PDI hierarchy. An optional view on the *left* shows the same hierarchy using a traditional tree view. Thumbnails can be zoomed using mouse-over. HiViewer contains a searching facility for finding specific thumbnails in large sets of data. In this screenshot, one blank thumbnail is shown, indicating a file that is currently being imported from another application

OMERO.insight also includes HiViewer, a tool that provides a graphical view of data hierarchies (Fig. 3.9). This tool is still very much experimental, but the goal is to provide a framework for viewing complex image sets and allowing rapid viewing and access to data organized into hierarchies. HiViewer provides views of thumbnails, either as a simple "flat" array or organized into "tiles" that indicate the properties of PDI or CG/C hierarchies. As in the DataManager, small icons overlaid on thumbnails indicate the presence of annotations or assignments to a CG/C hierarchy in the database. A magnification function allows a user to browse large collections of thumbnails with a mouse-over zoom. Multiple thumbnails can be selected, either using a familiar shift–click function or a click on a paperclip icon on each thumbnail. In the future, this selected set will be a "working set" that can be defined as a new dataset, annotated in batch, or sent for batch processing or analysis.

3.3.2.5 OMERO and Image Analysis

Currently OMERO has a much scaled-down implementation of a data model describing image analysis. The concepts of MEX and CHEX that are central to interactions with the OME server database have not yet been implemented. Instead, all annotations and image processing steps are stored as "events," which are simply defined as a database transaction that includes a modification of image data or metadata. The model for image analysis and processing inside OMERO is still being developed, but in general the OMERO server aims to define image manipulations and processing steps as clients of the server.

3.3.3 Developing Usable Tools for Imaging

With the increasing use of quantitative tools for biological discovery, there is an increasing need for software tools for data management and analysis. This has coincided with the maturing of open-source software, and there are now a number of open-source tools available for image visualization and analysis (ImageJ; ImageJ 2007) and data analysis (R; The R Project 2007). While these tools are often useful, they are often not fully polished software and are less user-friendly than many comparable commercial products. This is usually because functionality is more important than distribution to the community. More importantly, expertise in user interface design and human–computer interaction is often not available to academic research laboratories that are developing data analysis tools. In an effort to improve usability of OME and OMERO software, we have initiated the Usable Image project, a collaborative effort with software design experts to improve the appearance and functionality of OME software (The Usable Image Project 2007). In our first efforts, we have established a regular release and testing process, where users (all based in the author's laboratory) download, install and test new versions of OMERO software. All user interaction is video-taped (users are anonymous), and any bugs, feature requests or other discussions are logged, annotated then converted to an identified work item and prioritized for future development (Open Microscopy Environment 2007l). This process has proven quite effective as it provides rapid software improvements and also ensures users see improvements and additions that they request, thus engaging them in the software development process. We hope to extend this approach to include other imaging laboratories and possibly other data intensive approaches (e.g., mass spectrometry, microarray analysis).

3.4 Conclusions and Future Directions

The OME Consortium has produced a series of tools for managing large image data sets. It is by no means a finished project, but the tools are increasingly in use in the community (Open Microscopy Environment 2007b). The development of both the OME server and the OMERO server provide a flexible set of tools for image data management. With this foundation, the OME project is now moving to enhance the usability of its tools, as well as to extend the functionality delivered with OME software. In 2007, we intend to provide a full suite of support tools for our OME-TIFF and will work with various parts of the imaging community to help make this file format a standard for sharing data. Our OME server is now in a stable, functional form, and can be extended to support complex analysis chains, for high-end image processing and analysis. Our OMERO project has demonstrated the power of remote client applications, which can be extended to handle more of the workflow of the biological microscopist.

Acknowledgements Software development in the author's laboratory on is funded by the Wellcome Trust and the BBSRC. J.R.S. is a Wellcome Trust Senior Research Fellow.

References

Andrews PD, Harper IS, Swedlow JR (2002) To 5D and beyond: quantitative fluorescence microscopy in the postgenomic era. Traffic 3:29–36

Castleman KR (1979) Digital image processing. Prentice Hall, Englewood Cliffs

Conrad C, Erfle H, Warnat P, Daigle N, Lorch T, Ellenberg J, Pepperkok R, Eils R (2004) Automatic identification of subcellular phenotypes on human cell arrays. Genome Res 14:1130–1136

Eils R, Athale C (2003) Computational imaging in cell biology. J Cell Biol 161:477–481

Free Software Foundation (2007a) GNU general public license. http://www.gnu.org/copyleft/gpl.html. Cited 19 April 2007

Free Software Foundation (2007b) GNU lesser general public license. http://www.gnu.org/copyleft/lgpl.html. Cited 19 April 2007

Giepmans BN, Adams SR, Ellisman MH, Tsien RY (2006) The fluorescent toolbox for assessing protein location and function. Science 312:217–224

Goldberg IG, Allan C, Burel J-M, D. Creager D, Falconi A, Hochheiser HS, Johnston J, Mellen J, Sorger PK, Swedlow JR (2005) The Open Microscopy Environment (OME) data model and XML File: open tools for informatics and quantitative analysis in biological imaging. Genome Biol 6:R47

Hibernate (2007) Red Hat Middleware, Rayleigh. http://www.hibernate.org/. Cited 19 April 2007

ImageJ (2007) ImageJ. http://rsb.info.nih.gov/ij/. Cited 19 April 2007
JBoss.org (2007) JBoss, Atlanta. http://www.jboss.org. Cited 19 April 2007
Kiger A, Baum B, Jones S, Jones M, Coulson A, Echeverri C, Perrimon N (2003) A functional genomic analysis of cell morphology using RNA interference. J Biol 2:27
Laboratory for Optical Computation and Instrumentation (2007a) LOCI. http://www.loci.wisc.edu/. Cited 19 April 2007
Laboratory for Optical Computation and Instrumentation (2007b) OME at LOCI – OME-TIFF – OME-TIFF specification. http://www.loci.wisc.edu/ome/ome-tiff-spec.html. Cited 19 April 2007
Laboratory for Optical Computation and Instrumentation (2007c) OME at LOCI – software – Bio-Formats library. http://www.loci.wisc.edu/ome/formats.html. Cited 19 April 2007
Lippincott-Schwartz J, Snapp JE, Kenworthy A (2001) Studying protein dynamics in living cells. Nat Rev Mol Cell Biol 2:444–456
MySQL.com (2007) MySQL, Uppsala. http://www.mysql.com/. Cited 19 April 2007
Open Microscopy Environment (2007a) About OME. http://openmicroscopy.org/about. Cited 19 April 2007
Open Microscopy Environment (2007b) OME examples. http://openmicroscopy.org/use. Cited 19 April 2007
Open Microscopy Environment (2007c) Schema doc. http://openmicroscopy.org/XMLschemas/OME/latest/ome_xsd. Cited 19 April 2007
Open Microscopy Environment (2007d) OME-XML evolution. http://cvs.openmicroscopy.org.uk/tiki/tiki-index.php?page=OME-XML+Evolution. Cited 19 April 2007
Open Microscopy Environment (2007e) Analysis chains. http://openmicroscopy.org/api/xml/ACs.html. Cited 19 April 2007
Open Microscopy Environment (2007f) Image analysis with MATLAB. http://openmicroscopy.org/howto/quantitative-image-analysis-MATLAB.html. Cited 19 April 2007
Open Microscopy Environment (2007g) Introduction to OME-XML schemas. http://open microscopy.org/api/xml/. Cited 19 April 2007
Open Microscopy Environment (2007h) OME Web client "Marino". http://openmicroscopy.org/getting-started/web-client.html. Cited 19 April 2007
Open Microscopy Environment (2007i) Source directory of /OME/src/Excel. http://cvs.openmicroscopy.org.uk/horde/chora/browse.php?f=OME%2Fsrc%2FExcel%2F. Cited 19 April 2007
Open Microscopy Environment (2007j) http://cvs.openmicroscopy.org.uk. Cited 19 April 2007
Open Microscopy Environment (2007k) OME-downloads. http://openmicroscopy.org/downloads. Cited 19 April 2007
Open Microscopy Environment (2007l) OMERO Trac. http://trac.openmicroscopy.org.uk/omero. Cited 19 April 2007
Open Microscopy Environment (2007m) Introduction to the OME-XML schema. http://open microscopy.org/api/xml/OME/. Cited 19 April 2007.
Oracle (2007) Oracle 10g. http://www.oracle.com. Cited 19 April 2007
Phair RD, Misteli T (2001) Kinetic modelling approaches to in vivo imaging. Nat Rev Mol Cell Biol 2:898–907
Platani M, Goldberg I, Lamond AI, Swedlow JR (2002) Cajal body dynamics and association with chromatin are ATP-dependent. Nat Cell Biol 4:502–508
PostgreSQL (2007) PostgreSQL Global Development Group. http://www.postgresql.org. Cited 19 April 2007
Schiffmann DA, Dikovskaya D, Appleton PL, Newton IP, Creager DA, Allan C, Nathke IS, Goldberg IG (2006) Open microscopy environment and FindSpots: integrating image informatics with quantitative multidimensional image analysis. BioTechniques 41:199–208
Schiffman DA, Appleton PL, Goldberg IG (2007) FindSpots (OME v2.5.1) user guide. Available at http://www.openmicroscopy.org/howto/FindSpots-v2.pdf. Cited 19 April 2007
Simpson JC, Wellenreuther R, Poustka A, Pepperkok R, Wiemann S (2000) Systematic subcellular localization of novel proteins identified by large-scale cDNA sequencing. EMBO Rep 1:287–292

Sun Microsystems (2007) Remote method invocation home. http://java.sun.com/javase/technologies/core/basic/rmi/index.jsp. Cited 19 April 2007

Swedlow JR, Goldberg I (2006) Data models across labs, genomes, space, and time. Nat Cell Biol 8:1190–1194

Swedlow JR, Sedat JW, Agard DA (1997) Deconvolution in optical microscopy. In: Jansson PA (ed) Deconvolution of images and spectra. Academic, New York, pp 284–309

Swedlow JR, Goldberg I, Brauner E, Sorger PK (2003) Informatics and quantitative analysis in biological imaging. Science 300:100–102

The R Project (2007) The R project for statistical computing. http://www.r-project.org/. Cited 19 April 2007

The Usable Image Project (2007) The usable image project. http://www.usableimage.com. Cited 19 April 2007

Wouters FS, Verveer PJ, Bastiaens PI (2001) Imaging biochemistry inside cells. Trends Cell Biol 11:203–211

Yarrow JC, Feng Y, Perlman ZE, Kirchhausen T, Mitchison TJ (2003) Phenotypic screening of small molecule libraries by high throughput cell imaging. Comb Chem High Throughput Screen 6:279–286

ZeroC (2007) Welcome to ZeroC™, the home of Ice™. http://www.zeroc.com. Cited 19 April 2007

XML.org (2007) XML.org. http://www.xml.org. Cited 19 April 2007

4 Design and Function of a Light-Microscopy Facility

Kurt I. Anderson, Jeremy Sanderson, and Jan Peychl

Abstract Modern biological research depends on a wide variety of specialized techniques, which collectively are beyond the grasp of a single research group. Research infrastructure, in the form of services and facilities, is therefore an increasingly important foundation for a competitive research institution. A light-microscopy facility is a place of dynamic interaction among users, staff, and equipment. Staff provide the organization, continuity, and expert knowledge required to manage the laser-safe interaction between demanding, selfish, high-performance users and delicate, expensive, high-performance equipment. They introduce novice users to fundamental principles of image acquisition and analysis, often beginning with fluorescence basics, but collaborate with advanced users in the development of new imaging techniques. Intimate knowledge of the experimental needs of the user research groups is required to maximize the effectiveness of equipment purchases, which are also informed by critical evaluation of local sales and support teams. Equipment management encompasses evaluation, purchase, installation, operation, and maintenance, and depends critically on good relations with competent local technical support. Special care should be given to the architectural design of an imaging facility to maximize the utility and comfort of the user environment and the long-term performance stability of the equipment. Finally, we present the details of a web-based equipment scheduling database as an essential organizational tool for running an imaging facility, and outline the important points underlying the estimation of hourly instrument costs in a fee-for-use setting.

4.1 Introduction

Specialization is a hallmark of evolution, both of living organisms and of scientific fields. Modern cell biological research depends on a wide variety of highly evolved techniques, including among many others DNA sequencing, mass spectroscopy, bioinformatics, production of transgenic animals and antibodies, electron microscopy, and light microscopy. This list grows daily, as today's cutting-edge approaches

S.L. Shorte and F. Frischknecht (eds.), *Imaging Cellular and Molecular Biological Functions.*

become essential components of tomorrow's research. Likewise, insight in cell biology often depends on results obtained using multiple techniques. Research infrastructure, in the form of services and facilities, promotes research by allowing all researchers to access important specialized techniques. In simple terms, techniques are "provided" to users by the service in the form of expensive hardware and the expert knowledge required to use it effectively.

The distinction can be made between a service, where staff generate data for the users, and a facility, where staff help users to generate their own data. This distinction has important organizational consequences. In a service environment the equipment can be better maintained and may perform to a higher standard, but higher levels of staff are required. For a large number of systems the service approach may become untenable. In this chapter we will consider implementation of a light-microscopy facility which includes advanced imaging systems such as laser scanning confocal microscopes and has the capacity to cover from a few dozen to a hundred users.

Despite the profusion of core imaging facilities, there is a dearth of literature giving any guidance on how to design, set up and manage these, and the literature written before the turn of the century tends to cover electron microscopy (e.g., Alderson 1975). Judy Murphy has written on facility design, primarily for electron microscopy (Murphy 1993, 2002) and on database selection (Murphy 2001). The most recent article specifically on setting up and running a confocal microscope facility is that of DeMaggio (2002). Another article, by Helm et al. (2001), deals with installing three multimodal microscopes capable of single-photon and multiphoton operation onto one optical table. A usefully illustrated Bio-Rad technical note on setting up a laser scanning microscopy resource was provided by White and Errington (2001) and can be obtained with a request to the confocal microscopy listserver (University at Buffalo 1991). The microscopy (Zaluzec 1993) and confocal (University at Buffalo 1991) listservers both offer a dynamic forum where microscopy-managers exchange views and solutions regarding the practical challenges of running a core imaging facility. The issue of cost management is most often aired. Two recent articles (Humphrey 2004; Sherman 2003) cover policy aspects of managing a core facility, as does the paper by Angeletti et al. (1999), similar to the issues described here.

Light microscopy entails image acquisition, processing (including deconvolution), and analysis. In our view it is best to keep these functions under one roof, as opposed to having separate image acquisition and analysis facilities. The reason for this is that acquisition and analysis are intimately related, and must be closely coordinated for the final outcome to be valid. Separation of these functions creates the potential for conflicting advice from separate acquisition and analysis teams. However, image acquisition, processing, and analysis are each specialities, which can comprise full-time jobs. It is important to remain realistic about the level and type of support which can be offered with a given number of staff positions.

Light microscopy has long been a fundamental technique in cell and developmental biology. The development of genetically encoded fluorophores has revolutionized these fields. Genetic techniques exist to label and manipulate the expression of virtually any gene product. In response to these genetic tools, there have been tremendous technological advances to more accurately visualize fluorescent protein dynamics in living cells, tissues, and organisms. Today there exist an often bewildering multitude of advanced imaging techniques, some of which are broadly useful and some of which are only optimal for a narrow range of applications. It often occurs that molecular biology specialists reach suddenly for advanced imaging equipment at the end of a long genetic experimental procedure, with predictably variable results! This chapter describes approaches we have found successful in setting up, running, and expanding the imaging facilities of the Max Planck Institute for Cell Biology and Genetics (MPI-CBG) in Dresden, the Beatson Cancer Research Institute in Glasgow, and the University of Sheffield. The information contained here is the result of practical experience, which necessarily reflects our preferences and opinions. Others facing similar issues may choose to address them differently.

An imaging facility comprises hardware, people, and organization. "Hardware" refers both to the equipment and to the space in which it resides. "People" refers to both the staff and the users. "Organization" ultimately determines the efficiency with which hardware and people interact. In our view a Web-based database is an essential tool for organizing efficient interactions among hardware and people.

4.2 Users

Users are what it is all about; they can be a blessing and a curse! Good users push staff for assistance with cutting-edge applications, provide valuable feedback about the state of equipment, and are key to recognizing new trends. All users are important for identifying system problems and giving feedback to the staff about the state of equipment. Bad users are selfish monsters who expect the equipment to be in top working condition when they want to use it, but give no thought to the condition in which they leave it. Good luck!

All users should receive a formal introduction prior to using a piece of equipment, no matter what their experience level. The introductory session is an important chance for facility staff to assess the user's actual (as opposed to reported) level of experience, and to identify the applications involved. Novice microscopists need to be educated about basics of fluorescence microscopy, such as matching fluorophores to excitation sources and emission filters, choosing fluorophores for use in multiple-label experiments, controlling cross-talk, balancing parameters such as resolution, acquisition speed, sensitivity, and sample longevity, and finally the proper use of equipment so as not to damage it. It is important to separate training from data collection, especially where live samples are concerned. While it may be beneficial to use real samples as training specimens, biological interest may overwhelm the user's attention span, i.e., the desire to see a certain result may interfere with general learning about the microscope. Users should be encouraged to consult staff at an early stage when embarking on new applications. This ensures that users' expectations of the capabilities of the existing equipment are realistic, and gives staff time to prepare equipment for new applications. It is also important for users to be aware of the use-load on various systems to have realistic expectations about instrument availability in order to balance this against their experimental needs. Advanced users require help with advanced applications, which forces staff to invest time in learning and preparation in order to provide assistance. This should be encouraged because the expert knowledge of the staff is a crucial asset of the facility, which benefits future users.

4.3 Staff

Staff organize and run the facility by managing users and equipment. Facility staff represent a pool of expert knowledge and experience available for users to consult, who transmit information among users, imaging specialists, and the producers of imaging technology. Their importance and the number of staff required to support an imaging facility are often underestimated. A general overview of staff responsibilities includes safety compliance, especially laser safety, teaching, training, and general user support, equipment maintenance and quality control, and administration. Imaging equipment requires the expert knowledge of competent, motivated staff to deliver maximum performance. In addition, staff maintain an overview of, and serve as a knowledge repository for, the critical imaging applications of the local users. This provides continuity of research results as students and postdocs leave, taking their expert knowledge with them. In this context it may be useful for users to brief staff after important imaging sessions, and for staff to attend research group meetings. Continuing training and education are also essential components of the job, including attending scientific meetings and/or trade exhibits. Staff are able to set unbiased priorities for the resource allocation of a facility based on overall use, not the urgent needs of one vocal group or user. They also provide a vision for future trends by monitoring new developments in the imaging field. The number of staff positions needed to run an imaging facility is determined by the number of imaging systems present, their level of weekly use, the level of support expected by users, and the number of users (Sect. 4.5.3).

4.3.1 Workplace Safety

Staff play an essential role in establishing a safe work environment, especially where laser safety is involved. Very briefly, this first involves identifying hazards and the people likely to be affected by them. Then risk reduction measures are established, including both physical protection measures and organizational measures such as standard operating procedures designed to minimize risk. Users must be made aware of risks and trained in the standard operating procedures designed to protect them. Finally, all of these steps must be documented and periodically reviewed. Laser safety is discussed in Sect. 4.4.4.1.

4.3.2 User Training

User training is an important part of the job, which may not be initially appreciated. Users must first be trained to a level of unassisted competence in equipment use. This can be effectively accomplished during one or two standard, one-on-one training sessions. These sessions last a couple of hours each, and cover fundamental principles such as fluorescence basics, image formation, and confocal detection. Remember that poorly trained users will only be frustrated in their attempts to get good results, and this frustration will ultimately be turned back on the facility. Stomping out the many little fires users encounter on a daily basis dramatically improves the efficiency of equipment use and the level of user happiness. It is also important to follow up on user questions, to have answers by the user's next session. User training may involve courses on specific systems or techniques, which may be taught directly by the staff, or organized by staff in conjunction with company application specialists or recognized experts.

4.3.3 Equipment Management

On the technical side, staff provide assurance to users that the system is performing to a high standard by monitoring and documenting parameters such as the cleanliness of optical components, their mechanical stability, and the power levels of illumination sources such as lamps or lasers. Such quality control may also include regular estimates of system performance, such as resolution and sensitivity, through the use of reliable test samples.

A further crucial function of the staff is to identify problems with equipment and manage solutions. Staff must have the technical competence to handle smaller problems directly. Some manufacturers offer advanced training courses to staff, which allow them to undertake a wide range of adjustments and repairs (for example, laser fiber alignment to recover excitation power in a laser scanning confocal microscope). The cost of such courses should be viewed as an investment, which can be offset quickly by repair bills and wasted user time. When larger problems occur, staff can speed up repairs by performing initial diagnostic tests in close coordination with company service staff. This helps to avoid the situation that the service engineer diagnoses the problem on the first visit but has not brought the necessary parts or tools to finish the job, thus requiring a second visit which may be days to weeks later depending on the engineer's schedule. The goal is to ensure that service engineers bring the necessary parts and knowledge to fix the problem in one visit. Staff then follow up on repairs to ensure they are complete.

Note that the easiest way to keep the equipment in perfect working condition would be to lock the doors and keep the users out! Although seemingly trivial, it is important to ensure that the user does not become the enemy. One way to promote this is for the facility leader to be a facility user as well, i.e., to conduct research which uses the facility as well as managing it. It must be emphasized that a position split between research and running a facility is no longer a full-time research job; however, running a facility does confer research advantages. Leading the facility goes hand in hand with developing an expertise in imaging. This expertise, as well as the facility itself, can be used to attract collaborations with imaging nonspecialists. As a research group leader, the facility leader might develop customized imaging systems for his/her own use, which would then be made available through the facility to benefit the local user community.

4.4 Equipment

The equipment of an imaging facility consists of the imaging systems and their many accessories (i.e., large equipment and small equipment), the space in which they are located, and the tools necessary to keep them running.

4.4.1 Large Equipment

Microscopes, cameras, lasers, and computers are the heart of an imaging facility, and ultimately determine what experiments can be performed. Great care must be taken in purchasing large equipment to ensure that the most benefit is obtained from precious funds. Here are some points to consider with respect to equipment.

You do not just buy a microscope, you also buy the company which supports it. Equipment purchase creates a relationship with sales teams, application specialists, and most importantly service teams. When evaluating equipment for purchase it is crucial to evaluate these company support teams in the context of your own specific facility. How much expert knowledge exists among the users? What is the background of the company application specialists? What is the service procedure when equipment breaks down? How near are the service engineers based? How big an area do they serve? How good are they? A good service engineer is worth his/her weight in gold; a poor engineer will drive you progressively to despondency, despair, and drink – insist on good servicing.

Salespeople will generally tell you about the weaknesses of their competitor's products, not their own. Evaluating a system prior to purchase depends on a host of small decisions and factors gleaned from many sources. By all means get the official sales pitch, but also ask for a list of previous purchasers and reference customers. Contact these people for their experiences concerning ease and stability of system use, especially software, and also how they rate after-sales service and support.

It is crucial for facility staff to direct purchasing of the right equipment based on familiarity with the research of the local user community and an overview of products on the market. The latest-greatest imaging technology may not be useful for the applications of the users the facility covers. Software support for a wide variety of hardware components is key, for both versatile use of existing equipment and future system upgrades. Hardware and software flexibility can be maximized through modular system design, for example, though the ability to mount the same optical fiber for epifluorescence illumination on a mercury lamp, xenon lamp, and monochromator. Avoid combining too many features (i.e., total internal reflection fluorescence, spinning disc, and microinjection) on a single imaging system. User access to the system will ultimately become a problem if each separate function is in high demand. Complexity also increases the likelihood of failure. Failure of a system with multiple special functions means that more users will be affected than if TIRF, spinning disc, and microinjection features were on three separate microscopes.

"Many different systems, or many of the same system?" is another basic question. It takes time to become familiar with a system and learn to use it effectively. Having two or more of the same system increases user and staff familiarity with that system (especially its bugs!). The more different systems present, the more time required for a user to get to know them, or conversely the less likely that a user will be an expert user of all of them. User access to "cloned" systems is more flexible, if a clone breaks down, users can work on one of the others. User familiarity with equipment and consistency with previous results are strong inhibitors of trying something new. However, spending precious funds on multiple copies of the same system ultimately limits user access to other, potentially useful technologies. In the case of laser scanning confocal microscopes, each of the major systems has unique advantages over the others. Some experiments may truly work better on one system compared with the others. Company representatives are usually keen to have a monopoly position within an institute, i.e., to see that only microscopes from their company are bought within the whole institute. But in our experience having multiple suppliers in-house generates competition which keeps all the suppliers on their toes.

Technology development must be carefully considered in the context of a user facility. The raison d'être of an imaging facility is to provide access to commercially available, advanced imaging systems. Acquisition of experimental data in an imaging facility requires that system configurations are stable from week to week and month to month. Developmental imaging systems may offer important performance advantages over existing systems, but spend much of their time in configuration flux, i.e., the configuration changes often and the system spends much of its time in pieces on the benchtop. At the point where system parameters stop changing, the system is no longer under development. Advanced users may appreciate the benefits to be had from new technology and therefore be eager to help in development. Other users may be scared away from developmental systems by the extra patience needed to obtain results. It is often useful to "shake down" new systems by first allowing their use only among advanced users, who can identify pitfalls and bugs with facility staff, so they can be corrected or worked around before turning the system over to general use.

4.4.2 Small Equipment

A wide variety of small equipment is required in conjunction with a microscope, especially when live-cell imaging is involved. It is important to budget for this in order to extract maximum value from a microscope system which is already expensive enough. To name but a few bits, microscope accessories include objectives, stages, condensors, lamps, power supplies, and sets of fluorescence filters. For components which occasionally fail, such as power supplies, it is a good idea to keep a backup unit. Having an extra \$100 power supply can keep a \$100,000 system running; this is where standardization and flexibility of components are important. Small equipment can further encompass computers and monitors, antivibration tables, heating chambers, CO_2 regulators, peristaltic pumps, and microinjection equipment, including needle pullers, micromanipulators, and pressure regulators. All of these bits enable the staff to flexibly cope with shifting user applications, especially the ability to "just try" something out to see if it is worth pursuing.

4.4.3 Tools

As with any undertaking, be it plumbing or molecular biology, good tools are essential for doing a job quickly and correctly. The tools required to support an imaging facility include various screwdrivers (flat, Phillips, and hexagonal heads), spanners, and socket sets with a good selection of small sizes, a razor knife, flashlights, a multimeter, a laser power meter, and an electronic thermometer with a fine probe. Furthermore, compressed air, lens paper, and a variety of cleaning solutions in dropper bottles are essential cleaning aids. Useful solutions include water, ethanol, 1:1 mixture of water and ethanol, and petroleum benzene.

4.4.4 Imaging Facility Layout

There are many considerations in designing the physical space of an imaging facility (Fig. 4.1). These include:

- Laser safety
- User environment
- Equipment environment

4.4.4.1 Laser Safety

For a thorough introduction to laser safety the reader is referred to Winburn (1989). A common example is provided here for discussion. Facility staff must be aware of the wavelengths and power levels associated with each laser built into an imaging system. Imaging systems, such as confocal laser scanning microscopes, generally have lower overall laser classification than the lasers they contain, i.e., a system containing a (hazardous) class 3B Kr–Ar laser may be classified as class 3A (safe) because safety features of the microscope protect the user from the full power of the class 3B laser. Some of these safety features may be defeated by the user, for example, by removing an objective and inserting a mirror into the laser beam path while scanning. Safe operating procedures and user training are important to prevent users from unintentionally exposing themselves to hazardous levels of laser radiation. The full power of the Kr–Ar laser may also be emitted when the system is serviced. For this reason it is important to restrict access to lasers, as discussed below.

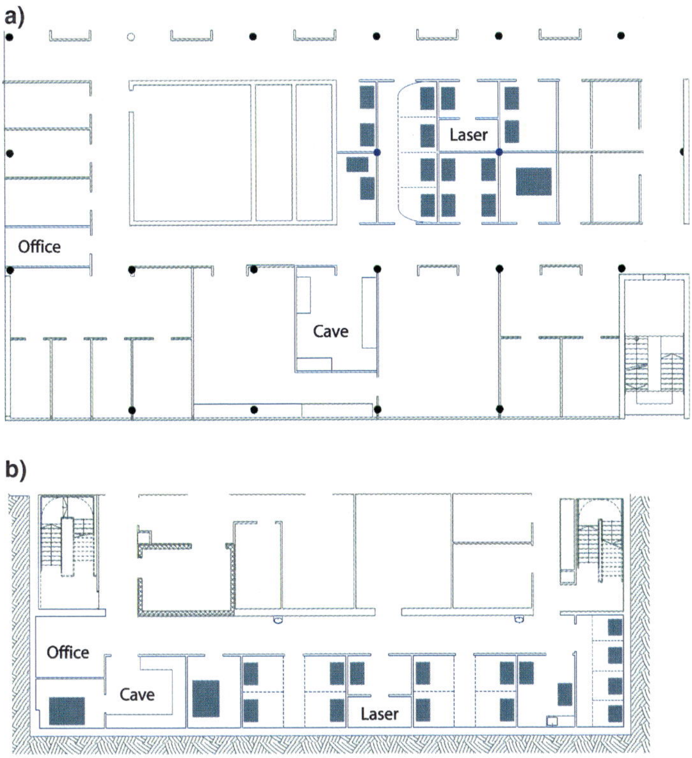

Fig. 4.1 Imaging facility floor plans. **a** Max Planck Institute for Cell Biology and Genetics, **b** Beatson Cancer Research Institute. *Dark lines* indicate space belonging to the imaging facility, *gray lines* delineate other laboratory space. Microscope workstations are indicated by *shaded rectangles*. Bench space for computers for image processing and analysis is indicated by *open rectangles* in the rooms marked *Cave*. Proximity of the cave to the acquisition stations is important for (1) allowing users to quickly check parameters associated with image acquisition and (2) allowing staff to assist with both image processing as well as acquisition. The office has a glass door and a partition overlooking the hallway, allowing staff to monitor events in the facility. Proximity of office to workstations is important for good staff oversight of users and equipment. *Broken lines* indicate sliding curtains used to flexibly partition large rooms into smaller workspaces. *Laser* indicates the laser room, which can accommodate lasers and other loud and/or hazardous equipment associated with microscopes in the adjoining rooms. Note that in **b** access to the imaging facility is controlled via doors with magnetic card locks at either end of the hallway. The facility in **a** is located on the first floor of the building. The facility in **b** is located in the basement, with a light-well providing daylight to the office

The manner and environment in which lasers are used are important aspects of laser safety. For example, it is generally safer to avoid using lasers in “free-space,” i.e., open on a benchtop. Fiber-optic coupling at the laser head prevents user interaction with the laser beam. Fiber-optic coupling has an additional advantage, introducing the opportunity to place the laser and the microscope in separate rooms. Additional advantages to user comfort and equipment stability are outlined below. The environment in which lasers are used is important for controlling user access to lasers, especially during the time of equipment servicing. Generally this requires locking all users out of the room in which the system is located while the service staff work on it. Alternatively, if lasers and microscopes can be placed in separate rooms via fiber-optic coupling, service staff can work on hazardous lasers in one room while users remain free to use neighboring systems in the next room.

4.4.4.2 User Environment

Live-cell fluorescence imaging typically requires darkness. For this reason it is important that the ambient lighting level be individually controllable and positionable at each imaging system, for example, by using an architect’s desk lamp. The microscope is not only a place for data acquisition, it should also be a place for users to present results to their colleagues and discuss experimental parameters. For this a quiet, private environment where two people can sit comfortably is optimal. The floor space required for an advanced imaging system is around 4–9 m^2 – enough space for a couple of tables and one or two people sitting at the microscope. Alternatively a simple upright microscope may require only 100 cm benchfront alongside other similar systems. Access to system components during installation and maintenance often requires a larger work space. For this reason we have chosen to compartmentalize larger rooms using sliding curtains as dividers. This flexible approach offers many advantages. When the curtains are closed, the user environment is small and private, with good control over ambient lighting. When the curtains are open, access to equipment is enhanced for installation and maintenance. Individual systems can be grouped together for teaching purposes. It is also easier for staff to oversee multiple users and for users to compare results on different systems. This approach is complemented by creation of a central equipment room to house all the lasers and other electronic equipment. This has the significant advantage of removing hot, loud, delicate, hazardous equipment from the user environment.

4.4.4.3 Equipment Environment

Optical components are extremely sensitive to fluctuations in temperature. Zeiss specifies $22 \pm 3°C$ and less than 65% humidity for the operating environment of an LSM 510. Fluctuations of $\pm 5°C$ can rapidly lead to misalignment of laser coupling and a drop in excitation power of the instrument, which generally requires a service visit for correction. High humidity, especially in the summertime, can destroy

Table 4.1 Utility considerations. Manufacturer's information on power and cooling requirements for selected instruments

	Power	Heat exhaust (kW)	Water cooling
Zeiss LSM 510 Vis Laser Module	230 VAC (Europe), 3 phase, 16 A per phase 115 VAC (USA), 2 phase, 25 per phase	4	–
Coherent Enterprise II	208–240 VAC, single phase, 29–34 A	–	7.6 l/min, 1.4–4.1 kg/cm^2, 10–60°C inlet temperature
Coherent Innova 70c	208 VAC, 3 phase with ground, 10 A per pulse	–	8.5 l/min, 1.8–4.23 kg/cm^2, 10–35°C inlet temperature
Spectra-Physics Chameleon	220 VAC/6 A 110 VAC/10 A	<2.4	Comes with closed loop chiller

expensive water-cooled lasers by causing condensation inside the laser head and power supply. Dust is an ever-present menace which interferes with the transmission of light through optical components, and in extreme cases can hinder equipment cooling. Control over these environmental parameters is easier to maintain by placing sensitive mechanooptical and electronic equipment in one small room, with restricted user access and a high concentration of utilities, such as heat ventilation, water cooling, special electrical supplies, compressed air, dehumidifier, and dust filter. High-power gas lasers, including the UV lasers used on many confocal microscopes, often have very specific power and cooling requirements, including input and output pressure, input temperature, and flow (Table 4.1). Restricting special utilities to one location can increase their effectiveness and decrease the cost compared with installing them in many rooms. Laser safety advantages with these approaches have already been mentioned. Additional advantages include increased operating stability and protection of sensitive components from user interference.

4.5 Organization

4.5.1 Equipment-Booking Database

The management of a service or facility lies somewhere between running a company and running a laboratory. It is not enough to concentrate on training users and maintaining the microscopes: precious time and effort are required to organize the operation of the facility in such a way as to minimize the administrative burden and maximize the efficiency of equipment and staff. Good organization will exponentially improve the efficiency of operation and contribute significantly to the happiness of users and staff. In our view, an equipment scheduling database is the central tool for managing a multiuser resource such as an imaging facility. It is the central interface with which all users interact in order to book equipment and make appointments with staff members. Paper calendars placed next to the equipment are simply insufficient for all but the smallest of operations.

To begin with, a booking database allows users to book equipment from anywhere on a given computer network, and allows both users and staff to monitor staff availability. This allows users to plan their experiments more conveniently and effectively, eliminating the need to run from one system to the next when checking availability. An online database also allows external users, guests, and collaborators to check equipment availability and confirm bookings ahead of their visit. Similarly, bookings can be cancelled at any time from anywhere. The booking database can also be used to introduce booking restrictions, such as the length of time a booking may last at certain times of the day. We use this feature to keep bookings short during the middle of the day, which increases user access to high-demand systems during peak-use hours. The booking database can also limit the number of days or weeks into the future when use may be booked. This increases the accuracy of user planning and minimizes unexpected cancellations. In the event that users are charged for instrument use, the booking database can automatically generate completely transparent and highly accurate bills to individual users or groups.

A booking database can also promote communication among users and staff. This can take the simple form of a text message appearing on the calendar next to slots which have been blocked or cancelled, explaining why the equipment is unavailable for use and possible effects on future availability. For example, the message "Microscope blocked for 543-nm laser service" has the dual function of informing users why the system is unavailable and reassuring users who had problems with the 543-nm laser that something is being done about it. Communication is further facilitated if the database can automatically generate e-mail to staff and users under appropriate conditions. For example if a system breaks down and staff are forced to cancel bookings, the booking database can automatically send e-mail to all affected users with a short explanation (written by staff) of why the cancellation was necessary. Another important moment for communication is when users book a member of staff for assistance. The database can incorporate a text field on the booking page where users enter comments which will be sent to the requested staff member about exactly what assistance is required (Fig. 4.2). This helps the staff prepare for the booking without having to track down the user and ask why they have been booked. Finally, a booking database can include error-reporting and problem-tracking features which allow users to report equipment problems and staff to record the actions taken to solve them.

Note that in addition to the booking database, mailing lists have an important role in the smooth running of daily events through increased communication among staff and users. For example, all users should belong to a facility users' mailing list, which can be used for important announcements. In addition, we have employed a mailing list for users to swap and fill cancelled bookings. This

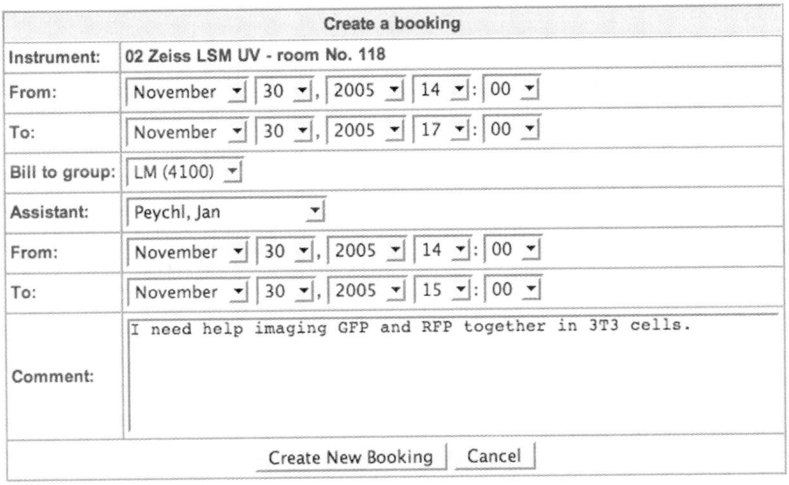

Fig. 4.2 Screenshot of booking database showing booking options. This window allows the user to book a microscope, in this case a 02 Zeiss LSM UV, as well as an assistant, Jan Peychl. The field "*Bill to group*" allows the charges for this booking to be directed to any groups or grants available to the user. Note that the equipment and assistant bookings begin at the same time but the assistant booking is only for 1 h. In this case the user expects assistance for the first hour, but will work independently thereafter. The text in the comment field along with the details of the booking are automatically sent in an e-mail to the assistant to aid in preparation for the booking. The time for which the assistant is booked is also automatically blocked out in the assistant's booking calendar

helps users to make use of last-minute cancellations by other users, and keeps the level of equipment use high.

Aside from scheduling and communication the booking database is an indispensable tool for planning and resource allocation, providing feedback about how the equipment is being used and the use patterns of individual users and groups (Fig. 4.3). Information such as the average number of hours a system is booked per week is essential to prioritize and rationally discuss future equipment purchases. Likewise it is important for staff to identify heavy users and groups, as opposed to vocal users and groups, when allocating resources. Finally, the sum total of all instrument-hours booked gives insight into the number of staff-hours needed to support the facility.

The only way to get a database which does exactly the things you want of it is to develop it in-house. Needless to say this is a costly and time-consuming process. There are a wide variety of free and commercially available calendars to be found on the Internet. Some of these may be adaptable for some purposes. The Institut Pasteur (2007) has developed a scheduling database which is provided free of charge. The database developed at the MPI-CBG light-microscopy facility is available for sale from Scionics.

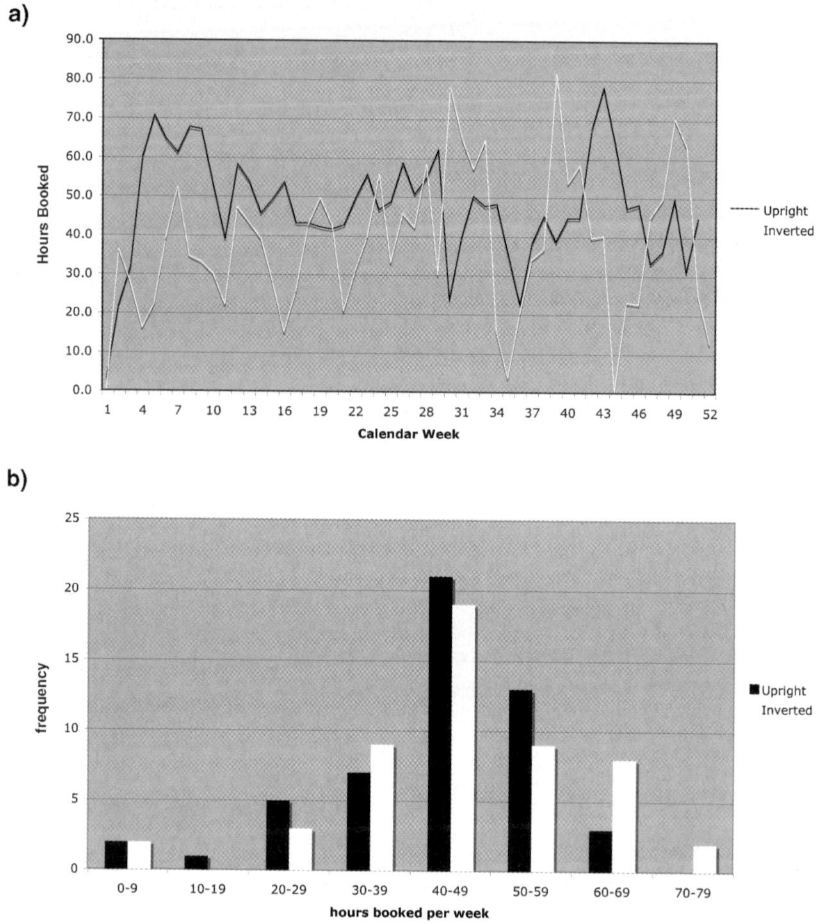

Fig. 4.3 Typical booking database output. Information from the booking database is important for showing that equipment is well used, and to justify purchase decisions. **a** Number of hours booked per week for two confocal microscopes based on upright (*green line*) or inverted (*blue line*) stands. Usage peaks often prompt complaints from users of insufficient microscope capacity (e.g., "The confocals are always booked, we need another one!") However, the average number of hours booked per week for the whole year is 42.7 for the upright microscope and 46.3 for the inverted microscope. Sharp drops in usage are often due to instrument breakdown or repair (inverted microscope, weeks 35 and 44). **b** The number of weeks having a given number of hours booked

4.5.2 *Fee for Service*

To bill or not to bill? That is the question. There are advantages and disadvantages to both approaches. One of the biggest effects of billing is to minimize frivolous equipment bookings. Frivolous bookings reduce the amount of time available for serious users, and in the extreme case (which we have witnessed) can lead to a hoarding mentality, in which people book instrument time defensively because they fear it is scarce. Assigning a cost to instrument use also raises the awareness of users and group leaders to the costs of running the facility. Billing for equipment time raises user awareness that equipment is expensive and must be treated carefully to promote a long lifetime, and that user mistreatment of equipment raises the cost of running it. In a fee-for-service environment there may be initial displeasure or even hostility on the part of users at having to pay for something they feel they deserve for free. In this context it is important to remember that nothing is free; either users are paying for their own equipment use or someone else is paying for it. In a fee-for-use environment there are no mysterious arrangements conferring preferential access to equipment on some groups; everyone has equal access rights. It can simplify access to facility equipment for outside users, if outside use is desired, and can thereby help to defray the costs for local users.

Alternatively, billing is extra work, requiring additional infrastructure and administration. It requires taking the time to devise and periodically update a cost matrix. This means keeping track of the staff and equipment costs associated with the number of hours of instrument use. It will periodically require explaining the hourly rates to the local group leaders and administrators to show they are getting value for money.

4.5.3 Cost Matrix

Accurate assessment of the hourly cost of running an imaging system requires careful consideration of the many different factors which contribute to cost, such as:

- Capital expenditure: cost of purchasing a system
- Evaluation cost: cost of reaching a purchase decision
- Operating cost: consumables, service contacts, replacement of parts
- Staff support cost: staff time required to keep a system running
- Overheads: building overheads, tools, secretarial support

These costs are summarized in Table 4.2. Operating and staff support costs are frequently underestimated; it costs money not just to buy equipment but also to keep it running. A facility may serve different types of user, as classified by their relationship to the parent or funding institution. For example, users may come from the same university department, different departments of the same university, or from companies outside the university. Depending on where the money to run the facility comes from, a price structure may be needed which reflects different obligations of the parent institution to the various user groups, as well as the ability of different groups to pay for use. For example, the cost matrix might distinguish between internal users, who do not pay the purchase cost or overheads, and external users, who contribute toward the full cost of running the facility.

Table 4.2 Hourly cost of instrument use. The total cost associated with 1 h of instrument use can be estimated by summing the different types of cost listed here. The internal price is the sum of the evaluation, operating, and support costs, which reflects only the cost of running the equipment. The external price reflects all costs associated with instrument use, including the purchase price and building/administrative overhead

	Upright system	Inverted system
Purchase cost	18.01	18.99
Evaluation cost	0.15	0.20
Operating cost	7.89	11.34
Support cost	5	6
Overhead cost	50	50
Internal price	13.05	17.54
External price	81.06	86.53

Table 4.3 Time basis for cost recovery. This is the number of hours (per week or year) which the system is used. The time basis for cost recovery must be estimated for a new system, but later may be set according to the actual number of hours a system has been booked

	Estimate	Upright system	Inverted system
Hours/day	10	–	–
Days/week	5	–	–
Hours/week	50	–	–
Less 10% downtime	45	42.7	46.3
Hours/year	2,340	2,220.4	2,407.6

4.5.3.1 Time Basis of Cost Recovery

The first point to consider in the design of a cost matrix is the time basis for cost recovery. Each cost must be recovered over a fixed length of time. For example, the purchase cost is recovered over the lifetime of the instrument, whereas service contracts run on yearly time intervals. Some repairs, such as the replacement of objective front lenses and cleaning of immersion oil from inside the objective, may be required every 2 years. The number of hours per year available for recovering these costs is essentially the number of hours per year the instrument is in use (Table 4.3). For a new instrument this number must be an estimate, but after the instrument has been used for a while the time basis for cost recovery can be set according to actual instrument use. If the time basis for cost recovery is set too high, the hourly cost to the user will be lower, but it will be impossible to completely recover costs. If the time basis for cost recovery is set too low, hourly costs to the user will be higher and more money will be recovered from the users than required to run the equipment. Increased use lowers the hourly cost by increasing the time basis of cost recovery, providing that increased use is not associated with increased *abuse*, which in turn increases the cost of repairs.

4.5.3.2 Purchase Cost

The purchase cost may or may not be passed on to the user depending on how the equipment was purchased, for example, through grants or directly through the budget of the parent institution. The time basis for recovery of the purchase cost is simply the lifetime of the instrument in years multiplied by the number of hours per year the instrument is used (Table 4.4). The anticipated lifetime of the instrument therefore has considerable impact on the hourly rate which must be charged to recover the purchase cost of the instrument. A longer lifetime results in a lower hourly rate to recover the purchase cost, but remember that an older system will become more expensive to run as time goes on owing to higher repair costs.

4.5.3.3 Evaluation Cost

The evaluation cost is the cost of making a purchase decision. This cost results primarily from the number of staff-hours spent on such tasks as instrument demonstrations and comparisons, discussions with users and company representatives, as well as time spent on equipment installation and testing (Table 4.5). This cost is generally small over the lifetime of an instrument; however, reaching a purchase decision can take up a substantial amount of staff time.

Table 4.4 Purchase cost. Recovery of the instrument purchase price (in arbitrary units) for two different systems over 5-, 7-, or 10-year time periods. Time basis refers to the number of hours per year the system is used, i.e., the number of hours per year available for cost recovery

	Upright system	Inverted system
Purchase price	280,000	320,000
Time basis	2,220.4	2,407.6
5-year recovery time	25.22/h	26.58/h
7-year recovery time	18.01/h	18.98/h
10-year recovery time	12.61/h	13.29/h

Table 4.5 Evaluation cost. The evaluation cost consists primarily of the staff time required to reach a purchase decision. Other costs, such as travel and accommodation for company or laboratory visits, might also be included

	Staff man-days required	
	Upright system	Inverted system
Company meetings	3	2
User meetings	2	2
Demonstrations	5	3
Installation	4	3
Total days	14	10
Staff cost per hour	30	30
Total cost	3,360	2,400
7-year recovery	0.22/h	0.14/h

4.5.3.4 Operating Cost

The operating cost includes the repair and replacement of parts, consumables, and service contracts (Table 4.6). Each repair or part replacement should have a length of time over which its cost must be recovered. Some repairs will be one-offs but others (such as laser tube replacement or objective lens repairs) can be expected to occur at regular intervals. Note that the estimation of operating costs for a new system will be very rough approximations at first, but will become more accurate as actual expenses accumulate. The operating cost requires periodic updating when costs have been completely recovered or new costs arise.

4.5.3.5 Staff Support Cost

The cost of staff support is determined by the desired level of staff support. How much help do the users expect to receive from staff? In our estimation this cost is time-consuming and difficult to objectively and accurately measure. One advantage of a booking database is that users can book staff directly for assistance and the cost can be accurately assigned back to the user who "consumed" it. However, much of

Table 4.6 Operating cost. The operating cost includes consumables, the repair and replacement of parts, and service contracts. The cost of each item is recovered according to its frequency of occurrence. Many events (replacement of lasers, repair of objectives) may be expected to occur at regular intervals

	Cost	Months to recover	Cost per hour
Upright system			
Mercury bulb	116	1	0.682
×40 objective repair	673.2	24	0.165
Service Contract	3,256.8	12	1.596
Maintenance visit	1,643.1	12	0.805
×60 objective repair	800	24	0.196
×100 objective repair	1,064	24	0.261
Scan-head overhaul	19,000	36	3.105
Argon laser	6,626	36	1.083
Total	–	–	7.89
Inverted system			
Mercury bulb	116	1	0.682
Service contract	3,727.12	12	1.827
Objective repair	633.27	24	0.155
Repair	6,948	24	1.703
Maintenance visit	1,891.25	12	0.927
Objective repair	809	24	0.198
Scan-head overhaul	15,375.1	36	2.512
Argon laser	6,626	24	1.624
UV laser	14,000	48	1.716
Total	–	–	11.34

Table 4.7 Cost of system support. The support ratio is the number of hours of system use requiring one general hour of staff support. The higher value (6) indicates the upright system requires less support that the inverted system. Staff cost is 30/h. Dividing the number of hours the system is used each week by the support ratio gives the number of staff-hours required to support each system. On the basis of the values in this example, general support for four confocal microscopes would require approximately 35.6h (=2×17.8) of staff support, or one full-time staff position. This is exclusive of the staff-hours needed to support direct booking by users for training and application assistance

System	Support ratio	Cost per hour	System-hours/week	Staff-hours/week
Upright	6	5	42.7	8.5
Inverted	5	6	46.3	9.3
Total	–	–	–	17.8

the staff's time is consumed by tasks which cannot be directly related back to any one user. General system support includes random trouble-shooting, sudden requests for user support, software maintenance, hardware maintenance, performance monitoring, and, of course, managing system repairs. The sudden user request deserves special mention. Such requests are typically associated with people using equipment who need help right there and then to be able to proceed with their day's experiment. Ten to 15 min of staff time can make the difference between a successfully used or a wasted booking. However, a series of sudden requests can easily consume the entire day. And because of their rapid-fire nature, it is extremely difficult to record how much time was spent with each user, and if the need for assistance was due to a fault in the system (for which all users should pay) or of the user (for which each user should pay). A simple and general estimation of the cost of staff support can be made by estimating the number of hours of equipment use which require 1 h of staff support (Table 4.7). This number will be lower for sophisticated systems requiring more support, and higher for robust systems requiring less support. These values can be adjusted according to the level of support desired by the local user community. Note than a partial estimation of the total number of staff-hours required to support the facility (i.e., the required number of staff positions) can be derived by dividing the number of system-hours booked by the support ratio.

4.5.4 Advisory Committees

Internal and external advisory committees can provide useful feedback and support for running a facility. An internal advisory committee composed of users can serve to keep relations with users smooth and give them insight into what it takes to keep the facility running. It provides an official conduit for user input on big issues such as new equipment purchases and the manner in which the facility is run, and a forum to air the many smaller issues which also arise. An external advisory committee, comprising other facility leaders and imaging specialists, provides an important reference point for the state of an imaging facility with respect to the status quo. An external advisory committee can provide proactive facility managers with crucial support for important "big picture" decisions, such as identifying trends and planning future spending.

4.6 Summary

An imaging facility integrates many functions within a research institute (Fig. 4.4) Many imaging facilities have evolved from a few microscopes to fill the available space in their local environment. In the future we expect that imaging facilities will be considered from the beginning as important components in the design of research buildings. This will allow imaging specialists to ensure that maximum utility is obtained from precious research funds through efficient planning of space, staff, and equipment. Staff play an important role in providing continuity and expertise above and beyond equipment maintenance. The flexibility of space use is of primary importance. We have emphasized that a laser room can improve laser safety while increasing the stability of equipment performance and comfort of equipment use. An equipment-scheduling database is vital for experiment planning, facilitating communication among users and staff, and establishing accountability.

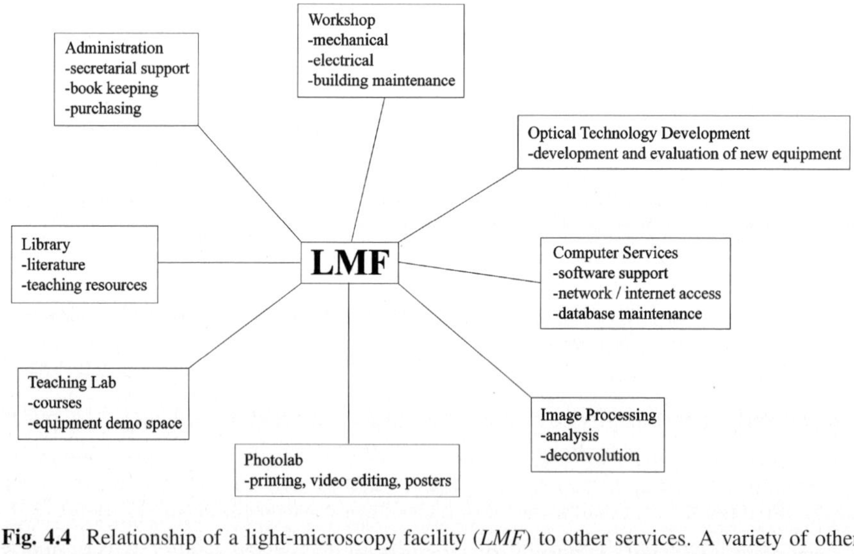

Fig. 4.4 Relationship of a light-microscopy facility (*LMF*) to other services. A variety of other support services are required to keep an LMF running. Depending on the local environment, these services may be completely external to the LMF or partly contained within it. For example, some imaging facilities also cover photolaboratory functions. Or image processing may exist as a separate facility if there are local specialists

References

Alderson RH (1975) Design of the electron microscope laboratory. Practical methods in electron microscopy, vol 4. Elsevier, New York

Angeletti RN, Bonewald LF, de Jongh K, Niece R, Rush J, Stults J (1999) Research technologies: fulfilling the promise. FASEB J 13:595–601

DeMaggio S (2002) Running and setting up a confocal microscope core facility. Methods Cell Biol 70:475–485

Helm JP, Haug F, Mogens S, Storm JF, Ottersen OP (2001) Design and installation of a multimode microscopy system. Proc SPIE 4262:396-406. http://www.math.uio.no/~atlej/LOV/files/HelmetalProcSPIE4262.pdf

Humphrey E (2004) How to promote a facility in order to increase use, acquire new equipment and, as a result, increase revenue. Microsc Today 12:32–36

Institut Pasteur (2007) PPMS: Pasteur platform management system for the PFID facility. http://www.pfid.org/html/ppms_agree/?fr. Cited 2 May 2007

Murphy JA (1993) Designing a microscopy facility: step by step procedure. In: Robards AW, Wilson AJ (eds) Procedures in electron microscopy, vol 7. Wiley, Chichester, pp 1:1.1–1:1.23, 1:2.1–1:2.2

Murphy JA (2001) Image management for a multi-instrument, multi-platform microscopy facility. Scanning May

Murphy JA (2002) Designing a microscopy/analytical instrumentation facility: step by step procedure. Microsc Today 10:36–39

Sherman D (2003) Core facility management session: maintaining major equipment in the core microscopy facility. Microsc Today 11:40–45

University at Buffalo (1991) Confocal@listserv.buffalo.edu – archives. http://listserv.acsu.buffalo.edu/archives/confocal.html. Cited 5 Dec 2006

White N, Errington R (2001) Setting up and managing a biological laser scanning microscope resource. Bio-Rad technical note 12 (reference no 9MRC50TN26). http://microscopy.bio-rad.com/reference/technical/Tech12.pdf

Winburn DC (1990). Practical laser safety. Dekker, New York

Zaluzec NT (1993) The Microscopy ListServer. http://www.microscopy.com/. Cited 5 Dec 2006

II
Advanced Methods and Concepts

5
Quantitative Colocalisation Imaging: Concepts, Measurements, and Pitfalls

Martin Oheim and Dongdong Li

Abstract Many questions in cell biology and biophysics involve the quantitation of the colocalisation of proteins tagged with different fluorophores and their interaction. However, the incomplete separation of the different colour channels due to the presence of autofluorescence, along with cross-excitation and emission 'bleed-through' of one colour channel into the other, all combine to render the interpretation of multi-band images ambiguous. Traditionally often used in a qualitative manner by simply overlaying fluorescence images ('red plus green equals yellow'), multicolour fluorescence is increasingly moving away from static dual-colour images towards more quantitative studies involving the investigation of dynamical three-dimensional interaction of proteins tagged with different fluorophores in live cells. Quantifying fluorescence resonance energy transfer efficiency, fluorescence complementation and colour merging following photoactivation or photoswitching provide related examples in which quantitative image analysis of multicolour fluorescence is required. Despite its widespread use, reliable standards for evaluating the degree of spectral overlap in multicolour fluorescence and calculating quantitative colocalisation estimates are missing. In this chapter, using a number of intuitive yet practical examples, we discuss the factors that affect image quality and study their impact on the measured degree of colocalisation. We equally compare different pixel-based and object-based descriptors for analysing colocalisation of spectrally separate fluorescence. Finally, we discuss the use of spectral imaging and linear unmixing to study the presence in a 'mixed pixel' of spectrally overlapping fluorophores and discuss how this technique can be used to provide quantitative colocalisation information in more complex experimental scenarios in which classic dual- or triple-colour fluorescence would produce erroneous results.

5.1 Introduction

During the past 15 years there has been a remarkable growth in the use of fluorescence imaging in biological microscopy. This development has been largely driven by the generation and widespread use of fluorescent protein chimeras (reviewed in

S.L. Shorte and F. Frischknecht (eds.), *Imaging Cellular and Molecular Biological Functions.*

Shaner et al. 2005; Giepmans2006). Also, three-dimensional imaging at the subcellular level has become possible for many researchers with the broad availability of confocal and two-photon-excited fluorescence (2PEF) microscopes to many laboratories and imaging platforms. After the identification of the key molecules and signalling pathways, many questions in cell biology and cell biophysics concern where and when these molecules interact (Schultz et al. 2005). As a consequence, microscopic multicolour imaging is moving away from classical confocal immunofluorescence (Miyashita 2004) towards studies that typically involve the quantification of the dynamics of the three-dimensional expression and –ideally – interaction of proteins tagged with different fluorophores in live cells. Information on molecular interaction could be derived by adding information derived from fluorescence resonance energy transfer (FRET) (Jares-Erijman and Jovin 2003), fluorescence complementation assays (Kerppola 2006), or colour merging following photoactivation or photoswitching (Betzig et al. 2006; Chudakov et al. 2006; Hess et al. 2006). Thus, larger and highly dimensional data sets must be handled.

Colocalisation studies using fluorescence imaging represent a powerful method for exploring putative associations between molecules and their targeting to discrete intracellular compartments. Ideally, several spectrally well-distinct fluorophores would be specifically addressed to their molecular-scale targets and imaged into distinct, spectrally separated detection channels so that the fluorescence intensity in each channel would contain spatial and concentration information exclusively derived from one fluorophore. These images could then be pseudocolour-coded displayed side by side or overlaid and the amount of colocalisation could be estimated from these intensity maps. Red and green equals yellow. The estimation of protein expression and colocalisation can be broken down to two steps: first, the selective labelling with and imaging of different fluorophores, followed by the quantification of their colocalisation from multicolour images. Both of these steps are based on hypotheses, for example that all collected fluorescence originates from endogenous label, and rely on the correct expression and subcellular targeting of fusion proteins, the existence of only negligible cross-talk between acquisition channels, or the linearity and spatial homogeneity of the analysed images.

In a real experiment, however, every one of the underlying assumptions is probably violated to some extent. Two simple questions arise:

1. To what extent are the spatial, spectral, and (to a lesser degree) temporal images truly independent?
2. How can we quantify the degree of colocalisation from such fluorescence images?

The aim of this chapter is to discuss the problems associated with the different techniques for dual-colour imaging and quantifying colocalisation and to evaluate their respective performance and limitations. Considering the microscope as a linear imaging device, we can describe the image of an arbitrary object as the linear superposition of point images of different intensity. We therefore restrict our discussion to imaging (subresolution) point objects. The generalisation to extended objects is straightforward.

The chapter is organised as follows: we first (Sect. 5.1.1) introduce a synthetic yet realistic dual-colour example that we will use throughout this chapter to study the impact of different image parameters on the colocalisation estimate and to evaluate different strategies to quantify colocalisation. This example has the virtue that we can know and control the true amount of colocalisation between probes and can vary their degree of spectral overlap and relative brightness, vary the image noise and background offset in a controlled manner, and study their impact on the colocalisation estimate. In Sect. 5.1.2, we stress the importance of matching dyes, filters, and intermediate optical components by regarding (Box 5.1) step by step the process of choosing appropriate combinations in a realistic experiment. Box 5.2 offers a swift review of optical sectioning techniques that can improve the colocalisation detection.

Box 5.1 Tracing spectral throughput along the excitation and emission optical path

The goal of dual-colour fluorescence microscopy is to simultaneously map the location and dynamics of two fluorescent vesicle markers from a dual-colour image pair. Multicolour maps are only as good as the raw images that are used to calculate them.

We here develop a simple rationale to chose optimal filters for a given fluorophore pair and to estimate the cross-talk that is engendered by filter mismatch. A more complete treatment can be found as an online resource on the Oheim laboratory Web site (Oheim et al. 2007). In the simplest case of only two fluorophores and negligible (or spatially and spectrally uniform) background, the fluorophore separation only depends on their excitation and emission spectral overlap and the filter bands used to isolate them. Neglecting higher-order effects (fluorophore saturation, self-absorption, aggregate formation and quenching, bleaching), the measured intensity (in analogue/digital [A/D] units) of a fluorophore on a microscopic image depends linearly on

1. The fluorophore spectral *extinction* $\varepsilon(\lambda)$, which describes the probability of absorbing a photon at wavelength λ and is typically given as $\varepsilon \times F_{abs}(\lambda)$, where ε is the *molar extinction coefficient* (mostly but not always specified at peak absorption) and $F_{abs}(\lambda)$ is the fluorophore absorption spectrum, with the peak absorption normalised to 1. ε is of the order of 61,000 mol^{-1} cm^{-1} for pEGFP-N1 (Clontech) and 48,000 mol^{-1} cm^{-1} for FM4-64 (Invitrogen). The curves for $F_{abs}(\lambda)$ are given in Fig. 5.6b. We next trace along the microscope light path (Fig. 5.6a, inset), starting from the light source[1], the fraction of the excitation light (black line) that is transmitted by the intermediate optical components and reaches the sample to excite the fluorophores.

(continued)

[1] The black line is the normalised experimental spectral distribution of the excitation light source used – here, a TILL polychrome II with about 20-nm excitation bandwidth. Alternatively, one could substitute here the transmission spectrum of an emission bandpass filter multiplied with the spectral emission of the Xe or Hg arc lamp.

Box 5.1 (continued)

Next, the dashed blue line shows the *reflectance* (1 minus transmission, neglecting absorption) of the primary dichroic beamsplitter.

2. The *fluorescence quantum yield* ϕ of a fluorophore defines the probability of an excited molecule relaxing to the ground state by emitting a fluorescence photon. The probability of emitting a photon at wavelength λ is obtained by multiplying ϕ with the normalised ($\int F_{em}(\lambda)\,d\lambda = 1$) fluorescence emission spectrum $F_{em}(\lambda)$. ϕ=0.60 for enhanced-green fluorescent protein (EGFP; Patterson et al. 1997). In the absence of specific information for FM4-64 in lipid membranes (Bill Betz, Denise Lo Invitrogen, personal communications), we equally assumed 0.6 for FM4-64 (Table 5.1). The product $\varepsilon\phi$ is sometimes referred to as the *fluorophore brightness.* The most absorptive fluorophores absorb more than 2 orders of magnitude more efficiently than the least. This is in stark contrast to the quantum efficiency, which typically falls in the range from 0.25 to 0.9 for most useful fluorophores, so they will not differ by more than a factor of 3–4 at most. As a result, ϕ will be of secondary importance in determining brightness.

Table 5.1 Spectroscopic properties of enhanced-green fluorescent protein (*EGFP*) and FM4-64

λ_{ex}, peak (nm)	ε (mol^{-1} cm)$^{-1}$	λ_{em}, peak (nm)	ϕ	Comment
EGFP 50 mM HEPES, pH 7.5	488	61,000[a]	509	0.6[b]
FM4-64 Measured in $CHCl_3$	558	48,000	734	0.6[c]

HEPES *N*-(2-hydroxyethyl)piperazine-*N*′-ethanesulfonic acid

[a] Cubitt et al. (1999)

[b] Patterson et al. (1997)

[c] FM dyes are almost non-fluorescent in water, but their quantum yield increases about 350 times when they partition into a hydrophobic environment. (Henkel et al. 1996). The actual ϕ value in a lipid is hard to determine, as the concentration is typically not known. We therefore simply assumed that FM4-64 and EGFP have the same quantum efficiency.

3. The fluorophore *concentration c,* in moles per litre, or its density, e.g. for membrane-resident dyes.
4. The camera exposure or, more generally, detector *integration time.*
5. The fluorescence excitation volume (related to the objective optical depth, along the *z*-axis and illuminated area (*xy*), and thus to the numerical aperture (NA); however, not in two-photon-excited fluorescence).
6. The fraction of fluorescence collected (i.e. the solid angle covered by the NA).
7. The fluorescence collection volume (in the case of confocal apertures or directional-emission detection; Axelrod 2001).
8. The A/D unit, i.e. the number of counts per detected electron/pixel.

9. Light flux (illumination intensity) and detector response (detector quantum yield).

We consider here a single-excitation side-by-side projection of two colour-channel images (Fig. 5.6a, inset); we only compare the relative contribution of isomolar EGFP and FM4-64 to each colour-channel image, *ceterum paribus*. Given the narrow spectral width of the excitation, we neglected the spectral profile of the light source, which is set unity for all λ. To estimate cross-excitation, we multiply for each dye the spectral profile of the excitation light source, the dichroic reflectance (i.e., 1 minus its transmittance, neglecting absorption) and $\varepsilon \times F_{abs}$ (λ) (Fig. 5.6b, bottom). Integration over λ shows that – under these conditions – EGFP is 1.3-fold more efficiently excited than FM4-64. Stated otherwise, the (intentional) cross-excitation is 43% (FM4-64) versus 57% (EGFP), so both fluorophores are excited with (roughly) equal efficiency.

On the *emission* side, to calculate bleed-through we proceed similarly by tracing back the transmitted fluorescence through both microscope detection arms. Hence, to calculate the contribution of each fluorophore to the 'red' and 'green' colour image we consecutively multiply their fluorescence emission spectra with ϕ and the transmission curve of the primary dichroic mirror, the transmission (for the 'red' channel) and reflection (for the 'green' channel) of the secondary dichroic mirror (Fig. 5.6a, inset), and, for each channel, the transmission curves of the respective emission bandpass filters. The corresponding curves are shown in Fig. 5.6c (top). Again, we neglected the transmission spectrum of the microscope intermediate optics as well as the spectral response of the detector, which we assume to be uniform in the wavelength range studied. The result is illustrated in the bottom panel of Fig. 5.6c, and integration over λ yields a 4% estimate of the contamination of the green-detection channel with FM4-64 signal and of less than 0.01% EGFP detected in the red colour channel.

Finally, the *total cross-talk* between the red and green channels is given by multiplying the excitation cross-talk and emission bleed-through and dividing through the sum of these products for both fluorophores (Oheim et al. 2007). In the specific case of 488-nm excitation and simultaneous dual-emission imaging of FM4-64 and EGFP, we obtain that 99.98% of the signal detected in the red channel comes from FM4-64 and 96.85% of the signal detected in the green channel results from EGFP.

Thus, from the spectral separability analysis we expect that both EGFP and FM4-64 largely dominate the green- and red-detection channels, respectively, with only negligible cross-talk.

To facilitate the comparison of colocalisation data across different studies and to evaluate the error of the colocalisation estimate, it should be good practice to explicitly state the amount of cross-talk between the different detection channels used. The spectral separability index defined here offers a convenient criterion for evaluating and comparing multicolour data sets (Oheim et al. 2007).

Box 5.2 Optical sectioning techniques to lower image background and increase image contrast

Confocal microscopy is a well-established optical sectioning technique that is based on the observation that a point source of excitation light (illumination pinhole) can be used to create a diffraction-limited focus in the specimen plane, which in turn corresponds to a confocal spot in the image plane. Thus, in-focus light at locations different from the illuminated spot as well as out-of-focus signal can be efficiently rejected by placing a small pinhole (roughly of the diameter of the Airy disc) in the confocal image plane. To create an image, the spot is scanned with respect to the specimen, in biological confocal microscopy typically by scanning the beam angle in the objective pupil. Although it (slightly) increases resolution and (substantially) reduces background, confocal microscopy is not particularly well suited for live-cell imaging, because – as most generated fluorescence is rejected at the confocal aperture – it makes very inefficient use of excitation photons.

Multiphoton excitation fluorescence (MPEF) microscopy is a laser-scanning technique as well. Here, the improved image contrast and background rejection are achieved by restricting the fluorescence excitation volume, rather than the fluorescence collection volume as in confocal microscopy. The technique is based on the near-simultaneous non-linear absorption of two or more photons that combine their energies to excite a fluorophore from the ground state to the first excited state. MPEF is restricted to a tiny volume near the focus, because high photon densities are required for the phenomenon of multiphoton absorption (one way to think about this is to realise that the trajectories of multiple photons must cross the excited molecule simultaneously). MPEF uses infrared light to excite ultraviolet (three-photon excited fluorescence) or visible (2PEF) fluorescence. This, together with the broad 2PEF absorption spectra (compared with one-photon excited fluorescence), and the availability of the entire visible-wavelength range for fluorescence detection, permits efficient filtering and facilitates multiband recordings.

Total internal reflection fluorescence microscopy (TIRFM) is a light-confinement technique that exploits the phenomenon of total internal reflection of a light (in most instances, a laser) beam at a dielectric interface to generate a thin (approximately $\lambda/5$), exponentially decaying evanescent field that skims the lower-refractive-index medium. This near-field perturbation can be used to create a near-surface fluorescence excitation. When cells are grown on the dielectric boundary (i.e. the glass–water interface) only fluorophores in a thin near-membrane space are excited, whereas the bulk of the cell is spared from fluorescence excitation and photobleaching. Owing to its extremely low background, evanescent-field microscopy is often used when studying single-fluorophore photodynamics. Because the image is restricted to a thin section, TIRFM is typically used in conjunction with epifluorescence excitation.

Deconvolution. Within the approximation of linear imaging theory, each point of the object can be described as a point-source of light that gives – depending on its intensity and precise focal position – rise to a shifted and weighted copy of the point-spread function (PSF). Conversely, with knowledge of the experimental PSF, the information contained in a three-dimensional image stack can – in principle – be used to back-calculate the initial fluorophore distribution that produced the blurred diffraction-limited image. Although the reassignment of detected photons to their original location is – in principle – possible, the mathematical algorithms to solve this 'inverse problem' are fairly noise sensitive and are somewhat notorious in generating artefacts. In laboratory practice, three-dimensional image restauration by 'deblurring' is often outperformed by confocal imaging.

Reversible saturable optical fluorescence transition (RESOLFT) concepts. Contrary to what one might expect from the optical diffraction limit, fluorescence microscopy is in principle capable of unlimited resolution. The necessary elements are spatially structured illumination light and a non-linear dependence of the fluorescence emission rate on the illumination intensity. In saturated structured-illumination microscopy, the non-linearity arises from saturation of the excited state (Gustafsson 2005). The diffraction barrier has equally been broken by a saturated depletion of the marker's fluorescent state by stimulated emission (Willig et al. 2006),but this approach requires picosecond laser pulses of gigawatt per square centimetre intensity. With use of much smaller intensities, subdiffraction resolution can be achieved from reversible photoswitching of a marker protein between a fluorescence-activated and a non-activated state, whereby one of the transitions is accomplished by means of a spatial intensity distribution featuring a zero (Betzig et al. 2006; Hofmann et al. 2005).

Section 5.2 revisits these examples to review different semiquantitative (colour merging, Sect. 5.2.1), pixel-based (Sect. 5.2.2), and object-based colocalisation estimates (Sect. 5.2.3) and discusses their performance. Box 5.3 extends the calculation of colocalisation coefficients to fluorophore abundance maps rather than fluorescence images. These maps are generated as a result of spectral imaging and linear unmixing (SILU) techniques in which the presence and relative contribution of fluorescent probes are analysed from a set of spectral images that is overdetermined, i.e., contains more planes than the sample contains fluorophores. We have recently introduced a variant of this technique specifically adapted for classifying resolution-limited point objects containing multiple fluorophores in live cells (Nadrigny et al. 2006).

Box 5.3 Multispectral and hyperspectral imaging

In addition to sequential or synchronous multiband recordings, many commercial laser scanning microscopes now permit multispectral or 'hyper'-spectral detection. Spectral detectors are based on a dispersion element (prism, grating) and the parallel detection of a range of wavelength, either on a linear photodiode array (Zeiss) or on an arrangement of multiple photomultiplier tubes with movable entry slits (Leica). These instruments generate a flexibility that filter-based multichannel acquisition cannot offer. Spectral imaging devices have the advantage over earlier integrative (photometric) devices of providing – for each pixel – spectral and localisation information in addition to fluorescence intensities.

Because microscope images are diffraction-limited, neighbouring pixels of fluorescent objects are not independent; therefore, *object-based* approaches to quantify colocalisation can take into account a priori knowledge of the imaged object. An advantage of such object-based colocalisation analysis is that one can make use of additional information, e.g. the size and shape of the subcellular object under study, or of correlations between neighbouring pixels (Nadrigny et al. 2006).

When thinking in terms of multidimensional histograms (Sect. 5.2.3), we can view linear unmixing as a projection of each spectral pixel vector onto an orthonormal basis. Thus, instead of parameterising the spectral vector in terms of N fluorescence detection channels, its coordinates are given in terms of a set of k (pure) fluorophore vectors. Already with a surprisingly low number of spectrally overlapping detection bands (Neher and Neher 2004a) spectral imaging and linear unmixing permits fingerprinting the expression of spectrally overlapping fluorescent proteins on single secretory vesicles in the presence of a spectrally broad autofluorescence. By making use of statistical tools and the knowledge of the microscope's PSF, this technique provides a robust alternative to error-prone dual-colour or triple-colour colocalisation studies in live cells (Nadrigny et al. 2006).

Most of the arguments used throughout this chapter rely on multiple-emission detection but symmetrically apply to experiments using multiple *excitation* wavelengths instead. We also note that although the optimal separation of fluorophore signal often requires both multiple-excitation *and* multiple-emission fluorescence imaging, we focus here on multiple-emission detection.

5.1.1 One Fluorophore, One Image?

Multicolour displays showing overlaid multichannel fluorescence images are increasingly being used in the cell and neurobiological literature to illustrate molecular colocalisation and interaction. It is generally assumed that one fluorescence channel contains a specific signal, exclusively related to one fluorophore, and that images are comparable among different acquisition channels. Figure legends will typically read '**a** Confocal fluorescence images of a … cell coexpressing a molecule X – enhanced-green fluorescent protein (EGFP) chimera (*top*), and protein Y fused to monomeric red fluorescent protein (mRFP, *bottom*). **b** Time series of EGFP (*top*) and mRFP images (*bottom*) reveals an increase in colocalisation after stimulation. **c** Pseudocolour overlay of EGFP and mRFP images. Note the increase in yellow indicating colocalisation (*arrowheads*)…' or similar. Often, the precise experimental conditions (illumination, filter and detection settings, fluorophore variants used, etc.) are not very explicit and controls are omitted. The critical evaluation of colocalisation data requires more information than is often given. The major problems encountered with quantitative multicolour microscopy are well identified. Their relative importance, however, can vary from one microscope to another, from one experiment to another, and probably even from one batch of cells to another, depending, e.g., on the level of protein expression, autofluorescence in the preparation, or detector noise. For each combination of fluorophores imaged, it is important to quantify to what degree the different detection bands really contain independent and fluorophore-specific information. Also, although it might appear tedious and time-consuming, understanding the physical limitations of what can be achieved with a given combination of flurophores, filters, and optical components is a useful exercise that lays the grounds for sensible instrument use. We stress this point specifically having the engineers and researchers in mind who are responsible for and run shared facilities and imaging platforms and can guide less experienced users to make judicious choices.

5.1.1.1 Spectral Overlap

Organic fluorophores typically display broad absorption bands that lead to considerable *cross-excitation*. Cross-excitation quantifies the amount of (usually unwanted[2]) excitation of fluorophores other than the one to be excited by this wavelength band. On the emission side, the problem is typically called *bleed-through* and relates to the amount of fluorescence that originates from other fluorophores detected in the fluorescence channel designed to view one specific fluorophore. Often, it is accentuated over cross-excitation because fluorescence tails off into the red owing to the decay into higher vibrational levels of the S_1 state and thermalisation of the excess vibrational energy, solvent effects, excited-state reactions, complex formation, or energy transfer. The total cross-talk will be proportional to the product of cross-excitation and bleed-through (Box 5.1).

[2] Special cases in which excitation cross-talk and emission bleed-through are not only tolerated but intentionally wanted are dual-colour emission detection with simultaneous excitation of two dyes emitting in different fluorescence bands, or dual-colour excitation with single-emission detection.

The ever-increasing generation of new fluorescent protein colour variants (reviewed in Shaner et al. 2005) and the expanding family of genetically encoded indicators (Griesbeck 2004) have not removed but rather accentuated the problem of fluorophore separation in multicolour fluorescence microscopy. Although a broader range of monomeric fluorescent proteins is becoming available, the choice of spectrally well separated variants is still very restricted. Also, for each new fusion protein and expression system, the specific targeting and lack of retention in the endoplasmic reticulum must be verified individually. For example, Hirrlinger et al. (2005) recently demonstrated that the formation of fluorescent precipitates limits the use of the spectrally attractive red-emitting reef coral proteins in transgenic animals. Those fluorescent proteins that work best have considerable overlap and cannot be separated using specific filter sets (Nadrigny et al. 2006; Zimmermann 2005).

We display in Fig. 5.1a two synthetic in-focus fluorophore maps. True fluorophore locations are represented by cross hairs (green) and circles (red), respectively. To model image formation, we convolved this high-resolution fluorophore map with an (experimentally determined interpolated high-resolution) point-spread function (PSF) of an objective with a numerical aperture (NA) of 1.45 and resampled the resulting diffraction-limited image with a pixelated imaging detector (Fig. 5.1b).The resulting red and green images were low-pass-filtered ($1\,\mu m^{-1}$) and the low-pass-filtered image was subtracted from the original image to remove high-frequency noise. The result was thresholded to exclude background, and binarised. We estimated the colocalisation in the red channel (index 1), by calculating the degree of overlap of the two binary masks of the dual-colour image relative to the red binary image squared (Lynch et al. 1991),

$$\rho_{\mathrm{bin}} = \frac{\sum\limits_{\forall(x,y)} w_1{}^*(x,y)\cdot w_2{}^*(x,y)}{\sum\limits_{\forall(x,y)} \left[w_1{}^*(x,y)\right]^2}, \tag{5.1a}$$

where

$$w_j{}^*(x,y) = \begin{cases} 255 & \text{if } w_1(x,y) \geq t_j \\ 0 & \end{cases}, \tag{5.1b}$$

and the sum runs over all pixels (x,y). $\frac{1}{255}\Sigma w_1{}^*(x,y)$ is the number of pixels above a threshold t_1. We chose $t_1=t_2$ but the threshold levels for the red (index 1) and green (index 2) channels can (in principle) be chosen independently. $w_1(x,y)$ and $w_2(x,y)$ are the pixel values of the red and green images, respectively. ρ_{bin} measures the fraction of pixels on the green binary image that are equally present on the red binary image, relative to the total area of red pixels. We chose the somewhat bulky notation to allow for later generalisation (see below).

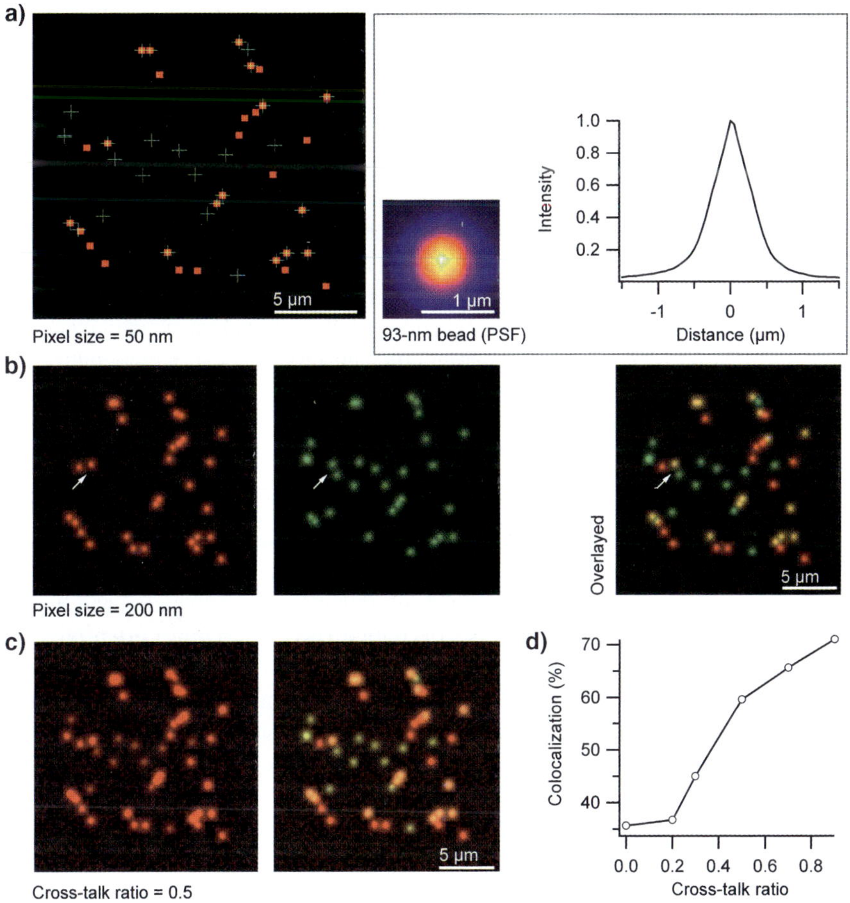

Fig. 5.1 Influence of cross-talk on colocalisation determination. **a** Artificially generated in-focus dual-colour images. To generate synthetic red and green high-resolution matrices 30 100-nm-diameter 'fluorophores' of each colour were placed on a grid of 50 nm. *Red squares* and *green cross hairs* indicate the 'true' particle positions. The actual degree of colocalisation was set as 50%, i.e. 15 particles colocalised, indicated by the overlap of both indicators, and the others were randomly distributed. Both high-resolution matrices were sequentially convolved with the experimental point-spread function (*PSF*), sampled by a low-resolution matrix with a pixel size of 200 nm, and shot noise was added to each pixel of the low-resolution matrix following a Poisson process. *Inset*: 'High-resolution' PSF. The experimental PSF determined by imaging a 93-nm fluorescent bead with an oil-immersion objective with a numerical aperture of 1.45 was radially averaged, interpolated, and resampled on a 50-nm grid. The line profile shows a linear cross-section of the interpolated in-focus PSF. **b** Green- and red-channel images and their pseudocolour overlap in the absence of spectral cross-talk. **c** Introduction of cross-talk. For clarity, only the cross-talk from the green channel into the red channel is evaluated. The cross-talk ratio is the amount of green fluorescence signal added to the red channel. An example (cross-talk ratio 0.5) is shown. **d** Increasing cross-talk adds a false colocalisation that can largely outnumber the 'true' amount of colocalisation. Colocalisation of 36% was estimated from the cross-talk-free fluorescent image pair using an object-based algorithm (see text for details)

Colocalisation in the green channel is estimated similarly, by dividing through $\Sigma[w_2{}^*(x,y)]^2$ instead. Owing to its use of binary image masks ρ_{bin} underestimates the true amount of colocalisation (36 vs. 50%), because even for a perfect match the added noise makes it impossible to delineate the real shape of the object. Thresholding favours the selection of high-intensity pixels, so the true particle size is underestimated when binarising the images and so is ρ_{bin}. See Sect. 52 for alternative descriptors of colocalisation.

What happens if we increase the cross-talk between both images? Figure 5.1c displays the red- and green-channel images and their pseudocolour overlay that result when increasing the fraction of the 'green' image leaking into the 'red' image. Whereas ρ_{bin} found 36% colocalisation on the original image pair, increasing the cross-talk ratio adds a false apparent colocalisation that can largely outnumber the true amount of overlap (Fig. 5.1d).

5.1.1.2 Low Signal

In order to provide meaningful estimates for fluorophore colocalisation, the fluorescence signal has to stick out of the image noise. In live-cell imaging, the sensitivity of the sample to high-intensity illumination (photodamage) and the loss of signal upon prolonged fluorophore excitation (photobleaching) often prescribe low excitation intensities. The resulting lower signal as well as the intrinsically dimmer fluorescence of fluorescent proteins (when compared with the commonly used organic fluorophores) requires additional precaution as to the interpretation of multicolour images. It might even be necessary to go back to fixed samples and use antibodies against the fluorescent proteins used so as to amplify the detectable signal above the noise level of the detector, however at the expense of losing quantitative intensity information (Martinez-Arca et al. 2003). Figure 5.2a shows, for the same image pair as shown previously, the effect of increasing the noise relative to a fixed signal in the green channel. The 'red' image is always the same, and we have assumed zero cross-talk between the two channels. Figure 5.2b displays the evolution of ρ_{bin} when the 'green' image fades away in the image noise. When always using the same fractional intensity for thresholding, the colocalisation becomes less and less apparent with increasing image noise, because fewer and fewer pixels remain after binarisation. However, other measures of colocalisation (Sect. 5.2) produce a different and even the opposite result. Median filtering (Demandolx and Davoust 1997) and deconvolution (Landmann 2002; Li et al. 2004) are two techniques that enhance the signal-to-noise ratio and colocalisation detection (not shown).

5.1.1.3 Three-Dimensional Spatial Resolution

Do two objects truly colocalise or do their images simply blur one into the other because the image resolution is low? Diffraction blurs the three-dimensional image

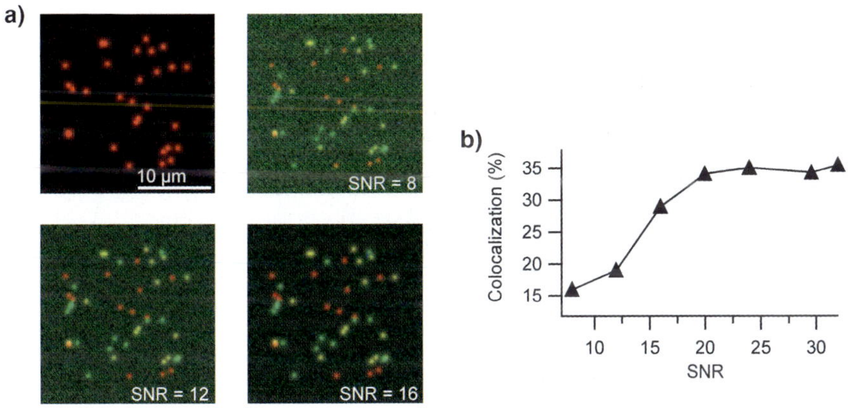

Fig. 5.2 Effect of decreasing the signal-to-noise ratio (*SNR*) in the green channel on colocalisation estimates. The 'red' image is always the same. In contrast, the noise was increased in the green channel for a constant signal of 500 counts. **a** Superimposed images of red- and green-channel images with different SNR levels as indicated. **b** Colocalisation is underestimated at low SNR. The amount of 'true' colocalisation in the absence of image noise is about 35% when using the object-based descriptor (Eq. 5.1) for estimating colocalisation. A minimal SNR is required to obtain reliable colocalisation estimates

of the object. In focus, subresolution objects appear with an apparent size much bigger than the true biological object given by the Airy disc, i.e. the in-focus plane of the three-dimensional PSF of the microscope (Fig. 5.1a, inset). Also, the microscope spatial resolution is not isotropic but is degraded along the microscope optical axis. Together with chromatic aberrations, diffraction results in spreading out a point object on the microscopic image, thereby creating a false apparent overlap between the images of proximal but not colocalised objects.

Defocus

Until now, we have assumed that all objects were located in focus ($z=0$) and the microscope had a perfect optical sectioning capacity. However, the objective will collect fluorescence from objects located above and below the nominal focal plane. These objects will contribute not with their in-focus image but with their respective (blurred) off-focus plane of the PSF. The spread of signal across multiple optical sections presents a significant source of false-positive artefact in the measurement of colocalisation (Fig. 5.3). To examine the effect of out-of-focus fluorescence on the colocalisation estimate, we moved the synthetic point objects to a random off-focus position drawn from a Gaussian distribution with mean dz and the width given by the approximately 1.9 μm effective depth of field of a NA-1.45 objective. Figure 5.3 shows an in-focus fluorescence image pair, along with the corresponding

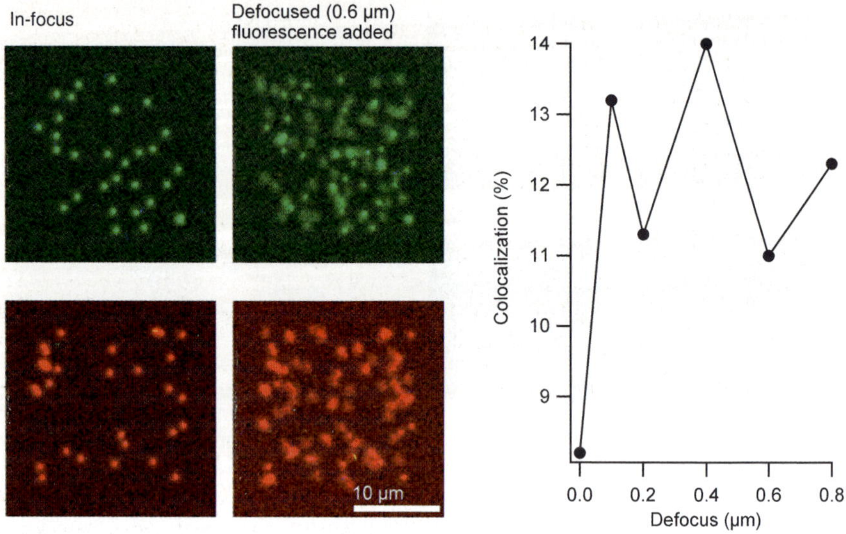

Fig. 5.3 Defocus results in false-positive colocalisation. Randomly distributed fluorophores were created for both channels. The colocalisation is low for the in-focus image pair, consistent with the small amount of random overlap resulting from the overlap of Airy patterns of proximal particles in the red and green channels. Moving objects slightly out of focus increases the apparent colocalisation and produces false positives (see text for details)

image in which we randomly introduced a mean defocus of 0.6 µm in both image channels. We equally plot the evolution of ρ_{bin} when increasing defocus from zero to 0.1, 0.2, 0.4, 0.6, or 0.8 µm. The amount of false-positive colocalisation resulting from overlapping Airy patterns for $dz=0$, and the effect resulting from the superposition of spatially unrelated signal give rise to an increasing false-positive colocalisation. Wide-field microscopy is only little suited for colocalisation analysis, because it suffers from out-of-focus blur. Image restoration by deconvolution can – in part – compensate for this problem, but is very sensitive to image noise and can generate bright pixels or grainy artefacts that are mistaken for fluorescent objects (Landmann 2002).

Lateral Resolution

The impact of lateral resolution on the apparent colocalisation of in-focus objects is illustrated in Fig. 5.4. We investigated this effect by placing red particles of $2r_1$=100-nm diameter and green particles of increasing size $2r_2$ randomly in the object plane, respecting their mutual size exclusion, i.e. the interparticle distance is r_1+r_2 or bigger. For object sizes below or close to the optical resolution, the estimated false-positive colocalisation is low and roughly constant. With increasing

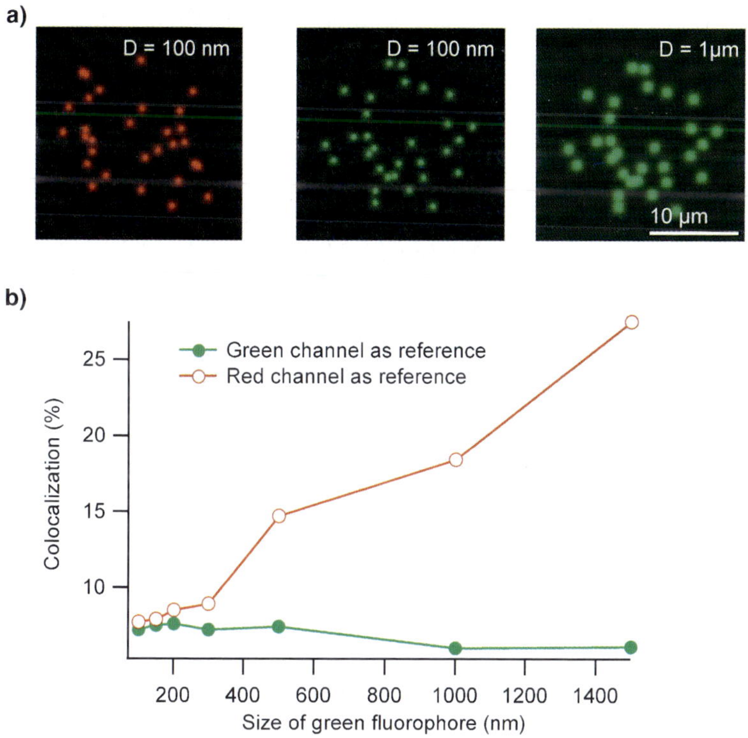

Fig. 5.4 The object size affects the colocalisation estimate. **a** Thirty red and green particles of 100-nm size each were created and randomly distributed in focus, respecting their mutual size exclusion, i.e. no two particle centres could get any closer than $r_{red}+r_{green}$, here 200 nm. The images were then blurred by convolution with the experimental PSF, Poisson noise was added, and the colocalisation was estimated as before. We then increased the diameter of green objects while keeping the red object size constant at 100 nm. The *leftmost panel* shows an example for 1-μm object size. **b** The size with which objects appear on fluorescence images is of critical importance in cases involving objects of near-resolution size. Evolution of the colocalisation estimate when using the red or the green channel as a reference (cf. Eq. 5.1a). The (false) apparent colocalisation rises with particle size in the red channel, because the number of pixels in which red and green overlap increases (owing to the bigger and bigger size of the green objects) while always dividing through a constant area or red pixels. Not the almost constant colocalisation estimate for particles with a size below or close to the optical resolution limit. In contrast, the colocalisation estimate is nearly constant when dividing the green and red overlap area by the concomitantly rising total area of green pixels. The slight drop results from the fact that the blurred edge creating the false overlap roughly grows as r_{green}, whereas the reference area grows as r_{green}^2

object size, ρ_{bin} (relative to the bigger-particle green image) slightly decreases, because the boundary effect becomes less and less prominent relative to the increasing green pixel area. As expected from the asymmetric three-dimensional resolution, ρ_{bin} is more sensitive to the axial than to the lateral resolution.

5.1.1.4 Non-rejected Background: Autofluorescence, Non-specific Protein Targeting, and Transmitted Excitation Light

Background is caused by electronic offset (Oshiro and Moomaw 2003), stray light, and blur from a specifically labelled or autofluorescent part of the object (Sheppard et al. 1995). Inefficient spectral filtering of ambient light or transmitted spurious excitation light can obscure faint signals too. Unfortunately, with sufficiently sensitive detection, virtually all cells contain detectable autofluorescence and the usually sufficient filter contrast ratios (i.e. the relative intensity of transmitted vs. reflected light) become limiting. Also, with intense ultraviolet and low visible excitation, glass components (cover slip, objective lens, etc.) can produce a background autofluorescence. Testing different substrates and immersion oils and performing appropriate controls can help to identify and reduce background. Belonging to the same class of problems, a fusion protein may not be exclusively directed to its target organelle but may be retained in the endoplasmic reticulum or in cytoplasmic and membranous locations as well. For example, appreciable amounts of the vesicle-associated membrane protein 2 (VAMP2) are found on the plasma membrane after expression of a fluorescent protein chimera (Nadrigny et al. 2006). Although these backgrounds have fundamentally different origins, they have in common that they produce a diffuse apparent spatial overlap outside areas of true colocalisation in a way similar to the effect observed at a low signal-to-noise ratio (Fig. 5.3). Figure 5.5 shows that spurious background impairs image quality in two respects. First, it buries low-intensity signal and obscures image detail (Fig. 5.5a). Second, it affects resolution, as the high-spatial-frequency image components are generally of less intensity than the low-frequency ones. Therefore, colocalisation analysis critically depends on the background level (Fig. 5.5b), which must be reduced to a minimum, e.g. by the use of optical sectioning techniques that reduce the fluorescence excitation or readout volume (Box 5.2).

5.1.1.5 Ultradeep Imaging in Intact Tissue with Two-Photon-Excitation Fluorescence Microscopy

Genetically encoded probes permit the targeting and tagging of subpopulations of cells in vivo and are emerging as a powerful tool for imaging cellular and molecular biological function in the living animal (reviewed in Miyawaki 2005). 2PEF imaging usually offers a more efficient filtering than its one-photon counterpart, owing to the large spectral separation between the infrared excitation and visible fluorescence, but this advantage is to some extent compensated for by the high instantaneous intensities required for efficient in vivo 2PEF and the faint fluorescent signals emerging from deep tissue sections. Also, in vivo 2PEF microscopy of optically thick tissue sections is associated with a number of problems that stem from the increasing diffusion (i.e. spatial redistribution) of photons with greater imaging depths (Beaurepaire and Mertz 2002; Oheim et al. 2001) and affect colocalisation measurements.

For example, with increasing depth, the scattering of excitation photons prohibits the formation of a tight focus, thereby degrading the resolution (through filtering out

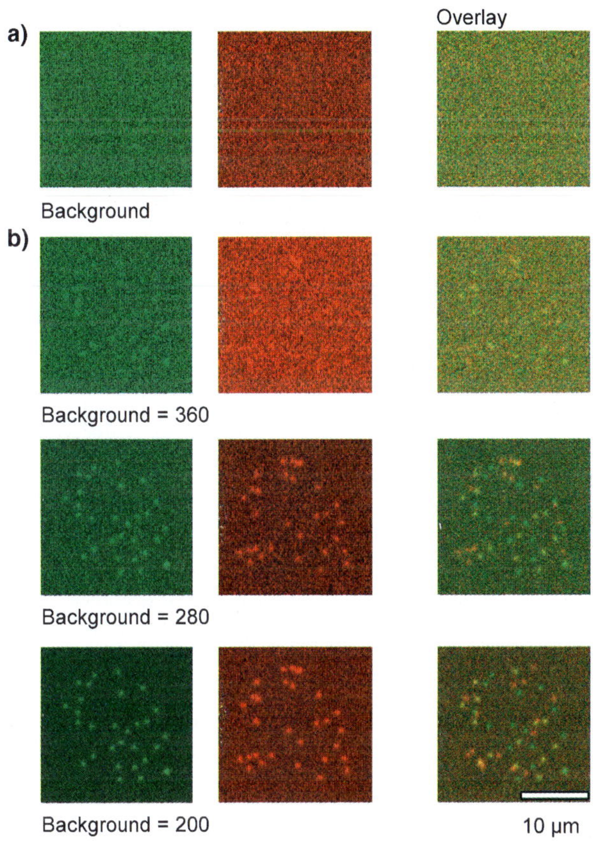

Fig. 5.5 Non-rejected background. **a** In the extreme case of colour-merging green- and red-colour background, the resulting overlap image displays a homogeneous yellow signal – featureless nearly perfect colocalisation. **b** To investigate the impact of non-rejected image background on the colocalisation estimate, the fluorophore intensity (useful signal) was kept constant at 500 counts, whereas the background level was chosen as indicated in the *leftmost panels*. Low background facilitates the detection of colocalisation

photons that travel at high NA) and reducing the signal (by decreasing the number of photons that arrive at the focal spot and contribute to 2PEF). Also, even when a detectable 2PEF is still generated at depths, fluorescence photons are scattered on their way to back the objective, so the collected fraction dwindles with increasing depth penetration, unless a high-NA large-field detector is used (Oheim et al. 2001). Finally, when compensating for these intensity losses by increasing the total laser power incident on the specimen, out-of-focus fluorescence and tissue autofluorescence are generated near the tissue surface. This is because the (diffuse) incident intensities are so high that the photon density in the unfocused pulsed beam is sufficiently high to generate out-of-focus 2PEF. This diffuse signal excited at the tissue surface is more efficiently collected and can attain the same order of magnitude and

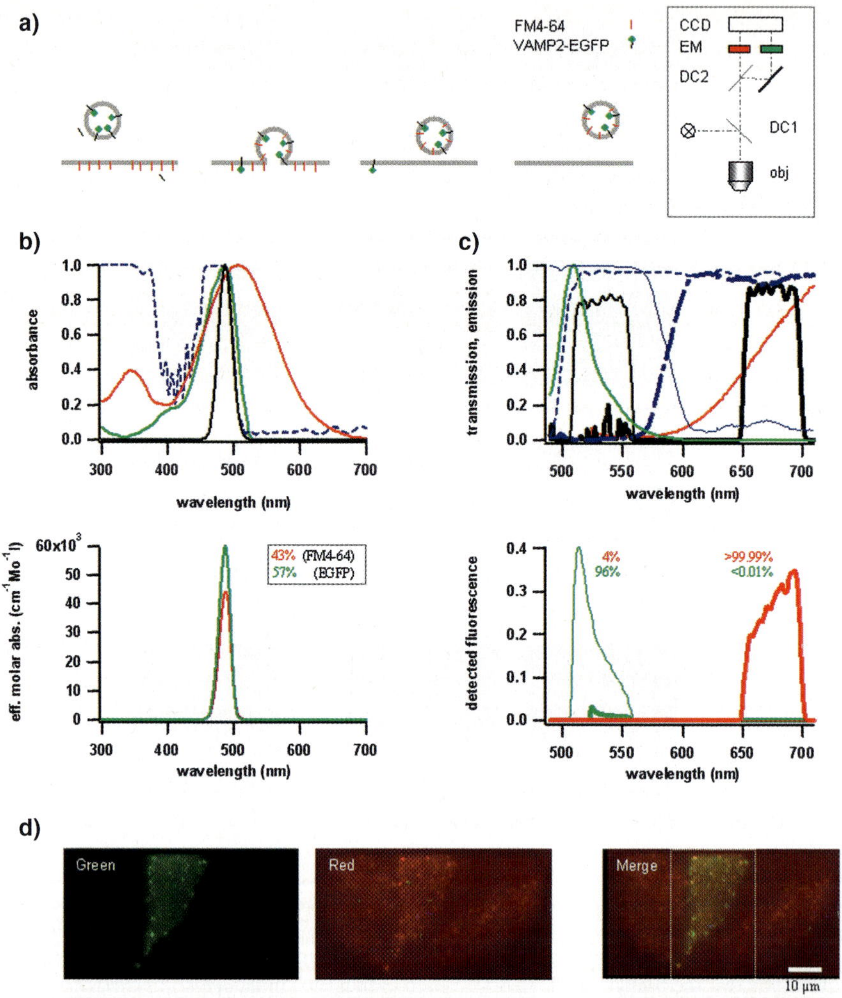

Fig. 5.6 Choosing appropriate filters for a given pair of fluorophores. **a** Endocytic uptake of the lipid membrane dye FM4-64 into secretory granules expressing a fusion protein of enhanced green fluorescent protein and vesicle-associated membrane protein 2 (EGFP–VAMP2). Note that FM4-64 is fluorescent when inserted in lipid membranes but virtually non-fluorescent in aqueous solution, as represented by *red sticks* and *black sticks*, respectively. *Inset:* The microscope light path with a dual viewer (500DCLP primary and 590DCXR secondary dichroic, HQ535/50m and HQ675/50m emitters). *obj* objective, *DC1* dichroic mirror, *DC2* secondary dichroic mirror, *EM* emission filters, *CCD* charge-coupled-device camera. **b** *Top*: Normalised spectra of the excitation band used (488t ± 12 nm, *solid black line*) and the reflection curve of the dichroic mirror (500DCLP, *dashed blue line*) as well the normalised absorption spectra of FM4-64 (red) and EGFP (green). *Bottom*: Product of the spectral excitation, dichroic reflectance, and molar extinction for FM4-64 (*red*) and EGFP (*green*), respectively. *Inset*: *Numbers* are the integral over wavelength, dλ. (Intentional) excitation cross-talk for 488-nm excitation is 43% (FM4-64) vs. 57% (EGFP), leading to roughly equal excitation of both dyes. **c** *Top*: Normalised emission spectra of FM4-64 (*red*) and EGFP (*green*), as well as the emission bandpass filters for their detection (*thick black line* and *thin black line*, respectively). *Blue lines*

obscure the faint collected in-focus signal from larger imaging depths (Oheim et al. 2001; Theer et al. 2003). In conclusion, for the same amount of physical colocalisation, the apparent measured colocalisation will vary with imaging depth.

5.1.2 A Practical Example of Dual-Band Detection

From the foregoing discussion it is clear that the quality of the colocalisation estimate critically depends on a variety of parameters. In practice, the 'quality' of the colocalisation measurement will depend on choosing appropriate combinations of dyes and filters in a given experimental situation, which requires knowledge about the level of autofluorescence in the sample and the sensitivity and spectral characteristics of the detector. Figure 5.6 shows a realistic experiment. We want to image the endocytic and exocytic dynamics of a defined subpopulation of secretory vesicles in cortical astrocytes, using a fluorescent marker of the styrylpyridinium dye family that is internalised upon endocytosis in an activity-dependent manner, and a monomeric fluorescent protein vesicle protein marker. A look at the available fluorescent probes suggests two possible combinations: (1) the use of the green fluorescent FM1-43 dye together with a monomeric red fluorescent protein (mRFP-1) chimera or (2) the (spectrally) inverse, i.e. the use of the red variant *N*-(3-triethylammoniumpropyl)-4-(6-(4-(diethylamino)phenyl)hexatrienyl)pyridinium dibromide (FM4-64) together with a green fluorescent protein construct (EGFP, pEGFP-N1, Clontech). When expressed in cells, the mRFP1–VAMP2 fusion protein produced a granular red fluorescence of spots; the size, dynamics, and response to intracellular calcium concentration elevations were compatible with secretory vesicles. Control experiments in which we expressed mRFP-1 displayed a clustered signal in front of the expected diffuse red cytoplasmic fluorescence (D. Li, F. Nadrigny, J. Hirrlinger, P.G. Hirrlinger, N. Ropert, F. Kirchhoff, and M. Oheim, unpublished data). EGFP–VAMP2 did not show a comparable mistargeting, so we rather opted for the FM4-64/EGFP pair (Blazer-Yost et al. 2001; Jomphe et al. 2005; Sharp and Pogliano 1999; Shoji et al. 2006; Tyler et al. 2006).

FM4-64 belongs to the same family of dyes as FM1-43 and is non-fluorescent in water but increases its fluorescence quantum yield by more than 2 orders of

Fig. 5.6 (continued) are the transmission of the primary (*dashed blue line*) and secondary (*dot-dashed blue line*) dichroic mirror as well as the transmission of the latter (*solid blue line*), respectively. *Bottom*: Effective contribution of FM4-64 (*thick line*) and EGFP (*thin line*) to the fluorescence detected in the red and green channels, respectively. See Box 5.1 for details. **d** Epifluorescence images of three mouse cortical astrocytes in culture, labelled with FM4-64 taken in the configuration shown in **a**. The cell in the *middle* equally expresses VAMP2-EGFP. *Left*: 'Green'-detection channel image (epi488|dic500|dic590|em535/50). *Middle*: 'Red'-detection channel image (epi488|dic500|dic590|em675/50). Note the partial colocalisation in the pseudocolour overlay, (*right*). The *boxed region* is analysed further in Fig. 5.12

magnitude upon insertion in lipid membranes (Brumback et al. 2004; Fig. 5.6a). Its peak excitation (558 nm) and emission (734 nm) are both redshifted (Table 5.1) compared with those of FM1-43 (Betz et al. 1996) owing to three double bonds linking the positively charged head and lipophilic tail group. Both EGFP and FM4-64 can be simultaneously and efficiently excited (with 99 and 93% of their peak absorption, respectively) with the quasimonochromatic (488 ± 12-nm) excitation band of a TILL Polychrome II (Messler et al. 1996) without further excitation filtering (Fig. 5.6b, top). Simultaneous EGFP/FM4-64 excitation removes the need to change filter cubes between acquisitions and offers the possibility of a simultaneous side-by-side projection of the green and red images on the same imaging detector. We used a 500DCLP primary dichroic mirror to reflect the excitation light onto the sample and a custom dual-viewer device in the two detection arms of the microscope (Fig. 5.6c, bottom).

Figure 5.6d displays the dual-channel fluorescence image pair of a group of cortical astrocytes labelled with FM4-64 and transfected with a plasmid encoding VAMP2–EGFP. As only the cell in the centre is expressing the fusion protein, we can directly estimate the amount of cross-talk from the images. Hardly any fluorescence is seen on the green-detection channel in the two cells flanking the transfected cell in the centre. Indeed, no FM4-64 signal is contaminating the green-detection channel. A comparison of the single-spot fluorescence and intensity line profiles on the red-detection channel reveals little if any increased intensities in the EGFP-expressing cell relative to its neighbours, indicating that EGFP signal is excluded from the red-detection channel (not shown).

However, even when choosing fluorophores of comparable brightness $\varepsilon\phi$ (Wessendorf and Brelje 1992), carefully balancing the detection efficiency, and avoiding spectral cross-talk, different (local) fluorophore concentrations can produce images of very different intensity that flaw colocalisation estimation. This is particularly a concern when fluorophore concentrations are unknown or only controlled in a very indirect manner, e.g. with acetoxymethyl ester loading or transfection (see Nadrigny et al. 2006 for a critical discussion).

Also, for many combinations of fluorophores and fluorescent proteins, fluorescence overlap is unavoidable. It can be dealt with, to some extent, with narrowband detection – at the expense of the collected signal. Dual-excitation dual-emission protocols can help to accentuate the differences between fluorophores while maintaining a constant signal, but changing filter cubes between acquisitions slows down the acquisition rate. Dichroic mirrors with multiple reflection bands can be employed to extend this dual-colour scheme to experiments with several dyes that excite and emit at different wavelengths. However, such multiband filters have reduced band-pass transmission and broadened reflection bands, leading to greatly reduced emitted photon collection efficiencies. To avoid such losses, a microscope design that allocates spatially separate portions of the objective lens aperture to excitation and emission beams without using a conventional dichroic mirror has been proposed (Friedman et al. 2006).

Finally with the exception of a few well-established cases (e.g. 4′,6-diamidino-2-phenylindole/fluorescein isothiocyanate/Texas Red), triple-band recordings of three labels are not feasible, particularly when autofluorescence is present or when the signals are faint (see, however, Finley et al. 2001; Lowy 1995; Xia et al. 2006). Instead, spectral 'oversampling' by acquiring a small number of spectral images larger than the number of fluorophores present in the sample can be a simple yet effective means to discern overlapping fluorophores (Nadrigny et al. 2006; Neher and Neher 2004a; Zimmermann 2005). In this case, the colocalisation estimate is not calculated from the fluorescence images but rather is derived from an 'unmixed' fluorophore abundance map (Box 5.3).

5.2 Quantifying Colocalisation

With a set of (near-) independent fluorophore maps at hand the next step is to calculate their relative degree of spatial overlap. Different techniques for measuring colocalisation are available. Although a clear distinction is not always straightforward, we group the various approaches in two principal families, those that:

1. Analyse *single-pixel intensity values* over the entire image area or subregions
2. First *segment* the image *and then detect objects* and compare their area and/or intensity

Whereas the former approach considers the acquired data set as a $5N$-dimensional vector [two lateral spatial dimensions (xy) plus one axial spatial dimension (z) plus time- plus N spectral dimensions] in which each pixel is treated as an independent element, the later approach identified objects and classifies them, by detecting fluorophore presence and colocalisation, and then tracks them over time. Both approaches have their distinctive advantages and drawbacks, but it is important to realise that they generally will not produce identical colocalisation estimates.

5.2.1 *'Colour Merging'*

Many published estimates of colocalisation are only qualitative and are based on an image-overlay method, i.e. the superposition of one fluorescence image, (pseudo-) coloured 'green' (Fig. 5.6d, left), on image two, coloured 'red' (Fig. 5.6, middle) to give 'yellow' (Fig. 5.6, right). A formally equivalent qualitative representation of the total overlap is obtained by subtracting one component image from the other (Akner et al. 1991; Oheim et al. 1999).

Colour merging is implemented in many commercial imaging software packages. The evaluation of the degree of colocalisation is generally visual-based and therefore prone to error and bias, as the 'amount of yellow' depends on the brightness of the merged images, the monitor settings, as well as the viewer's perception. Also, the displayed image triplet only reports one anecdotic observation and does not account for the variance in colocalisation observed for an ensemble of image pairs studied.

5.2.1.1 The Dos and Don'ts for Measuring Colocalisation

Any colocalisation measurement should:

- Use maximally independent image pairs (Sect. 5.1), and quantify the degree of cross-talk that is present (and tolerated) in the experiment.
- Define the parameter used for measuring co-localisation.
- Report a distribution of this parameter for a statistically significant sample of objects, cells, fields of view, etc., rather than a single value.
- Normalise the reported degree of colocalisation to some commonly accessible reference standard, e.g. multispectral subresolution beads, or dual-colour conjugates of nanocrystals. In practice, the absence of this type of normalisation is one of the main difficulties that prevents that data from the literature being directly compared.
- Show the same type of analysis for a negative control, i.e., an experiment, in which the same pair of fluorophores is directed to two molecules or subcellular compartments that can be assumed to be distinct. Sometimes, such negative controls have been performed by rotating one colour-channel image by 90° with respect to the other and recalculating the amount of colocalisation. A better strategy that takes into account the object size and density as well as the cross-talk is to place randomly the same number of solid (i.e. finite-extend) objects, convolved with the experimental PSF, on an area equal to the available area on the colocalisation image studied, and then to recalculate the colocalisation estimate over 30 trials. This procedure inherently takes into account the non-zero random overlap (Fig. 5.4b) resulting from the diffraction-limited resolution as well as trial-to-trial variability. In order to be called 'colocalised', the experimental distribution should be statistically different from the control distribution that gives a lower bound of meaningless (random) colocalisation.
- Report a positive control, i.e. an experiment, in which the same pair of fluorophores labels the same molecule or subcellular compartment. Again, and depending on the precise parameter calculated (see below), the calculated colocalisation estimate will not be 100%, even in the case of perfect spatial overlap. Alternatively, a synthetic image can be generated, convolved with the PSF, and image noise added as shown earlier. The resulting parameter distribution defines an upper bound and – together with the negative control - identifies the interval for meaningful colocalisation estimates.
- Use statistical tests to decide if the observed colocalisation distribution differs from a random (or perfect) control situation.

The result will probably be one of *partial colocalisation* (Fig. 5.6d). Even careful controls cannot remove the ambiguity of how to interpret the calculated parameter. Absolute values of colocalisation are always instrument-dependent. The best one can hope for is to arrive at statements like 'the measured amount of colocalisation was significantly higher than what is expected from a random distribution of the same organelles/molecules/…' or 'however, not all molecules X colocalised with Y, as the measured colocalisation significantly differed from a perfect match…'. A consistent yet laborious strategy to circumvent such problems is to compare the amount of colocalisation between the molecule of interest and different (spectrally identical) markers, e.g. the fluorescent proteins fused to different proteins, and to report their relative degree of colocalisation. When the different constructs are of about the same size and are expressed under the control of the same promoter, this procedure can also, at least in part, compensate for artefacts resulting from protein overexpression.

5.2.2 *Pixel-Based Techniques*

A more rigorous quantification of the overlap region than the colour merging method requires the simultaneous evaluation of spatial and intensity data in both colour channels (Garcia Peñarrubia et al. 2005). The first aspect, looking for the spatial correspondence of data pixels, is easily achieved once the two images have been brought in register. This can be done with the help of subresolution multiple-emission point sources as reference points (e.g. MultiSpeck™ beads, Invitrogen) as positive controls for true colocalisation at different image locations. If the sensitivity of the detector used is too low, imaging larger (and brighter) beads that contain one fluorophore on their surface and are labelled throughout with the other are a good choice. Next, after images have been aligned and corrections for chromatic aberrations have been made (Kozubek and Matula 2000) the spatial correspondence is investigated by collecting all pixel-intensity data pairs at corresponding x,y locations and plotting them as a two-dimensional histogram. In Figs. 5.7–5.11 we plot this type of analysis for the synthetically generated images previously shown. Figure 5.12 displays the scatterplot for the FM4-64/EGFP double-labelled cell shown in Fig. 5.6d.

5.2.2.1 Scatterplots, Multidimensional Histograms, and Spectral Angle Matching

The result is a two-dimensional scatterplot (or 'cytofluorogram'; Fig. 5.7), where the axes represent the intensity ranges of the two images and one point is one pixel. The point density in the pixel cloud is indicative of the frequencies of the intensity data pairs. In a noise-free, background-subtracted image, pixels containing pure fluorophores populate the paraxial regions (Fig. 5.7, top right). Extended, multifluorescent objects with their pixels having roughly a fixed intensity ratio between both channels result in a diagonal pixel cloud, the thickness of which reflects noise contributions of background noise. As noise-reduction techniques, low-pass and median filtering as well as image deconvolution can help to thin down the cloud, particularly at the low-intensity end (Landmann 2002). Cross-talk between detection channels brings the lobes of pure and mixed pixels closer to each other.

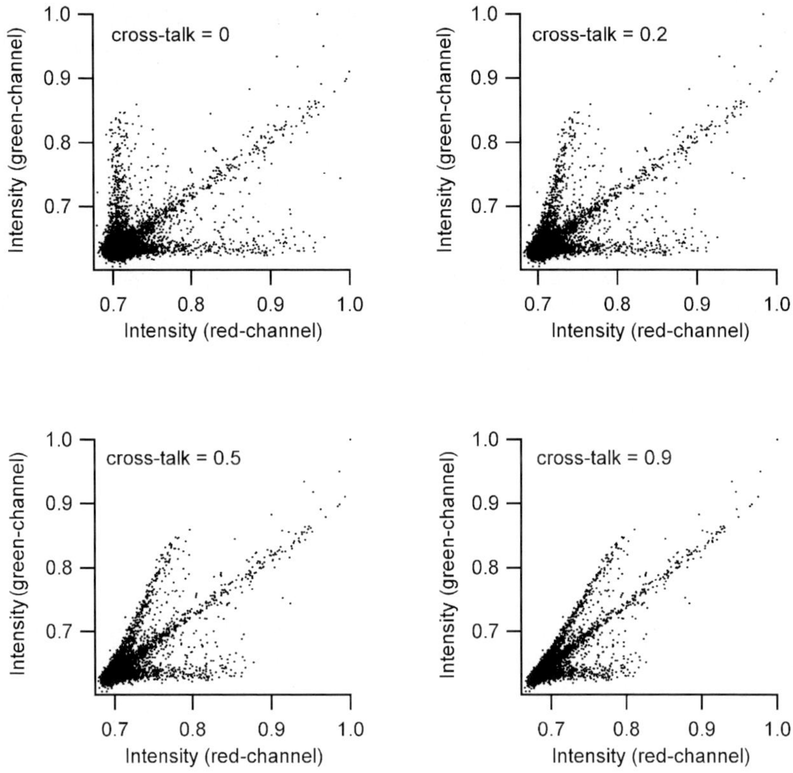

Fig. 5.7 Pixel analysis for the influence of cross-talk on colocalisation determination. Each image pixel is represented as a spot on the two-dimensional histogram. Axes are normalised to maximal intensity in each component image. The graphs illustrate the impact of increasing the cross-talk from the green into the red channel for the same images as shown in Fig. 5.1. Pure fluorescent species are found in paraxial regions, whereas mixed pixels that appear at roughly equal intensity on both component images populate the diagonal. Brightest pixels are located in the distant corners. Cross-talk brings the lobes representing pure pixels closer to the region of mixed pixels, reducing the spectral angle between 'pure' and mixed pixels, thereby increasing the apparent colocalisation

Therefore, the second (and trickier) part involves the determination of the zone of the scatterplot in which 'true' colocalisation occurs. Only pixels inside these boundaries will be considered for the colocalisation analysis. In practice, rectangular or conical colocalisation areas are the most frequent selections. The intensity-based scatterplot analysis is frequent in photometric devices like flow cytometers and cell counters (where labelled cells are relatively easily distinguished from autofluorescent and unlabeled ones and one cell analysed gives rise to one point on the scatterplot). In microscopic imaging, heterogeneity occurs at the subcellular level and the segmentation and interpretation of scatterplots is less straightforward (Garcia Peñarrubia et al. 2005). The lack of general criteria to select an area of colocalisation on the scatterplot makes this task both difficult and ambiguous. Typically, selecting

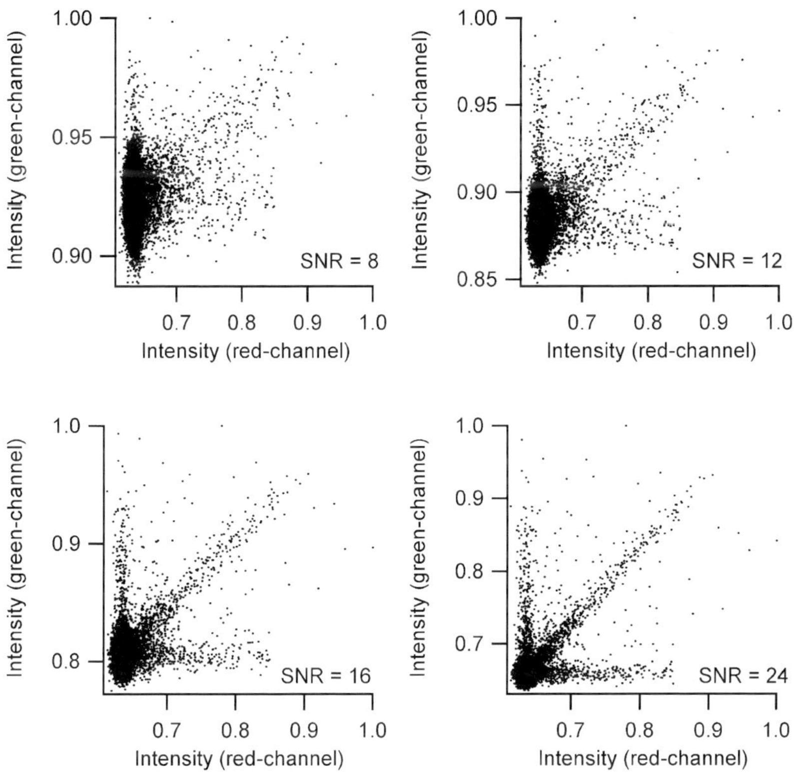

Fig. 5.8 Pixel analysis of the effect of the SNR on the colocalisation estimation. The graphs illustrate the impact of decreasing the SNR in the green channel, as shown on the images in Fig. 5.2. The red-channel image is always constant. Decreasing the SNR in the green-detection channel produces an increasingly featureless pixel cloud. The number of pixels that can be extracted and used for the calculation of the colocalisation coefficient becomes lower. With decreasing SNR, the interpretation and segmentation of scatterplots becomes less and less straightforward

a scatterplot region is done by thresholding each component image individually, which again leaves room for user-bias. One way to interactively control the choice is to produce a binary mask on the analysed images so that the user can see the position on the image of the selected pixel and compare this choice with some a priori knowledge about the fluorescent label, e.g. object size, shape, or localisation relative to other visible subcellular structures.

The *multidimensional histogram* extends the scatterplot type of analysis to data sets containing more than two fluorophores and prepares the ground for SILU approaches (Box 5.3). Each pixel (x,y) is represented by a spectral vector:

$$\mathbf{w}(x,y) = \left[w_1(x,y), w_2(x,y), \ldots, w_N(x,y)\right]^{T}. \tag{5.2a}$$

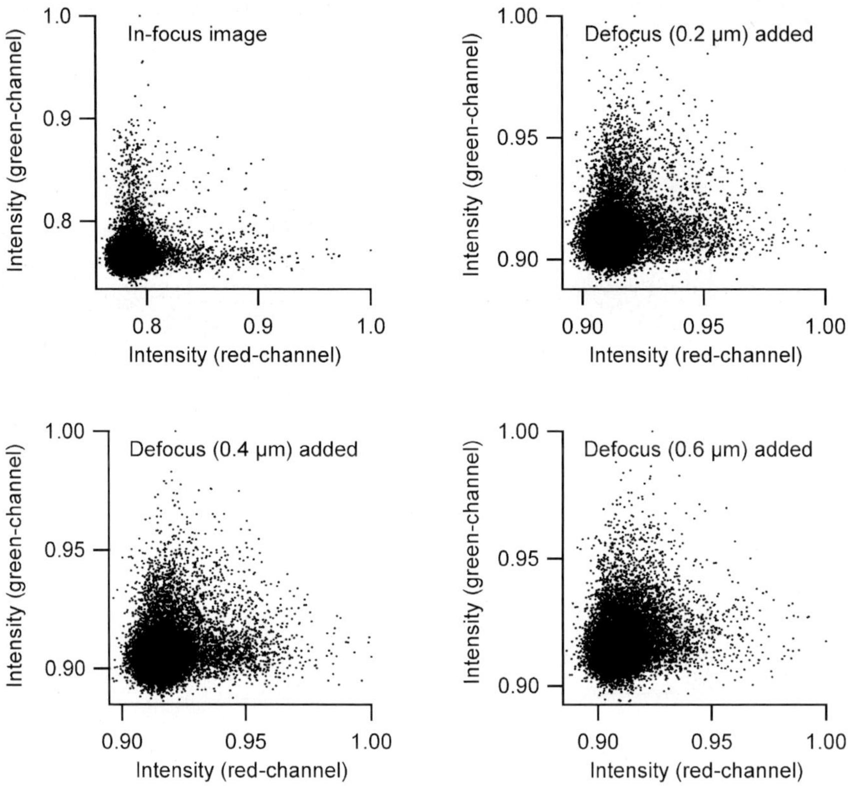

Fig. 5.9 Pixel analysis of the effect of out-of-focus fluorescence. The graphs illustrate the impact of moving some of the red and green objects out of focus, as shown in Fig. 5.3. On the two-dimensional scattergram, the initially well-separated lobes of unrelated fluorescent spots of the red and green component images increasingly merge upon addition of defocus. The spectral angle decreases and more and more pixels populate the diagonal and are (falsely) mistaken for colocalising pixels

Its N components w_i denote the intensity of pixel (x,y) in each detection band. For a more convenient notation, we replace the two-dimensional pixel index (x,y) by a running index $i \in [1,\ldots, n]$, where $n = x_{max} \cdot Y_{max}$,

$$\mathbf{w}(i) = [w_1(i), w_2(i), \ldots, w_N(i)]^{\mathrm{T}} \quad (5.2b)$$

In the case of zero cross-talk between the N detection channels, the N unit vectors form an orthonormal basis that spans the N-dimensional fluorophore space. Otherwise, the projection of $\mathbf{w}(i)$ on the pure fluorophore vectors defines a k-dimensional subspace ($N > k$, linear unmixing). Each pixel is represented by a vector and the projection on the axes measures the relative amount of fluorophore(s) present in that pixel. As in the two-dimensional case, above, the tricky part consists of delineating the N-dimensional

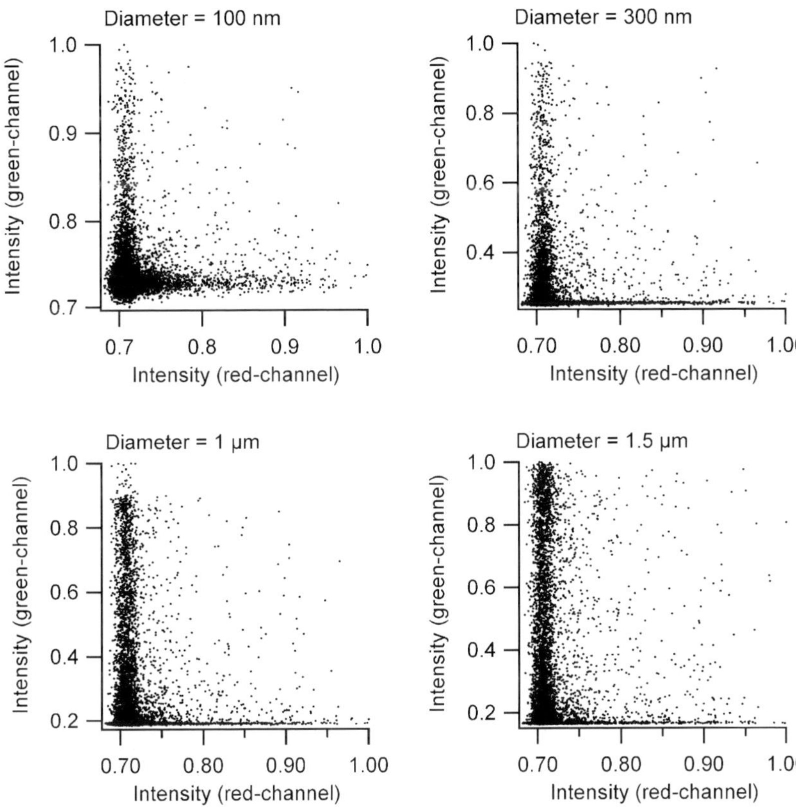

Fig. 5.10 Pixel analysis of the influence of the change in the object size on the colocalisation estimate. Diffraction adds a blur to the single-particle image that creates a false-positive colocalisation, even in the case of only proximal, non-overlapping (i.e. in-focus) particles. The graphs illustrate the impact of increasing the physical size of the green objects, as shown in Fig. 5.4. Diameters specify the 'true' physical size of the green particles before convolution with the PSF and addition of noise. Red particles always measured 100 nm in diameter. As green particles get bigger and bigger, more and more green pixels (that correspond to the luminous centre of the spherical particles) populate the high-intensity end of the two-dimensional scattergram

volumes that identify pure or coexisting fluorophores. Therefore, a different and more intuitive strategy classifies pixels by measuring their spectral similarity, based on *spectral angle* mapping (SAM) (Kruse et al. 1993; for examples see Neteler et al. 2004; Shrestha et al. 2005) or by using statistical descriptors that test for outliers among the spectral vectors (Nadrigny et al. 2006). The spectral angle,

$$\theta_i = \arccos\left(\frac{\mathbf{w}(i)\cdot\mathbf{r}}{\|\mathbf{w}(i)\|\cdot\|\mathbf{r}\|}\right) = \arccos\left(\frac{\sum_{j=1}^{N} w_j r_j}{\sqrt{\sum_{j=1}^{N} w_j^2}\cdot\sqrt{\sum_{j=1}^{N} r_j^2}}\right), \tag{5.3}$$

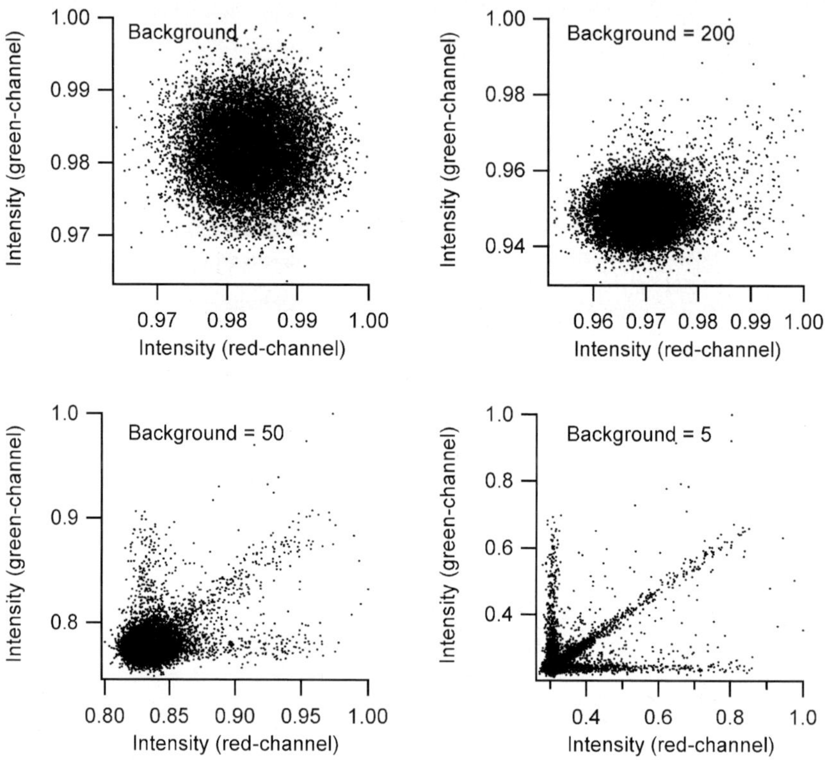

Fig. 5.11 Effect of non-rejected background on the two-dimensional scatterplot. The graphs illustrate the impact of adding a homogenous image background to the green image (mean intensity indicated) on the colocalisation estimate. The corresponding images are shown in Fig. 5.5. *Top left*: Two-dimensional histogram for pure red and green background images. *From top right to bottom right*: Effect of decreasing the relative contribution of a constant background. The peak signal of the 'true' signal was 500 counts. With decreasing background, the three-lobe signal (pure red, pure green, and colocalising pixels) gradually emerges from the centrosymmetric background scattergram

measures the resemblance of the pixel vector **w**(i) with a reference vector **r**, which will typically represent a known fluorophore, e.g. cytoplasmically expressed EGFP. Since SAM compares only the spectral angle between pixels containing known fluorophores and pixels containing unknown (potentially colocalised) fluorophores and not the length of the vector, the method is fairly insensitive to intensity differences. Also, no a priori knowledge about the exact shape of **w**(i) is required, so SAM is useful in situations where strong autofluorescence is present. An intuitive way to represent colocalisation is to measure the average vector $\langle \mathbf{w}(i) \rangle_{\text{coloc}}$ from an image region (or control experiment) where colocalisation occurs and to compare this reference vector with each pixel vector **w**. θ is then determined for each pixel i and the result is plotted as a pseudocolour map $\theta_i \in [0,1]$. This type of analysis bears resemblance to the classification problem in the satellite imaging and remote sensing literature (cf. multispectral and hyperspectral imaging).

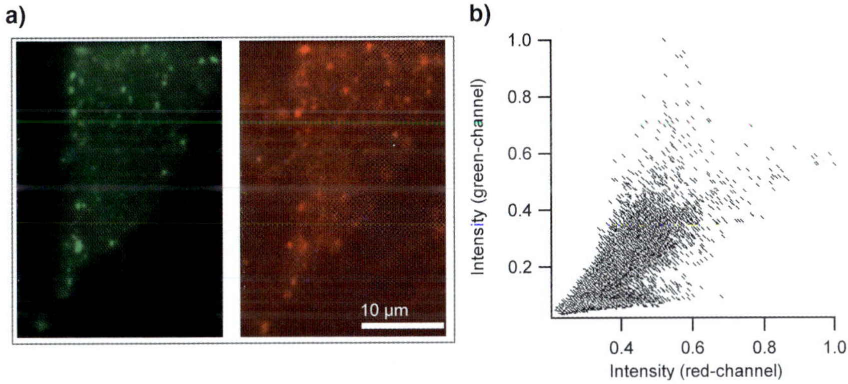

Fig. 5.12 Pixel analysis of the FM4-64/EGFP double-labelled astrocyte. **a** Excised region of interest taken from the centre of the dual-colour image shown in Fig. 5.6, showing a cortical astrocyte labelled with FM4-64 and expressing VAMP2–EGFP, viewed through the 'red' (HQ675/50m) and 'green' (HQ535/50m) microscope detection arms. See Box 5.1 for details. **b** Two-dimensional scattergram of the image pair shown in **a**. Three lobes can be distinguished in the pixel cloud. The lobe structure reveals that the intensity and contrast are higher in the green channel and that the signal-to-background and SNR are lower in the red-detection channel. The spectral angles apparent from the lobes are often misleading, owing to the non-uniform pixel density obscured by the finite symbol size.

5.2.2.2 Cross-Correlation Analyses: Pearson's Correlation Coefficient and Overlap Coefficients

A different method that uses the information of all pixels as well but calculates the degree of correlation between the intensity grey values of the pixels in a dual-colour image is the estimate provided by Pearson's correlation coefficient ρ_p (Manders et al. 1992, 1993). Pearson's correlation coefficient is one of the standard measures in pattern recognition (Gonzales and Wintz 1987) for matching one image with another and provides information about the similarity of shape without regard to the average intensity of the signals in both component images. It is calculated for two component images 1 and 2 as

$$\rho_p = \frac{\sum_i^N \left[w_1(i) - \langle w_1 \rangle_{\forall i}\right]\left[w_2(i) - \langle w_2 \rangle_{\forall i}\right]}{\sqrt{\sum_i^N \left[w_1(i) - \langle w_1 \rangle_{\forall i}\right]^2 \cdot \sum_i^N \left[w_2(i) - \langle w_2 \rangle_{\forall i}\right]^2}}, \quad (5.4a)$$

where $w_j(i)$ and $\langle w_j \rangle$ represent the fluorescence intensity of each pixel i and the average over all pixels of the component image j, respectively. N is the total number of pixels in each image. ρ_p is formally and mathematically equivalent to the cross-correlation coefficient (Stauffer and Meyer 1997), in which fluorophore colocalisation is measured by

$$\rho_c = \frac{\frac{1}{N}\sum_i^N \left[w_1(i) - \langle w_1 \rangle_{\forall i}\right]\left[w_2(i) - \langle w_2 \rangle_{\forall i}\right]}{\sqrt{\frac{1}{N}\sum_i^N \left[w_1(i) - \langle w_1 \rangle_{\forall i}\right]^2}\sqrt{\frac{1}{N}\sum_i^N \left[w_2(i) - \langle w_2 \rangle_{\forall i}\right]^2}}. \tag{5.4b}$$

A tool for automating this process in ImageJ has been published (Rodgers 2002).

In this type of correlation analysis, the average grey values of the two analysed images are subtracted from the respective pixel value, so pixels contribute to the colocalisation coefficient in proportion to their intensity difference to the average rather than their absolute intensities. As a consequence, both ρ_p and ρ_c vary from −1 to 1, i.e. perfect negative or positive correlation, perfect mutual exclusion, or perfect overlap of both fluorophores. However, the interpretation of the intermediate values is not straightforward. Therefore, Manders et al. (1993) proposed a slightly different formulation that takes the *overlap coefficient*

$$\rho_o = \frac{\sum_i^N w_1(i) \cdot w_2(i)}{\sqrt{\sum_i^N \left[w_1(i)\right]^2 \cdot \sum_i^N \left[w_2(i)\right]^2}} \tag{5.5}$$

as the starting point. ρ_o can assume values from 0 to 1. Also, Eq. 5.5 is insensitive to differential photobleaching of both fluorophores, as is readily seen by substituting $w_j(i) = \alpha \cdot w_j'(i)$. However, ρ_o will create biased estimates for component images with very different intensities and very different densities of fluorescent particles. This effect can be cancelled out by splitting ρ_o into two different (but interdependent) coefficients,

$$\rho_o^2 = k_1 k_2, \tag{5.6a}$$

where

$$k_1 = \frac{\sum_i^N w_1(i) \cdot w_2(i)}{\sum_i^N \left[w_1(i)\right]^2} \tag{5.6b}$$

and

$$k_2 = \frac{\sum_i^N w_1(i) \cdot w_2(i)}{\sum_i^N \left[w_2(i)\right]^2}. \tag{5.6c}$$

The degree of colocalisation is expressed using two different parameters, the first measuring intensity differences relative to channel 1, the second relative to channel 2. Two new colocalisation coefficients can be defined from this which are proportional to the amount of fluorescence of the colocalising objects in each component image, relative to the total fluorescence in that component (Manders et al. 1993):

$$M_1 = \frac{\sum_i^N w_1'(i)}{\sum_i^N w_1(i)} \tag{5.7a}$$

and

$$M_2 = \frac{\sum_i^N w_2'(i)}{\sum_i^N w_2(i)}, \tag{5.7b}$$

where $w_j'(i) = w_j(i)$ if $w_{i \neq j}(i) > t$, and is zero otherwise. As before, t defines some intensity threshold. Alternatively, a spectral-angle map (Sect. 5.2.2.1) can be used as the basis for selecting a threshold. Thus, only pixels in the second component image that contribute some appreciable intensity (or display a certain degree of spectral resemblance with image $k \neq j$) contribute to the numerator of M_1 and do so in proportion to the total fluorescence in image 1. M_1 and M_2 can even be determined when the intensity differences between the component images are very large and can be thought of as a the generalisation of Eq. 5.1, with the major difference that only their numerator is thresholded, but to the true intensity value, and is zero otherwise. Thus, instead of the overlapping pixel *area* alone, M_1 and M_2 weigh the area with the colocalised pixel *intensity*, i.e. they are – in some way –a hybrid between a pixel-based and an object-based measurement. The degree of colocalisation is defined as the ratio of the integral of the intensity distribution of colocalising pixels and the total intensity in the component image studied. When the number of pixels carrying an intensity above the threshold t is very different in images 1 and 2, M_1 and M_2 are a proper choice. Yet, the problems of thresholding, background subtraction, and treating outlier pixels remain, as with the other coefficients. A qualitative analysis of the factors that affect Manders's and Person's colocalisation coefficients is found in Garcia Peñarrubia et al. (2005).

Other methods for quantifying fluorophore colocalisation on a pixel-by-pixel basis have been described (Smallcombe 2001).

5.2.2.3 Regions of Interest and Segmenting Tools

Object-based colocalisation estimates, i.e. the segmentation of labels into distinct pixel clusters in three-dimensional space, followed by colocalisation of these clusters yield more reliable and sensitive measures of colocalisation than a simple determination of the number (or summed intensities) of colocalised pixels (pixel-based analysis). This is because object-based techniques utilise information about object shape and size in addition to intensity-information to sharpen the criteria used to designate colocalising pixels (Silver and Stryker 2000).

5.2.3 *Object-Based Techniques*

5.2.3.1 Threshold-Based Object Recognition

The simplest technique that does not rely on global intensity analysis was introduced by Lynch et al. (1991). Binary masks are created for both component images by thresholding and the overlap between the thresholded *areas* is calculated (cf. Eq. 5.1). A similar approach is implemented in many imaging software packages. For example, the MetaMorph (Molecular Devices) COLOCAL drop-in allows the user to chose between different descriptors of overlap (area, average or integrated intensities in the region of overlap) in thresholded image (sub-) regions of interest. These parameter measurements can be transformed into a true quantitative colocalisation estimate using a trick (Becherer et al. 2003; Rappoport et al. 2003); by introducing an artificial pixel shift of one component image relative to the other and recalculating the parameters, one obtains a modified parameter. This is repeated, one pixel at a time, for, e.g. ten pixels in each direction and averaged over, e.g., the eight cardinal directions. The plot of the parameter measured with increasing deliberate misalignment of the two images allows the determination of a characteristic length scale on which both fluorophores colocalise. Irrespective of the choice of the intensity threshold made for each channel, the procedure is inherently *pixel-based*, i.e. within the regions of interest created the data are processed without introducing further assumptions for the object that is being imaged.

5.2.3.2 Localising Spots Using Segmented Image Subegions

Until now, we have treated images or even three-dimensional image stacks as large planar or cubic assemblies of independent pixels. One can – and indeed should – use the information contained in the image rather than treating each pixel individually. Although – mathematically speaking – the presence of image noise makes each pixel statistically independent of its neighbours, the intensity envelope, i.e. the low-spatial-frequency signal extending over an ensemble of nearby pixels, is not independent, owing to the diffraction limitation.

One example using correlated multipixel information is the accurate determination of the two-dimensional (or three-dimensional) *position* of point objects in a fluorescence image (or *z*-stack of images) (Ghosh and Webb 1994) by fitting a small region of the intensity image with a centre of mass (centroid) or two-dimensional Gaussian (Cheezum et al. 2001; Gennerich and Schild 2005) to locate the spot. Its position is calculated on the basis of all pixels that belong to the domain of the spot, so a meaningful contour must be delineated that defines the region on the image that belongs to the spot, e.g., using largest (Manders et al. 1996) or active (Dufour et al. 2005) contour spatial segmentation.

The object coordinate (rather than its intensity distribution) can then be used for the investigation of colocalisation. Spots are localised in *independent* image channels so that the accuracy of the particle position is not resolution-limited but rather depends on the signal-to-noise ratio of the fitted image and the measured PSF (Churchman et al. 2005; Karakikes et al. 2003; Morrison et al. 2003); therefore, the term *precision* rather than resolution is often used in this context. With bright molecular fluorophores, molecular distances can be measured with an accuracy better than 10 nm using conventional far-field optics (Lacoste et al. 2000; Michalet et al. 2001) and less than 2 nm using total internal reflection fluorescence microscopy (Yildiz et al. 2003). Of course, for this precision to be attained the component images of the different colour channels must be truly independent, stressing the importance of eliminating cross-talk between images. Although this calculation is simple, its error analysis is demanding and has generally not been correctly applied (Churchman 2006).

When spectral overlap cannot be avoided, using the spatial distribution of fluorescence lifetimes instead of intensities can be an alternative (Berezovska et al. 2003; Brismar and Uifhake 1997; Heilemann et al. 2002; Wahl et al. 2004). In fluorescence lifetime imaging microscopy (FLIM), several (picosecond) time-resolved images of a sample are obtained at various time delays after pulsed laser excitation of the microscope field of view. Lifetimes are calculated pixel by pixel from these time-resolved images, and the spatial variations of the fluorescence lifetime are then displayed in a two-dimensional pseudocolour-coded map. Combining FLIM with polarisation-modulated excitation allows one to obtain, simultaneously, information about the relative orientation of fluorophores (Heinlein et al. 2005).

5.2.3.3 Studying Single-Pair Colocalisation and Interaction with Single-Molecule Fluorescence

Probably the most intuitive way of establishing colocalisation with object-based techniques is single-particle tracking. When two molecular fluorophores consistently move together, they are probably attached one to the other (Yang and Musser 2006). Dual-colour fluorescence cross-correlation spectroscopy (FCCS) is capable of measuring interacting fluorescently tagged macromolecules via temporal cross-correlation analysis of fluorescence intensity fluctuations collected from a small observation volume defined by the excitation beam focus (Schwille et al. 1997). Intensity fluctuations arising from changes in fluorophore concentration within the beam focus are recorded simultaneously in two channels and correlated in time to reveal transport properties and number densities of interacting and non-interacting species (reviewed in Bacia et al. 2006). Employing simultaneous two-photon excitation of three distinct dye species, Heinze et al. (2004) demonstrated their successful discrimination on a single-molecule level. This enables the direct observation of higher-order molecular complex formation in the confocal volume. Image cross-correlation spectroscopy (ICCS) relies on the same principles as FCCS, but utilises spatial correlation analysis of intensity fluctuations in fluorescence images (Brown et al. 1999). A quantitative comparison between the standard, fluorescence microscopy colocalisation algorithms and spatial ICCS has been published (Comeau et al. 2006). A similar double labelling and coincidence fluorescence detection method has been used to enhance the sensitivity of single-molecule detection and observe individual DNA molecules labelled with two different fluorophores in solution (Li et al. 2003). Single-molecule single-pair FRET (spFRET) experiments (Ha et al. 1996; Yang et al. 2006) extend these measurements to studying true molecular interaction (Allen et al. 2003; reviewed in Yeh et al. 2005). Finally, bimolecular fluorescence complementation assays (Hu et al. 2002) in which two non-fluorescent protein fragments are combined to give a functional flurophore may soon attain single-molecule sensitivity (reviewed in Hebert et al. 2006; Kerppola 2006; Piehler 2005).

All these techniques have in common that they rely on the ultrasensitive detection and identification, in extremis, of single molecular species. However, because of the faint signals involved, single-molecule techniques are particularly vulnerable to the incomplete separation of the different colour channels owing to the presence of autofluorescence, along with cross-excitation and emission bleed-through (see earlier). A SILU technique that uses the statistical correlations between pixels on the image single-diffraction-limited spots has been used to quantify the expression and colocalisation of about 15 copies of fluorescent protein molecules on single secretory vesicles (Nadrigny et al. 2006). Using classification and feature extraction techniques borrowed from multispectral and hyperspectral imaging techniques (see Box 5.3) and applied to microscopic imaging (reviewed in Zimmermann 2005), spectral unmixing improves FRET detection (Ecker et al. 2004; Gu et al. 2004; Neher and Neher 2004b).

5.3 Conclusions

- The resolving power of the instrument defines a three-dimensional minimal volume that gives the ‘unit cell’ for meaningful colocalisation analysis. For objects smaller than this volume, it is conceivable that both fluorophores are present in the same voxel accidentally without being associated.
- Colocalisation of intensity images is restricted to data sets with high signal-to-noise ratios and cannot provide colocalisation information at the low-intensity end.
- Image processing (filtering, deconvolution, unmixing) improves the colocalisation estimate, at the expense of spatial resolution. Appropriate controls must ascertain that artefacts that can be generated by image processing do not influence the estimate.
- Depending on the technique, the results of the colocalisation analysis differ qualitatively and quantitatively. Therefore, to allow data to be compared or reproduced, a detailed protocol must complement the colocalisation analysis.
- Irrespective of the precise technique used for estimating fluorophore presence and colocalisation, the reduction of a high-dimensional data set with millions of image elements to one or two numbers necessarily implies a considerable loss of information. Therefore, it is important to use a colocalisation measurement that extracts and preserves the information from the images that should be retained. Also, in analysing colocalisation, absolute numbers are often not terribly meaningful. Reporting relative parameter distributions and comparing the amount of colocalisation between different – spectrally equivalent – fluorescent markers can often be a sensible compromise.
- Single-molecule techniques are increasingly being used to localise and colocalise single fluorescently labelled biomolecules and, combined with FRET or fluores-

cence complementation analyses, to trace out molecular interactions. Owing to the faint intensities, these techniques are particularly vulnerable to spectral cross-talk and benefit from multispectral imaging and unmixing techniques.

While this chapter was in proof, Adler and Parmryd presented a normalised Pearson's coefficient for calculating co-localisation while taking into account image noise in the two detection channels. This approach based on comparing first the frame-to-frame variations within one color channel on replicate images and then calculating the corrected co-localisation estimate between the two channels. See I. Palmryd and J. Adler, *Making Accurate Measurement of Colocalization by Correcting for Image Noise*, 2007 Biophysical Society Meeting Abstracts, Biophysical Journal, Supplement, Abstract. p321a for details.

References

Akner GM, K., Wilkström A, Sundqvist K, Gustafsson J (1991) Evidence for co-localisation of glucocorticoid receptor with cytoplasmic microtubules in human gingival fibroblasts, using two monoclonal anti-GR antibodies, confocal microscopy and image analysis. J Steroid Biochem Mol Biol 39:419–432

Allen MW, Bieber Urbauer RJ, Zaidi A, Williams TD, Urbauer JL, Johnson CK (2003) Fluorescence labeling, purification, and immobilization of a double cysteine mutant calmodulin fusion protein for single-molecule experiments. Anal Biochem 325:273–284

Axelrod D (2001) Selective imaging of surface fluorescence with very high aperture microscope objectives. J Biomed Opt 6:6–13

Bacia K, Kim SA, Schwille P (2006) Fluorescence cross-correlation spectroscopy in living cells. Nat Methods 3:83–89

Beaurepaire E, Mertz J (2002) Epifluorescence collection in two-photon microscopy. Appl Opt 41:5376–5982

Becherer U, Moser T, Stuhmer W, Oheim M (2003) Calcium regulates exocytosis at the level of single vesicles. Nat Neurosci 6:846–853

Berezovska O, Ramdya P, Skoch J, Wolfe MS, Bacskai BJ, Hyman BT (2003) Amyloid precursor protein associates with a nicastrin-dependent docking site on the presenilin 1-γ-secretase complex in cells demonstrated by fluorescence lifetime imaging. J Neurosci 23:4560–4566

Betz WJ, Mao F, Smith CB (1996) Imaging exocytosis and endocytosis. Curr Opin Neurobiol 6:365–371

Betzig E, Patterson GH, Sougrat R, Lindwasser OW, Olenych S, Bonifacino JS, Davidson MW, Lippincott-Schwartz J, Hess HF (2006) Imaging intracellular fluorescent proteins at nanometer resolution. Science 313:1642–1645

Blazer-Yost BL, Butterworth M, Hartman AD, Parker GE, Faletti CJ, Els WJ, Rhodes SJ (2001) Characterization and imaging of A6 epithelial cell clones expressing fluorescently labeled ENaC subunits. Am J Physiol Cell Physiol 281:C624–632

Brismar H, Uifhake B (1997) Fluorescence lifetime measurements in confocal microscopy of neurons labeled with multiple fluorophores. Nat Biotechnol 15:373–377

Brown CM, Roth MG, Henis YI, Petersen NO (1999) An internalization-competent influenza hemagglutinin mutant causes the redistribution of AP-2 to existing coated pits and is colocalized with AP-2 in clathrin free clusters. Biochemistry 38:15166–15173

Brumback AC, Lieber JL, Angleson JK, Betz WJ (2004) Using FM1-43 to study neuropeptide granule dynamics and exocytosis. Methods 33:287–294

Cheezum MK, Walker WF, Guilford WH (2001) Quantitative comparison of algorithms for tracking single fluorescent particles. Biophys J 81:2378–2388

Chudakov DM, Chepurnykh TV, Belousov VV, Lukyanov S, Lukyanov KA (2006) Fast and precise protein tracking using repeated reversible photoactivation. Traffic 7:1304–1310
Churchman LS, Okten Z, Rock RS, Dawson JF, Spudich JA (2005) Single molecule high-resolution colocalization of Cy3 and Cy5 attached to macromolecules measures intramolecular distances through time. Proc Natl Acad Sci USA 102:1419–1423
Churchman LS, Flyvbjerg H, Spudich JA (2006) A non-Gaussian distribution quantifies distances measured with fluorescence localization techniques. Biophys J 90:668–671
Comeau JWD, Costantino S, Wiseman PW (2006) A guide to accurate fluorescence microscopy colocalization measurements. Biophys J 91:4611–4622
Cubitt AB, Heim R, Wollenweber LA (1999) Understanding structure-function relationships in the Aequorea victoria green fluorescent protein. In: Sullivan KF, Kay SA (eds) Green fluorescent proteins, vol. 58. Academic, San Diego, pp 19–33
Demandolx D, Davoust J (1997) Multicolour analysis and local image correlation in confocal microscopy. J Microsc 185:21–36
Dufour A, Shinin V, Tajbakhsh S, Guillen-Aghion N, Olivio-Marin JC, Zimmer C (2005) Segmenting and tracking fluorescent cells in dynamic 3-D microscopy with coupled active surfaces. IEEE Trans Image Proc 14:1396–1410
Ecker RC, de Martin R, Steiner GE, Schmid JA (2004) Application of spectral imaging microscopy in cytomics and fluorescence resonance energy transfer (FRET) analysis. Cytometry A 59:172–181
Finley KR, Davidson AE, Ekker SC (2001) Three-color imaging using fluorescent proteins in living zebrafish embryos. BioTechniques 31:66–70
Friedman LJ, Chung J, Gelles J (2006) Viewing dynamic assembly of molecular complexes by multi-wavelength single-molecule fluorescence. Biophys J 91:1023–1031
Garcia Peñarrubia P, Férez Ruiz X, Galvez J (2005) Quantitative analysis of the factors that affect the determination of colocalization coefficients in dual-color confocal images. IEEE Trans Image Proc 14:1–8
Gennerich A, Schild D (2005) Sizing-up finite fluorescent particles with nanometer-scale precision by convolution and correlation image analysis. Eur Biophys J 34:181–199
Ghosh RN, Webb WW (1994) Automated detection and tracking of individual and clustered cell surface low density lipoprotein receptor molecules. Biophys J 66:1301–1318
Giepmans BNG, Adams SR, Ellisman MH, Tsien RY (2006) The fluorescent toolbox for assessing protein location and function. Science 312:217–224
Gonzales RC, Wintz P (1987) Digital image processing. Adisson-Wesley, Reading
Griesbeck O (2004) Fluorescent proteins as sensors for cellular functions. Curr Opin Neurobiol 14:636–641
Gu Y, Di WL, Kelsell DP, Zicha D (2004) Quantitative fluorescence resonance energy transfer (FRET) measurement with acceptor photobleaching and spectral unmixing. J Microsc 215:162–173
Gustafsson MGL (2005) Nonlinear structured-illumination microscopy: wide-field fluorescence imaging with theoretically unlimited resolution. Proc Natl Acad Sci USA 102:13081–13086
Ha T, Enderle T, Ogletree DF, Chemla DS, Selvin PR, Weiss S (1996) Probing the interaction between two single molecules: fluorescence resonance energy transfer between a single donor and a single acceptor. Proc Natl Acad Sci USA 93:6264–6268
Hebert TE, Gales C, Rebois RV (2006) Detecting and imaging protein-protein interactions during g protein-mediated signal transduction in vivo and in situ by using fluorescence-based techniques. Cell Biochem Biophys 454:85–109
Heilemann M, Herten D-P, Heintzmann R, Cremer C, Müller C, Tinnefeld P, Weston KD, Wolfrum J, Sauer M (2002) High-resolution colocalization of single dye molecules by fluorescence lifetime imaging microscopy. Anal Chem 74:3511–3517
Heinlein T, Biebricher P, Schlüter C, Roth M, Herten D-P, Wolfrum J, Heilemann M, Müller C, Tinnefeld P, Sauer M (2005) High-resolution colocalization of single molecules within the resolution gap of far-field microscopy. ChemPhysChem 6:949–955

Heinze K, Jahnz M, Schwille P (2004) Triple-color coincidence analysis: One step further in following higher order molecular complex formation. Biophys J 86:506–516

Henkel AW, Lübke J, Betz WJ (1996) Fm1-43 dye ultrastructural localization in and release from frog motor nerve terminals. Proc Natl Acad Sci USA 93:1918–1923

Hess ST, Girirajan TPK, Mason MD (2006) Ultra-high resolution imaging by fluorescence photoactivation localization microscopy (FPALM). Biophys J 91:4258–4272

Hirrlinger PG, Scheller A, Braun C, Quintela-Schneider M, Fuss B, Hirrlinger J, Kirchhoff F (2005) Expression of red coral fluorescent proteins in the central nervous system of transgenic mice. Mol Cell Neurosci 30:291–303

Hofmann M, Eggeling C, Jakobs S, Hell SW (2005) Breaking the diffraction barrier in fluorescence microscopy at low light intensities by using reversibly photoswitchable proteins. Proc Natl Acad Sci USA 102:17565–17569

Hu C-D, Chinenov Y, Kerppola TK (2002) Visualization of interactions among bZIP and Rel family proteins in living cells using bimolecular fluorescence complementation. Mol Cell 9:789–798

Jares-Erijman EA, Jovin TM (2003) FRET imaging. Nat Biotechnol 21:1387–1395

Jomphe C, Bourque M-J, Fortin GD, St-Gelais F, Okano H, Kobayashi K, Trudeau L-E (2005) Use of TH-EGFP transgenic mice as a source of identified dopaminergic neurons for physiological studies in postnatal cell culture. J Neurosci Methods 146:1–12

Karakikes I, Barber RE, Morrison IEG, Fernandez N, Cherry RJ (2003) Co-localization of cell surface receptors at high spatial resolution by single-particle fluorescence imaging. Biochem Soc Trans 31:1453–1455

Kerppola TK (2006) Visualization of molecular interactions by fluorescence complementation. Nat Rev Mol Cell Biol 7:449–456

Kozubek M, Matula P (2000) An efficient algorithm for measurement and correction of chromatic aberrations in fluorescence microscopy. J Microsc 200:206–217

Kruse FA, Lefkoff AB, Boardman JW, Heiderecth KB, Shapiro AT, Barloon JP, Goetz AF (1993) The spectral image processing system (SIPS) – interactive visualization and analysis of imaging spectrometer data. Remote Sens Environ 44:145–163

Lacoste TD, Michalet X, Pinaud F, Chemla DS, Alivisatos AP, Weiss S (2000) Ultrahigh-resolution multicolor colocalization of single fluorescent probes. Proc Natl Aacd Sci USA 97:9461–9466

Landmann L (2002) Deconvolution improves colocalization analysis of multiple fluorochromes in 3D confocal data sets more than filtering techniques. J Microsc 208:134–147

Li D, Xiong J, Qu A, Xu T (2004) Three-dimensional tracking of single secretory granules in live PC12 cells. Biophys J 87:1991–2001

Li H, Ying L, Green JJ, Basasubrananian S, Klenerman D (2003) Ultrasensitive coincidence fluorescence detection of single DNA molecules. Anal Chem 75:1664–1670

Lowy RJ (1995) Evaluation of triple-band filters for quantitative epifluorescence microscopy. J Microsc 178:240–250

Lynch RM, Fogarty KE, Fay FS (1991) Modulation of hexokinase association with mitochondria analyzed with quantitative three-dimensional confocal microscopy. J Cell Biol 112:385–395

Manders EM, Stap J, Brakenhoff GJ, van Driel R, Aten JA (1992) Dynamics of three-dimensional replication patterns during the S-phase, analysed by double labelling of DNA and confocal microscopy. J Cell Sci 103:857–862

Manders EM, Verbeek FJ, Aten JA (1993) Measurement of co-localization of objects in dual-colour confocal images. J Microsc 169:375–382

Manders EM, Hoebe R, Strackee J, Vossepoel AM, Aten JA (1996) Largest-contour segmentation: a tool for the localization of spots in cofocal images. Cytometry 23:15–21

Martinez-Arca S, Rudge R, Vacca M, Raposo G, Camonis J, Proux-Gillardeaux V, Daviet L, Formstecher E, Hamburger A, Filippini F, D'Esposito M, Galli T (2003) A dual mechanism controlling the localization and function of exocytic v-SNAREs. Proc Natl Acad Sci USA 100:9011–9016

Messler P, Harz H, Uhl R (1996) Instrumentation for multiwavelengths excitation imaging. J Neurosci Methods 69:137–147

Michalet X, Lacoste TD, Weiss S (2001) Ultrahigh-resolution colocalization of spectrally separable point-like fluorescent probes. Methods 25:87–102

Miyashita T (2004) Confocal microscopy for intracellular co-localization of proteins. In: Fu H (ed) Protein-protein interactions methods and applications, vol 261. Humana, Totowa, pp 399–410

Miyawaki A (2005) Innovations in the imaging of brain functions using fluorescent proteins. Neuron 48:189–199

Morrison IEG, Karakikes I, Barber RE, Fernandez N, Cherry RJ (2003) Detecting and quantifying colocalization of cell surface molecules by single particle fluorescence imaging. Biophys J 85:4110–4121

Nadrigny F, Rivals I, Hirrlinger PG, Koulakoff A, Personnaz L, Vernet M, Allioux M, Chaumeil M, Ropert N, Giaume C, Kirchhoff F, Oheim M (2006) Detecting fluorescent protein expression and colocalisation on single secretory vesicles with linear spectral unmixing. Eur Biophys J 35:533–547

Nadrigny F, Li D, Kemnitz K, Ropert N, Koulakoff A, Rudolph S, Vitali M, Giaume C, Kirchhoff F, Oheim M. Systematic Co-localization Errors between Acridine Orange and EGFP in Astrocyte Vesicular Organelles. Biophys J. 2007 Apr 6; [Epub ahead of print]

Neher R, Neher E (2004a) Optimizing imaging parameters for the separation of multiple labels in a fluorescence image. J Microsc 213:46–62

Neher RA, Neher E (2004b) Applying spectral fingerprinting to the analysis of fret images. Microsc Res Tech 64:185–195

Neteler M, Grasso D, Michelazzi I, Miori L, Merler S, Furlanello C (2004) New image processing tools for grass. In: Proceedings of FOSS/GRASS user conference 2004, Bangkok, Thailand

Oheim M, Loerke D, Stühmer W, Chow RH (1999) Multiple stimulation-dependent processes regulate the size of the releasable pool of vesicles. Eur Biophys J 28:91–101

Oheim M, Beaurepaire E, Chaigneau E, Mertz J, Charpak S (2001) Two-photon microscopy in brain tissue: Parameters influencing the imaging depth. J Neurosci Methods 111:29–37

Oheim M, Li D, Luccardini C, Yakovlev A (2007) Online resource for calculating the spectral separabilty index Xijk. http://www.biomedicale.univ-paris5.fr/neurophysiologie/Groups/oheimropertgroup.php. Cited 4 Apr 2007

Oshiro M, Moomaw B (2003) Cooled vs. intensified vs. electron bombardment CCD cameras – applications and relative advantages. Methods Cell Biol 72:133–156

Patterson GH, Knobel SM, Sharif WD, Kain SR, Piston DW (1997) Use of the green fluorescent protein and its mutants in quantitative fluorescence microscopy. Biophys J 73:2782–2790

Piehler J (2005) New methodologies for measuring protein interactions in vivo and in vitro. Curr Opin Struct Biol 15:4–14

Rappoport JZ, Taha BW, Lemeer S, Benmerah A, Simon SM (2003) The AP-2 complex is excluded from the dynamic population of plasma membrane-associated clathrin. J Biol Chem 278:47357–47360

Rodgers W (2002) An automated method for quantifying fluorophore colocalization in fluorescence double-labeling experiments. BioTechniques 32:28–34

Schultz C, Schleifenbaum A, Goedhart J, Gadella TW Jr (2005) Multiparameter imaging for the analysis of intracellular signaling. Chembiochem 8:1323–1330

Schwille P, Meyer-Almes FJ, Rigler R (1997) Dual-color fluorescence cross-correlation spectroscopy for multicomponent diffusional analysis in solution. Biophys J 72:1878–1886

Shaner N, Steinbach PA, Tsien RY (2005) A guide to choosing fluorescent proteins. Nat Methods 2:905–909

Sharp MD, Pogliano K (1999) An in vivo membrane fusion assay implicates spoiiie in the final stages of engulfment during Bacillus subtilis sporulation. Proc Natl Acad Sci USA 96:14553–14558

Sheppard CJR, Gan X, Gu M, Roy M (1995) Signal-to-noise in confocal microscopes. In: Pawley JB (ed) Handbook of confocal microscopy Plenum, New York

Shoji J-Y, Arioka M, Kitamoto K (2006) Vacuolar membrane dynamics in the filamentous fungus Aspergillus oryzae. Eukaryot Cell 5:411–421
Shrestha DP, Margate DE, van der Meer F, Anh HV (2005) Analysis and classification of hyper-spectral data for mapping land degradation: an application in southern Spain. Int J Appl Earth Obs Geoinf 7:85–96
Silver MA, Stryker MP (2000) A method for measuring colocalization of presynaptic markers with anatomically labelled axons using double label immunofluorescence and confocal microscopy. J Neurosci Methods 94:205–215
Smallcombe A (2001) Multicolor imaging: the important question of co-localization. BioTechniques 30:1240-1246
Stauffer TP, Meyer T (1997) Compartmentalized IgE receptor-mediated signal transduction in living cells. J Cell Biol 139:1447–1457
Theer P, Hasan MT, Denk W (2003) Two-photon imaging to a depth of 1000 microns in living brains by use of a Ti:Al2O3 regenerative amplifier. Opt Lett 28:1022–1024
Tyler WJ, Zhang X-L, Hartman K, Winterer J, Muller W, Stanton PK, Pozzo-Miller L (2006) Bdnf increases release probability and the size of a rapidly recycling vesicle pool within rat hippocampal excitatory synapses. J Physiol 574:787–803
Wahl M, Koberling F, Patting M, Rahn H, Erdmann R (2004) Time-resolved confocal fluorescence imaging and spectrocopy system with single molecule sensitivity and sub-micrometer resolution. Curr Pharm Biotechnol 5:299–308
Wessendorf MW, Brelje TC (1992) Which fluorophore is brightest? A comparison off the staining obtained using fluorescein, tetramethylrhodamine, lissamine rhodamine, Texas Red, and cyanine 3.18. Histochemistry 98:81–85
Willig KI, Kellner RR, Medda R, Hein B, Jakobs S, Hell SW (2006) Nanoscale resolution in GFP-based microscopy. Nat Methods 3:721–723
Xia J, Kim SHH, Macmillan S, Truant R (2006) Practical three color live cell imaging by wide-field microscopy. Biol Proc Online 8:63–68
Yang J, Chen H, Vlahov IR, Cheng J-X, Low PS (2006) Evaluation of disulfide reduction during receptor-mediated endocytosis by using FRET imaging. Proc Natl Acad Sci USA 103:13872–13877
Yang W, Musser SM (2006) Visualizing single molecules interacting with nuclear pore complexes by narrow-field epifluorescence microscopy. Methods 39:316–328
Yeh HC, Chao SY, Ho YP, Wang TH (2005) Single-molecule detection and probe strategies for rapid and ultrasensitive genomic detection. Curr Pharm Biotechnol 6:453–461
Yildiz A, Forkey JN, McKinney SA, Ha T, Goldman YE, Selvin PR (2003) Myosin v walks hand-over-hand: Single fluorophore imaging with 1.5-nm localization. Science 300:2061–2065
Zimmermann T (2005) Spectral imaging and linear unmixing in light microscopy. Adv Biochem Eng Biotechnol 95:245–265

6
Quantitative FRET Microscopy of Live Cells

Adam D. Hoppe

Abstract Quantitative fluorescence resonance energy transfer (FRET) microscopy is a powerful tool for analyzing dynamic protein–protein interaction within living cells. FRET microscopy is increasingly employed to access the molecular mechanisms governing diverse cellular processes such as vesicular transport, signal transduction and the regulation of gene expression. However, evaluation of experimental approaches for FRET microscopy and the data they produce requires an appreciation of the techniques at the photophysical, molecular and data-acquisition level. This chapter aims to provide a conceptual framework for comparing FRET technologies and interpreting the data they produce. We begin with a qualitative discussion of FRET physics and the molecular interactions that can be probed by FRET. The discussion then shifts to the aspects of quantitative microscopy necessary for FRET-based measurements. With this foundation, we move to an overview of the current techniques in FRET microscopy, including acceptor photobleaching, spectral fingerprinting, FRET stoichiometry, and polarization FRET. Lastly, we discuss interpretation of FRET data and emerging applications to protein network analysis. Altogether, this chapter provides a progressive overview of FRET microscopy, beginning with fluorescent excited states, moving to detection methods and ending with interpretation of cell biology data.

6.1 Introduction

Proteins, lipids and nucleic acids form organized and dynamic chemical networks within the three-dimensional space of the living cell. Many of the protein–protein interactions that make up these networks have been identified. Until recently, our ability to analyze these networks in living cells was limited by a lack of microscopic methods for observing these interactions within their native context. Quantitative fluorescence resonance energy transfer (FRET) microscopy is emerging as a powerful instrument to meet this need. Here I describe the fundamentals of FRET microscopy and the future of this technique. My goal in this chapter is to provide the reader with a conceptual framework for understanding how current FRET technologies work and for evaluating the images they produce.

S.L. Shorte and F. Frischknecht (eds.), *Imaging Cellular and Molecular Biological Functions.*

FRET is the transfer of energy from an excited donor fluorophore to an acceptor fluorophore by a dipolar interaction that occurs over distances in the range of 1–10 nm. As such, in vitro FRET spectroscopy has been used extensively in biology to study molecular structure, conformational changes and molecular associations (reviewed in Lakowicz 1999). Over the last decade, FRET methods have been developed for microscopic analysis of molecular associations and conformations within living organisms. This growth in popularity has been fueled by systems biology and proteomics interests in understanding protein interactions in living cells. Furthermore, the development of the spectral variants of green fluorescent protein (GFP) has greatly accelerated FRET microscopy in living cells by allowing tagging of proteins by genetic manipulation. Together, the tools of GFP combined with quantitative FRET microscopy promise to provide unprecedented new insights into understanding the dynamics and localization of molecular interactions inside living cells.

This chapter builds a conceptual framework for understanding FRET and the microscopic methods that use FRET to measure molecular interactions. We begin with a description of the physics of FRET, paying attention to key parameters that govern FRET and how FRET manifests itself in the fluorescence of the donor and acceptor. We then cover fluorescence microscopy image formation and the concepts behind quantitative image analysis for FRET. At the heart of the chapter, we discuss the major approaches to FRET imaging, including photobleaching, sensitized emission, polarization and fluorescence lifetime. Lastly, this chapter will deal with data display and data interpretation, paying attention to the advantages and disadvantages of using fluorescent proteins in FRET imaging.

6.2 Introductory Physics of FRET

Perrin postulated mechanisms for FRET in 1927, and a physical formalism for FRET was published by Theodore Förster in 1948 (Lakowicz 1999). A detailed description of Förster's original derivation, an alternative derivation and a comparison with a quantum mechanical derivation are reviewed by Clegg (1996). Here I will recapitulate only the main concepts and results from Förster's original derivation as they pertain to the nature of the fluorescence signals created by FRET. My goal is to convey an intuitive understanding of FRET that the biologist can use for contextualizing FRET imaging results.

Förster's original derivation considers two classical charged oscillators coupled to each other by electrostatic interactions with dipole moments of magnitude μ. In other words, the fluorophores are modeled as molecular-scale antennae. These two oscillators can exchange energy only if they have the same resonant frequency. Initially, the donor oscillator is vibrating and the acceptor is not. This can be thought of as a transmitting antenna (donor) and a receiving antenna (acceptor). The donor oscillator can either give up its energy to its surroundings by emission of a photon (or other nonradiative processes) or transfer energy to the acceptor. The likelihood of that energy transfer depends on how strongly the two dipoles are coupled, which in turn depends on their relative orientation and distance and on the likelihood that they are at resonance (i.e., the energy of the donor's excited state matches the energy that can be absorbed by the acceptor). Förster realized that most fluorescent molecules have broad excitation and emission spectra owing to their interactions with solvent. In short, he deduced that the probability that the donor and acceptor will be in perfect resonance depends on the overlap of the donor emission energy levels with the acceptor absorption energy levels, and on the bandwidth of resonant energies. Combining this condition with results from electromagnetics describing the distance and orientation dependencies yields the fundamental prediction of Förster's theory, the rate of energy transfer (reproduced from Clegg (1996),

$$k_{\mathrm{T}} \approx \left(\frac{\kappa^2}{R^6}\right)\left(\frac{\Omega'}{\Omega^2}\right)\left(\frac{\mu^4}{\hbar^2 n^4}\right). \tag{6.1}$$

The first grouping contains the distance and orientation components; κ is a geometrical factor describing the relative orientation of the two dipoles and R is the distance between the dipoles. The second group of terms indicates the degree of spectral overlap. The final grouping of terms contains Planck's constant $\hbar$, the index of refraction n and the magnitude of the dipole moment μ. For most biological studies, the choice of fluorophores (and medium) dictates the last two groupings of terms. The leftmost grouping says that the rate of energy transfer depends on the relative orientation of the dipoles (κ^2) and the distance between them (R^6).

For fluorophores with nonzero spectral overlap that are brought close enough together and have nonzero κ^2 orientations, the energy transfer process will occur with rate k_{T} and will compete with the rate of radiative emissions (τ_{D}^{-1}) of the donor. Thus, the FRET efficiency (E) can be defined as the number of quanta transferred from the donor to the acceptor divided by the number of quanta absorbed by the donor. Generally, E is expressed in terms of these rate constants as (Lakowicz 1999)

$$E = \frac{k_{\mathrm{T}}}{\tau_{\mathrm{D}}^{-1} + k_{\mathrm{T}}}. \tag{6.2}$$

The Förster distance (R_0) is defined as the distance between randomly oriented donors and acceptors for which the FRET efficiency is 50%. Typical Förster distances are on the order of 2–7 nm. By comparing the equation for k_{T} and the equation for E, we see that E depends on the distance between the fluorophores to the sixth power (R^6). Likewise the influences of the orientation of the fluorophores on E are relayed through the term κ^2. κ^2 can range between 0 and 4, where orthogonal dipoles give a value of 0, parallel dipoles give a value of 4 and randomly oriented dipoles give a value of 2/3. Hence, from the above description, the efficiency of FRET depends on the choice of fluorophores, the distance between them and the relative orientation of their dipole moments.

6.3 Manifestations of FRET in Fluorescence Signals

In the absence of FRET, the donor and acceptor fluorescence will have characteristic spectral properties including excitation and emission spectrum, quantum yield, fundamental polarization and natural fluorescence lifetime. All fluorescence signals detected in the absence of FRET will simply be a sum of the characteristic donor and acceptor fluorescence. If the donor and acceptor are brought close together, FRET can alter their spectral properties in ways that introduce new spectroscopic components to the system. In general, FRET results in four spectroscopic changes affecting: the net fluorescence spectrum, polarization, fluorescence lifetime and photobleaching rates.

6.3.1 Spectral Change (Sensitized Emission)

As FRET efficiency increases, the donor emits fewer photons and the acceptor emits proportionally more photons (Fig. 6.1). This hallmark reduction in donor fluorescence and consequent increase in acceptor fluorescence seems to provide an obvious mechanism by which to measure FRET. However, sensitized emission can

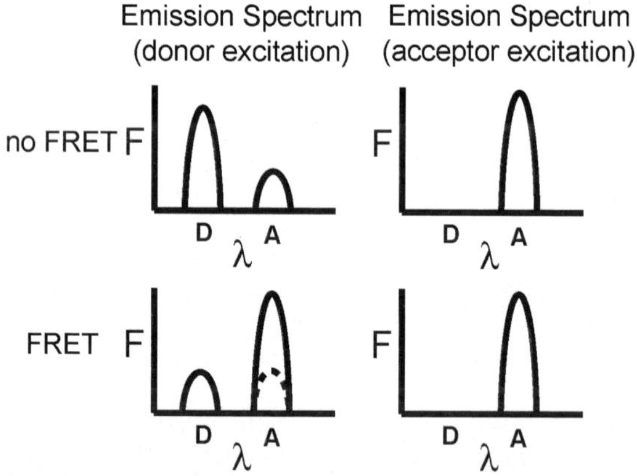

Fig. 6.1 The spectral changes associated with fluorescence resonance energy transfer (*FRET*). The fluorescence emission from the donor (D) and acceptor (A) under donor excitation are altered by FRET, whereas the directly excited acceptor fluorescence is unperturbed by FRET. For most donor–acceptor pairs, light used to excite the donor also excites the acceptor. Conversely, excitation of the donor when exciting the acceptor can also occur; however, this is rarely encountered. When FRET occurs, the sensitized emission from the acceptor increases, the donor fluorescence decreases and the directly excited acceptor remains unchanged (the *dotted line* indicates fluorescence from directly excited acceptor is still present). For clarity, this example describes the spectral relationships assuming that the donor and acceptor emissions do not overlap. For most real fluorophores there will be some emission overlap.

be difficult to quantify because of spectral overlap between donor and acceptor emission and excitation. This means that analysis of FRET by spectral change requires either observing the donor and acceptor fluorescence with and without FRET or a calibration scheme to separate and scale the sensitized emission and the direct fluorescence of the donor and acceptor. Methods that utilize sensitized emission in an appropriately calibrated microscope are very powerful and have become the mainstay of FRET microscopy.

6.3.2 *Fluorescence Lifetime*

A characteristic property of fluorescence is the long time it takes for a molecule to relax from the excited state to the ground state when compared with the time needed for other molecular transitions. Following excitation, the donor or acceptor will relax from the excited state exponentially by emission of a photon or by nonradiative processes. The typical decay rate constant for fluorophores to emit a photon is on the order of 1–10 ns after excitation. This long lifetime means that the fluorescence decay of a population of fluorophores can be measured using pulsed lasers and special detectors (Lakowicz 1999). As mentioned above, FRET is a dynamic rate process that competes with the rate of radiative relaxation. In other words, the rate of FRET depends on the time required by the donor and acceptor to encounter a resonant energy level. Thus, FRET creates an alternative route for the donor to reach the ground state, resulting in a shortened fluorescence lifetime decay of the donor. Furthermore, the fact that the rate of transfer competes with the donor lifetime means that acceptors will be excited by FRET at various times during the donor's time in the excited state, resulting in a positive growth and protracted decay of the acceptor fluorescence (Fig. 6.2).

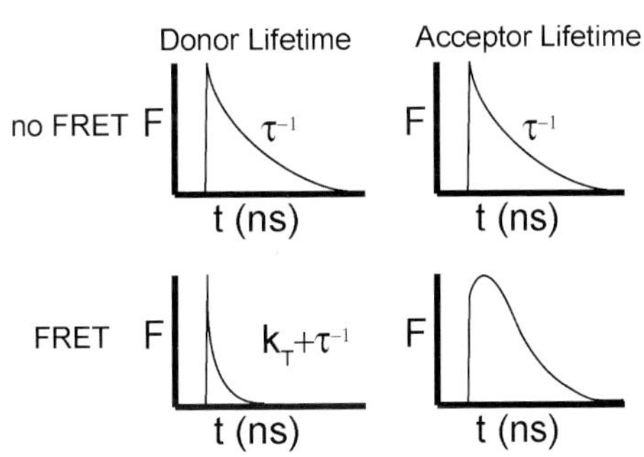

Fig. 6.2 The fluorescence lifetime of the acceptor and donor are affected by FRET. Following a very brief excitation pulse in the absence of FRET, the donor and acceptor will display their natural fluorescence lifetime decays. When engaged in FRET, the donor's lifetime is shortened and the acceptor's lifetime is protracted

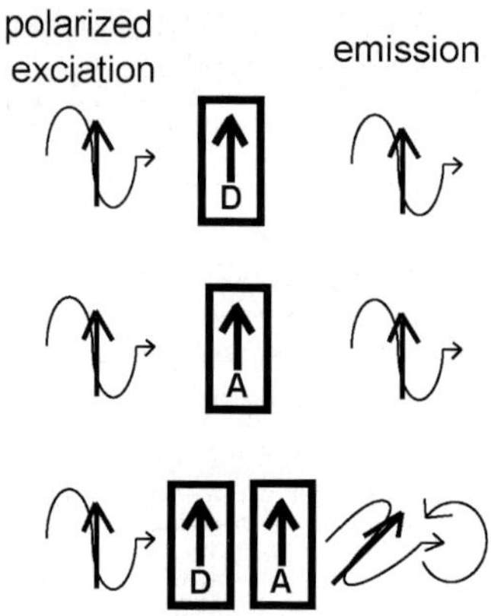

Fig. 6.3 Fluorescence polarization is affected by FRET. Polarized excitation of a slowly rotating donor or acceptor molecule (such as cyan fluorescent protein and green fluorescent protein) results in a highly polarized fluorescence emission. Energy transferred from a donor excited by polarized light and then emitted from an acceptor is largely depolarized owing to small angles between the dipole moments of the donor and acceptor

6.3.3 Polarization

Polarization is the property of light that describes the direction of light's electric field vector. This vector is perpendicular to the direction the light is traveling. Excitation of fluorescent molecules is most efficient when the polarization of the incident photon is aligned with the molecule's dipole moment. Likewise, the emission of a fluorophore is also polarized, usually in the same orientation as the excitation dipole. For fluorophores that rotate slowly relative to their fluorescence lifetime (e.g., GFP), the fluorescence will have polarization similar to that of the excitation light. FRET, however, requires that energy be passed between dipoles that usually have some angular displacement and therefore FRET has a strong propensity to scramble the polarization of the sensitized emission (Clegg 1996; Fig. 6.3). The depolarization caused by FRET has formed the basis for several methods for analyzing FRET in living cells.

6.3.4 Accelerated Photobleaching

FRET accelerates the photobleaching of the accepter while reducing the rate of donor photobleaching in a FRET efficiency dependent manner (Kubitscheck et al. 1991; Young et al. 1994; Jares-Erijman and Jovin 2003; Fig. 6.4). This effect originates from the fact that the rate of photobleaching of a fluorophore depends on its excited-state lifetime. As mentioned in Sect. 6.3.2, FRET shortens the time that the donor spends in the excited state and prolongs the time that the acceptor spends in the excited state, and consequently altering the photobleaching rates of the donor and acceptor. Both the enhanced acceptor and the diminished donor bleaching rates have been used for measurement of FRET efficiency (reviewed in Jares-Erijman and Jovin 2003).

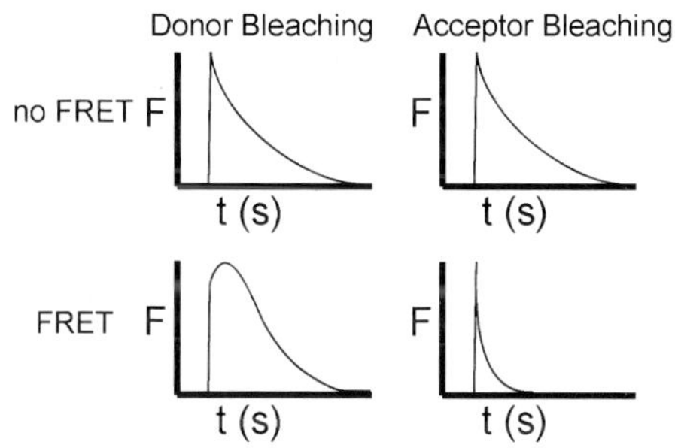

Fig. 6.4 Acceptor and donor photobleaching rates are affected by FRET. In the absence of FRET, the donor and acceptor will bleach exponentially during illumination. When the donor and acceptor are engaged in FRET, the donor bleaching rate is reduced and the acceptor bleaching is accelerated. Rapid bleaching of the acceptor results in an enhanced fluorescence from the liberated donors

6.4 Molecular Interaction Mechanisms That Can Be Observed by FRET

The strong dependence of FRET efficiency on distance and orientation allows for analysis of changes in molecular structure and interactions. The molecular mechanisms amenable to FRET analysis can be grouped into three categories: conformational change, molecular association and molecular assembly (Fig. 6.5).

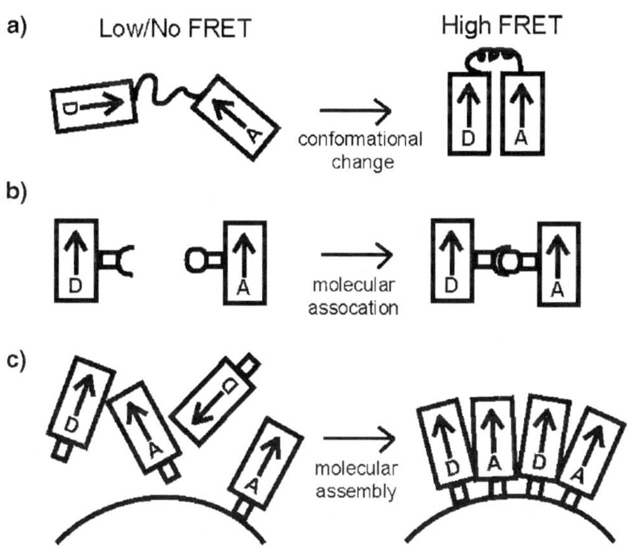

Fig. 6.5 The molecular mechanisms of observed by FRET. **a** Conformational change, **b** molecular association and **c** molecular assembly

6.4.1 Conformational Change

Analysis of changes in molecular conformation requires that the molecule be labeled with both donor and acceptor fluorophores. For proteins, this attachment is usually accomplished by placing the donor and acceptor at the N-terminus or the C-terminus of the molecule (e.g., genetic fusion with cyan fluorescent protein, CFP, or yellow fluorescent protein, YFP). Thus, when the molecule undergoes a conformational change due to an interaction or binding of analyte, the distance and orientation between the donor and acceptor are changed, resulting in a change in the FRET efficiency. Conformational change has formed the basis for all of the biosensor-type FRET constructs discussed later.

6.4.2 Molecular Association

The binding of two or more proteins can be monitored by FRET. If two different proteins with affinity for each other are labeled with donor and acceptor and these two molecules are allowed to associate, they may bring the donor and acceptor close enough together that they can undergo FRET. An example of FRET analysis of molecular associations in live cells is the interaction of activated Rho-GTPases with effector domains (Kraynov et al. 2000; Hoppe et al. 2002). Detection of a nonzero FRET efficiency is a good indication that the molecules are very close together (within about 10 nm); however, it does not necessarily indicate direct binding. It is possible for the two proteins to interact via an intermediate binding partner and still create a FRET signal. Conversely, lack of a FRET signal does not rule out an interaction as the protein complex may result in a distance and orientation unfavorable for FRET.

6.4.3 Molecular Assembly

Assembly of molecules into higher-order structures or onto surfaces can also be monitored by FRET. In this case, donor- and acceptor-labeled molecules that accumulate in the structure are brought close enough for FRET to occur despite the lack of a specific association between the two molecules. Examples of this effect include the density of GPI-anchored proteins on the surface of cells (Kenworthy and Edidin 1999), assembly of phosphoinositide binding domains on the inner leaflet of the plasma membrane (van der Wal et al. 2001) and assembly of molecules into the spindle pole body in yeast cells (Muller et al. 2005).

6.5 Measuring Fluorescence Signals in the Microscope

Before describing methods for FRET microscopy, it is helpful to review the parameters that effect detection of fluorescence. Measuring FRET in the microscope requires accurate determination of fluorescence intensities and accounting for acquisition parameters. Fluorescence microscopy has long suffered from its reputation as a qualitative tool. In part, this reputation originates from the many parameters that determine the magnitude of fluorescence signals. For quantitative image analysis such as FRET microscopy, it is important that these parameters be held constant, or be correctly scaled to allow calibration of the microscope and quantitative comparison between fluorescence signals. Here, I briefly describe the important parameters involved in acquisition of a fluorescence image.

If we neglect blurring in the microscope, we can describe formation of a fluorescence image by excitation and emission processes. The number of molecules excited (per second) in the focal volume of the microscope (F^*) is given by the product of the concentration of molecules $[F]$ with the excitation light intensity $L(\lambda)$ and the fluorophore's extinction coefficient $\varepsilon(\lambda)$ integrated over the excitation wavelengths λ :

$$F^* = [F]\int L(\lambda)\varepsilon(\lambda)\mathrm{d}\lambda. \tag{6.3}$$

The fluorescence from these molecules (F, per second) is given by the product of the quantum yield Q (e.g., the fraction of excitations resulting in emissions), and the integral of the product of the fluorophore's emission spectrum $S(\lambda)$, the emission filter's bandpass $B(\lambda)$ and the spectral response of the camera $C(\lambda)$ over λ :

$$F = F^* Q\int B(\lambda)S(\lambda)C(\lambda)\mathrm{d}\lambda. \tag{6.4}$$

Combining the excitation and emission equations gives the fluorescence intensity impinging on the camera or detector during the exposure time Δt,

$$I = F\Delta t = \Delta t Q\int B(\lambda)S(\lambda)C(\lambda)\mathrm{d}\lambda [F]\int L(\lambda)\varepsilon(\lambda)\mathrm{d}\lambda.$$

From this equation, we can see that the measured fluorescence intensity depends on fluorophore-specific parameters $\varepsilon(\lambda)$, $S(\lambda)$ and Q and on microscope-specific parameters $B(\lambda)$, $L(\lambda)$, $C(\lambda)$ and Δt. All quantitative microscopy methods, including quantitative colocalization, ratio imaging and FRET microscopy, require that these parameters either remain constant or can be accounted for from one sample to the next. Fortunately, many of these parameters are constants and depend only on fluorophore choice and the microscope configuration. In particular, the emission bandpass $B(\lambda)$ and camera response $C(\lambda)$ are usually constant for a given microscope setup. The excitation intensity, however, can be more difficult to keep constant because aging mercury arc lamps, variable power supplies, adjustable lasers and adjustable neutral density filters affect $L(\lambda)$. Such changes need to be corrected for by frequent calibration or by adjustments to the microscope to minimize their effect (e.g., using a long-lived xenon lamp instead of a mercury lamp). Perhaps the most frequently varied microscope parameter is the exposure time Δt. Fortunately, this parameter is easy to account for since the relative intensity of two images will simply scale with the ratio of exposure times over a wide range (approximately 2×10^{-3}–2 s for most CCD cameras).

In FRET microscopy, the fluorophore-specific parameters $\varepsilon(\lambda)$, $S(\lambda)$ and Q are determined by selection of the donor and acceptor fluorophores. Switching from one FRET pair to another means that these parameters will change and the microscope/fluorophore system will have to be recalibrated. Importantly, not all of these parameters need to be measured in order to calibrate a microscope for FRET imaging. Rather, FRET imaging techniques for sensitized emission describe the parameters as ratios (Sect. 6.6.2).

In fluorescence microscopy, the fluorescence image is the spatial distribution of I, e.g., $I(x,y)$, whose values we expect will correspond with the concentration of fluoropores in the two-dimensional optical section of the cell [$F(x,y)$]. For this correspondence to hold, microscope images must be preprocessed. In particular, images collected with a CCD camera have an offset such that even when no light reaches the camera the image is made up of nonzero values (Fig. 6.6). This arbitrary value is often referred to as the camera bias level and can be placed into our intensity equation (now modified to represent a two-dimensional image with x,y-dependent parameters),

$$\begin{aligned} I(x,y) &= F(x,y)\Delta t \\ &= \Delta t Q \int B(\lambda)S(\lambda)C(\lambda)\mathrm{d}\lambda \left[F(x,y)\right] \int L(\lambda)\varepsilon(\lambda)\mathrm{d}\lambda + \mathrm{bias}(x,y). \end{aligned} \tag{6.5}$$

This bias value must be subtracted from the image to obtain the fluorescence intensity. The bias image can be obtained by collecting images while blocking all light from reaching the camera. Most CCD cameras will produce a somewhat uneven bias image, bias (x,y), that often has larger values on one side than the other. The best approach for correcting for the camera bias is to subtract a mean bias image obtained from a 10 or 20 frame average with no light reaching the camera (Fig. 6.6).

A second, but very important consideration is the spatial variation in the illumination field across the image and between different excitation wavelengths. Above, we said it was important that $L(\lambda)$ remain constant. If $L(\lambda)$ varies across the image, e.g., $L(\lambda,x,y)$, then corrections need to be applied to compare fluorescence signals in various regions of the image. Furthermore, for two images created with excitation wavelength $L(\lambda_1,x,y)$ that has a specific spatial pattern in x,y and excitation wavelength $L(\lambda_2,x,y)$ that has its own x,y distribution, the ratio $L(\lambda_1,x,y)/L(\lambda_2,x,y)$ (and hence any calibration depending on these two illuminations) will not be uniform over the image. This situation is often encountered in microscopes with directly attached mercury arc lamps and must be corrected. Liquid light guide illumination systems tend to create more even illumination fields. The simplest way to correct images for differences in illumination patterns is to collect an image of the illumination (or "shading") pattern. This can be accomplished by making a thin solution (about

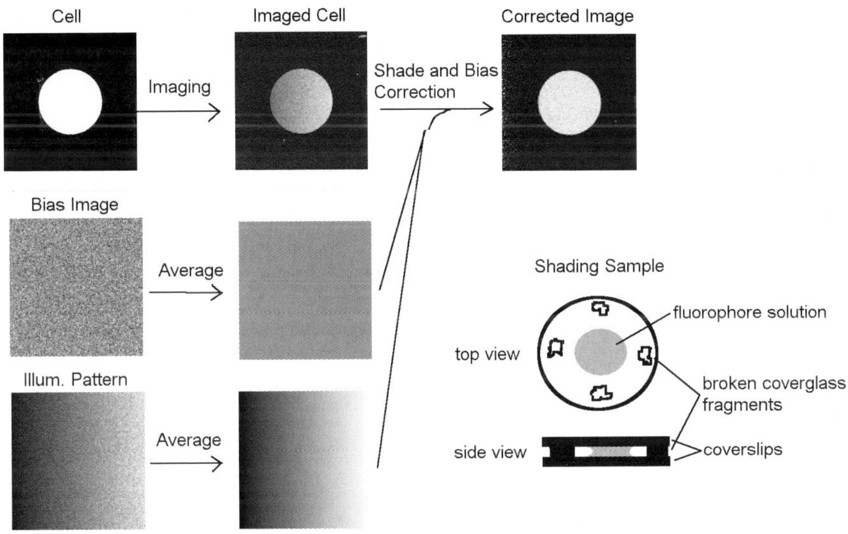

Fig. 6.6 Image preprocessing to correct for uneven illumination and camera bias/background. This is a simulation of the imaging of a model cell with uniform fluorophore distribution for a microscope with an uneven illumination. Similar, but less exaggerated illumination heterogeneities are seen in real microscopes. When the model cell is imaged, this microscope creates an image that displays uneven fluorescence intensity across the cell with a nonzero offset or bias. The bias image can be measured by collecting an image with the camera shutter closed. Averaging multiple bias images will produce a low-noise estimate of the offset, which can be subtracted from the image of the cell. Likewise, a good estimate of the illumination pattern can be obtained by averaging noisy images collected from a thin solution of fluorophore sandwiched between two cover slips separated with broken cover-glass fragments. The averaged bias and shading image can then be used to correct the cellular image

100 μm thick) of fluorophore between two cover glasses supported by cover-glass fragments (Hoppe et al. 2002). An average of ten to 20 frames of this solution can be used to correct the raw data according to the following equation:

$$I_{\text{corrected}} = \left(I_{\text{raw}} - \langle I_{\text{bias}}\rangle\right) / \left(\langle I_{\text{shade}}\rangle - \langle I_{\text{bias}}\rangle\right) \times \text{Max}\left(I_{\text{shade}} - \langle I_{\text{bias}}\rangle\right), \tag{6.6}$$

where <> indicates a ten to 20 frame averaged image (Fig. 6.6). Once images have been corrected for bias and shading, they will then display the parameter dependencies described above and will be ready for use in FRET calculations.

6.6 Methods for FRET Microscopy

A number of fluorescence imaging techniques for FRET have been developed that exploit the spectral properties of FRET. However, no single technique has become a standard for FRET imaging (although sensitized emission microscopy is gaining ground). Rather than attempt to review the evolution of these techniques, I will break down FRET microscopy into four types of approaches: (1) photobleaching approaches, (2) sensitized emission, (3) fluorescence polarization and (4) fluorescence lifetime. These approaches constitute two categories of microscopy: (1) those that quantify FRET in terms of concentrations of interacting proteins and FRET efficiency and (2) those that measure a "FRET signal" in terms of an arbitrary value. The approaches in the first category have the advantage of being instrument-independent and yield direct interpretation of molecular association data. Their principal disadvantage is the requirement of more sophisticated calibration procedures. The approaches in the second category have a number of disadvantages compared with those on the first category. In particular, they can be nonlinear, and they have arbitrary units that are instrument- and fluorophore-dependent and do not clearly describe molecular associations. In general, the arbitrary methods are simplified (e.g., in which certain parameters and calculations are omitted) versions of the quantitative methods. As such, I will focus the discussion on the quantitative methods. Also, the reader should be cautioned regarding this distinction as there are a significant number of publications that erroneously present arbitrary data labeled as "FRET efficiency." FRET efficiency is a clearly defined, fundamental parameter that should not be confused with a method or particular approach. In other words FRET efficiency and fundamental interaction parameters (such as the fraction of donors or acceptors in a complex) are independent of the instrument or method by which they are measured.

Here we will focus on FRET techniques that allow analysis of associations of independently labeled proteins and that therefore better mimic the function of their native counterparts in cellular networks and pathways. Simplified versions of these FRET methods can be used when conformational biosensors with fixed stoichiometry of donor and acceptors are used for FRET-based detection of analyte binding or phosphorylation without instrument calibration. Biosensors and their uses and limitations will be discussed later in the chapter.

6.6.1 Photobleaching Approaches

There are mainly two types of photobleaching approaches for FRET microscopy: dynamic bleaching and acceptor annihilation, with the latter seeing the lion's share of application in FRET experiments nowadays.

Dynamic photobleaching FRET experiments take advantage of the accelerated acceptor and reduced donor bleaching rates imparted by FRET. In dynamic bleaching experiments, the FRET efficiency is inferred from the rate of photobleaching of either the donor (Kubitscheck et al. 1991, 1993; Young et al. 1994) or the acceptor (Jares-Erijman and Jovin 2003; Van Munster et al. 2005), or both. In any case, the samples are imaged over time and the (apparent) FRET efficiency can be estimated by fitting the exponential or multiexponential bleaching curves from control (no-FRET) samples vs. experimental (with FRET) samples. The apparent donor efficiency (which is the product of the fraction of donors in complex and the FRET efficiency) can be estimated by

$$E_{\mathrm{D}} = Ef_{\mathrm{D}} = \left(\tau_{\mathrm{DIDA}} - \tau_{\mathrm{DID}}\right)/\tau_{\mathrm{DIDA}} \tag{6.7}$$

where τ_{DID} is the bleaching rate of the donor alone and τ_{DIDA} is the bleaching rate of the donor in the presence of acceptor. A similar though slightly more complicated equation can generated for the acceptor (Mekler et al. 1997). When used together, these approaches could allow measurement of the apparent donor and acceptor FRET efficiencies from dynamic photobleaching data; however, this has not been reported. Both approaches are fundamentally limited to photobleaching measurements that are much faster than changes in the FRET efficiency or fractions of molecules in complex. Often in live-cell imaging, the rate of molecular complex formation and dissolution is on the order seconds to minutes and may be too fast for dynamic photobleaching approaches. A second limitation to this approach is that photobleaching rates may be influenced by other factors, such as the local cellular environment, which may complicate the interpretation. Control experiments that directly assess the bleaching rates of donor and acceptors in the correct compartment should overcome this limitation.

A more direct alternative to dynamic photobleaching is acceptor annihilation or acceptor photobleaching. This approach measures the apparent FRET efficiency by quantifying the increase in donor fluorescence after intense illumination is used to destroy the acceptor fluorophore while preserving as many donors as possible (Kenworthy and Edidin 1998, 1999; Kenworthy et al. 2000; Jares-Erijman and Jovin 2003). The result is an apparent donor FRET efficiency which is the true FRET efficiency times the fraction of donors in complex with acceptors (f_{D}):

$$E_{\mathrm{D}} = Ef_{\mathrm{D}} = \left(I_{\mathrm{D}} - I_{\mathrm{DA}}\right)/I_{\mathrm{D}}, \tag{6.8}$$

where, I_{DA} is the fluorescence of the donor in the presence of the acceptor and I_{D} is the fluorescence of the donor after the acceptor has been photobleached. These measurements require background and camera bias corrections as described in Sect 6.5. This technique allows for comparisons of the fraction of donors engaged in the complex, although it does not directly measure f_{D}. One caveat of using only E_{D} to detect an interaction is that E_{D}~0 could mean that the majority of donors were not associated with acceptors or that they were associated but the FRET efficiency was very low (e.g., they were at an unfavorable orientation for FRET or they were too far apart). This ambiguity is particularly difficult to resolve by the acceptor photbleaching method because the molar ratio of acceptors to donors is not determined.

The photobleaching approach has the advantage of not requiring calibration. The experimenter simply measures the donor fluorescence before and after the acceptor is bleached. This makes the photobleaching approach perhaps the simplest quantitative FRET measurement. The appearance of many faulty sensitized emission FRET measurements in the literature has led to an opinion in the field that all FRET measurements should be validated by the photobleaching approach. Recent evidence however, suggests that during the photobleaching of YFP some of the

YFP molecules photoconvert to a form that has a CFP-like fluorescence (Valentin et al. 2005). This photoconversion could be detrimental to the quantitative nature of the photobleaching method and could lead to false-positive results and would be complicated by the fact that the magnitude of the distortion would depend on the molar ratio of acceptors to donors. Importantly, the degree of perturbation caused by the acceptor photconversion during the bleach has not yet been determined.

Acceptor photobleaching has some intrinsic limitations. Perhaps the most significant limitation is that this measurement can only be performed once on a given cell, thereby precluding measurement of protein interaction dynamics. Furthermore, live cells and diffusing proteins can move too quickly, causing intensities in the bleached region to lose correspondence with the prebleached regions, resulting in false-positive and false-negative value for E_D. This leads to the temptation to perform photobleaching on fixed cells despite potential artifacts that can arise from fixation, such as contaminating background fluorescence, perturbation of the protein structure and destruction of some unknown fraction of fluorescent protein labels. Another potential problem with this approach is that the acceptor bleach often results in destruction of some donors. Overall this is not a serious problem, because incidental bleaching of the donor will not produce a false positive, rather it will simply reduce the measured E_D value (or even give rise to a negative one) and approximate corrections can be applied by assessing donor-bleaching rates from donor-only samples. Lastly, without calibration, the photobleaching approach provides no information on the concentration of the acceptor relative to that of the donor or no information on the fraction of acceptors that are participating in FRET.

6.6.2 *Sensitized Emission*

Perhaps the most direct way to overcome the limitations of the photobleaching method is to measure FRET by sensitized emission. Here, the objective is to isolate the sensitized emission signal from overlapping donor emission and directly excited acceptor emission (Fig. 6.1). This can be achieved by microscope calibration to allow mathematical separation of the sensitized emission (Youvan 1997; Gordon et al. 1998). Further calibration allows scaling of the donor, acceptor and sensitized emissions and calculation of apparent FRET efficiency (Erickson et al. 2001; Hoppe et al. 2002) and acceptor-to-donor molar ratios (Hoppe et al. 2002).

To quantify molecular interactions by sensitized emission we must separate the three spectral components from the fluorescence signals. These spectral components (Fig. 6.1) are the (1) direct donor fluorescence – fluorescence from donor molecules due to excitation from the light source, (2) direct acceptor fluorescence – fluorescence due to acceptor molecules that are excited by the light source and (3) the sensitized emission – fluorescence from the acceptor due to energy transfer from the donor. Measurement of the apparent FRET efficiencies requires that these three signals be isolated and scaled. Here I will describe two strategies for accomplishing this, where the first strategy is really a special case of the second, more general approach.

First and foremost to a sensitized emission FRET calculation is the isolation of the sensitized emission signal. In principle, sensitized emission can be directly measured by simply exciting the donor and measuring the acceptor emission. In practice, however, fluorophores used for FRET have significant spectral overlaps. This implies that unless the acceptor and donor both have large Stoke's shifts there will be some excitation of the acceptor by the donor illumination and that the donor emission will spill into the acceptor emission channel. In many cases (e.g., CFP and YFP), it is possible to choose donor and acceptor excitation and emission filter combinations such that part of the donor emission and directly excited acceptor emission can be obtained independent of sensitized emission. In this case, two calibration constants can be defined that allow subtractive isolation of the sensitized emission (Youvan 1997; Gordon et al. 1998),

$$\mathrm{SE} = I_{\mathrm{F}} - \beta I_{\mathrm{D}} - \alpha I_{\mathrm{A}}, \tag{6.9}$$

where SE is the sensitized emission and the three images are I_{D}, the fluorescence from the donor (i.e., the CFP image, using filters for donor excitation and donor emission), I_{A}, the FRET-independent fluorescence from the acceptor (i.e., the YFP image, using filters for acceptor excitation and acceptor emission), and I_{F}, the mixture of signals from directly excited donors and acceptors, plus acceptors excited by FRET (i.e., the FRET image, using filters for donor excitation and acceptor emission). α and β are coefficients that reflect the fraction of fluorescence crossing over between channels. These parameters are simply defined for samples containing donor or acceptor only:

$$\begin{aligned} \alpha &= I_{\mathrm{F}} / I_{\mathrm{A}} \qquad \text{(for acceptor only)} \\ \beta &= I_{\mathrm{F}} / I_{\mathrm{D}} \qquad \text{(for donor only)} \end{aligned} \tag{6.10}$$

Some fluorophores, however, overlap to the extent that the donor's emission bandpass contains acceptor emission and the acceptor's excitation bandpass excites the donor. In this case at least two more coefficients must be defined. FRET in arbitrarily overlapping systems has been discussed theoretically (Neher and Neher 2004), and will be discussed below.

The sensitized emission image is sufficient to detect FRET; however, it does not directly describe the relation between bound and free donors and acceptors. We have defined three terms that can be used to interpret FRET data in the context of a molecular interaction and allow the estimation of the FRET efficiency (Hoppe et al. 2002). First, the sensitized emission can be compared with the directly excited acceptor's emission to allow estimation of the apparent acceptor efficiency E_{A} (which is the FRET efficiency times the fraction of acceptors in complex). Since the sensitized emission came from the acceptor it will have the acceptor's emission spectrum. The difference between the sensitized emission and the direct acceptor emission is that excitation energy was absorbed by the donor; therefore, the relationship between direct acceptor emission and the sensitized emission is described by the ratio of the acceptor's extinction coefficient and the donor's extinction coefficient at the donor's excitation wavelength,

$$\gamma = \varepsilon_A(\text{donor ex})/\varepsilon_D(435 \text{ donor ex}). \quad (6.11)$$

Using γ and α, we can define the apparent acceptor efficiency as the fraction of acceptors in complex times the FRET efficiency and written in terms of SE and I_A as (Erickson et al. 2001; Hoppe et al. 2002)

$$E_A = E[\text{DA}]/[A_T] = \gamma \text{SE}/\alpha I_A. \quad (6.12)$$

Here, αI_A estimates the directly excited acceptor in the FRET image and γ relates the sensitized emission to the direct acceptor emission.

We have also defined a second term called the apparent donor efficiency (E_D) which measures the FRET efficiency and the fraction of donors in complex (Hoppe et al. 2002). For this definition, the units of the sensitized emission must be scaled to the units of donor fluorescence. This was accomplished by noting that the difference between the sensitized emission and the donor fluorescence is the ratio of the donor and acceptor emission spectra and natural quantum yields, giving

$$\xi = S_D(\text{donor em})\, Q_D / S_A(\text{acceptor em}) Q_A. \quad (6.13)$$

With ξ, the apparent donor efficiency is

$$E_D = E[\text{DA}]/[\text{D}_T] = \xi \text{SE}/(\xi \text{SE} + I_D). \quad (6.14)$$

Furthermore, thet molar ratio of acceptors to donors can also be defined (Hoppe et al. 2002):

$$R_M = [A_T]/[D_T] = \xi \alpha I_A / \gamma(\xi \text{SE} + I_D). \quad (6.15)$$

Together E_A, E_D and R_M provide the fundamental information needed to interpret a protein interaction. Note that in Hoppe et al. (2002) ξ was originally defined as ξ/γ, however this is unnecessary, and it has been replaced by simply ξ (Beemiller et al. 2006).

Although the parameters α, β, γ and ξ have fundamental physical meanings, they are rarely calculated from first principles. Rather, these values are determined using calibration molecules consisting of covalently linked donor and acceptor with a predetermined FRET efficiency. With the linked calibration molecule, the system can be calibrated simply by collecting data from samples consisting of each individual fluorophore species: donor only, acceptor only and linked molecule with known FRET efficiency.

How is this linked molecule created and calibrated? Is it really necessary? The linked molecule is a simple calibration standard created by fusing CFP and YFP together in an expression plasmid with some amino acid spacer. The FRET efficiency of the calibration molecule can be measured by a number of approaches, such as fluorescence lifetime spectroscopy (Hoppe et al. 2002), or by the photobleaching method (Zal and Gascoigne 2004). We typically measure the FRET efficiency by time-domain fluorescence lifetime either in live cells or from purified proteins. Importantly, the FRET efficiency of this protein is an intrinsic parameter which should be independent of the method of measurement. Furthermore, for a well-designed fusion protein, the FRET efficiency should be independent of environmental factors and can be used as a calibration standard in any microscope.

Some confusion has arisen in the literature about the use of a calibration molecule and how values obtained from it influence the above equations. In particular, one research group has erroneously stated that calibration of the linked molecule by photobleaching will give a linear result when used in the above equations, whereas calibration with a molecule calibrated by fluorescence lifetime will not (Zal and Gascoigne 2004). Given the complications with the photobleaching approach described above, it is more likely that a linked molecule calibrated by fluorescence lifetime will be more accurate.

6.6.3 Spectral Fingerprinting and Matrix Notation for FRET

Recently, Neher and Neher (2004) proposed a "spectral fingerprinting" method for FRET analysis. In this notation the parameters for FRET are expressed as a system of equations described by matrix multiplication. This approach currently suffers from a lack of calibration formalism: however, experimental implementation should be a tractable problem.

For the three-image approach described in Sect. 6.6.2, the matrix formalism can be written as

$$\mathbf{I}=\mathbf{M}\times\mathbf{C}, \tag{6.16}$$

where **I** is a vector containing the data, **M** is the mixing matrix and **C** is a vector containing the spectral components. In conventional linear unmixing algorithms, **C** is thought of as a vector containing the concentrations of each species and **M** is a matrix that contains the emission spectra for each fluorophore. For FRET, however, the values in the mixing matrix take on new meaning. In particular, they contain information about the extinction coefficients and quantum yield of each fluorophore. This method is precisely equivalent to the three-image approach described in Sect. 6.6.2 and can written explicitly as

$$\begin{bmatrix} I_{\mathrm{D}} \\ I_{\mathrm{A}} \\ I_{\mathrm{F}} \end{bmatrix} = \begin{bmatrix} a_{11} & a_{12} & a_{13} \\ a_{21} & a_{22} & a_{23} \\ a_{31} & a_{32} & a_{33} \end{bmatrix} \times \begin{bmatrix} D \\ A \\ \mathrm{EDA} \end{bmatrix} \tag{6.17}$$

where **C** represents the apparent concentration of the three species: D – donor fluorescence, A – acceptor fluorescence and EDA – E times the DA complex concentration or the sensitized emission. Note that these are relative concentrations. The true concentrations of each of these are masked by the optical blurring of the microscope.

Substituting in the constants for FRET stoichiometry and noting that each a_{ij} is given by

$$a_{ij} = I_i \varepsilon_{ij} Q_j S_{ij}, \tag{6.18}$$

where is the intensity of the *i*th illumination condition, Q_j is the quantum yield of the *j*th fluorophore, ε_{ij} is the extinction coefficient of the *j*th fluorophore integrated over the excitation wavelengths of illumination *i*, and S_{ij} is the emission of the *j*th fluorophore integrated over the *i*th emission wavelengths, and rewriting the equations of FRET stoichiometry (Hoppe et al. 2002) gives

$$\begin{bmatrix} I_{\mathrm{D}} \\ I_{\mathrm{A}} \\ I_{\mathrm{F}} \end{bmatrix} = \begin{bmatrix} \xi & 0 & -\xi \\ 0 & \gamma/\alpha & 0 \\ \xi\beta & \gamma & 1-\beta\xi \end{bmatrix} \times \begin{bmatrix} D \\ A \\ \mathrm{EDA} \end{bmatrix}. \tag{6.19}$$

The solutions for the three spectral components can be regrouped to give the identical solutions for E_{A}, E_{D} and R_{M}. For example, $E_{\mathrm{A}}=\mathrm{SE}/A$.

The principal advantage of this approach over the conventional formalism for FRET is that it provides a convenient mathematical formalism (particularly in the case of donor and acceptor fluorophores that display large excitation and emission overlaps). Furthermore, this mathematical convention should open new avenues for exploring systems in which FRET can occur between three or more fluorophores.

6.6.4 Polarization

As described already, polarization is a fluorescence property that is modulated by FRET. It has been known for some time that polarized excitation of the donor produces a sensitized emission that is largely depolarized (Lakowicz 1999). The depolarizing nature of FRET has been utilized in fluorescence imaging as a qualitative indicator of FRET between fluorophores of the same type (called homotransfer) (Varma and Mayor 1998). In this technique, data collection consists of exciting the donor with polarized light and collecting the fluorescence at two polarizations, e.g., $I_{\parallel}$ and $I_{\perp}$ (for differently colored donor and acceptors, any crossover of donor fluorescence would have to be subtracted from these signals before analysis). With these two images, the anisotropy (r) can be used as a FRET indicator:

$$r = I_{\parallel} - I_{\perp} / \left(I_{\parallel} + 2I_{\perp}\right). \tag{6.20}$$

Unfortunately, it is difficult to define a relationship between anisotropy and FRET efficiency.

A more quantitative approach was recently described (Mattheyses et al. 2004) that allows for simultaneous data acquisition of all data required for FRET calculations. This approach called polFRET accomplishes its goal by excitation of the donor and acceptor with orthogonally polarized light at their respective excitation maxima. The fluorescence of the donor and that of the acceptor are then collected via an image splitter that projects onto a CCD camera the donor emission, and the acceptor emission is split into two polarizations, resulting in three images I_D, $I_{A\parallel}$ and $I_{A\perp}$. Although it is possible to split the donor image into two polarizations, this is unnecessary in the case of CFP and YFP. Furthermore by not splitting the donor image, more signal from the dimmer CFP molecule is collected on the CCD camera. These three images contain all of the information needed to analyze the FRET images.

Analysis of polFRET data is carried out by a matrix operation similar to the multispectral method described in Sect. 6.6.2. In particular, images of the donor only, acceptor only and a linked calibration molecule are collected from a number of cells and are used to generate a matrix. This matrix contains the fractional contributions of fluorescence to each of the I_D, $I_{A\parallel}$ and $I_{A\perp}$ channels. The matrix formalism for polFRET is given in terms of the concentration of free donor [D], free acceptor [A] and donor–acceptor complex [DA] as (Mattheyses et al. 2004)

$$\begin{bmatrix} I_D \\ I_{A\perp} \\ I_{A\parallel} \end{bmatrix} = \begin{bmatrix} \alpha_{11} & \alpha_{12} & \alpha_{13} \\ \alpha_{21} & \alpha_{22} & \alpha_{23} \\ \alpha_{31} & \alpha_{32} & \alpha_{33} \end{bmatrix} \times \begin{bmatrix} [\mathrm{D}] \\ [\mathrm{A}] \\ [\mathrm{DA}] \end{bmatrix} \tag{6.21}$$

in which the α values represent the fractional fluorescence components arriving in each of the I_D, $I_{A\parallel}$ and $I_{A\perp}$ images. These α values were obtained from free donor, free acceptor and a linked molecule of known efficiency. Importantly, in its current form, this method can only be used to measure the fractions of interacting molecules when the complex has approximately the same FRET efficiency as the calibration molecule. Thus, the current version of polFRET is considered to be semiquantitative. Nonetheless, this is the only method which allows for simulations acquisition of all of the data required for a FRET calculation.

6.7 Fluorescence Lifetime Imaging Microscopy for FRET

A fluorophore's fluorescence lifetime, τ, describes the rate at which a fluorophore relaxes to its ground state by emission of a photon or nonradiative processes. Fluorescence decays are generally exponential or multiexponential decays. FRET provides an alternative path for the donor to reach the ground state. As such, the donor's fluorescence lifetime is shorter when it is engaged in FRET. In fact, this shortened donor lifetime can be used to directly estimate the FRET efficiency by the following equation:

$$E = 1 - \frac{\tau_{DA}}{\tau_D}, \tag{6.22}$$

where τ_D is the fluorescence lifetime of the donor in the absence of acceptor and τ_{DA} is the fluorescence lifetime of the donor in the presence of acceptor. It should be noted that this equation is a special case that only applies when (1) all of the donor is in complex with acceptor, (2) the donor displays a single-exponential fluorescence lifetime and (3) there is a single, fixed distance between the donor and the acceptor. Data not adhering to these criteria require a more complex expression for FRET efficiency. Fluorescence lifetime imaging microscopy (FLIM) uses information about the relaxation of the excited state. Current FLIM techniques aim to quantify FRET signals by measuring the reduction of the donor fluorescence lifetime. These approaches are often based on pulsed lasers and time-resolved detection schemes; these are beyond the scope of this chapter. For more information on FLIM approaches to FRET imaging the reader is referred to Lakowicz (1999).

6.8 Data Display and Interpretation

What information do we hope to learn from our FRET experiments? This question ultimately defines the mechanism we will use to analyze and display our data. If the microscope were a perfect imaging device, then it could be calibrated to return the concentration or number of fluorescent molecules in an arbitrarily small volume with arbitrarily fast temporal resolution; however, no microscopes have this capacity. Rather, we are limited to using instruments that collect light from cells with imperfect optical sectioning, thereby resulting in images that are influenced by excluded subresolution volumes within the plane of focus and fluorescence from other planes of focus. The magnitude of this effect will depend on the type of microscope (widefield vs. confocal) and its settings (e.g., pinhole size). Thus, we must define our measurement of FRET in accordance with this limitation. If the question at hand is simply "Do protein A and protein B interact?" then we can simply calculate a sensitized emission image, but we must be careful in interpreting this image. Bright regions in the cell do not necessarily mean a higher FRET efficiency or that even a greater fraction of molecules form a complex in that region, rather they may simply reflect more total molecules are in that region. If, however, if we want to draw inference about the quantities of molecules participating in an interaction or the FRET efficiency, then we need to display our data in a way that removes the confounding intensity due to cell thickness. This can be accomplished by splitting out the fractional components into images such as those of E_A, E_D and R_M (Fig. 6.7). These images are ratios, and are therefore not confounded by cell thickness, therefore simplifying data interpretation.

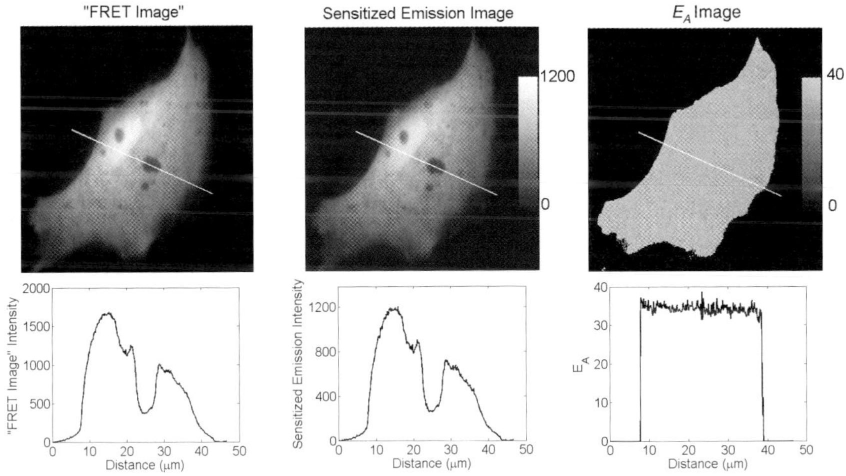

Fig. 6.7 Visualization of FRET data. These are data taken from a sensitized emission FRET experiment of a cell expressing a "linked" donor and acceptor molecule that displays FRET and distributes uniformly throughout cytoplasm and nucleoplasm. The FRET image shows the raw data collected in the I_F channel (e.g., excite donor and measure acceptor emission). This image, and a linescan across the cell, shows that the fluorescent molecules are distributed throughout the cytosol of the cell, and are displaced from some organelles and cellular structures. The variation in intensity along the linescan is largely due to contributions of fluorescence from above and below the plane of focus, consistent with the nuclear region being thicker than the rest of the cell body. Calculation of the sensitized emission image by Eq. 6.9 produces an image that looks nearly identical to that of the FRET image. This image is sufficient to detect the presence of FRET; however, the intensity of the image is challenging to interpret. One may examine a brighter region (e.g., the nuclear region) and mistakenly conclude that there is more FRET in this region than in other parts of the cell, when in reality the intensity reflects cell thickness and concentration of molecules. To create an image that better represents the fraction of interacting molecules (which is 100% in this case), the E_A or E_D image can be calculated. As expected, the E_A image (calculated from Eq. 6.12) for this sample displays a uniform apparent acceptor FRET efficiency as seen in both the E_A image and the linescan across the cell body

6.9 FRET-Based Biosensors

FRET-based biosensors are molecules labeled with both donor and acceptor that change conformation in response to signaling events or changes in analyte concentration. This change in conformation alters the donor/acceptor distances and orientations, resulting in changes in FRET efficiency. The first example of a genetically encoded biosensor made use of FRET between blue fluorescent protein (BFP) and GFP linked by a peptide from myosin light chain kinase (MLCK) (Romoser et al. 1997). This molecule was an indicator of Ca^{2+}-mediated signaling by calmodulin such that when calmodulin bound to the MLCK peptide it altered the distance and orientation between BFP and GFP, thereby changing the efficiency of FRET. A very similar sensor was also constructed that was composed of CFP and YFP separated by calmodulin and a peptide that calmodulin binds in the presence of high calcium concentration (Miyawaki et al. 1997).

In the years that followed, more linked biosensors were developed to detect signals other than those of calcium. Development of these sensors generally made use of the experimenter's knowledge of signal transduction to combine an activatable component and a domain that binds the activated component to alter the conformation of the molecule. Recent examples include phosphorylation sensors, small G-protein-activation sensors for Ras and Rac, and sensors for protein kinase C activity (Sato et al. 2002; Zacharias et al. 2002; Yoshizaki et al. 2003).

Linked biosensors have advantages and disadvantages. The principal advantage of the linked-biosensor approach is that analysis is simple. Second, the molar ratio of donor to acceptor is always 1, which means that FRET can be measured by the ratio of donor emission to acceptor emission, exciting at the donor's excitation maximum (Miyawaki et al. 1997). In the nomenclature of this chapter, data from a linked biosensor can be calculated as a simple ratio,

$$\text{Simple ratio} = \frac{I_F}{I_D}. \tag{6.23}$$

Unfortunately, this goes to infinity as E approaches 1 and results in slight skewing of data owing to the relationship between I_D and I_F.

A better expression is the FRET ratio,

$$\text{FRET ratio} = \frac{I_F}{I_F + I_D}, \tag{6.24}$$

which, although this is not the FRET efficiency, at least has the same functional form as E_D.

The principal disadvantage of linked biosensors is that they often lack structural components that are important for proper localization, and they are often singling-deficient molecules, which may disrupt certain pathways. Furthermore, these sensors are often difficult to construct, requiring many iterations to produce a functional sensor. Lastly, the change in FRET signal from these sensors is usually small.

6.10 FRET Microscopy for Analyzing Interaction Networks in Live Cells

How well does measurement of binding events by FRET represent the magnitude and kinetics of what is really happening in the living cell? A large portion of cell biology and medically relevant research requires that FRET measurements be made in genetically intractable cells. As such we are generally limited to working with overexpressed fluorescent protein chimeras that produce distorted reflections of the normal interactions in a protein network. The degree of distortion depends on three interrelated effects: (1) the degree to which the fluorescent chimeras function like their endogenous counterparts, (2) the degree to which overexpression of fluorescent signaling molecules perturbs the balance of the interaction network and (3) the extent of competitive binding of fluorescently tagged donor or acceptor with endogenous (unlabeled) molecules. These effects depend on the biology and must be addressed on a case-by-case basis. In general, assays need to be on hand to measure if the fluorescently tagged molecules behave like their unlabeled counterparts and to determine if the labeled molecules perturb the normal set of interactions.

These distortions can be put into perspective by considering an ideal system for measuring the magnitudes and kinetics of molecular associations. This is a useful exercise because it provides a framework for understanding parameters that define the interplay between measurement and the protein associations. In the ideal system, the endogenous singling proteins would be replaced with GFP-labeled proteins, be produced at exactly the correct level and function exactly as the endogenous proteins. In such a system, it should then be possible to determine the apparent affinity and kinetics of the interaction to ultimately allow quantitative modeling of interaction networks.

Only in genetically tractable organisms such as *Saccharomyces cerevisiae* and the social amoeba *Dictyostelium discoideum* can endogenous proteins be replaced with fluorescent chimeras with reasonable efficiency. For systems where the endogenous proteins cannot be removed, an alternative, as suggested by others (Chamberlain et al. 2000), is to have GFP chimeras expressed at low levels, such that they participate in the binding events but do not significantly increase the overall level of interaction components. In this case, the FRET approach will track the binding dynamics correctly; but the magnitude will be decreased by some unknown fraction owing to interactions with unlabeled binding partners. This low expression level approach is limited because interactions between tagged proteins would become quite rare. For example, consider an endogenous molecule S that binds to two endogenous effectors E1 and E2 both present at equal concentrations. We then introduce fluorescently tagged S* and E1*, where S and S* are expressed at low concentration relative to E1, E2 and E1*. For E1* expressed at 1/100th the concentration of E1 and E2, the maximum fraction of E1* bound to S* is only 0.5%. The detection limits of FRET microscopy are not nearly good enough to allow the measurement of such a small interaction, indicating that low level expression approaches are not plausible.

Thus, we are required to measure fluorescent chimeras expressed at concentrations that compete with the endogenous molecules. In this scenario, overexpressed molecules are at concentrations greater than or equal to those of endogenous molecules. This means that a smaller fraction of total molecules (endogenous plus fluorescent) will bind the same number of effectors. Furthermore, one usually does not know the proportion of labeled to unlabeled molecules, which means that the relationship between the FRET signal and the true concentration of the complex is ambiguous. Nonetheless, as long as the presence of the overexpressed molecules does not significantly perturb the normal network of interactions, then FRET microscopy can still provide a valid measurement of the dynamics and localization of the binding event.

6.11 Conclusion

FRET microscopy is a powerful tool for examining molecular interactions within living cells. The technologies for FRET microscopy are as varied as the spectroscopic effects of FRET and truly quantitative analysis of FRET microscopy requires consideration of a menagerie of parameters. In this chapter I hope to have conveyed the basic principles behind these FRET approaches and to have provided the reader with a conceptual framework for evaluating quantitative FRET imaging.

References

Beemiller P, Hoppe AD, Swanson JA (2006) A phosphatidylinositol-3-kinase-dependent signal transition regulates ARF1 and ARF6 during Fcγ receptor-mediated phagocytosis. PLoS Biol 4:e162

Chamberlain CE, Kraynov VS, Hahn KM (2000) Imaging spatiotemporal dynamics of Rac activation in vivo with FLAIR. Methods Enzymol 325:389–400

Clegg RM (1996) Fluoresence resonance energy transfer. In: Wang F, Herman B (eds) Fluorescence imaging spectroscopy and microscopy. Wiley, New York, pp 179–252

Erickson MG, Alseikhan BA, Peterson BZ, Yue DT (2001) Preassociation of calmodulin with voltage-gated Ca(2+) channels revealed by FRET in single living cells. Neuron 31:973–985

Gordon GW, Berry G, Liang XH, Levine B, Herman B (1998) Quantitative fluorescence resonance energy transfer measurements using fluorescence microscopy. Biophys J 74:2702–2713

Hoppe A, Christensen K, Swanson JA (2002) Fluorescence resonance energy transfer-based stoichiometry in living cells. Biophys J 83:3652–3664

Jares-Erijman EA, Jovin TM (2003) FRET imaging. Nat Biotechnol 21:1387–1395

Kenworthy AK, Edidin M (1998) Distribution of a glycosylphosphatidylinositol-anchored protein at the apical surface of MDCK cells examined at a resolution of <100 A using imaging fluorescence resonance energy transfer. J Cell Biol 142:69–84

Kenworthy AK, Edidin M (1999) Imaging fluorescence resonance energy transfer as probe of membrane organization and molecular associations of GPI-anchored proteins. Methods Mol Biol 11:37–49

Kenworthy AK, Petranova N, Edidin M (2000) High-resolution FRET microscopy of cholera toxin B-subunit and GPI- anchored proteins in cell plasma membranes. Mol Biol Cell 11:1645–1655

Kraynov VS, Chamberlain C, Bokoch GM, Schwartz MA, Slabaugh S, Hahn KM (2000) Localized Rac activation dynamics visualized in living cells. Science 290:333–337

Kubitscheck U, Kircheis M, Schweitzer-Stenner R, Dreybrodt W, Jovin TM, Pecht I (1991) Fluorescence resonance energy transfer on single living cells. Application to binding of monovalent haptens to cell-bound immunoglobulin E. Biophys J 60:307–318

Kubitscheck U, Schweitzer-Stenner R, Arndt-Jovin DJ, Jovin TM, Pecht I (1993) Distribution of type I Fc epsilon-receptors on the surface of mast cells probed by fluorescence resonance energy transfer. Biophys J 64:110–120

Lakowicz JR (1999) Principles of fluorescence spectroscopy. Kluwer/Plenum, New York

Mattheyses AL, Hoppe AD, Axelrod D (2004) Polarized fluorescence resonance energy transfer microscopy. Biophys J 87:2787–2797

Mekler VM, Averbakh AZ, Sudarikov AB, Kharitonova OV (1997) Fluorescence energy transfer-sensitized photobleaching of a fluorescent label as a tool to study donor-acceptor distance distributions and dynamics in protein assemblies: studies of a complex of biotinylated IgM with streptavidin and aggregates of concanavalin A. J Photochem Photobiol B 40:278–287

Miyawaki A, Llopis J, Heim R, McCaffery JM, Adams JA, Ikura M, Tsien RY (1997) Fluorescent indicators for Ca2+ based on green fluorescent proteins and calmodulin. Nature 388:882–887
Muller EG, Snydsman BE, Novik I, Hailey DW, Gestaut DR, Niemann CA, O'Toole ET, Giddings TH Jr, Sundin BA, Davis TN (2005) The organization of the core proteins of the yeast spindle pole body. Mol Biol Cell 16:3341–3352
Neher RA, Neher E (2004) Applying spectral fingerprinting to the analysis of FRET images. Microsc Res Tech 64:185–195
Romoser VA, Hinkle PM, Persechini A (1997) Detection in living cells of Ca2+-dependent changes in the fluorescence emission of an indicator composed of two green fluorescent protein variants linked by a calmodulin-binding sequence. A new class of fluorescent indicators. J Biol Chem 272:13270–13274
Sato M, Ozawa T, Inukai K, Asano T, Umezawa Y (2002) Fluorescent indicators for imaging protein phosphorylation in single living cells. Nat Biotechnol 20:287–294
Valentin G, Verheggen C, Piolot T, Neel H, Coppey-Moisan M, Bertrand E (2005) Photoconversion of YFP into a CFP-like species during acceptor photobleaching FRET experiments. Nat Methods 2:801
van der Wal J, Habets R, Varnai P, Balla T, Jalink K (2001) Monitoring agonist-induced phospholipase C activation in live cells by fluorescence resonance energy transfer. J Biol Chem 276:15337–15344
Van Munster EB, Kremers GJ, Adjobo-Hermans MJ, Gadella TW Jr (2005) Fluorescence resonance energy transfer (FRET) measurement by gradual acceptor photobleaching. J Microsc 218:253–262
Varma R, Mayor S (1998) GPI-anchored proteins are organized in submicron domains at the cell surface. Nature 394:798–801
Yoshizaki H, Ohba Y, Kurokawa K, Itoh RE, Nakamura T, Mochizuki N, Nagashima K, Matsuda M (2003) Activity of Rho-family GTPases during cell division as visualized with FRET-based probes. J Cell Biol 162:223–232
Young RM, Arnette JK, Roess DA, Barisas BG (1994) Quantitation of fluorescence energy transfer between cell surface proteins via fluorescence donor photobleaching kinetics. Biophys J 67:881–888
Youvan DC, Silva CM, Bylina EJ, Coleman WJ, Dilworth MR, Yang MM (1997) Calibration of fluorescence resonance energy transfer in microscopy using genetically engineered GFP derivatives on nickel chelating beads. Biotechnology 3:1–18
Zacharias DA, Violin JD, Newton AC, Tsien RY (2002) Partitioning of lipid-modified monomeric GFPs into membrane microdomains of live cells. Science 296:913–916
Zal T, Gascoigne NR (2004) Photobleaching-corrected FRET efficiency imaging of live cells. Biophys J 86:3923–3939

7
Fluorescence Photobleaching and Fluorescence Correlation Spectroscopy: Two Complementary Technologies To Study Molecular Dynamics in Living Cells

Malte Wachsmuth and Klaus Weisshart

Abstract Fluorescence recovery or redistribution after photobleaching (FRAP) and fluorescence correlation or fluctuation spectroscopy (FCS) are probably the most widely used techniques employed to study the transport and diffusion as well as the interaction and immobilisation of biological molecules inside the cellular environment. This has been promoted by the emergence of fluorescent proteins for in vivo labelling and the development of confocal laser scanning microscopes that also allow for photobleaching and fluctuation spectroscopy experiments. FRAP represents a family of methods which are all based on the photoinduced bleaching (or activation) of marker molecules in selected areas of a cell followed by the relaxation back to equilibrium. FCS stands for another and complementary set of relaxation methods which are based on the observation and analysis of thermal fluctuations of sparse labelled molecules in a microscopic observation volume. Being conceptually different, these techniques taken together and combined with confocal imaging give access to a wide time and concentration range and can yield qualitatively and quantitatively biochemical and biophysical data such as concentrations, reaction rates, free and bound fractions, or diffusion coefficients. We present aspects of the biological and physical background, outline typical fields of applications, and give some guidance on how to carry out FRAP, FCS, and continuous photobleaching experiments with an emphasis on practical aspects and pitfalls.

7.1 Introduction

Fluorescence recovery after photobleaching (FRAP; Axelrod et al. 1976; Edidin et al. 1976; Peters et al. 1974) and fluorescence correlation spectroscopy (FCS; Elson and Magde 1974; Magde et al. 1972, 1974) were first introduced in the 1970s and are now probably the most common techniques applied to study transport and diffusion as well as interactions of biological molecules in living cells. FRAP is also known as fluorescence redistribution after photobleaching, fluorescence photobleaching recovery (FPR), or fluorescence microphotolysis (FM) and represents a

S.L. Shorte and F. Frischknecht (eds.), *Imaging Cellular and Molecular Biological Functions.*

whole family of related methods all based on photoinduced bleaching (or activation) of marker molecules in selected areas of a cell. This imposes an unbalanced distribution on a large number of fluorescent molecules and allows observation of the relaxation back to equilibrium. FCS stands for another, complementary set of similar methods, which can be summarised as fluorescence fluctuation spectroscopy or microscopy (FFS, FFM) and which are also relaxation techniques. Here, however, there is no need to disturb a steady-state fluorescence distribution because thermally induced fluctuations of sparse labelled molecules in a microscopic observation volume around equilibrium are analysed statistically.

Based initially on custom-designed microscope systems and the sometimes difficult introduction of fluorescently labelled biomolecules into cells, FRAP and FCS became more and more popular in the second half of the 1990s as a result of two developments:

1. Commercially available confocal laser scanning microscopes (CLSMs) were improved significantly with respect to resolution, speed, sensitivity, and flexibility, making photobleaching experiments easier to perform. Owing to the almost identical optical setup, these improvements also led to the launch of dedicated commercial FCS systems.
2. The emergence of the green fluorescent protein (GFP) and an increasing number of other autofluorescent proteins often derived from GFP has revolutionised fluorescent labelling of proteins in vivo.

7.1.1 FRAP and Other Photobleaching Methods

A typical FRAP experiment measures the equilibrium signal from a fluorescently labelled molecule in a region of interest (ROI; Fig. 7.1). The labelled molecules in this region are then rapidly bleached and subsequently the fluorescence signal from the ROI is recorded over time using the same conditions as before bleaching. Owing to diffusion and other transport processes as well as association to and dissociation from immobilised structures, the bleaching-induced unbalanced distribution of labelled molecules gradually reequilibrates and the fluorescence signal from the ROI recovers. A quantitative analysis may yield parameters such as the diffusion coefficient, the fraction of fully free, transiently immobilised and fully immobilised molecules or the residence time at the binding site responsible for immobilisation. Depending on the application, the required time resolution and the setup used, intensity monitoring and bleaching is either carried out as imaging FRAP by confocal imaging, e.g. of a whole cell, with switching the laser to high intensity within the selected ROI only for the bleaching sequence (Wedekind et al. 1994) or performed as spot FRAP with a focused and fixed laser beam, the intensity of which is modulated appropriately (Axelrod et al. 1976).

In most photobleaching applications the fluorescence intensities in the bleaching ROI and other regions are averaged and plotted versus time. However, integrating

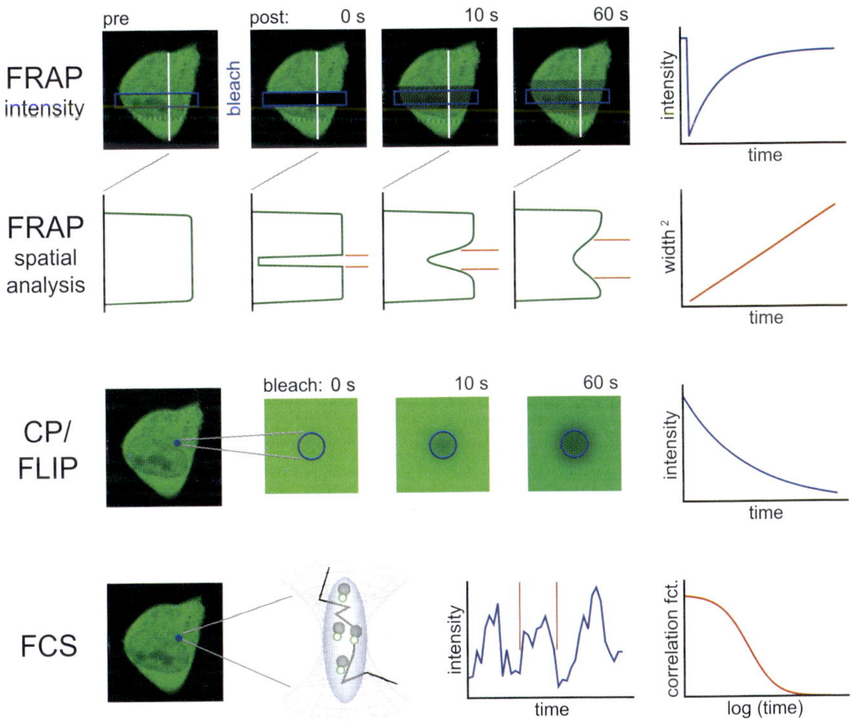

Fig. 7.1 Overview of the methods outlined in this chapter. FRAP intensity: cell with bleaching region of interest (ROI; *blue*) and line for profile analysis (*white*) at different pre- and postbleach time points and the averaged intensity in the ROI versus time. FRAP spatial analysis: intensity profile along the white line at different pre- and postbleach time points and the squared width of the bleached strip versus time. CP/FLIP: cell with bleach spot (*blue*), magnified view at different time points, and the spot intensity versus time. FCS: cell with measurement spot (*blue*), magnified view of the focus, time course of intensity, and the intensity autocorrelation function

intensities always means a loss of spatial information; therefore, an additional spatially resolved analysis of the fluorescence distribution over time (Fig. 7.1) combined with analytical or numerical models can provide more detailed and more reliable interpretations of photobleaching data (Houtsmuller et al. 1999; Phair and Misteli 2001).

In another set of techniques both photobleaching and imaging/measuring of the fluorescence intensity distribution are carried out simultaneously. These methods are called continuous photobleaching (CP; Wachsmuth et al. 2003), fluorescence loss in photobleaching (FLIP; Cole et al. 1996; White and Stelzer 1999), or continuous fluorescence microphotolyis (CFM; Peters et al. 1981). A typical FLIP experiment (Fig. 7.1) consists of repeated photobleaching of a certain sample region and simultaneous imaging of the whole sample. Thus, the bleached region is depleted of the fluorescent molecules that either reside in it or can enter it from the nonbleached areas due to diffusion or other transport processes. FLIP is therefore especially suited to studying the connectivity of cellular compartments, regions, or cells and the corresponding transport processes. While confocal imaging is generally used for FLIP, in a typical CP experiment a laser beam with constant intensity is parked at the position of interest and used at the same time for bleaching and monitoring of the fluorescence. Applied to an assay with molecules transiently bound to immobilised structures, a dynamic equilibrium between association, dissociation, diffusion/transport, and photobleaching is reached, from which the binding properties can be derived quantitatively.

Another set of methods, which will not be discussed in detail in this chapter, is based on generating an excess of fluorescent molecules in a small ROI covering only a minor fraction of the whole sample area, as opposed to FRAP and FLIP/CP, which involve depletion of fluorescence in the ROI. Examples for these techniques are inverse FRAP (iFRAP; Dundr et al. 2002) where fluorescence is bleached in the whole sample except the ROI, and uncaging or photoactivation experiments (Patterson and Lippincott-Schwartz 2002; Politz 1999) where appropriate synthetic dyes or GFP variants are switched to a fluorescent state in the ROI usually with UV light. Diffusion as well as association/dissociation kinetics can be analysed from the time course of the fluorescence distribution.

7.1.2 FCS and Other Fluctuation Analysis Methods

In a typical FCS experiment, the focus of a confocal laser illumination and fluorescence detection system like that of a confocal microscope defines a small observation volume, the confocality ensuring that only light from the focal volume is detected (Rigler et al. 1993). It is fixed at a position of interest (Fig. 7.1), since FCS is not an imaging method. Owing to their diffusion, i.e. thermally induced Brownian motion, fluorescently labelled molecules can enter and leave the focus, resulting in signal fluctuations at the detector. The average lengths and amplitudes of the fluctuations are determined by so-called temporal autocorrelation analysis. Appropriate models allow, for example, quantification of concentrations and diffusion coefficients and distinction of small free proteins from large complexes to which they are bound. Other sources of fluctuations are photophysical effects such as molecular blinking, interactions with cellular structures, or the motion of these structures themselves. Initially, FCS was mostly applied to in vitro assays, but in recent years more and more in vivo applications have emerged.

In a two-colour setup, potential binding partners are labelled with spectrally different fluorophores (Ricka and Binkert 1989; Schwille et al. 1997). When they associate, diffusion generates synchronised fluctuations in different detection channels, which can be quantified by cross-correlation analysis featuring high sensitivity for biomolecular interactions and yielding the fractions of bound and free molecules and other association/dissociation parameters.

For another approach called photon counting histogram (PCH; Chen et al. 1999), or fluorescence intensity distribution analysis (FIDA; Kask et al. 1999), which will not be discussed in detail in this chapter, the amplitude of diffusion-induced fluctuations rather than the time course is analysed statistically. Thus, fluorescently labelled molecules can be distinguished in terms of their intensity rather than their diffusion properties. This proves to be useful for multimerisation processes.

7.1.3 Comparing and Combining Techniques

Imaging-based FRAP requires high concentration gradients of fluorescent molecules and a high imaging rate or beam scanning speed for a good signal-to-noise ratio and reasonable time resolution. For classic spot FCS, on the other hand, no scanning but precise fixing of the beam in the sample is necessary and concentrations should be as low as possible in order to obtain pronounced fluctuations. It is worth noting that the signal relevant for FCS is precisely the noise to be avoided in imaging and FRAP. In practical terms, the concentration range for FCS lies between picomolar and micromolar and the time range of observable processes between microseconds and seconds. FRAP, in turn, can be employed in the nanomolar to millimolar range and in the millisecond range and above. Conceptually, FCS is "blind" for sufficiently immobilised molecules but ideally suited to measure fast diffusion especially on a molecular level, whereas FRAP is much more suited to slow diffusion resulting from transient immobilisation. A method lying in-between is CP, which shares sensitivity and time resolution with FCS but also allows one to assess slower processes in the same experiment.

The seemingly contradictory requirements are fulfilled by modern CLSMs (Fig. 7.2; Zeiss LSM 510 ConfoCor 3; Leica TCS SP5 FCS2) where the beam is both scanned and parked with the same galvanometer-driven mirrors (Wachsmuth et al. 2003; Wedekind et al. 1994). Imaging and FRAP use conventional photomultiplier tubes (PMTs) with a broad dynamic range to read the fluorescence signal, while FCS and CP employ single photon counting detectors with maximal quantum efficiency and minimal readout noise, mainly avalanche photodiodes (APDs).

It is very advantageous that these combined methods not only cover a wide dynamic and concentration range but that they also have sufficient overlap to be applied simultaneously. In this chapter, we want to outline the usefulness as well as the limitations of these technologies for cellular work. We focus mainly on practical aspects limiting mathematical derivations and expressions as much as possible – they are given more explicitly in a separate section. We will also point out pitfalls and place emphasis on how to carry out experiments and evaluate the data. For a deeper understanding, many reviews are available for photobleaching (Carmo-Fonseca et al. 2002; Carrero et al. 2004; Houtsmuller and Vermeulen 2001; Kimura et al. 2004; Klonis et al. 2002; Lippincott-Schwartz et al. 2001; Misteli 2001; Peters and Kubitscheck 1999; Phair et al. 2004a; Rabut and Ellenberg 2005; Reits and Neefjes 2001; Stavreva and McNally 2004; Verkman 2002, 2003; White and Stelzer 1999) as well as for

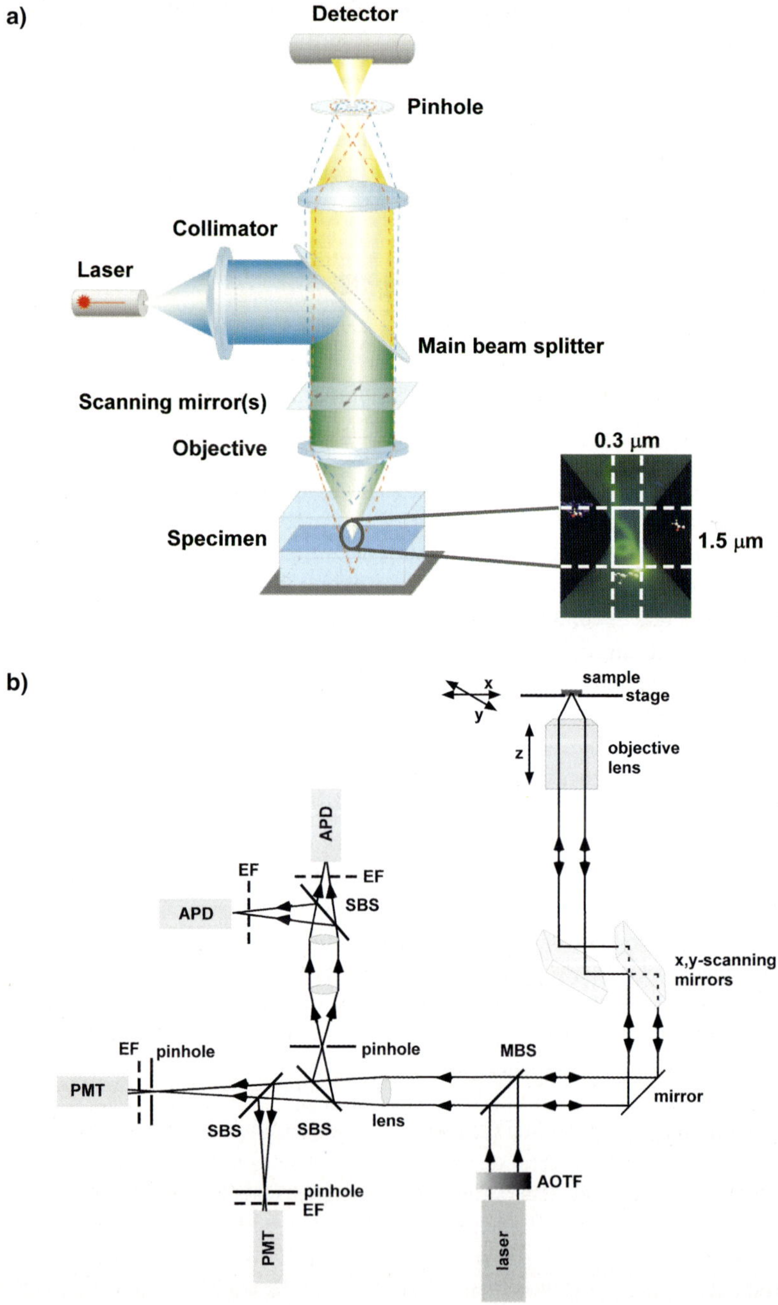

Fig. 7.2 Confocal principle and setup. **a** Advanced confocal optics limit the detection of fluorescence to an ellipsoid of approximately 1.5 μm along the optical axis and 0.3 μm laterally. This confocal volume amounts to less than a quarter of a femtolitre (1 fl = 10^{-15} l) and is either scanned

FCS-related methods (Bacia and Schwille 2003; Elson 2004; Gösch and Rigler 2005; Grünwald et al. 2005; Hess et al. 2002; Hink et al. 2002; Jahnz and Schwille 2004; Kohl and Schwille 2005; Medina and Schwille 2002; Rigler and Elson 2001; Sanchez and Gratton 2005; Thompson et al. 2002; Vukojevic et al. 2005; Weidemann et al. 2002).

7.2 Fundamentals

7.2.1 Fluorescent Labelling

The demands for fluorescent probes for photobleaching and FCS-like applications are manifold and partly contradictory: in general, they should exclusively label the desired molecules or structures and must not impede the biological function of the tagged biomolecules. They should have a large absorption cross-section and a high quantum yield in order to reduce unwanted deposition of illumination energy in the sample. At the low intensities as used for FCS and FRAP before and after bleaching, they should be very photostable, but at the higher illumination intensities used for the bleaching sequence in FRAP or for CP, they should rapidly and irreversibly switch to a nonfluorescent state without much photoinduced toxicity. But even for close-to-ideal fluorophores, a high local concentration of free oxygen radicals may still be produced (Sect. 7.2.4), which cause strong bleaching artefacts. This holds for FCS and CP with long-term low- to medium-intensity illumination of single spots with low fluorophore concentrations as well as for imaging-based FRAP with short-term high-intensity illumination of many spots in a cell with higher fluorophore concentrations.

One can choose between chemical fluorophores, including quantum dots, genetically encoded fluorescent proteins, and hybrid systems between them with respective pros and cons. Fluorophores from each of the groups can show unwanted photophysical effects such as reversible photobleaching and blinking, which can be explained as delayed recovery of the fluorescence capability or transient occupation of a nonfluorescent state (Sect. 7.2.4).

Fig. 7.2 (continued) for imaging and FRAP or fixed for FCS and CP. **b** The intensities of the laser lines are modulated with an acousto-optical tuneable filter (*AOTF*) before the light is fed into the confocal scan head. Here, it is reflected by the main dichroic beam splitter (*MBS*) and the scanning mirrors into the microscope optics, where it is focused with the objective lens into the sample. The emitted light returns on the same path and passes the MBS because of its longer wavelength. The light passes the confocal pinhole(s) and is distributed spectrally with secondary beam splitters (*SBS*) and emission filters (*EF*) onto suitable detectors, which are usually avalanche photodiodes (*APD*) for FCS and CP, and photomultiplier tubes (*PMT*) for imaging. The scanning mirrors are used for imaging and FRAP with synchronised intensity switching in the AOTF as well as for fixing of the focus in the sample for FCS and CP. Relative positioning of focus and sample could also be achieved using a motorised stage

7.2.1.1 Chemical Fluorophores

Numerous chemical fluorophores suitable for both FCS and photobleaching methods are available, including fluoresceine, rhodamine, cyanine, or Alexa dyes. All of them can be obtained with various reactive groups that allow their coupling to carboxy, amino, or sulfhydryl groups of amino acids, for example. They should not exhibit too much hydrophobicity, otherwise they can bind unspecifically to intracellular structures such as membranes. A high labelling ratio per molecule is desirable as long as the biological function is not impeded and quenching artefacts are avoided. For small ligands, dyes can be included in the synthesis; however, the biological activity of these modified ligands should be carefully checked in control experiments (Briddon et al. 2004). Uniform labelling is very important for FCS, otherwise molecules generated with different brightnesses render interpretation of the data difficult (Sect. 7.4). In order to circumvent random labelling of proteins, the one of interest can be modified genetically so that it contains one cystein and a thiol reactive group is used that guarantees one label per molecule. Labelling of nucleic acids is more straightforward since strands can be labelled specifically at one end or the other. A major disadvantage of biomolecules labelled with chemical fluorophores is that they must be purified and labelled with the selected chemical fluorophore and then introduced into cells by microinjection or other invasive methods.

Quantum dots are semiconductor nanocrystals that show bright fluorescence owing to the combination of semiconductor properties and quantum confinement (Medintz et al. 2005; Michalet et al. 2005). Depending on the material, they have relatively broad excitation spectra, whereas the emission spectra are restricted to narrow bands of 20–30 nm and can be tuned by changing the size of the crystal. In addition to spectral flexibility, they show high biochemical and photostability, but also blinking. Their relatively large size and high molecular weight and the sometimes difficult 1:1 coordination with target molecules might limit biological compatibility.

7.2.1.2 Fluorescent Proteins

GFP and its variants and relatives (Shaner et al. 2005; Tsien 1998; Zhang et al. 2002) are in general very suitable for FCS, FRAP, and related methods. They provide a much more convenient way to label proteins in living cells since they are genetically attached to the protein of interest. However, they can all display more or less distinct spontaneous reversibility of photobleaching or blinking, they tend to multimerise at high concentrations, and/or they can impede the labelled proteins' biological function.

The enhanced version of GFP, EGFP, is probably the most widely used fluorescent protein for in vivo labelling and is optimised for mammalian cells at 37°C in terms of quantum yield, folding, maturation, and stability (Patterson et al. 1997). It shows pronounced blinking in the time range of 10–100 μs owing to reversible internal and external protonation of the chromophore, as well as minor reversible photobleaching (Haupts et al. 1998). This can also be observed with the yellowish and the bluish variants EYFP and ECFP (Schwille et al. 2000), respectively, where especially ECFP is more susceptible to photobleaching (Rizzo et al. 2004). The red-shifted relatives *Discosoma striata* red fluorescent protein (DsRed) and *Heteractis crispa* red fluorescent protein (HcRed; Zhang et al. 2002), which have strong oligomerisation tendencies, and the recently introduced monomeric red fluorescent protein (mRFP; Campbell et al. 2002) all seem to have comparable photophysical properties. The list of available fluorescent proteins is constantly growing by isolation of new species and genetic engineering of existing species (Shaner et al. 2005).

More recent approaches include the addition of a short polypeptide (FlAsH; Griffin et al. 2000) or a larger protein (Promega HaloTag system) supplying a binding site specific for appropriate reactive groups of synthetic fluorophores and potentially more spectral flexibility, provided that the biological function is not affected.

7.2.2 *Microscope Setup*

Many different systems have been developed for the combination of FCS, FRAP, and imaging (Brock et al. 1998; Gennerich and Schild 2000; Verkman 2003; Wachsmuth et al. 2003). Here, we will describe the schematic setup as it is accessible for many users, e.g. in imaging facilities in the form of commercial CLSM and FCS systems (Fig. 7.2). Usually, an inverted research microscope forms the base. Due to the refractive index of the cellular interior, water immersion objective lenses are very suitable. A high numerical aperture (NA) provides laser light illumination of the focus with high intensity and efficient collection of fluorescence photons. Such lenses are frequently equipped with a correction collar that permits accounting for thickness variations of the cover-glass. The objective should also be well corrected for chromatic aberrations to use in quantitative cross-correlation experiments and multicolour imaging. Most manufacturers offer objectives with a magnification of 40–63x and a NA of 1.2, which are well suited especially for combined FRAP/FCS experiments. As excitation sources, a variety of continuous-wave or pulsed lasers can be used in the UV to visual range for one-photon excitation and with IR emission for two-photon excitation. Laser diodes are an inexpensive alternative. For example, a multiline argon ion gas laser with an output power of 25 mW at 488 nm provides enough intensity in the focus to bleach EGFP efficiently – with a 63x 1.2 NA water lens and full illumination of the back aperture, the maximum power in the focus might be around 400 μW and the intensity about 200 kW cm^{-2}. The laser light is collimated to a parallel beam, reflected by an excitation or main beam splitter (usually a dichroic mirror), passes optional additional elements, and is focused with the objective lens to a diffraction-limited spot in the sample. Fluorescent substances in the sample are excited by the laser light and emit photons of longer wavelengths, which are collected by the same optics and are transmitted through the main beam splitter in order to separate them from the excitation light. The light passes additional secondary beam splitters and emission filters or a spectrometer in order to distribute the fluorescence photons over several detection channels and to suppress any reflected and Rayleigh- or Raman-scattered light. Commonly used detectors are PMTs with a broad dynamic range for confocal imaging and APDs for FCS, which have a high quantum efficiency and are capable of counting single photons. A crucial element for a confocal setup is the detection pinhole in the image plane of the microscope, which ensures that fluorescence is detected from only the illuminated focal volume and rejected from everywhere else. There are two ways to place the pinhole: either a common one in the path before the light is spectrally separated, or individual ones in front of every detector. The former solution is more stable and allows one to change beam splitters and filters without needing to realign the pinhole. The confocality of different detection channels is inherently guaranteed. The latter setup benefits from more flexibility for different detection channels because the pinholes can be aligned and the size can be changed individually. For two-photon excitation, the pinhole can be omitted.

Combined photobleaching-adapted CLSM and FCS systems are generally implemented in two ways. In one design concept a confocal scanning head is attached to one port of the microscope stand and a separate fixed-beam detection module for spectroscopy is mounted to another port. This proves to be advantageous because particularly the optical path for FCS does not have to pass additional elements such as scan optics. As a drawback, stage-based positioning of the sample is necessary, including the risk of an offset between the positions of the foci as defined by the stage and the scanning head, and requiring calibration. In another design concept the same beam is used for imaging, FRAP, and FCS, i.e. it is moved at high speed for imaging and is fixed for FCS. Thus, no alignment between the imaging and the FCS focus is needed, and intensities for imaging and FCS are easily comparable. Depending on the application, one can switch between PMTs and APDs as detectors.

In order to record images and for imaging-based FRAP, the focus is scanned pixel by pixel, line by line, and frame by frame while collected photons are integrated over the time the laser beam dwells on each pixel and are displayed as grey values. The laser intensity can be changed from pixel to pixel and with microsecond time resolution by means of an acousto-optical tunable filter (AOTF) providing intensity ratios of up to 1,000:1 between bleaching and nonbleaching. For CP, FCS, and related methods, the beam is set to constant intensity and fixed at the desired position. The photons collected are recorded over time and subjected to further processing, especially computing time correlation functions in real-time or offline using hardware or software correlators.

Commercial CLSM setups are equipped with large and versatile software systems for multidimensional image acquisition, including complex time series and "application and evaluation suites" for FRAP and similar experiments. For FCS, too, dedicated software modules are available that provide, for example, an online correlation and intensity display (allowing optimisation of experimental parameters), data acquisition, processing, and evaluation or fitting functionality. Usually, raw data – time series of images for imaging FRAP, time series of intensity values for FCS and CP – can also be imported into third-party or custom-designed software for further, more detailed, or alternative processing and evaluation steps.

7.2.3 *Diffusion and Binding in Living Cells*

Diffusion or Brownian motion is thermally induced stochastic movement, for example, of fluorescently labelled molecules in a solution or inside a cell. The concept can be generalised to all thermally driven stochastic processes such as fluctuations in protein structure. On the macroscopic scale of a whole cell where large numbers of molecules are involved, diffusion attempts to equilibrate macroscopic concentration gradients, for example, generated by depleting the fluorescence from a certain area of a cell – experimentally accessible with FRAP. The diffusion coefficient characterises the relation between the gradient and the resulting flux. On a microscopic scale, single molecules roam randomly through their environment, thus generating concentration fluctuations when they enter and leave a small observation volume – experimentally accessible with FCS. The area covered by roaming or the mean squared displacement (MSD) grows linearly in time, the proportionality factor being the diffusion coefficient, which is frequently given in square microns per second.

Various processes can effect a change in the diffusive behaviour:

1. When molecules associate with larger yet mobile complexes, their movements are slowed down and the actual diffusion coefficient is reduced.
2. In complex environments like the cell nucleus or heterogeneous membranes, the diffusion process can deviate significantly from free diffusion, i.e. the MSD grows less than linearly in time, resulting in anomalous, obstructed, or confined Brownian motion (Görisch et al. 2004; Saxton and Jacobson 1997; Schwille et al. 1999; Wachsmuth et al. 2000).
3. Specific binding to cellular structures and thus immobilisation also leads to a reduced mobility.

It is useful to discriminate between two cases (Carrero et al. 2003; Phair et al. 2004b; Sprague et al. 2004). When there is always a significant fraction of free molecules they redistribute rapidly, most of the binding sites are occupied, and the corresponding kinetics is dominated by the dissociation process that can be called a pseudo-first-order reaction. A free and a bound fraction can be distinguished. In the regime of limited diffusion, a sufficient number of binding sites are not occupied and when a molecule is released, it is rapidly recaptured by another site. The molecules appear as one fraction with a reduced effective diffusion coefficient.

7.2.4 Fluorescence, Blinking, and Photobleaching

All processes around fluorescence can be represented in a so-called Jablonski diagram (Lakowicz 1999; Pawley 1995; Fig. 7.3). A fluorophore molecule can occupy different electronic states with discrete energy values (thick horizontal lines in Fig. 7.3) and each electronic state splits up into vibrational levels (thin lines in Fig. 7.3). Without excitation, a synthetic fluorophore molecule rests in the vibrational ground state of the electronic ground state S_0, S standing for singlet according to the electron configuration. Absorption of a photon with appropriate energy raises the molecule into higher vibrational levels of the first excited electronic state S_1 within around 10^{-15} s. After vibrational relaxation (around 10^{-12} s), the molecule drops back spontaneously within a few nanoseconds (around 10^{-9} s, the so-called fluorescence lifetime) into higher vibrational levels of S_0 with emission of a fluorescence photon. Further vibrational relaxation leads it back to the starting point and the molecule is ready for another excitation and emission cycle.

Alternatively, with a probability of a few tens of percent, the molecule can undergo a so-called intersystem crossing from S_1 to the triplet state T_1. The repeated occupation and relatively long lifetime of this state where the fluorophore is excluded from the fluorescence cycle makes this "dark state" detectable, for example, as triplet blinking in FCS. After a few microseconds (around 10^{-6} s), the molecule again undergoes intersystem crossing back to S_0 and to the fluorescence cycle, either nonradiatively or with emission of a phosphorescence photon.

Proposed mechanisms for photobleaching are the irreversible (photoinduced) oxidation or reduction into nonfluorescent conformations when the molecules are

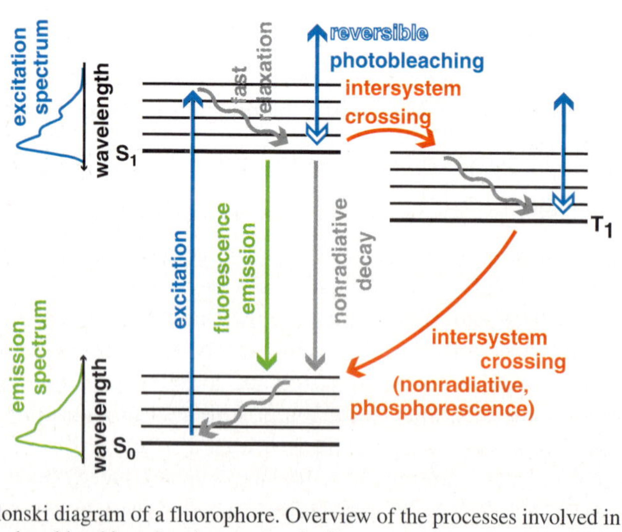

Fig. 7.3 Jablonski diagram of a fluorophore. Overview of the processes involved in fluorescence, blinking, and photobleaching. See the text for details

in the electronically excited S_1 or T_1 or higher states (Peters et al. 1981; Song et al. 1995, 1996, 1997). It is generally assumed that the photobleaching probability grows linearly with illumination intensity. The photochemical reactions frequently generate free radicals that may show significant toxicity.

Many fluorophores ranging from synthetic dyes (Periasamy et al. 1996; Stout and Axelrod 1995) to fluorescent proteins (Swaminathan et al. 1997; Umenishi et al. 2000) or quantum dots (Doose et al. 2005; Nirmal et al. 1996) show a phenomenon called reversible photobleaching, molecular blinking, or emission intermittence. As a general mechanism, it is assumed that especially at high illumination intensities as used for FRAP, a fluorophore can absorb a second photon while still in an excited state, which drives it into another excited state. This state can be metastable with a lifetime in the range from a few tens of microseconds to several seconds (around 10^{-5}–10 s) before the molecule returns to S_0 and the fluorescence cycle.

Although GFP and its variants as well as quantum dots usually have more intricate Jablonski diagrams, the basic principle is the same: the fluorescence cycle can be aborted – photobleaching – or transiently interrupted, either for several tens to hundreds of microseconds – blinking/flickering as observed with FCS – or for several hundred milliseconds – reversible photobleaching as observed with FRAP. Isomerisation or protonation of the chromophore and trapping in surface states were proposed to create the switching between fluorescent and dark states in fluorescent proteins and quantum dots, respectively.

7.2.5 *Two-Photon Excitation*

The principle of a confocal setup with continuous-wave one-photon fluorescence excitation implies that good axial resolution of the whole system is a result of reasonable axial resolution on both the illumination and the detection side. However, excitation and thus photobleaching also take place outside the focal plane. With a two-photon setup, on the other hand, excitation and thus photobleaching only occur in the focal volume (Helmchen and Denk 2005; Svoboda and Yasuda 2006). It is based on the almost simultaneous (within 10^{-15} s) absorption of two photons of twice the wavelength and thus requires a pulsed high-power IR laser. While it is well established for imaging, there are a limited number of published two-photon FRAP (Brown et al. 1999; Van Keuren and Schrof 2003) and FCS studies (Sanchez and Gratton 2005).

The advantages of two-photon excitation are deeper sample penetration and suppressed off-focus excitation and bleaching. In addition, one laser line can be used for simultaneous excitation of spectrally different fluorophores, for example for two-colour FCS. The pulsing enables additional fluorescence lifetime imaging and spectroscopy. Due to the underlying process, the effective focal volume for bleaching as well as for FCS is quite different from that for single photon excitation, something that must be taken into account in the design and analysis of FRAP, FCS, and CP experiments (Sects. 7.3.2, 7.4.2).

As major drawbacks, the absorption cross-section for two-photon excitation is very small, it features more localised photodamage (Hopt and Neher 2001), and the equipment is very expensive.

7.3 How To Perform a FRAP Experiment

7.3.1 The Principle of Imaging-Based FRAP

A FRAP experiment in a living cell consists of (at least) three steps: (1) a time series of prebleach images of the cell in order to record the bona fide equilibrium distribution of the fluorescently labelled molecules of interest; (2) the bleach sequence, in which the ROI of the cell is depleted of these molecules as fast and as efficiently as possible by imaging this ROI with high-intensity illumination; (3) postbleach imaging, i.e. one or several time series under the same conditions as used in the prebleach sequence in order to record the redistribution of labelled molecules back to equilibrium owing to transport, diffusion, and more or less transient immobilisation.

Ideally, there is no photobleaching at all during the pre- and postbleach sequence, while during the bleach sequence all fluorescent molecules in the ROI are instantaneously and completely depleted. The share of bleached molecules amongst all molecules, on the other hand, should be negligibly small, and no diffusion should take place during bleaching. The postbleach sequence should be long enough to cover the whole redistribution of molecules. The time steps should be an order of magnitude shorter than the recovery process in order to obtain good resolution. For a real experiment it is necessary to optimise setup and parameters to come as close as possible to the ideal conditions (Sect. 7.3.2).

7.3.1.1 Intensity-Based Evaluation

The most common quantification method is to average the fluorescence intensity over the ROI and plot it as a function of time. The "amount" of bleaching (Axelrod et al. 1976) or degree of incomplete bleaching, the half time or time constant of recovery, and the immobile fraction can be determined as empirical parameters. They characterise the recovery, independent of the actual processes behind the redistribution of molecules. This is useful when only changes in mobility or binding properties are studied. In order to obtain the actual physical parameters of the processes involved, such as the diffusion coefficient or the dissociation rate, models must be applied that take into consideration Brownian motion, binding and immobilisation, cellular topology, and the geometry of the ROI. There are quite a few examples of more or less explicit expressions for the intensity recovering over time (Axelrod et al. 1976; Braga et al. 2004; Calapez et al. 2002; Periasamy and

Verkman 1998; Potma et al. 2001; Saxton 2001; Soumpasis 1983; Sprague et al. 2004; Starr and Thompson 2002). When recovery is dominated by either diffusion or binding/immobilisation, the quantitative analysis yields the diffusion coefficient or the dissociation rate, respectively. Very often, however, none of the processes dominate the mobility and a mere intensity analysis is not sufficient.

7.3.1.2 Spatial Analysis

It is adequate to refer to the evaluation as spatial analysis when the time course of the intensity is used in more spatial detail than only averaged over the bleach ROI. Different levels of spatial resolution and differentiation are conceivable and have been implemented. The first step beyond intensity-based analysis in the ROI includes intensities averaged over a few nonbleached areas (Carrero et al. 2003, 2004; Starr and Thompson 2002). This can be extended by generating continuous one-dimensional profiles through bleached and nonbleached areas (Berk et al. 1993; Houtsmuller et al. 1999; Kubitscheck et al. 1994; Oancea et al. 1998; Subramanian and Meyer 1997; Fig. 7.5, panels H–K). Finally, one can employ the complete two- or three-dimensional distribution as a function of time (Elsner et al. 2003; Siggia et al. 2000). What all these approaches have in common is that in order to describe the data quantitatively they require biophysical models that cover binding-induced immobilisation and diffusional exchange on the length scales used, leading to analytical or semiempirical functions or even the need for fully numerical modelling (Phair and Misteli 2001; Phair et al. 2004b).

7.3.2 Choosing and Optimising the Experimental Parameters

7.3.2.1 Numerical Aperture and Pinhole Size

The most prominent advantage of a CLSM is high resolution in three dimensions for imaging as well as for photobleaching, which is well suited for three-dimensional objects like cells. However, for most FRAP applications the acquisition of three-dimensional image stacks for the prebleaching, postbleaching, and bleaching sequence is too slow, restricting data acquisition to two-dimensional images. As an additional drawback, cells tend to move in z and the stage and the focus of the microscope tend to drift over time; therefore, it is very often useful and necessary to sacrifice axial resolution both for bleaching and imaging. This aims at (1) making the FRAP experiment less sensitive to cellular movements and mechanical drift, (2) increasing bleaching efficiency and detection sensitivity, and (3) reducing the diffusion or transport process to effectively two dimensions and thus simplifying the analysis.

How can this be achieved on the illumination side? The z resolution of the illumination is mainly determined by the effective NA of the objective lens. As shown in Fig. 7.4, illumination with a high-NA lens bleaches mainly in the focal plane and less strongly above or below it. In order to obtain an even axial bleach profile, one can either use an objective lens with a smaller NA, or reduce the illumination radius at the entrance aperture of the high-NA lens with an appropriate telescope or beam expander as is implemented in some commercial systems. Moreover, the latter makes more light available in the focus for more efficient depletion. By bleaching homogeneously in z through a whole cellular compartment, for example, the system is simplified to a two-dimensional problem.

On the detection side, an additional loss of z resolution can be achieved by expanding the pinhole diameter above that of one Airy disc, the optimised value for resolution and intensity yield. In this way, more light also falls onto the detector and the signal-to-noise ratio is improved, especially during the low-light pre- and postbleach sequences.

On the other hand, in many applications, a high spatial definition of the bleach ROI is desired so that the axial resolution is limited to a few microns (Fig. 7.4).

7.3.2.2 Laser Power and Detector Settings

Photobleaching is faster and more efficient when more laser intensity is available in the sample; therefore, the laser should be operated at high output power, and for the bleach sequence, transmission of the AOTF should be set to 100%, also in order

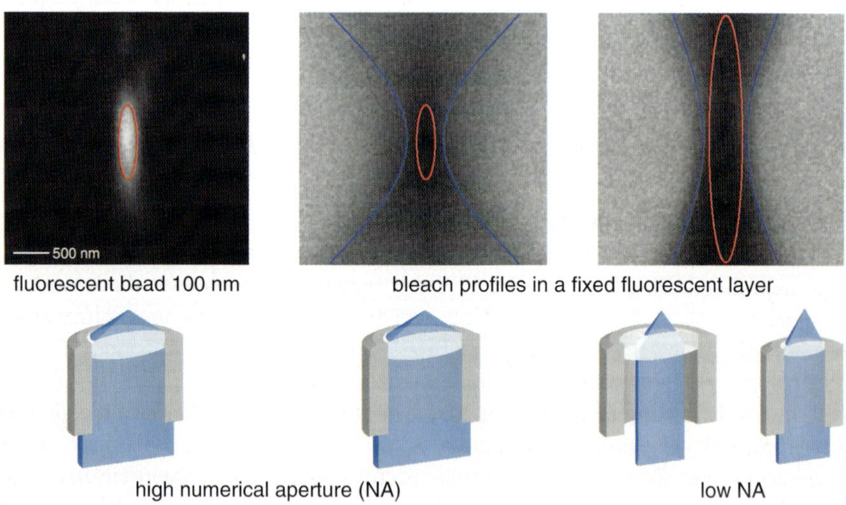

Fig. 7.4 Size and geometry of the focus. *Left*: Effective focal volume (*red profile*) for FCS represented by a confocal image of a fluorescent bead with a diameter of 100 nm taken with a 63× 1.2 NA water immersion objective lens with overfilled back aperture. *Middle*: Effective bleach volume (*red*) created under the same conditions as for the image on the *left* with 1-s bleaching in a fluorescent layer; the illumination cone is denoted in *blue*. *Right*: Same as for the image in the *middle*, but bleaching carried out with a sixfold reduced diameter of the laser at the entrance aperture of the objective lens

to exploit the full dynamic range of the AOTF. Additional laser lines can be activated, too. As an example, when EGFP and an Argon laser are used, alongside the main excitation at 488 nm, the lines at 458 or 514 nm can be set to 100% for stronger bleaching. In order to minimise photobleaching during pre- and post bleaching sequences, laser intensity is best set to the lowest possible value. An intensity ratio of 100:1 or more is desirable.

The PMT offset and gain must be set so that the full dynamic range is used. Background signal should be at pixel values above zero in order to facilitate background correction, whereas even the brightest pixels must not show saturation, otherwise they would bias the quantitative analysis.

7.3.2.3 Imaging Geometry, Speed, and the Bleach ROI

Usually, in a CLSM, the scanned area or physical image size can be chosen flexibly, ranging for a high magnification like 63× from a few hundred microns for the full field of view down to a few microns for maximal zoom. Other user-defined parameters are the number of pixels per line, the number of lines, and the line scanning frequency or time required per line including averaging. They all determine the acquisition speed (typically given in frames per second, fps), the pixel size, and the detection sensitivity, and a tradeoff between them must be found.

For mammalian cells, the whole cell including all potentially involved regions should be imaged. An additional neighbouring reference cell can be very useful to determine photobleaching, cellular movements, and hardware drift during imaging. When the processes under investigation are slow enough, one can choose parameters as for conventional confocal imaging. On the other hand, for very fast processes such as free diffusion, it might be better to image only a part of the cell or even the ROI because at a higher zoom factor, the scanned area is smaller and the galvanometer movement can be faster, permitting a higher acquisition speed and more efficient photobleaching by delivering the same laser power in a smaller area. Oversampling, i.e. a pixel size smaller than the resolution defined by the optics, and averaging should be avoided and the number of lines minimised to speed up acquisition since the required time per line remains constant. A tradeoff can be found by zooming into the ROI only for the bleach sequence.

Commercial CLSMs allow an almost arbitrary definition of two-dimensional ROIs, which should be adapted in size and shape to the cellular topology and the mobility of the molecules of interest. When a bleach ROI with dimensions close to the resolution limit is chosen, e.g. a circle with a lateral diameter of 0.5 μm and diffraction-limited z resolution, mobility can be studied with very high spatial resolution and the share of bleached molecules amongst all molecules is very small. However, a high acquisition speed might be necessary. For very fast processes or connectivity studies, it can be advisable to define a whole organelle or a strip across it as the ROI so that the effective dimensionality is reduced and the redistribution is sufficiently slow to be resolved (Fig. 7.5b).

7.3.2.4 The Prebleach, Bleach, and Postbleach Sequences

In order to determine the bona fide equilibrium distribution and the photobleaching during the nonbleach sequences, several frames, e.g. 10, should be taken as the prebleach sequence. The parameters must be the same as for the postbleach sequence, optimised as mentioned in the previous section (Fig. 7.5a).

Photobleaching during the second sequence should be carried out as fast and as completely as possible. Test experiments in separate cells can be used to optimise all the aforementioned parameters and the number of bleach frames, starting with one. The whole sequence should remain around 10 times shorter than the recovery half time. Nevertheless, incomplete bleaching can still occur owing to diffusion of fluorescent molecules into the ROI during bleaching and reversible photobleaching.

The postbleach sequence, taken with the same parameters as the prebleach images, is ideally acquired on the one hand round 10 times faster and on the other hand around 10 times longer than the observed processes. Since a logarithmic time scale is usually not supported, one could define several subsequences with increasing repetition times.

7.3.3 Quantitative Evaluation

For any quantitative analysis several correction and normalisation steps must or can be carried out. This requires the definition of additional regions over which the intensity is averaged (Fig. 7.5b, c; Sect. 7.6.1). In addition to the CLSM software packages for acquisition and processing, ImageJ (NIH) or Metamorph (Molecular Devices), for example, are suitable tools for the quantification steps.

7.3.3.1 Background Correction

In order to account for background signals, a nonstained area in the prebleach/postbleach images is defined. The average intensities from this region are then subtracted from all other average intensities, time step by time step, before further processing (Fig. 7.5d).

bl from **d** normalised to the total intensity tot. **f2** Intensity bl from **e2** normalised to the prebleach intensity. **g** A corrected and normalised bleaching ROI intensity plot from an imaging FRAP experiment and the empirical parameters that can be obtained. **h** The same first postbleach image as in **b** now with the directions over which the intensity profile is plotted and averaged, respectively. **i** Averaged intensity profiles for the prebleach (*pre*), the first postbleach (*0 s*), and a later postbleach (*48 s*) image. **j** The postbleach profiles from **i** normalised to the prebleach profile. **k** The squared fitted full width at half maximum from **j** versus the corresponding postbleach time with a linear fit

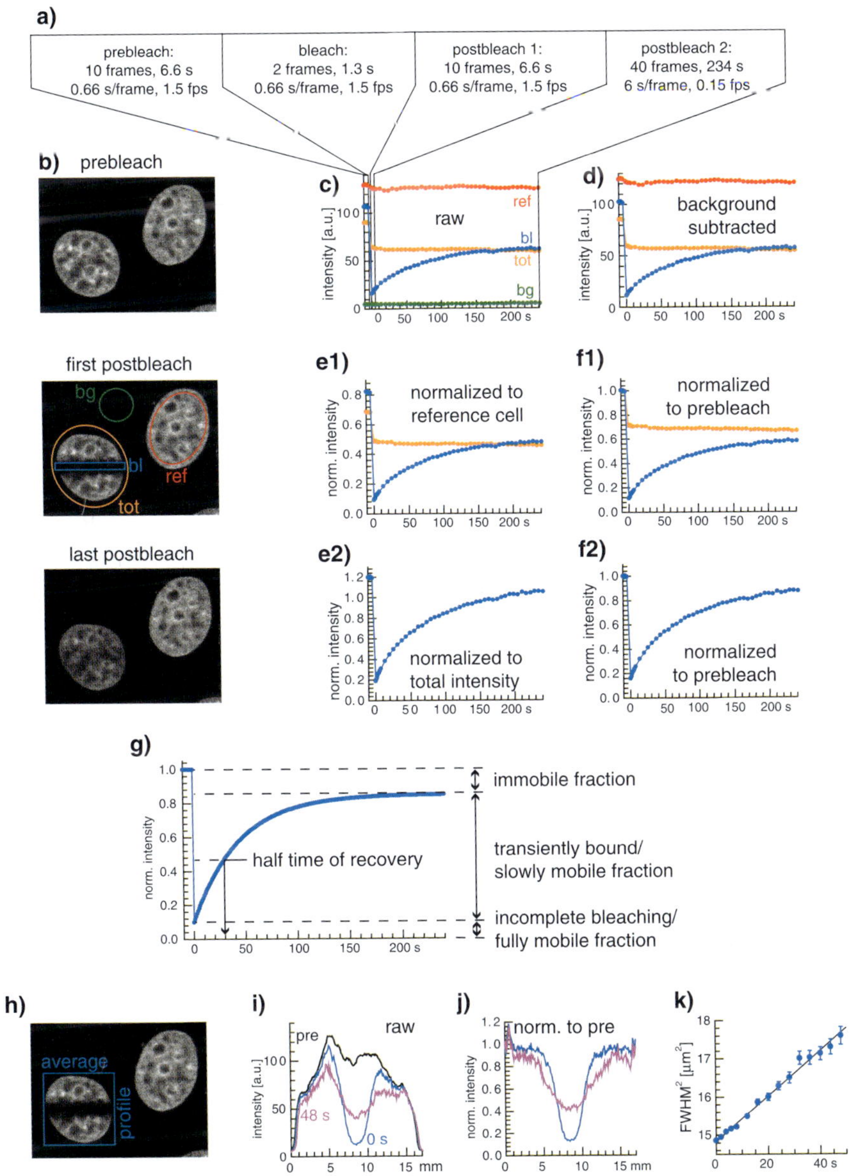

Fig. 7.5 A FRAP experiment. **a** Exemplary timing of an imaging FRAP experiment. **b** Prebleach as well as first and last postbleach image with the bleaching ROI (*bl*, *blue*), a ROI for a background signal (*bg*, *green*), a ROI for a reference signal from a neighbouring cell (*ref*, *red*), and a ROI for the total signal from the bleached cell (*tot*, *yellow*). **c** The averaged intensities from the ROIs shown in **b** versus time; the zero time point corresponds to the first postbleach image. **d** Intensities bl, ref, and tot from **c** after background correction. **e1** Intensities bl and tot from **d** normalised to the reference intensity ref. **f1** Intensities bl and tot from **e1** normalised to the prebleach intensity. **e2** Intensity

7.3.3.2 Correction for Bleaching and Other Slow Fluctuations

Even with the most careful choice of parameters, photobleaching during pre- and postbleach imaging is likely to occur. When the images also cover a neighbouring reference cell (at least partially), the average intensities of all other regions can be divided by the average intensity from this cell, time step by time step for both the prebleach and the postbleach sequence (Fig. 7.5e1). If no reference cell or area that is not affected by the bleaching step and subsequent transport/diffusion is available, the prebleach intensities of the bleached cell or region can be fitted with an exponential decay and this model function, extrapolated to the whole experiment, can be used for division. However, this does not correct for other fluctuations, e.g. due to laser instabilities. Alternatively, the average intensity from the bleached cell can be used for correction (Fig. 7.5e2), inherently normalising to the amount of fluorescence available in the focal plane (see next section). This gives more precise data for amplitudes and fractions, while correction with a reference cell results in a more precise time course of the recovery.

7.3.3.3 Normalisation

As mentioned already, one can normalise the ROI intensities to the amount of fluorescence available in the whole bleached cell; however, this requires that the intensity seen in the image represents the whole fluorescence available, which can be achieved with an extended z resolution (Sect. 7.3.2).

Finally, the intensities in the bleach ROI and other regions can be normalised to the respective prebleach average values (Fig. 7.5f1, f2) so that an incomplete recovery is indicative of a real immobilised fraction. In addition, when only the time course of recovery is of interest, the first postbleach value can be set to zero (Axelrod et al. 1976).

7.3.3.4 Data Analysis and Fitting

A first basic and model-independent analysis step can be applied to all fully corrected and normalised ROI intensities as shown in Fig. 7.5g: the degree of incomplete recovery stands for a fully immobilised fraction, the recovery amplitude for a transiently bound/slowly mobile fraction, the degree of incomplete bleaching gives an upper limit for a fast, fully mobile fraction, and the time course can be quantified by the half time of recovery. These parameters can be obtained from a good fit of any reasonable function to the data (Sect. 7.6.1). In the binding limit (Sect. 7.2.3) and for some experimental designs in the diffusion limit, analytical model functions are known that reflect the actual physical processes (Ellenberg et al. 1997; Soumpasis 1983; Sprague et al.v 2004; Sect. 7.6.1).

The experimental function is fitted by a model equation with nonlinear least-squares methods, which are included in a number of data analysis packages such as Origin (OriginLab), MATLAB (MathWorks), SigmaPlot (Systat), KaleidaGraph

(Synergy), Igor Pro (WaveMetrics), Excel (Microsoft), Maple (Maplesoft), or Mathematica (Wolfram), to name just a few. The fit procedure minimises χ^2, the squared difference between the measured data and the model function summed over all data points, by varying the parameter values of the model equation. However, it is possible that the fitting algorithm finds different, apparently optimised sets of parameters for different sets of start values, even when those are provided by the software. Moreover, good fits can be found for different model functions and it is only possible to rule out the "wrong" ones. An estimation of which fit might be better can be obtained from comparing the χ^2 values. Care has also to be taken regarding the number of free parameters to be fitted – the fewer there are the more accurate the fit will become and any parameter that is known a priori should be fixed. For example, if an immobilised fraction was determined independently, its value should be fixed for further fitting. Global fitting where parameters from several measurements are linked can help to improve the statistical significance (Skakun et al. 2005).

7.3.3.5 Spatial Analysis and Modelling

In many cases, however, numerical modelling is necessary that includes the spatial information available in the data. In general, a set of coupled differential equations in space and time covering diffusion/transport and binding/immobilisation must be defined and solved. With increasing spatial resolution, this ranges from a small number of spatially separated compartments to a two- or three-dimensional solution of the spatiotemporal distribution of fluorescent molecules. In addition to writing customized software, several modelling software packages are available (Phair and Misteli 2001; Rabut and Ellenberg 2005).

Under sufficiently simplified experimental conditions, diffusion and binding properties can be determined directly from the time series of effectively one-dimensional profiles through the bleach region (McGrath et al. 1998; Oancea et al. 1998; Tardy et al. 1995). Figure 7.5h–k shows an example where a rectangular ROI was bleached into the distribution of a nuclear protein tagged with EGFP. After the profiles had been averaged as indicated (Fig. 7.5h, i) and had been normalized to the prebleach distribution (Fig. 7.5j), the widths of the profiles were squared and plotted as a function of postbleach time (Fig. 7.5k). The diffusion coefficient can be determined from the slope (Sect. 7.6.1).

7.3.4 Controls and Potential Artefacts

7.3.4.1 Reference Experiments

In order to assess the limits of permanently immobilised and fully mobile molecules, respectively, it is helpful to carry out the same FRAP experiments on fixed cells containing the molecules of interest or cells expressing fluorescent fusion proteins that are known to be (quasi) permanently bound/immobilised, such as core histones (Kimura and Cook 2001; Weidemann et al. 2003), as well as on cells containing fully mobile molecules like free GFP or fluorescently labelled dextrans. The quantitative analyses described in previous sections should be applied to check their validity.

7.3.4.2 Measuring the Bleached Region

The geometry of the bleached region in three dimensions crucially influences the amplitude and time course of the redistribution process. Bleaching in an immobilised fluorescence distribution under the same conditions as used for the experiments helps to determine and optimise the geometry of the actual bleached region. Figure 7.4 shows the extreme case where a pointlike ROI results in an extended bleached area with a circular shape in x and y and a NA-dependent more conical or more cylindrical profile in z. In the focal plane, the bleached region can differ from the chosen bleach ROI owing to diffusion into and out of the region, a pixel size smaller than the resolution limit, and instrumental inaccuracies. Along the optical axis, the effective NA of the objective lens determines size and shape in addition to possible diffusion.

7.3.4.3 Reversible Photobleaching

As mentioned in Sect. 7.2.4, especially a fast recovering component can result from reversible photobleaching. Although this effect can hardly be avoided, continuous imaging under identical conditions, but with varying frame repetition rates, can help identify and even quantify the effect, supported by additional FCS experiments (Sect. 7.4). Additionally, FRAP measurements with different ROI sizes allow one to distinguish it from diffusive recovery because it does not depend on the ROI size. Although fixation would allow one to distinguish it from mobility-related recovery in a FRAP experiment, fixation modifies the microenvironment of the fluorophore (e.g. a fluorescent protein) and can thus change the photophysical properties significantly.

7.3.4.4 Cell Movements and Hardware Drift

Another frequent problem is the movement of cells and the microscope stage and/or focus drift during the experiment. Axial movements can be covered to some extent with an extended z resolution, while lateral displacement can be compensated by tracking the translation (and rotation) of the cells and realigning the images afterwards with appropriate software tools – the ImageJ plugin TurboReg (EPFL) or Metamorph – before starting quantification steps.

7.3.4.5 Phototoxicity

During FRAP experiments, the energy delivered is transformed into photons or heat or is used to split chemical bonds. Although the first is desired for fluorescence, it can also affect photosensitive intracellular processes, making all methods based on photoinduced fluorescence problematic. Heat generation typically results in an almost negligible local warming of 0.5 Kelvin or less (Rabut and Ellenberg 2005). A more dramatic effect is due to the photochemical creation of free radicals and reactive oxygen that interact with other molecules and impair their function, an effect used in protocols like CALI (Surrey et al. 1998). Although this effect is not very efficient in fluorescent proteins, it is necessary to check for phototoxicity especially for the processes and functions of interest as well as for cell viability in general by carrying out bleaching experiments and subsequently monitoring morphology, cell division, and other relevant biological functions with independent methods.

7.4 How To Perform an FCS Experiment

7.4.1 The Principle of FCS

FCS measures the length and amplitude of stochastic fluctuations of the fluorescence signal from labelled molecules in a microscopic observation volume (Elson and Magde 1974) defined by the confocal optical setup. These fluctuations are mainly generated by Brownian motion, but also by photophysical processes where the fluorophores switch between bright and dark states. Diffusion of fluorescent molecules results in dwell times in the focus of typical setups ranging from around 30 μs for pure fluorophores in water to several tens to hundreds of milliseconds for membrane-based complexes, whereas the stochastic occupation of triplet and other dark states results in molecular blinking or flickering with time constants between 1 μs and several hundred microseconds (Sect. 7.2.4). For data analysis it is very important to distinguish the sources of fluctuations (Schwille 2001).

As pointed out earlier, the concentration range practically accessible with FCS lies between picomolar and micromolar, constrained as follows. Generally, a low concentration is desirable for FCS because for fewer molecules in the focus, the relative signal fluctuations upon a molecule entering or leaving the focus are more pronounced: compare the signal difference between three and four molecules – 33% – and 30 and 31 molecules – 3.3%. At a concentration around 1 μM, corresponding to a few hundred molecules, all diffusional fluctuations average out. Below 1 nM, on the other hand, the number of molecules is so small that often the fluorescence signal cannot be distinguished from detector noise and background signals. The time range of observable processes lies between microseconds and seconds. The lower limit is defined by the time resolution of the hardware (e.g. 50–100 ns as determined by the dead time of the APDs), which should be about 10 times smaller than the correlation time of the fastest process of interest. The upper constraint is given by the recommendation that for good statistics the measurement should last at least 1,000 times longer than the slowest process of interest, while the whole system must be stable, i.e. no bulk photobleaching, no slow movements of a cell, and no variations of laser intensity should occur.

The photons as collected by the detectors, typically APDs, are transformed into digital electronic pulses, which are recorded over time and further processed. The mathematical function or algorithm mainly used for this is a correlation analysis where the signal obtained at a time point is multiplied with the signal from the same channel (autocorrelation) or another channel (cross-correlation) recorded at a fixed time lag. Averaging over all available time points generates a correlation function with values for a whole range of lag times (Fig. 7.6). In most cases, the lag times follow a so-called multiple τ scheme with a pseudologarithmic scale where the raw data are rebinned for larger lag times. This is in order to reduce noise and computation time while covering several orders of magnitudes. When the signal fluctuations result from a diffusion process, the correlation function stands for the probability that given a molecule is in the focus, it will still be there after the lag time. Therefore, the correlation function drops to zero for large lag times and only mobile molecules contribute.

Not only for diffusion, but for all other molecular relaxation processes such as blinking, the corresponding correlation function is a decaying curve. For independent processes – either different ones at the level of one molecule such as diffusion and blinking, or the same process for different molecules – the resulting correlation function is just the sum of the respective single correlation functions. For a number of diffusional and photophysical processes, and under appropriate assumptions for the optics, a mathematical treatment yields analytical expressions (Sect. 7.6.2). The experimental function is fitted by a model equation with nonlinear least-squares methods, which are included in a number of data analysis packages (Sect. 7.3.3.4). The suppliers of FCS systems also provide appropriate fitting software.

7.4.1.1 Autocorrelation

For an autocorrelation function (ACF), the signal is compared with itself. The amplitude of the ACF is inversely proportional to the number of molecules in the focus, i.e. increasing the concentration decreases the amplitude. When looking at diffusion-induced fluctuations, the inflection point of the sigmoidal ACF is determined by the mean dwell time of the molecules in the focus, also referred to as the diffusion correlation time (Fig. 7.6). Moreover, the shape of the curve changes with different contributions and a triplet transition, for example, adds a short exponential shoulder to the diffusion ACF (Widengren et al. 1995). When several species contribute to the fluctuations, their contributions to the ACF add up. As an example, for two molecules with the same label but significantly different diffusion coefficients, the resulting ACF bears two "steps" and two inflection

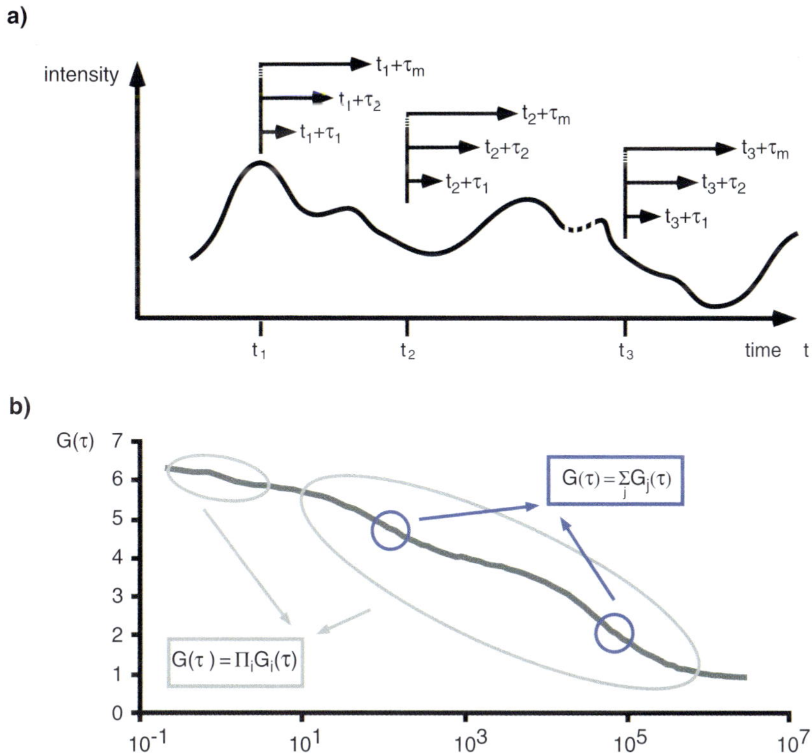

Fig. 7.6 The principle of FCS. **a** Fluctuation analysis. In a correlation analysis of a fluctuating signal, the intensity at different measurement times t_i is compared with signal intensities at different lag times τ_j later. See the text for details. **b** A correlation curve shows exponential decay for photophysical processes and sigmoidal decay for diffusion processes. Contributions from independent components add up and the contributions of dependent processes multiply to result in the total correlation. This is a simulated correlation curve for two components – the first component with a diffusion correlation time of 100 μs and a fraction of 40% and the second component with a diffusion correlation time of 50 ms and a fraction of 60% – and a number of diffusing particles of 0.2, a structure parameter of 5, a triplet fraction of 10%, and a triplet correlation time of 1 μs. The contributions of the two components are added to obtain the total diffusional correlation, which is multiplied with the triplet contribution to obtain the total correlation function

points. Due to the broad sigmoidal decay of the ACF, diffusion correlation times must differ by at least a factor of 1.7–2 in order to be distinguishable (Meseth et al. 1999), and sometimes much more, especially for noisy data. It should be noted that for more globular molecules, which are mobile in three dimensions, the effective radius and the diffusion correlation time (both inversely proportional to the diffusion coefficient) scale with the second to third root of the mass. Hence, autocorrelation experiments that aim at distinguishing different species based on the diffusion coefficient are frequently restricted to small ligands binding to substantially larger molecules or complexes. This holds true especially in living cells, where multiple interactions and the inhomogeneity of the cellular environment hamper the interpretation of binding reactions. For diffusion studies in cells it might be beneficial to analyse distributions of diffusion times rather than single ones (Modos et al. 2004; Sengupta et al. 2003).

7.4.1.2 Cross-Correlation

An important extension to the autocorrelation analysis from a single-colour experiment is a two-colour approach where the temporal cross-correlations of different signals are computed (Ricka and Binkert 1989; Schwille et al. 1997). It was developed in order to resolve interactions of molecules of similar sizes where the decrease in the diffusion coefficient is too small to be resolved in the ACF. This method requires potentially interacting molecules to be labelled with spectrally different fluorophores that can be distinguished by detection in separate channels. When the molecules do not interact, signal fluctuations in the two channels are totally independent and the cross-correlation function (CCF), which is calculated in the same way as the ACF, is ideally zero. When all the molecules form dimers, for example, Brownian motion and resulting fluctuations are synchronised or correlated and the CCF is ideally identical to the ACF in each of the channels. The relative fractions of free and bound molecules can be quantified from the CCF and the total concentrations from the ACFs (Strohner et al. 2005; Weidemann et al. 2002). In a real system, fluorescence cross-talk from one channel to the other generates an offset cross-correlation signal even in the absence of binding. In addition, the size of a diffraction-limited focus depends on the wavelength. Therefore, the observation volumes in spectrally different detection channels differ in size and – in the presence of chromatic errors – in their position. Due to this imperfect overlap, the CCF never aligns perfectly with the ACF. These deviations must be borne in mind and corrected for (Sect. 7.4.2).

7.4.2 Instrument Alignment and Calibration

The basic intrinsic readouts from a correlation measurement and analysis for every channel used are the fluorescence signal (given as count rates in counts per second, CPS, or kilohertz, kHz), the average number of molecules, the diffusion correlation time, and the signal coincidence in the case of a cross-correlation measurement. In order to transform them into instrument-independent physical parameters such as concentrations and diffusion coefficients, the size, geometry, and overlap of the focal volumes must be well defined and known. The most accurate way to determine them would be probing the intensity distribution in the focal volume using near-field optics (Hecht 2004; Rigler and Elson 2001); however, the necessary instrumentation is very expensive and usually not available. Therefore, in an indirect approach the aforementioned intrinsic parameters are employed using photophysically and hydrodynamically well characterised fluorophore molecules with high quantum yield and photostability.

7.4.2.1 Cover-Glass Correction

For FCS and in vivo imaging, high-NA water immersion objective lenses are recommended. They are frequently equipped with a correction collar that allows one to account for thickness variations of the cover-glass and thus to obtain an aberration-free focus over a wide range of sample depths in aqueous solutions as well as the cellular interior, which both feature optical properties similar to those of water. Using a fluorescent solution, the focus should be fixed more than 20 µm above the cover-glass. After the laser has been activated, the correction collar should be turned carefully until the signal readout at the detector reaches a maximum. Alternatively, in a combined FCS/CLSM system, vertical scanning in reflection mode allows one to monitor the reflection of the laser light on the upper surface of the cover-glass. The correction collar should be turned carefully until the reflection is as sharp as possible.

7.4.2.2 Numerical Aperture and Pinhole Size

As already outlined in Sect. 7.3.2 for FRAP and as shown in Fig. 7.4, the effective NA and the pinhole size crucially influence the size and geometry of the focus. A small number of molecules, a high probability of detecting them, and an aberration-free focus are crucial for FCS. While it is frequently recommended to fill the back aperture of the objective to only two thirds with the $1/e^2$ diameter of the illumination beam in order to guarantee a three-dimensional Gaussian intensity distribution in the focus (Haustein and Schwille 2003), with modern high-quality lenses the focus size can be minimised by overfilling the back aperture while the approximate Gaussian profile is preserved (Rigler and Elson 2001). For similar reasons, a pinhole size well above one Airy disc is used in many setups (Rigler et al. 1993). However, modern optics allow one to use a resolution-optimised pinhole size of one Airy disc. In general, the same diffraction-limited conditions can be used for resolution-oriented CLSM and for FCS.

Pinhole size and NA determine the size of the focus and thus the dwell time of the molecules, i.e. their diffusion correlation time. In contrast, photophysical sources of fluctuations (occupation of triplet and other nonfluorescent states) do not depend on the focus size; therefore, varying the pinhole size and NA is a tool to distinguish between diffusional and photophysical contributions to the correlation function. Moreover, this can be used to increase the focal size for the (rare) cases when the number of molecules is too small or the diffusional processes are too fast.

7.4.2.3 Focal Size and Geometry

When the combined illumination intensity and detection probability, the so-called molecular detection efficiency (MDE; Rigler et al. 1993), is approximated as a three-dimensional Gaussian function, the focus is fully characterised by the lateral and the axial dimension. As an open volume, the focus is arbitrarily defined as the volume within which the MDE values are above $1/e^2$ (13.5%) of the maximum. From the autocorrelation analysis of a solution of freely diffusing fluorophores with a known diffusion coefficient, the lateral and axial radii can be calculated on the basis of the diffusion correlation time and the structure parameter (the ratio of axial to lateral radius) as fitted to the data (Weisshart et al. 2004; Sect. 7.6.2). The fluorophore should feature a high quantum yield and photostability and should not be prone to too much blinking and triplet-state occupation in order to obtain a good signal-to-noise ratio. Only then are the fitting parameters accurate and meaningful. Although theoretical calculations come up with structure parameters of 2–3, typical experimental values lie between 5 and 8 owing to optical aberrations of the system (Muller and Gratton 2003).

The amplitude of a correlation function is given by a so-called geometric factor divided by the number of molecules in the focus (Sect. 7.6.2). This factor accounts for the intensity profile and dimensionality of the focus (e.g. Gaussian or circular, two- or three-dimensional), and the excitation process, e.g. one- or two-photon (Thompson 1991). In order to determine it, one can take correlation measurements of a dilution series (around 10 nM) of a fluorophore solution with a known concentration. Plotting the diffusional contribution to the correlation amplitude separated from the photophysical part (Sect. 7.6.2) versus the concentration reveals a linear range, from which the geometric factor is derived together with the linear dimensions of the focus.

Once the dimensions and the geometric factor of the observation volume are known, diffusion coefficients can be calculated from the measured diffusion correlation times and concentrations from the correlation amplitudes, respectively.

7.4.2.4 Volume Overlap, Cross-Talk, and Maximal Amplitude for Cross-Correlation Experiments

Ideally, the two focal volumes of a two-colour setup illuminated by two different laser lines and observed by two different detectors should match perfectly in size and position. However, since the linear dimensions of the illuminated focus are proportional to the laser wavelength, the first requirement can be met from the illumination side only when the illumination of the back aperture of the objective, and thus the effective NA, can be adjusted separately for each wavelength (Schwille et al. 1997; Sect. 7.3.2). This is usually not possible in commercial systems. In terms of detection, when two independent pinholes are implemented, the more redshifted detection volume can be decreased by closing and/or the more blueshifted volume can be increased by opening the respective pinhole.

More important, however, is a minimal chromatic displacement between the channels, which is most likely along the optical axis. Misaligned volumes will result in cross-correlation diffusion times longer than the slower component of the autocorrelations (Bacia and Schwille 2003). The overlap depends mainly on the colour correction of the objective lens and to a smaller extent of other optical elements in the microscope – some commercial systems are equipped with correction optics in the detection head. In general, it is advisable to use highly corrected lenses because on the illumination side, the user has virtually no means to minimise chromatic displacements. In systems with independently movable pinholes the misalignment can be measured and to a minor extent also minimised by maximising the cross-correlation amplitude.

Excitation of spectrally different fluorophores with one laser line reduces the volume mismatch to the detection side. Recently, red fluorophores with a long Stokes shift have become available that allow for excitation of both the more blueshifted and the more redshifted fluorophore with the same laser line (Hwang and Wohland 2005) and also quantum dots of different colours have very similar excitation spectra.

Cross-talk especially, but not exclusively, due to emission of the shorter-wavelength fluorophore into the longer-wavelength channel generates a contribution to the CCF that must be corrected for (Weidemann et al. 2002; Sect. 7.6.2). While this is the most dominant component, all other combinations (cross-excitation and cross-emission, subsumed in a matrix) also play a role. In order to minimise cross-talk, the fluorophores should be spectrally well separated and optimal filters and dichroics should be chosen. However, this might deteriorate the volume overlap; see above. Not all matrix elements must be determined to quantify the cross-talk contribution to the cross-correlation. It is sufficient to measure an equimolar mixture of the two types of fluorophores and to determine the ratio of cross-correlation to autocorrelation amplitude, which should not exceed 0.1 significantly. Cross-talk in a CCF can also be identified on the basis of triplet contributions because they are not correlated between different fluorophores.

7.4.2.5 Sensitivity and Background

Channel sensitivity can be defined as counts per molecule per second (CPMS, also referred to as count rate per molecule, CPM). It is calculated from the count rate divided by the number of molecules as obtained from the correlation function and represents the brightness of a molecule species. It depends not only on the properties of the fluorophore, but also on the transmission of optical filters, the quantum yield of the detectors, and the laser power.

Especially in living cells, autofluorescence and scattered light can form some background signals. They are best assessed quantitatively with cells not containing the labelled molecules of interest. A noncorrelated background signal generally decreases correlation amplitudes, but this can be corrected for quantitatively after acquisition (Sect. 7.6.2). When in the sequence of an FCS measurement followed by whole-cell bleaching and another FCS measurement the correlation amplitude does not rise, this is strongly indicative of a background signal. When the background signal is correlated it has to be treated as an additional component, provided that the diffusion correlation times can be distinguished.

7.4.2.6 Position Calibration

In intracellular FCS experiments performed in combined CLSM and FCS systems, the interesting biological information, such as concentrations or diffusion and interaction properties, are of course strongly dependent on the position of the measurement spot – quantitatively measuring these parameters with diffraction-limited spatial resolution is actually a major benefit of such a system. Both beam scanner and motorised stages can show an offset and/or hysteresis. This should be calibrated before intracellular experiments in order to ensure that the physically addressed position corresponds to the desired position. Usually, the software allows one to choose an FCS measurement spot in a prerecorded confocal image. After fixing the beam in the image of a fixed homogeneous fluorescent layer and bleaching for a short time with high intensity, the subsequently repeated image should show a darker bleached spot whose position can be compared with the desired position. Commercial systems provide functionalities to compensate for a potential spatial offset.

7.4.3 Setting Up an Experiment

After the required calibration steps, setting up experiments is fairly straightforward in solution. The laser power should not lead to bleaching of the dye. The length of data acquisition should be 1,000 times longer than the processes of interest for good statistics. If binding constants are being investigated, one autocorrelation or cross-correlation measurement can be enough, but it is recommended to set up experiments by varying the concentration of one partner and analysing the data by Scatchard analysis. If dissociation constants are too high, appropriate concentrations could be too high for FCS. In this case, unlabelled molecules can be added. Whenever possible, control experiments should be performed to assess as many parameters as possible beforehand. For example, the structure parameter, which is instrument-dependent, and the diffusion correlation time of a small ligand can be determined independently and the values fixed for the fittings performed on the following measurements.

Cellular work is more demanding. Autofluorescence, background fluorescence, and scattering from immobile molecules as well as cell or organelle movement can all disturb the correlation signal. Optimisation of the laser power is often restricted owing to bleaching problems as is the measurement time; hence, in cells, one has to work under suboptimal conditions. For cell work it is best to measure consecutively for shorter times and to reduce the laser power as much as possible. The count rate per molecule should be at least around 1 kHz, otherwise the measurements should be discarded. Most commercial systems are well equipped for measurements taken in cells when they provide a combined CLSM and FCS platform. The cell is imaged, the site of interest designated, and the measurement triggered. In a cellular environment a short prebleach can be beneficial to remove the contribution of immobile molecules. If concentrations and fluorescence levels are to be reduced because of overexpression, it is advisable not to bleach in one spot by FCS but rather to bleach the whole cell by imaging. This will reduce the overall load of oxygen radicals and keep cells healthier. Data analysis in cells is also more demanding, since free three-dimensional diffusion will apply in rare cases only and the user has to decide on the model; hence, previous knowledge of the system, e.g. from imaging and FRAP, is very helpful. Also, if diffusion times are interpreted it would make a stronger case if mutants are available to help discriminate them. For example, when a fast and a slow diffusion time are observed, one can only reach the conclusion that the faster belongs to the monomer and the slower to a complex if a mutant known not to be able to incorporate into the complex lacks only the longer diffusion time. For membrane studies care should be taken regarding the position of the confocal volume with respect to the membrane. Diffusion times and concentrations will differ depending on whether the membrane lies just in the centre (of the height) or is offset (Bacia and Schwille 2003). Also one should note that the volume is 0.3 μm in diameter and 1.5 μm in height; hence, not only the membrane is sampled but also the external medium and the internal cytoplasm.

While for free two- or three-dimensional diffusion, the data-fitting modules integrated into the acquisition software often can be used, for other or more complex fitting models and approaches, third-party software must be used (Sect. 7.3.3.4).

7.4.4 Types of Applications

There are five intrinsic results that can be obtained from an (intracellular) correlation measurement (Haustein and Schwille 2003): the low concentration localisation of molecules (if an autocorrelation signal can be obtained), their concentration (from the amplitude), their mobility properties (binding and diffusion, from the decay) a spatiotemporal coincidence (in the case of signals obtained simultaneously in two channels), and intramolecular and photophysical dynamics (from the decay).

7.4.4.1 Localisation Measurements

In general, the localisation of fluorescently labelled molecules is studied by imaging with high spatial resolution. However, if only trace amounts are present imaging technologies are often not sensitive enough for detection. In this case, one can prove the existence of a molecule in a certain compartment or the cell medium with 100-fold improved sensitivity using FCS. As such, uptake of sparse molecules can be efficiently followed by taking FCS measurements in solution, on a membrane, or in different cellular compartments. If high-precision positioning is required, it might be useful to counterstain the structure of choice with a different fluorophore, preferentially a red-shifted one that does not cross-talk into the channel of interest.

7.4.4.2 Measuring and Imaging Concentrations

Since the amplitude of the ACF is inversely proportional to the average number of molecules in the focus, the concentration of the molecule under study can be exactly determined once the effective volume and the geometric factor are known from calibration measurements. The computation in cross-correlation experiments is straightforward as well. Control experiments should be performed to validate the numbers and to rule out artefacts such as surface adsorption that might influence the results when working with minute amounts of proteins. This can be overcome by coating the vessels with bovine serum albumin. Owing to the sensitivity of the method all reagents, even water, have to be checked for trace amounts of fluorescent particles. Especially in binding assays one should also check for quenching and dequenching of the fluorophore because brightness differences have to be taken into account.

In an intracellular application, the ACF is measured at a certain position in a prerecorded image. Using the number of molecules in the focus or the concentration, one can transform the intensity value of the corresponding image pixel into a concentration and thus the whole intensity image into a concentration image (Weidemann et al. 2003).

7.4.4.3 Mobility Measurements

The submicrometre resolution of the confocal volume allows FCS to be used not only to assess mobility-related parameters in solution, but also in cells. It is ideally suited to distinguish different forms of molecular mobility and to understand the underlying transport process. For example, it is often important to distinguish between directed motion, free diffusion, anomalous subdiffusion, or confined diffusion. In this respect FCS offers a higher dynamic performance and sensitivity than imaging methods but can suffer substantially from photobleaching problems, so FRAP and CP might be the methods of choice. The mobility of the molecules depends to a great extent on the environment, and FCS can help to identify the underlying processes. Care has to be taken if chemical dyes are employed. They can be quite lipophilic, e.g. rhodamine dyes, and tend to associate with intracellular membranes, altering the diffusion characteristics of the molecules they are attached to. This problem is not encountered with fluorescent fusion proteins. However, nonphysiological overexpression can cause nonspecific binding to subcellular structures or accumulation in certain cellular compartments.

Membrane-bound receptors or nuclear proteins often display anomalous subdiffusion (Schwille et al. 1999; Wachsmuth et al. 2000). Environmental heterogeneities in membranes can often be better accessed in model systems such as giant unilamellar vesicles (GUVs), where it is possible to alter membrane composition in a controlled way.

7.4.4.4 Intermolecular Interactions

Binding studies can be performed in autocorrelation experiments, provided that the change in diffusion time is large enough (1.7–2-fold corresponding to a factor of 3–8 in molecular weight); hence, this kind of scenario works mostly only for small labelled ligands binding to large partner molecules, as is the case with receptor–ligand interactions. In the course of the reaction, the small and thus fast species would be increasingly diminished until a chemical equilibrium is reached or the species is completely consumed. In such a setup it is advisable to determine separately the diffusion time of the pure ligand, which can be fixed in later fit procedures.

If the ligand only becomes fluorescent after binding, the binding efficiencies over time are easily assessed by taking only the amplitudes of the correlation into account. In general, analysis is more complicated if the ligand changes its brightness states. If potential binding partners are of similar size, cross-correlation experiments should be considered where mainly the amplitudes are of interest, which makes the analysis more reliable.

7.4.4.5 Intramolecular and Structural Dynamics

Changes in diffusion are not only triggered by binding to larger molecules, but also by alteration of conformation, inducing, in turn, changes in the so-called hydrodynamic radius (the radius of a sphere with the same diffusion properties). Again, a factor of 1.7–2 is required. This is sometimes accompanied by a change in the brightness of the label, which can be followed with FCS as well.

As mentioned already, especially at higher laser intensities fluorophores can show blinking or emission intermittence, i.e. occupation of dark states of the molecule that can endure for up to several seconds. This is expressed as an exponentially decaying shoulder in the ACF. Triplet-state parameters depend not only on the laser power but also on environmental conditions such as the oxygen concentration. Similarly, some blinking behaviour of fluorescent proteins depends on the pH. Hence, environmental conditions can be assessed in terms of photophysical dynamics.

When fluorescent molecules are attached to or close to dynamic structures such as membranes and polymers, their mobility is strongly influenced by the thermal fluctuations and relaxations of these structures, often resulting in confined diffusion (Fradin et al. 2003; Lumma et al. 2003; Wachsmuth et al. 2003). This can be identified in FCS experiments and used to quantify structural properties.

7.4.5 Potential Artefacts

7.4.5.1 Photobleaching

Photobleaching can induce severe artefacts in FCS experiments. Depending on the extent, two effects can arise in autocorrelation measurements. When photobleaching occurs mainly in the confocal volume (spot bleaching) it causes an apparent fast movement of the molecules since loss in fluorescence is misinterpreted as diffusional exit from the volume. Especially in a solution with a large reservoir of molecules, bleached ones are easily replaced, and the count rate trace will not show any pronounced decrease even at high laser intensities. The best way to check for spot bleaching is to increase the laser power gradually and monitor whether diffusion times become faster.

On the other hand, bleaching also happens outside the confocal volume in the excitation cones (bulk bleaching). In this case, an overall loss of fluorescent molecules will be observed, which is expressed as a drop in the count rate trace. This leads to a decay in the correlation to an offset well above or below zero, resulting in a so-called divergent curve. In most cases, this corrupts proper curve evaluation and the data should be disregarded. If the curves are analysed there seems to be an increase in diffusion time because the drop seen at high lag times is misinterpreted as slow diffusion interfering with the real one. If the curve is divergent, i.e. it continuously decreases even below zero, this can no longer be accounted for in the fit by using an artificial slow component.

Bleaching is most likely for slowly moving or immobile molecules. In fact in a cell, especially when the protein of interest is overexpressed, a distribution of diffusion times can be observed (Jankevics et al. 2005) owing to nonspecific binding to cellular structures. In this case a prebleach can remove this slow and immobile fraction and at least the faster fraction can be measured with accuracy. However, when the slow movement or immobilisation is of physiological significance, this should be accessed by FRAP or CP. Another option would be scanning FCS, where the motion of the particle is replaced by the motion of the beam (Digman et al. 2005; Levi et al. 2003).

In the case of cross-correlation experiments, an artificial rise in the cross-correlation amplitude will occur if both molecules are simultaneously photobleached. In the case of bulk bleaching this is easily observed by a decay of the count race trace in both channels. Again, scanning FCS, where the position-dependent signal is converted into a time-dependent signal, can help asses the molecules that are immobile on an FCS time scale.

7.4.5.2 Differences in Molecular Brightness

In a binding reaction where a small labelled ligand binds to a larger receptor, the fluorophore might be quenched or dequenched depending on its microenvironment. The contribution of a molecular species to a correlation function is weighted with the square of the molecular brightness, for example a fourfold brighter ligand molecule will have 16 times more weight than the dimmer bound form. In this case, the fit would not yield the true fractional contribution, which can be calculated when the molecular brightness of each species (or at least their ratio) is known from independent experiments such as a PCH, FCS, or bulk spectroscopy (Weidemann et al. 2002; Sect. 7.6.2). In the worst case, the dimmer part is fully lost.

In the presence of larger aggregates of fluorescent molecules, singular bright events will dominate the correlation function. Luckily, they are easily spotted as huge peaks in the count rate trace. If they only occur rarely it is advisable to take several shorter measurements and exclude the affected ones from the average curve. The alternative would be postacquisition processing, were the photon counting trace is reanalysed after cutting out the contribution of the aggregates. This requires storage of the raw data.

7.4.5.3 Cell Movements and Hardware Drift

As for FRAP experiments, for intracellular FCS the movement of cells and drift of hardware during measurement are serious problems. Therefore, it is always advisable to acquire an image of the same area under identical conditions before and after a correlation measurement. In the case where the cell has changed its position and/or shape significantly, one should disregard the data because a retroactive correction is not possible.

7.5 How To Perform a CP Experiment

7.5.1 The Principle of CP

In a CP experiment, the focus of a confocal system, e.g. an FCS setup, is fixed at a desired position in a cell and the decrease of fluorescence in the so-defined observation volume is recorded under continuous illumination. Gradually, a dynamic equilibrium between association to and dissociation from immobile structures, diffusion/transport, and photobleaching is established. This is represented by a characteristic decay of the fluorescence signal, which often shows biphasic behaviour: a fast initial decay mainly from focal bleaching of an immobilised fraction and a slow asymptotic decay from focal as well as off-focus bleaching in the illumination cone (Fig. 7.4) of the whole pool of freely mobile molecules exchanging with the bound fraction. Already qualitatively, one can distinguish the case of fully diffusive from fully immobilised molecules or a mixture (Fig. 7.7) from the curve. A quantitative analysis yields properties of the binding interaction. However, very often, it is advisable or necessary to combine CP with other independent experiments such as FRAP or FCS in order to obtain more reliable values.

CP studies are based on imaging (FLIP; Cole et al. 1996; White and Stelzer 1999), line-scanning (Cutts et al. 1995; Kubitscheck et al. 1996; Wedekind et al. 1996), or more general geometries (Schulten 1986). Here we will focus on point CP (Brünger et al. 1985; Peters et al. 1981; Wachsmuth et al. 2003) where only one position is addressed so that spatially differentiated measurements must be carried out sequentially. On the other hand, the bleaching and observation volume is sufficiently well defined so that quantitative analysis is feasible. Moreover, the focus is so small and hence diffusion so fast that the conditions for a pseudo-first-order approximation of the immobilisation reaction (Sect. 7.2.3) are met in most cases when the bleaching and the dissociation rate are of the same order of magnitude.

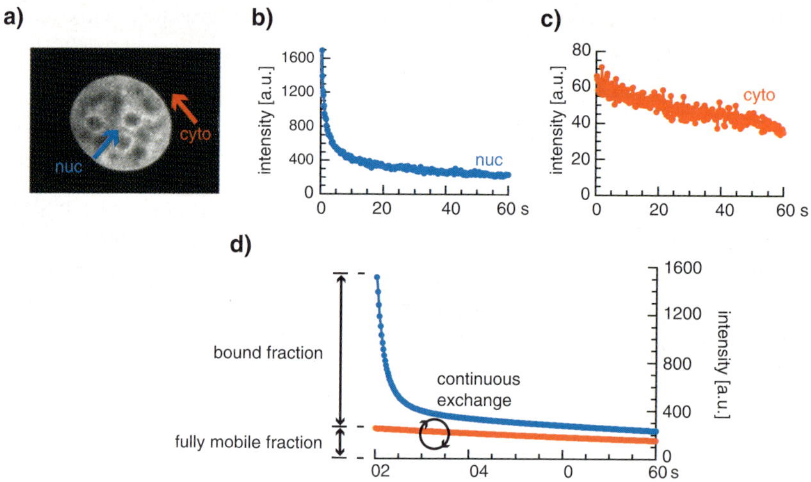

Fig. 7.7. A CP experiment. **a** Intensity image of a cell with a fluorescently labelled nuclear protein and a nuclear (*nuc*, *blue*) and a cytoplasmic (*cyto*, *red*) measurement point. **b** Decaying intensity taken at position *nuc*. **c** Decaying intensity taken at position *cyto*. **d** Schematic representation of the intensity time course in a typical CP experiment with the contribution from a bound and a mobile fraction that can be obtained quantitatively

Since CP can easily be carried out with an FCS system, the theoretical time resolution is as good. In practice, for a quantitative approach one assumes that diffusion of the free fraction through the focus is faster than the photobleaching rate in the centre of the focus (around 0.1–10 s^{-1}) or the association and dissociation rates to/from immobilised structures. Even for slow membrane-based receptors diffusion across the size of an FCS focus is below a few tens of milliseconds, which defines a lower limit for the usable time resolution and an upper limit for the dissociation rate. Thus, CP can assess residence times well below 1 s, which is significantly less than imaging FRAP can cover (Peters 1983). At the other end, measurement time is not limited but should not be much shorter than the residence time of the molecules at their binding sites (frequently above 1 s). When dissociation is significantly faster than bleaching and binding-related and diffusional residence times in the focus become comparable, one can consider the molecules as one mobile fraction with an apparently reduced diffusion coefficient. Especially when an FCS system is used the accessible concentration range is also comparable with that for FCS conditions, i.e. above a few nM.

7.5.2 Choosing and Optimising the Experimental Parameters

Since it is recommended to use an FCS setup for CP, the preparatory steps for cover-glass correction as well as for NA and pinhole size are the same as for FCS. Comparably detailed knowledge of the focal geometry is not necessary as long as the MDE can be approximated as a three-dimensional Gaussian function. In combined FCS and CLSM systems, the position calibration should be carried out in the same way as described before.

7.5.2.1 Timing

It is very important for quantitative analysis of CP data that the start time of illumination is known exactly with respect to the start time of data acquisition. Ideally, they should coincide. Even for commercial systems, this is not always precisely defined but should be determined beforehand.

7.5.2.2 Laser Power

The models applied to describe CP data are based on the assumption that the bleaching rate is smaller than the diffusion rate and has a certain ratio to the dissociation rate (Sects. 7.5.3, 7.6.3). By changing the laser intensity appropriately, this can be fulfilled and a test experiment including evaluation can help to find suitable settings. As an example, at around 10 kW cm^{-2} the bleaching rate for GFP is expected to be of the order of 1 s^{-1} (Harms et al. 2001; Wachsmuth et al. 2003).

7.5.2.3 Background Signal

As mentioned in the Sect. 7.4.2.5 for FCS, especially in cells autofluorescence and scattered light can form some background signal. It is best assessed quantitatively under the same conditions as the actual experiment in cells not containing the labelled molecules of interest. A nonbleaching contribution can be considered as a constant offset that should be subtracted from the CP curve before further analysis, while bleaching autofluorescence should be subtracted as a time-dependent signal. As described before, FCS can also be used to identify a background signal.

7.5.3 *Quantitative Evaluation*

After background correction and determination of the starting point of illumination, the data can be fitted with a model function (Sect. 7.6.3). As a starting point, one should fit an exponential decay to the slowly decreasing tail and subsequently keep the resulting time constant fixed. The amplitude is proportional to the sum of the concentration of free molecules and a contribution of bound molecules, which depends on the dissociation rate from the binding sites. Extending the fit to the whole curve and to a CP model equation gives the photobleaching rate, the dissociation rate, and the relative free and bound fractions. Since the last of these fitting parameters can show some dependence on the other ones, it is advisable to obtain it independently, for example in FRAP experiments.

There are two approximations for CP model equations depending on the ratio of dissociation and bleaching rates (Sect. 7.6.3). After fitting with one of them, one can tell from the ratio obtained whether to switch to the other one. As for fitting of FRAP and FCS data, a different model should be applied and inappropriate ones ruled out. On the other hand, the CP model equations can easily be extended to more types of binding/immobilisation sites that differ in association and dissociation rates.

7.5.4 Controls and Potential Artefacts

7.5.4.1 Postbleach Image, Cell Movements, and Hardware Drift

As for FRAP and FCS experiments, the movement of cells and drift of hardware during the measurement are a serious problem with intracellular CP. Therefore, it is always advisable to acquire an image of the same area under identical conditions before and after a bleaching measurement. In cases where the cell has changed its position and/or shape significantly, one should disregard the data because retroactive correction is not possible. If this can be ruled out, the postbleach fluorescence distribution can help with choosing the right model.

7.5.4.2 Reference Experiments

As for FRAP, it is helpful to carry out the same CP experiments on fixed cells containing the molecules of interest or cells expressing fluorescent fusion proteins that are known to be (quasi) permanently bound/immobilised as well as on cells or areas containing fully mobile molecules (Fig. 7.7) in order to assess the limits of permanently immobilised and fully mobile molecules. In addition, quantitative results such as the immobilised fraction or the dissociation rate from CP and FRAP on the same cell should correspond to each other.

7.5.4.3 Reversible Photobleaching and Phototoxicity

Since in a CP experiment fluorophores are illuminated with a constant intensity, reversible photobleaching is expected to result in only a reduced effective bleach rate when the lifetime of the nonfluorescent state is not too long. Moreover, for CP the bleaching intensity is much smaller than for FRAP, so all photoinduced processes, including reversible photobleaching, occur less frequently (see Sect. 7.2.4 and the corresponding sections for FRAP).

Just as during a FRAP experiment, a significant amount of energy is deposited in a cell – except with a smaller intensity and more localised. See Sect. 7.3.4.5 for more details about potential cytotoxic effects and how to avoid them.

7.6 Quantitative Treatment

7.6.1 *Fluorescence Recovery After Photobleaching*

7.6.1.1 Correction and Normalisation Steps for FRAP Intensity Evaluation

The time course of intensities averaged over the bleaching ROI, the background, reference, and total fluorescence regions are named $I_{bl}(t)$, $I_{bg}(t)$, $I_{ref}(t)$, and $I_{tot}(t)$, respectively. The time point $t = 0$ corresponds to the acquisition time of the first postbleach image. Several steps must be taken to obtain the fully corrected and normalised signal for further analysis, starting with the background subtraction

$$I_{\mathrm{bl,corr}}(t) = I_{\mathrm{bl}}(t) - I_{\mathrm{bg}}(t), \quad I_{\mathrm{ref,corr}}(t) = I_{\mathrm{ref}}(t) - I_{\mathrm{bg}}(t),$$
$$I_{\mathrm{tot,corr}}(t) = I_{\mathrm{tot}}(t) - I_{\mathrm{bg}}(t). \tag{7.1}$$

Then, two normalisation strategies can be applied. First, the bleaching ROI and reference intensities can be normalised to the intensity from a reference cell, followed by a normalisation to the prebleach values, $\langle \ldots \rangle_{\mathrm{prebleach}}$ representing averaging over the prebleach data points/images:

$$I_{\mathrm{bl,ref}}(t) = \frac{I_{\mathrm{bl,corr}}(t)}{I_{\mathrm{ref,corr}}(t)}, \quad I_{\mathrm{tot,ref}}(t) = \frac{I_{\mathrm{tot,corr}}(t)}{I_{\mathrm{ref,corr}}(t)};$$
$$I_{\mathrm{bl,pre}}(t) = \frac{I_{\mathrm{bl,ref}}(t)}{\langle I_{\mathrm{bl,ref}}(t)\rangle_{\mathrm{prebleach}}}, \quad I_{\mathrm{tot,pre}}(t) = \frac{I_{\mathrm{tot,ref}}(t)}{\langle I_{\mathrm{tot,ref}}(t)\rangle_{\mathrm{prebleach}}}. \tag{7.2}$$

Alternatively, the bleaching ROI intensity can be normalised to the total intensity from the same cell, followed by a normalisation to the prebleach values:

$$I_{\mathrm{bl,tot}}(t) = \frac{I_{\mathrm{bl,corr}}(t)}{I_{\mathrm{tot,corr}}(t)};$$
$$I_{\mathrm{bl,pre}}(t) = \frac{I_{\mathrm{bl,tot}}(t)}{\langle I_{\mathrm{bl,tot}}(t)\rangle_{\mathrm{prebleach}}}. \tag{7.3}$$

7.6.1.2 Evaluation of the Corrected and Normalised In tensity

From the corrected and normalised bleaching ROI intensity, several empirical parameters can easily be obtained: the immobile fraction F_{immo}, corresponding to the degree of incomplete recovery, the fully diffusive fraction and/or degree of incomplete bleaching F_{diff}, and the transiently bound or slowly mobile fraction F_{trans} according to

$$F_{immo} = 1 - I_{bl,pre}(\infty), \quad F_{diff} = I_{bl,pre}(0), \quad F_{trans} = 1 - F_{immo} - F_{diff} \tag{7.4}$$

as well as the half time of recovery $t_{1/2}$ with

$$I_{bl,pre}(t_{1/2}) = \frac{I_{bl,pre}(0) + I_{bl,pre}(\infty)}{2}. \tag{7.5}$$

Another way to determine these basic parameters would be a fit of an exponential recovery according to

$$I_{bl,pre}(t) = I_{bl,pre}(0) + \left[I_{bl,pre}(\infty) - I_{bl,pre}(0)\right]\left[1 - \exp\left(-\frac{t \ln 2}{t_{1/2}}\right)\right]. \tag{7.6}$$

In the binding-dominant case, the diffusional contribution is equilibrated very rapidly and recovery is dominated by the dissociation rate k_{off} from binding sites. The recovery curve can be fitted with

$$I_{bl,pre}(t) = I_{bl,pre}(0) + \left[I_{bl,pre}(\infty) - I_{bl,pre}(0)\right]\left[1 - \exp(-k_{off} t)\right]. \tag{7.7}$$

In the diffusion-dominant case, recovery depends crucially on the geometry of the bleaching ROI. Examples are

$$I_{bl,pre}(t) = I_{bl,pre}(0) + \left[I_{bl,pre}(\infty) - I_{bl,pre}(0)\right] \\ \text{x} \exp\left(-\frac{w^2}{2Dt}\right)\left[I_0\left(\frac{w^2}{2Dt}\right) + I_1\left(\frac{w^2}{2Dt}\right)\right] \tag{7.8}$$

in the case of a two-dimensional circular ROI, with w being the diameter and D the effective diffusion coefficient. For a two-dimensional strip ROI of width w with effective one-dimensional diffusion, we have

$$I_{bl,pre}(t) = I_{bl,pre}(0) + \left[I_{bl,pre}(\infty) - I_{bl,pre}(0)\right]\left(1 - \frac{w^2}{w^2 + 4\pi Dt}\right)^{1/2}. \tag{7.9}$$

7.6.1.3 Profile Analysis

When the intensity profile averaged along the desired direction of a two-dimensional strip bleaching ROI, for example, is given as I_{pre} (x) averaged over the prebleach images and I_{post} (x,t) for the postbleach images, the normalised profile is given as

$$I_{norm}(x,t) = I_{post}(x,t)/I_{pre}(x). \tag{7.10}$$

In the diffusion-dominant case, the time course of the intensity depends on the geometry. However, it is sufficient to determine the effective width w, e.g. by fitting a Gaussian function to the profile. Plotting the squared width versus time, w^2 (t), describes the diffusion process on the space and time scales considered.

7.6.2 *Fluorescence Correlation Spectroscopy*

7.6.2.1 Definition of the ACF

The ACF is defined as

$$\begin{aligned} G^{\delta I}(\tau) &= \frac{\langle \delta I(t)\cdot \delta I(t+\tau)\rangle}{\langle I(t)\rangle^2} = \frac{\frac{1}{T}\int_0^T [\delta I(t)\cdot \delta I(t+\tau)]dt}{\left[\frac{1}{T}\int_0^T I(t)dt\right]^2} = \frac{T\int_0^T [\delta I(t)\cdot \delta I(t+\tau)]dt}{\left[\int_0^T I(t)dt\right]^2}, \\ G^{I}(\tau) &= \frac{\langle I(t)\cdot I(t+\tau)\rangle}{\langle I(t)\rangle^2} = \frac{\frac{1}{T}\int_0^T [I(t)\cdot I(t+\tau)]dt}{\left[\frac{1}{T}\int_0^T I(t)dt\right]^2} = \frac{T\int_0^T [I(t)\cdot I(t+\tau)]dt}{\left[\int_0^T I(t)dt\right]^2}, \end{aligned} \tag{7.11}$$

where <...> denotes the time average and $\delta I(t) = I(t) - < I(t) >$ describes the fluctuations around the mean intensity. For a long time average of I without bleaching, the following relation exists:

$$G^{I}(\tau) = 1 + G^{\delta I}(\tau). \tag{7.12}$$

7.6.2.2 Definition of the CCF

The formalism for the CCF is identical to that for the ACF, with the exception that the signal in one channel is not compared with itself, but with a signal in a second channel. If one assigns the indices r and b to the red and the blue channel, respectively, then the CCF reads

$$G_X^{\delta I}(\tau) = \frac{\langle \delta I_\mathrm{b}(t) \cdot \delta I_\mathrm{r}(t+\tau) \rangle}{\langle I_\mathrm{b}(t) \rangle \cdot \langle I_\mathrm{r}(t) \rangle} = \frac{\langle \delta I_\mathrm{r}(t) \cdot \delta I_\mathrm{b}(t+\tau) \rangle}{\langle I_\mathrm{b}(t) \rangle \cdot \langle I_\mathrm{r}(t) \rangle},$$
$$G_X^{I}(\tau) = \frac{\langle I_\mathrm{b}(t) \cdot I_\mathrm{r}(t+\tau) \rangle}{\langle I_\mathrm{b}(t) \rangle \cdot \langle I_\mathrm{r}(t) \rangle} = \frac{\langle I_\mathrm{r}(t) \cdot I_\mathrm{b}(t+\tau) \rangle}{\langle I_\mathrm{b}(t) \rangle \cdot \langle I_\mathrm{r}(t) \rangle}. \tag{7.13}$$

7.6.2.3 General Expression for the ACF

$$G_\mathrm{tot}^{I}(\tau) = 1 + C + B^2 \cdot A \cdot G_k(\tau), \tag{7.14}$$

with offset C; with background correction $B^2 = (1 - I_\mathrm{bg}/I_\mathrm{tot})^2$, where I_bg and I_tot are the background and the total intensity, respectively; with the amplitude $A = \gamma / N$, where γ and N are the geometric factor and the number of molecules, respectively – the geometric factor has different values for different experimental situations: $\gamma_\mathrm{C} = 1$ (cylindrical), $\gamma_\mathrm{2DG} = 0.35$ (two-dimensional Gaussian), $\gamma_\mathrm{3DG} = 0.75$ (three-dimensional Gaussian), $\gamma_\mathrm{GL} = 0.035$ (Gaussian–Lorentzian); and with a term $G_k(\tau)$ for the correlated processes, for example $G_k(\tau) = \sum_{i=1}^{n} \Phi_i \, g_{\mathrm{diff},i}(\tau)$, which is the weighted sum of n different components, the number of which is usually restricted to 3.

7.6.2.4 Analytical Expressions

The term for translational or lateral diffusion is defined as

$$\sum_{i=1}^{n} \Phi_i g_{\mathrm{diff},i}(\tau) = \sum_{i=1}^{n} \Phi_i \left[1 + \left(\frac{\tau}{\tau_{\mathrm{d},i}} \right)^{\alpha_i} \right]^{-e_{\mathrm{d}1}} \left[1 + \left(\frac{\tau}{\tau_{\mathrm{d},i}} \right)^{\alpha_i} \frac{1}{S^2} \right]^{-e_{\mathrm{d}2}}. \tag{7.15}$$

The contribution of each component is defined by the fractional intensity $\Phi_\mathrm{i} = f_i \eta_i^2 / \sum_{i=1}^{n} f_i \eta_i^2$, with f_i and η_i being the fraction and the relative molecular brightness of each component, respectively, and with the constraint $\sum_{i=1}^{n} \Phi_i = 1$. For equal brightness, i.e. $\eta_1 = \eta_2 = \eta_3$, the fractional intensity will become the fraction $\Phi_i = f_i$. The observed or apparent brightness η is determined by the contribution of each brightness,

$$\eta = \frac{\sum_{i=1}^{n} f_i \eta_i^2}{\sum_{i=1}^{n} f_i \eta_i}. \tag{7.16}$$

The fraction is defined as $f_i = N_i / N = N_i / \sum_{i=1}^{n} N_i$, with N_i and N being the number of molecules of the ith species and the total number of molecules and with the constraint $\sum_{i=1}^{n} f_i = 1$. The actual number of diffusing molecules N_{diff} is derived from the fitted number N with the relation

$$N_{\mathrm{diff}} = N \frac{\left(\sum_{i=1}^{n} f_i \eta_i\right)^2}{\sum_{i=1}^{n} f_i \eta_i^2}. \tag{7.17}$$

The dimensionality of diffusion determines the fixed exponents: $e_{\mathrm{d1}} = 1/2$, $e_{\mathrm{d2}} = 0$ in one dimension; $e_{\mathrm{d1}} = 1$, $e_{\mathrm{d2}} = 0$ in two dimensions; $e_{\mathrm{d1}} = 1$, $e_{\mathrm{d2}} = 1/2$ in three dimensions. The fractionality of diffusion determines the anomaly parameter or temporal coefficient: $\alpha < 1$ for anomalous subdiffusion; $\alpha = 1$ for free diffusion; $\alpha > 1$ for superdiffusion.

The shape of the confocal volume is defined by the structure parameter $s = w_z / w_r$, with w_z and w_r being the axial and lateral $1/e^2$ radius of the focus, respectively. The number of molecules can be converted to a concentration by $c = N/V\,N_A$, with $V = \pi^{3/2} w_r^3\, s$ being the confocal volume and $N_A = 6.023 \times 10^{23}\ \mathrm{mol}^{-1}$ the Avogadro number.

The average dwell time of the molecule in the confocal volume is defined by $\tau_D = w_r^2/4D$ for free diffusion, with D being the diffusion coefficient. In the case of two-photon excitation, it is $\tau_D = w_r^2/8D$. For anomalous diffusion with the transport coefficient Γ, it is defined by $\tau_D^{\alpha} = w_r^2/\Gamma$ and $\tau_D^{\alpha} = w_r^2 / 2\,\Gamma$ for one- and two-photon excitation, respectively.

7.6.2.5 Evaluation of Cross-Correlation Data

The analytical expression for cross-correlations is formally identical to the ones for autocorrelations. The confocal cross-correlated volume and diffusion time are defined by

$$V_{\mathrm{rb}} = (\pi/2)^{3/2}\left(w_{\mathrm{r,r}}^2 + w_{\mathrm{r,b}}^2\right)\left(w_{z,\mathrm{r}}^2 + w_{z,\mathrm{b}}^2\right)^{1/2}$$

and

$$\tau_{\mathrm{D,rb}} = \left(\tau_{\mathrm{D,r}} + \tau_{\mathrm{D,b}}\right)/2 = \left(w_{\mathrm{r,r}}^2 + w_{\mathrm{r,b}}^2\right)/8D\,.$$

In general, one has a mixture of free blue, free red, and bound blue-red molecules. From the ratio of cross-correlation to autocorrelation amplitudes, which corresponds to a dynamic correlation coefficient, the fraction of bound and free blue and red molecules can be obtained, depending more or less intricately on the stoichiometry of the binding reaction:

$$\mathrm{ratioG} = \frac{G_X^{\delta I}(0)}{\left[G_{\mathrm{b}}^{\delta I}(0) G_{\mathrm{r}}^{\delta I}(0)\right]^{1/2}} = F\left(N_{\mathrm{b}}, N_{\mathrm{r}}, N_{\mathrm{br}}\right). \tag{7.18}$$

7.6.3 Continuous Fluorescence Photobleaching

First, the time-dependent intensity $I(t)$ from the measurement spot is corrected by subtracting the background intensity $I_{bg}(t)$, resulting in $I_{corr}(t) = I(t) - I_{bg}(t)$.

In order to find an analytical model equation, the differential equations of the concentration of a free and a bound fraction must be solved taking into consideration the photobleaching with a rate α (the reciprocal mean time of illumination until a fluorophore is bleached), the dissociation rate k_{off} from the binding/immobilisation sites, the diffusion, which is assumed to be much faster than the other processes, and the illumination profile, which is approximated as a Gaussian function. Also using a Gaussian profile for detection, one obtains for the diffusive/mobile fraction

$$I_{diff}(t) = I_{diff}(0)\exp(-\beta t), \tag{7.19}$$

with β characterising the depletion of diffusive molecules due to the continuous bleaching of a constant fraction of the entire pool in the focus as well as off focus. For the bound fraction exchanging with the diffusive molecules, we have

$$I_{bound}(t) = I_{bound}(0) \left(\left[G(\alpha t) - Q\left(\frac{k_{off}}{\alpha}\right)H(\alpha t)\right]\exp(-k_{off}t) + Q\left(\frac{k_{off}}{\alpha}\right)\exp(-\beta t)\right), \tag{7.20}$$

with the dimensionless functions

$$G(\alpha t) = \left(1 + \frac{\alpha t}{2} + \frac{\alpha^2 t^2}{6}\right)^{-1},$$

$$H(\alpha t) = \begin{cases} \left(1 + \dfrac{3\alpha t}{7}\right)^{-1} & \dfrac{k_{off}}{\alpha} \le 0.5 \\ \left(1 + \dfrac{\alpha t}{2} + \dfrac{2\alpha^2 t^2}{15}\right)^{-1} & \dfrac{k_{off}}{\alpha} > 0.5 \end{cases},$$

$$Q\left(\frac{k_{off}}{\alpha}\right) = \begin{cases} \dfrac{12 k_{off}/\alpha}{5 + 14 k_{off}/\alpha} & \dfrac{k_{off}}{\alpha} \le 0.5 \\ \dfrac{2 k_{off}/\alpha}{1 + 2 k_{off}/\alpha} & \dfrac{k_{off}}{\alpha} > 0.5 \end{cases}. \tag{7.21}$$

The different approximations are valid for small and medium dissociation rates, respectively. For large dissociation rates, one obtains pure diffusion with a reduced apparent diffusion coefficient.

From the fit of the sum $I_{diff}(t)+I_{bound}(t)$ of the diffusive and the bound contribution to $I_{corr}(t)$, the diffusive and the bound fraction as well as k_{off} and α are determined, a fully immobilised fraction being the special case where $k_{off} = 0$:

$$F_{diff} = \frac{I_{diff}(0)}{I_{diff}(0)+I_{bound}(0)}, \quad F_{bound} = \frac{I_{bound}(0)}{I_{diff}(0)+I_{bound}(0)}. \tag{7.22}$$

7.7 Conclusion

Confocal fluorescence microscopy and spectroscopy have experienced dynamic developments over the last few years. In combination with genetic methods that allow fluorescent labelling in vivo, many new noninvasive techniques have emerged to measure the dynamics of biological processes in living cells on a molecular level, providing good spatial and temporal resolution and high biological specificity. Among them, FRAP and other photobleaching methods, as well as FCS and other fluctuation spectroscopy methods are becoming more and more popular because they can be performed easily on commercial systems. Combining these techniques with each other and with rigorous quantitative analyses, e.g. based on numerical modelling of cellular processes, can cover a broad dynamic range and yield quantitative information about molecular diffusion and interactions, structural dynamics, or intracellular/intercellular topology. Nevertheless, the increasing demand for quantitative results in the field of systems biology requires good reproducibility and the development of standards.

References

Axelrod D, Koppel DE, Schlessinger J, Elson E, Webb WW (1976) Mobility measurement by analysis of fluorescence photobleaching recovery kinetics. Biophys J 16:1055–1069

Bacia K, Schwille P (2003) A dynamic view of cellular processes by in vivo fluorescence auto- and cross-correlation spectroscopy. Methods 29:74–85

Berk DA, Yuan F, Leunig M, Jain RK (1993) Fluorescence photobleaching with spatial Fourier analysis: measurement of diffusion in light-scattering media. Biophys J 65:2428–2436

Braga J, Desterro JM, Carmo-Fonseca M (2004) Intracellular macromolecular mobility measured by fluorescence recovery after photobleaching with confocal laser scanning microscopes. Mol Biol Cell 15:4749–4760

Briddon SJ, Middleton RJ, Cordeaux Y, Flavin FM, Weinstein JA, George MW, Kellam B, Hill SJ (2004) Quantitative analysis of the formation and diffusion of A1-adenosine receptor-antagonist complexes in single living cells. Proc Natl Acad Sci USA 101:4673–4678

Brock R, Hink MA, Jovin TM (1998) Fluorescence correlation microscopy of cells in the presence of autofluorescence. Biophys J 75:2547–2557

Brown EB, Wu ES, Zipfel W, Webb WW (1999) Measurement of molecular diffusion in solution by multiphoton fluorescence photobleaching recovery. Biophys J 77:2837–2849

Brünger A, Schulten K, Peters R (1985) Continuous fluorescence microphotolysis to observe lateral diffusion in membranes. Theoretical methods and applications. J Chem Phys 82:2147–2160

Calapez A, Pereira HM, Calado A, Braga J, Rino J, Carvalho C, Tavanez JP, Wahle E, Rosa AC, Carmo-Fonseca M (2002) The intranuclear mobility of messenger RNA binding proteins is ATP dependent and temperature sensitive. J Cell Biol 159:795–805

Campbell RE, Tour O, Palmer AE, Steinbach PA, Baird GS, Zacharias DA, Tsien RY (2002) A monomeric red fluorescent protein. Proc Natl Acad Sci USA 99:7877–7882

Carmo-Fonseca M, Platani M, Swedlow JR (2002) Macromolecular mobility inside the cell nucleus. Trends Cell Biol 12:491–495

Carrero G, McDonald D, Crawford E, de Vries G, Hendzel MJ (2003) Using FRAP and mathematical modeling to determine the in vivo kinetics of nuclear proteins. Methods 29:14–28

Carrero G, Crawford E, Th'ng J, de Vries G, Hendzel MJ (2004) Quantification of protein-protein and protein-DNA interactions in vivo, using fluorescence recovery after photobleaching. Methods Enzymol 375:415–442

Chen Y, Muller JD, So PT, Gratton E (1999) The photon counting histogram in fluorescence fluctuation spectroscopy. Biophys J 77:553–567

Cole NB, Smith CL, Sciaky N, Terasaki M, Edidin M, Lippincott-Schwartz J (1996) Diffusional mobility of Golgi proteins in membranes of living cells. Science 273:797–801

Cutts LS, Roberts PA, Adler J, Davies MC, Melia CD (1995) Determination of localized diffusion coefficients in gels using confocal scanning laser microscopy. J Microsc 180:131–139

Digman MA, Brown CM, Sengupta P, Wiseman PW, Horwitz AR, Gratton E (2005) Measuring fast dynamics in solutions and cells with a laser scanning microscope. Biophys J 89:1317–1327

Doose S, Tsay JM, Pinaud F, Weiss S (2005) Comparison of photophysical and colloidal properties of biocompatible semiconductor nanocrystals using fluorescence correlation spectroscopy. Anal Chem 77:2235–2242

Dundr M, Hoffmann-Rohrer U, Hu Q, Grummt I, Rothblum LI, Phair RD, Misteli T (2002) A kinetic framework for a mammalian RNA polymerase in vivo. Science 298:1623–1626

Edidin M, Zagyansky Y, Lardner TJ (1976) Measurement of membrane protein lateral diffusion in single cells. Science 191:466–468

Ellenberg J, Siggia ED, Moreira JE, Smith CL, Presley JF, Worman HJ, Lippincott-Schwartz J (1997) Nuclear membrane dynamics and reassembly in living cells: targeting of an inner nuclear membrane protein in interphase and mitosis. J Cell Biol 138:1193–1206

Elsner M, Hashimoto H, Simpson JC, Cassel D, Nilsson T, Weiss M (2003) Spatiotemporal dynamics of the COPI vesicle machinery. EMBO Rep 4:1000–1004

Elson EL (2004) Quick tour of fluorescence correlation spectroscopy from its inception. J Biomed Opt 9:857–864

Elson EL, Magde D (1974) Fluorescence correlation spectroscopy. I. Conceptual basis and theory. Biopolymers 13:1–27

Fradin C, Abu-Arish A, Granek R, Elbaum M (2003) Fluorescence correlation spectroscopy close to a fluctuating membrane. Biophys J 84:2005–2020

Gennerich A, Schild D (2000) Fluorescence correlation spectroscopy in small cytosolic compartments depends critically on the diffusion model used. Biophys J 79:3294–3306

Görisch SM, Wachsmuth M, Ittrich C, Bacher CP, Rippe K, Lichter P (2004) Nuclear body movement is determined by chromatin accessibility and dynamics. Proc Natl Acad Sci USA 101:13221–13226

Gösch M, Rigler R (2005) Fluorescence correlation spectroscopy of molecular motions and kinetics. Adv Drug Deliv Rev 57:169–190

Griffin BA, Adams SR, Jones J, Tsien RY (2000) Fluorescent labeling of recombinant proteins in living cells with FlAsH. Methods Enzymol 327:565–578

Grünwald D, Cardoso MC, Leonhardt H, Buschmann V (2005) Diffusion and binding properties investigated by fluorescence correlation spectroscopy (FCS). Curr Pharm Biotechnol 6:381–386

Harms GS, Cognet L, Lommerse PH, Blab GA, Schmidt T (2001) Autofluorescent proteins in single-molecule research: applications to live cell imaging microscopy. Biophys J 80:2396–2408.

Haupts U, Maiti S, Schwille P, Webb WW (1998) Dynamics of fluorescence fluctuations in green fluorescent protein observed by fluorescence correlation spectroscopy. Proc Natl Acad Sci USA 95:13573–13578

Haustein E, Schwille P (2003) Ultrasensitive investigations of biological systems by fluorescence correlation spectroscopy. Methods 29:153–166

Hecht B (2004) Nano-optics with single quantum systems. Philos Trans R Soc Lond Ser A 362:881–899

Helmchen F, Denk W (2005) Deep tissue two-photon microscopy. Nat Methods 2:932–940

Hess ST, Huang S, Heikal AA, Webb WW (2002) Biological and chemical applications of fluorescence correlation spectroscopy: a review. Biochemistry 41:697–705

Hink MA, Bisselin T, Visser AJ (2002) Imaging protein-protein interactions in living cells. Plant Mol Biol 50:871–883

Hopt A, Neher E (2001) Highly nonlinear photodamage in two-photon fluorescence microscopy. Biophys J 80:2029–2036

Houtsmuller AB, Vermeulen W (2001) Macromolecular dynamics in living cell nuclei revealed by fluorescence redistribution after photobleaching. Histochem Cell Biol 115:13–21

Houtsmuller AB, Rademakers S, Nigg AL, Hoogstraten D, Hoeijmakers JH, Vermeulen W (1999) Action of DNA repair endonuclease ERCC1/XPF in living cells. Science 284:958–961

Hwang LC, Wohland T (2005) Single wavelength excitation fluorescence cross-correlation spectroscopy with spectrally similar fluorophores: resolution for binding studies. J Chem Phys 122:114708

Jahnz M, Schwille P (2004) Enzyme assays for confocal single molecule spectroscopy. Curr Pharm Biotechnol 5:221–229

Jankevics H, Prummer M, Izewska P, Pick H, Leufgen K, Vogel H (2005) Diffusion-time distribution analysis reveals characteristic ligand-dependent interaction patterns of nuclear receptors in living cells. Biochemistry 44:11676–11683

Kask P, Palo K, Ullmann D, Gall K (1999) Fluorescence-intensity distribution analysis and its application in biomolecular detection technology. Proc Natl Acad Sci USA 96:13756–13761

Kimura H, Cook PR (2001) Kinetics of core histones in living human cells: little exchange of H3 and H4 and some rapid exchange of H2B. J Cell Biol 153:1341–1353

Kimura H, Hieda M, Cook PR (2004) Measuring histone and polymerase dynamics in living cells. Methods Enzymol 375:381–393

Klonis N, Rug M, Harper I, Wickham M, Cowman A, Tilley L (2002) Fluorescence photobleaching analysis for the study of cellular dynamics. Eur Biophys J 31:36–51

Kohl T, Schwille P (2005) Fluorescence correlation spectroscopy with autofluorescent proteins. Adv Biochem Eng Biotechnol 95:107–142

Kubitscheck U, Wedekind P, Peters R (1994) Lateral diffusion measurement at high spatial resolution by scanning microphotolysis in a confocal microscope. Biophys J 67:948–956

Kubitscheck U, Heinrich O, Peters R (1996) Continuous scanning micro-photolysis: a simple laser scanning microscopic method for lateral transport measurements employing single- or two-photon excitation. Bioimaging 4:158–167

Lakowicz JR (1999) Principles of fluorescence spectroscopy. Kluwer/Plenum, New York

Levi V, Ruan Q, Kis-Petikova K, Gratton E (2003) Scanning FCS, a novel method for three-dimensional particle tracking. Biochem Soc Trans 31:997–1000

Lippincott-Schwartz J, Snapp E, Kenworthy A (2001) Studying protein dynamics in living cells. Nat Rev Mol Cell Biol 2:444–456

Lumma D, Keller S, Vilgis T, Radler JO (2003) Dynamics of large semiflexible chains probed by fluorescence correlation spectroscopy. Phys Rev Lett 90:218301

Magde D, Elson EL, Webb WW (1972) Thermodynamic fluctuations in a reacting system – measurement by fluorescence correlation spectroscopy. Phys Rev Lett 29:705–708

Magde D, Elson EL, Webb WW (1974) Fluorescence correlation spectroscopy. II. An experimental realization. Biopolymers 13:29–61

McGrath JL, Tardy Y, Dewey CF Jr, Meister JJ, Hartwig JH (1998) Simultaneous measurements of actin filament turnover, filament fraction, and monomer diffusion in endothelial cells. Biophys J 75:2070–2078

Medina MA, Schwille P (2002) Fluorescence correlation spectroscopy for the detection and study of single molecules in biology. BioEssays 24:758–764

Medintz IL, Uyeda HT, Goldman ER, Mattoussi H (2005) Quantum dot bioconjugates for imaging, labelling and sensing. Nat Mater 4:435–446

Meseth U, Wohland T, Rigler R, Vogel H (1999) Resolution of fluorescence correlation measurements. Biophys J 76:1619–1631

Michalet X, Pinaud FF, Bentolila LA, Tsay JM, Doose S, Li JJ, Sundaresan G, Wu AM, Gambhir SS, Weiss S (2005) Quantum dots for live cells, in vivo imaging, and diagnostics. Science 307:538–544

Misteli T (2001) Protein dynamics: Implications for nuclear architecture and gene expression. Science 291:843–847

Modos K, Galantai R, Bardos-Nagy I, Wachsmuth M, Toth K, Fidy J, Langowski J (2004) Maximum-entropy decomposition of fluorescence correlation spectroscopy data: application to liposome-human serum albumin association. Eur Biophys J 33:59–67

Muller JD, Gratton E (2003) High-pressure fluorescence correlation spectroscopy. Biophys J 85:2711–2719

Nirmal M, Dabbousi BO, Bawendi MG, Macklin JJ, Trautman JK, Harris TD, Brus LE (1996) Fluorescence intermittency in single cadmium selenide nanocrystals. Nature 383:802–804

Oancea E, Teruel MN, Quest AF, Meyer T (1998) Green fluorescent protein (GFP)-tagged cysteine-rich domains from protein kinase C as fluorescent indicators for diacylglycerol signaling in living cells. J Cell Biol 140:485–498

Patterson GH, Knobel SM, Sharif WD, Kain SR, Piston DW (1997) Use of the green fluorescent protein and its mutants in quantitative fluorescence microscopy. Biophys J 73:2782–2790

Patterson GH, Lippincott-Schwartz J (2002) A photoactivatable GFP for selective photolabeling of proteins and cells. Science 297:1873–1877

Pawley JB (ed) (1995) Handbook of biological confocal microscopy. Plenum, New York

Periasamy N, Verkman AS (1998) Analysis of fluorophore diffusion by continuous distributions of diffusion coefficients: application to photobleaching measurements of multicomponent and anomalous diffusion. Biophys J 75:557–567

Periasamy N, Bicknese S, Verkman AS (1996) Reversible photobleaching of fluorescein conjugates in air-saturated viscous solutions: singlet and triplet state quenching by tryptophan. Photochem Photobiol 63:265–271

Peters R (1983) Fluorescence microphotolysis. diffusion measurements in single cells. Naturwissenschaften 70:294–302.

Peters R, Kubitscheck U (1999) Scanning microphotolysis: Three-dimensional diffusion measurement and optical single-transporter recording. Methods 18:508–517

Peters R, Peters J, Tews KH, Bahr W (1974) A microfluorimetric study of translational diffusion in erythrocyte membranes. Biochim Biophys Acta 367:282–294.

Peters R, Brünger A, Schulten K (1981) Continuous fluorescence microphotolysis: a sensitive method for study of diffusion processes in single cells. Proc Natl Acad Sci USA 78:962–966

Phair RD, Misteli T (2001) Kinetic modelling approaches to in vivo imaging. Nat Rev Mol Cell Biol 2:898–907

Phair RD, Gorski SA, Misteli T (2004a) Measurement of dynamic protein binding to chromatin in vivo, using photobleaching microscopy. Methods Enzymol 375:393–414

Phair RD, Scaffidi P, Elbi C, Vecerova J, Dey A, Ozato K, Brown DT, Hager G, Bustin M, Misteli T (2004b) Global nature of dynamic protein-chromatin interactions in vivo: three-dimensional genome scanning and dynamic interaction networks of chromatin proteins. Mol Cell Biol 24:6393–6402

Politz JC (1999) Use of caged fluorochromes to track macromolecular movement in living cells. Trends Cell Biol 9:284–287

Potma EO, Boeij WP, Bosgraaf L, Roelofs J, van Haastert PJ, Wiersma DA (2001) Reduced protein diffusion rate by cytoskeleton in vegetative and polarized dictyostelium cells. Biophys J 81.2010–2019

Rabut G, Ellenberg J (2005) Photobleaching techniques to study mobility and molecular dynamics of proteins in live cells: FRAP, iFRAP, and FLIP. In: Spector D, Goldman D (eds) Live cell imaging: a laboratory manual. Cold Spring Harbor Laboratory Press, Cold Spring Harbor, pp 101–126

Reits EA, Neefjes JJ (2001) From fixed to frap: measuring protein mobility and activity in living cells. Nat Cell Biol 3:E145–147

Ricka J, Binkert T (1989) Direct measurement of a distinct correlation function by fluorescence cross correlation. Phys Rev A 39:2646–2652

Rigler R, Elson ES (eds) (2001) Fluorescence correlation spectroscopy: theory and applications. Springer, Berlin

Rigler R, Mets Ü, Widengren J, Kask P (1993) Fluorescence correlation spectroscopy with high count rate and low background: analysis of translational diffusion. Eur Biophys J 22:169-1-75

Rizzo MA, Springer GH, Granada B, Piston DW (2004) An improved cyan fluorescent protein variant useful for FRET. Nat Biotechnol 22:445–449

Sanchez SA, Gratton E (2005) Lipid-protein interactions revealed by two-photon microscopy and fluorescence correlation spectroscopy. Acc Chem Res 38:469–477

Saxton MJ (2001) Anomalous subdiffusion in fluorescence photobleaching recovery: a Monte Carlo study. Biophys J 81:2226–2240

Saxton MJ, Jacobson K (1997) Single-particle tracking: applications to membrane dynamics. Annu Rev Biophys Biomol Struct 26:373–399

Schulten K (1986) Continuous fluorescence microphotolysis by a $\sin^2$ kx grating. Chem Phys Lett 124:230–236

Schwille P (2001) Fluorescence correlation spectroscopy and its potential for intracellular applications. Cell Biochem Biophys 34:383–408

Schwille P, Meyer-Almes FJ, Rigler R (1997) Dual-color fluorescence cross-correlation spectroscopy for multicomponent diffusional analysis in solution. Biophys J 72:1878–1886

Schwille P, Korlach J, Webb WW (1999) Fluorescence correlation spectroscopy with single-molecule sensitivity on cell and model membranes. Cytometry 36:176–182

Schwille P, Kummer S, Heikal AA, Moerner WE, Webb WW (2000) Fluorescence correlation spectroscopy reveals fast optical excitation-driven intramolecular dynamics of yellow fluorescent proteins. Proc Natl Acad Sci USA 97:151–156

Sengupta P, Garai K, Balaji J, Periasamy N, Maiti S (2003) Measuring size distribution in highly heterogeneous systems with fluorescence correlation spectroscopy. Biophys J 84:1977–1984

Shaner NC, Steinbach PA, Tsien RY (2005) A guide to choosing fluorescent proteins. Nat Methods 2:905–909

Siggia ED, Lippincott-Schwartz J, Bekiranov S (2000) Diffusion in inhomogeneous media: theory and simulations applied to whole cell photobleach recovery. Biophys J 79:1761–1770

Skakun VV, Hink MA, Digris AV, Engel R, Novikov EG, Apanasovich VV, Visser AJ (2005) Global analysis of fluorescence fluctuation data. Eur Biophys J 34:323–334

Song L, Hennink EJ, Young IT, Tanke HJ (1995) Photobleaching kinetics of fluorescein in quantitative fluorescence microscopy. Biophys J 68:2588–2600

Song L, Varma CA, Verhoeven JW, Tanke HJ (1996) Influence of the triplet excited state on the photobleaching kinetics of fluorescein in microscopy. Biophys J 70:2959–2968

Song L, van Gijlswijk RP, Young IT, Tanke HJ (1997) Influence of fluorochrome labeling density on the photobleaching kinetics of fluorescein in microscopy. Cytometry 27:213–223

Soumpasis DM (1983) Theoretical analysis of fluorescence photobleaching recovery experiments. Biophys J 41:95–97

Sprague BL, Pego RL, Stavreva DA, McNally JG (2004) Analysis of binding reactions by fluorescence recovery after photobleaching. Biophys J 86:3473–3495

Starr TE, Thompson NL (2002) Fluorescence pattern photobleaching recovery for samples with multi-component diffusion. Biophys Chem 97:29–44

Stavreva DA, McNally JG (2004) Fluorescence recovery after photobleaching (FRAP) methods for visualizing protein dynamics in living mammalian cell nuclei. Methods Enzymol 375:443–455

Stout AL, Axelrod D (1995) Spontaneous recovery of fluorescence by photobleached surface-adsorbed proteins. Photochem Photobiol 62:239–244

Strohner R, Wachsmuth M, Dachauer K, Mazurkiewicz J, Hochstätter J, Rippe K, Längst G (2005) A 'loop recapture' mechanism for ACF-dependent nucleosome remodeling. Nat Struct Mol Biol 12:683–690

Subramanian K, Meyer T (1997) Calcium-induced restructuring of nuclear envelope and endoplasmic reticulum calcium stores. Cell 89:963–971

Surrey T, Elowitz MB, Wolf PE, Yang F, Nedelec F, Shokat K, Leibler S (1998) Chromophore-assisted light inactivation and self-organization of microtubules and motors. Proc Natl Acad Sci USA 95:4293–4298

Svoboda K, Yasuda R (2006) Principles of two-photon excitation microscopy and its applications to neuroscience. Neuron 50:823–839

Swaminathan R, Hoang CP, Verkman AS (1997) Photobleaching recovery and anisotropy decay of green fluorescent protein GFP-S65T in solution and cells: cytoplasmic viscosity probed by green fluorescent protein translational and rotational diffusion. Biophys J 72:1900–1907

Tardy Y, McGrath JL, Hartwig JH, Dewey CF (1995) Interpreting photoactivated fluorescence microscopy measurements of steady-state actin dynamics. Biophys J 69:1674–1682

Thompson NL (1991) Fluorescence correlation spectroscopy. In: Lakowicz JR (ed) Topics in fluorescence spectroscopy, vol 1. Plenum, New York, pp 337–378

Thompson NL, Lieto AM, Allen NW (2002) Recent advances in fluorescence correlation spectroscopy. Curr Opin Struct Biol 12:634–641

Tsien RY (1998) The green fluorescent protein. Annu Rev Biochem 67:509–544

Umenishi F, Verbavatz JM, Verkman AS (2000) cAMP regulated membrane diffusion of a green fluorescent protein-aquaporin 2 chimera. Biophys J 78:1024–1035

Van Keuren E, Schrof W (2003) Fluorescence recovery after two-photon bleaching for the study of dye diffusion in polymer systems. Macromolecules 36:5002–5007

Verkman AS (2002) Solute and macromolecule diffusion in cellular aqueous compartments. Trends Biochem Sci 27:27–33

Verkman AS (2003) Diffusion in cells measured by fluorescence recovery after photobleaching. Methods Enzymol 360:635–648

Vukojevic V, Pramanik A, Yakovleva T, Rigler R, Terenius L, Bakalkin G (2005) Study of molecular events in cells by fluorescence correlation spectroscopy. Cell Mol Life Sci 62:535–550

Wachsmuth M, Waldeck W, Langowski J (2000) Anomalous diffusion of fluorescent probes inside living cell nuclei investigated by spatially-resolved fluorescence correlation spectroscopy. J Mol Biol 298:677–689

Wachsmuth M, Weidemann T, Muller G, Hoffmann-Rohrer UW, Knoch TA, Waldeck W, Langowski J (2003) Analyzing intracellular binding and diffusion with continuous fluorescence photobleaching. Biophys J 84:3353–3363

Wedekind P, Kubitscheck U, Heinrich O, Peters R (1996) Line-scanning microphotolysis for diffraction-limited measurements of lateral diffusion. Biophys J 71:1621–1632

Wedekind P, Kubitscheck U, Peters R (1994) Scanning microphotolysis: a new photobleaching technique based on fast intensity modulation of a scanned laser beam and confocal imaging. J Microsc 176:23–33

Weidemann T, Wachsmuth M, Tewes M, Rippe K, Langowski J (2002) Analysis of ligand binding by two-colour fluorescence cross-correlation spectroscopy. Single Mol 3:49–61

Weidemann T, Wachsmuth M, Knoch TA, Muller G, Waldeck W, Langowski J (2003) Counting nucleosomes in living cells with a combination of fluorescence correlation spectroscopy and confocal imaging. J Mol Biol 334:229–240

Weisshart K, Jungel V, Briddon SJ (2004) The LSM 510 META–ConfoCor 2 system: an integrated imaging and spectroscopic platform for single-molecule detection. Curr Pharm Biotechnol 5:135–154

White J, Stelzer E (1999) Photobleaching GFP reveals protein dynamics inside live cells. Trends Cell Biol 9:61–65

Widengren J, Mets Ü, Rigler R (1995) Fluorescence correlation spectroscopy of triplet states in solution: a theoretical and experimental study. J Phys Chem 99:13368–13379

Zhang J, Campbell RE, Ting AY, Tsien RY (2002) Creating new fluorescent probes for cell biology. Nat Rev Mol Cell Biol 3:906–918

8
Single Fluorescent Molecule Tracking in Live Cells

Ghislain G. Cabal*, Jost Enninga*, and Musa M. Mhlanga*

Abstract Biological macromolecules, such as DNA, RNA, or proteins, display a distinct motility inside living cells that is related to their functional status. Understanding the motility of individual biological macromolecules under specific physiological situations gives important clues regarding the molecular role of these macromolecules. Therefore, it is important to track individual biological macromolecules in the context of living cells. Often, biological macromolecules are constituents of larger multimacromolecular assemblies that can be denominated as macromolecular *particles*. During the last few years, various approaches based on fluorescent imaging have been developed to allow the tracking of single particles inside living cells. In this chapter, we present several such approaches to track individual (1) DNA loci, (2) ribonucleoprotein particles, or (3) membrane proteins. Particularly, we focus on the practical aspects of these approaches, allowing the reader to adopt the methods presented to their specific scientific problems. The methods presented are based on different principles: Tracking of chromosomal DNA loci is achieved via operator/repressor recognition using fluorescent repressor molecules. Individual ribonuleoprotein particles can be followed with small oligonucleotide sensor molecules called molecular beacons. Individual membrane proteins can be tracked via their specific labeling with antibody–quantum dot conjugates. Subsequently, we outline the principles of single particle tracking algorithms that have been developed in the field of bioinformatics, and that are crucial for a rapid, unbiased analysis of the tracked particles.

8.1 Introduction

Looking at living cells constituted of biological macromolecules and large cities inhabited by millions of people, we can draw an intriguing analogy: the dimension of an average, soluble protein (about 2–5 nm in diameter) in comparison to the dimension to an entire eukaryotic cell (about 5–10 μm in diameter) exhibits the same

*All authors contributed equally

spatial ratio as the dimension of an individual (about 1.75 m in height) inside a large city such as Paris (10 km in diameter). Generally, the role of an individual is important for a community; this becomes evident in a negative way if a group of individuals with a specific occupation is on strike. Similarly, the billions of different biological macromolecules constituting each living cell have to work in concert to allow proper cell function and viability. Finding out about the function of the individual cellular constituents is one of the main goals of cell biologists. In the macroscopic world, following an individual in a city from an airplane reminds us of secret-agent movies from the time of the cold war. Similarly, peering at individual macromolecules in the context of living cells has become possible in recent years by novel light-microscopy approaches and is giving us important insights into how the individual components of a cell render it functional.

In this chapter, we will describe several new technologies that allow the imaging of individual biological macromolecules in living cells. Overall, biological macromolecules can be classified into DNAs, RNAs, proteins, and lipids. Moreover, such macromolecules often exist as complexes made of various macromolecular constituents (e.g., messenger RNA, mRNA, complexed with proteins). Therefore, we will generally refer to the tracked macromolecules as *particles*. Evidently, choosing the right approach allowing the tracking of single particles requires specific consideration of the various classes of the cellular macromolecular constituents. The size, the abundance, the localization, and the accessibility of a specific particle are the determining parameters that dictate the approaches used to track them. For example, small macromolecules diffuse more than large ones, and they require rapid image acquisition. High abundance of a specific particle is not necessarily advantageous for tracking because a high concentration of independently moving particles in a defined volume may complicate the adequate tracking of individual ones.

Technically, the method of choice for tracking single macromolecules in living cells is based on fluorescent labeling owing to its sensitivity. In this chapter, we will follow the central dogma of molecular biology, firstly focusing on the dynamics of single DNA loci, secondly presenting approaches to track the mobility of RNA particles, and thirdly describing ways to track proteins with a focus on membrane proteins.

8.2 Tracking of Single Chromosomal Loci

8.2.1 General Remarks

Most of the eukaryotic, cellular DNA content in a specific organism is found in the chromosomes – encoding the genetic information, and thus constituting its genome. While genomes are defined by their primary sequence, their functional properties depend on multiple layers of regulatory processes. To understand these processes, it is of importance to consider the highly dynamic nature of the chromatin fibers. In this regard, it has been proposed that chromosomes are not randomly distributed in the nucleus, and that one newly recognized level of regulation involves the information encoded by gene positioning within a given nuclear volume (Cremer and Cremer 2001; Misteli et al. 2004). However, how specific gene positioning is achieved, what nuclear components affect chromatin dynamics, and how these dynamics influence gene transcription are still open questions. Therefore, it is important to understand the chromatin fiber dynamics of individual chromosomal domains. But it has been difficult to measure these owing to the limits of indirect in situ visualization. Just very recently, cell biologists have been successful in developing novel tools devoted to single chromosomal locus tracking. First, we will describe several approaches that have been used to tackle this problem. Generally, they are based on the interaction between repeats of a bacterial operator integrated at a precise site in the genome, and on factors fused to fluorescent proteins that specifically recognize such sequence repeats. Subsequently, we will focus on a specific method based on this approach allowing the in vivo visualization of a single chromosomal locus in yeast.

8.2.2 In Vivo Single Loci Tagging via Operator/Repressor Recognition

The operator/repressor recognition method was initially developed for the visualization of a single locus in the budding yeast *Saccharomyces cerevisiae* (Straight et al. 1996) before being adapted for use in mammalian cells (Robinett et al. 1996). A track of 256 tandem repeats of the bacterial *lac* operator (LacO) was integrated into the chromatin fiber and then recognized by a hybrid protein that consisted of green fluorescent protein (GFP) fused to the amino terminus of the Lac inhibitor (LacI). In a similar fashion, another group used the *tet* operator (TetO) repeats and the *tet* repressor (TetR) to label chromosomal loci in yeast (Michaelis et al. 1997). In both methods the operator sequences recruit with high specificity the GFP-coupled repressor and appear as a very bright dot via fluorescence microscopy (Fig. 8.1, panels A, B). So far, such operators have been integrated in the genomes of various organisms via (1) homologous recombination (yeast, Straight et al. 1996; Michaelis et al. 1997; and bacteria, Webb et al. 1997), via (2) transposable elements (*Drosophila*, Vazquez et al. 2001), or via (3) nonhomologous recombination (*Caenorhabditis elegans*, Gonzalez-Serricchio and Sternberg 2006), and human cells (Chubb et al. 2002).

The ability to follow a single chromosomal locus in space and in real time provides significant information. For example, the Belmont and Sedat laboratories have used the LacO/LacI labeling system in yeast to track a single chromosomal locus over time, and have applied analytical tools from physics, such as motion analysis, to measure the parameters of chromatin dynamics (Marshall et al. 1997). More recently, the Gasser group and the Nehrbass laboratory have combined this strategy and basic genetic approaches to relate the physical properties of the chromatin fiber with biological function (Heun et al. 2001; Bystricky et al. 2005; Cabal et al. 2006; Taddei et al. 2006).

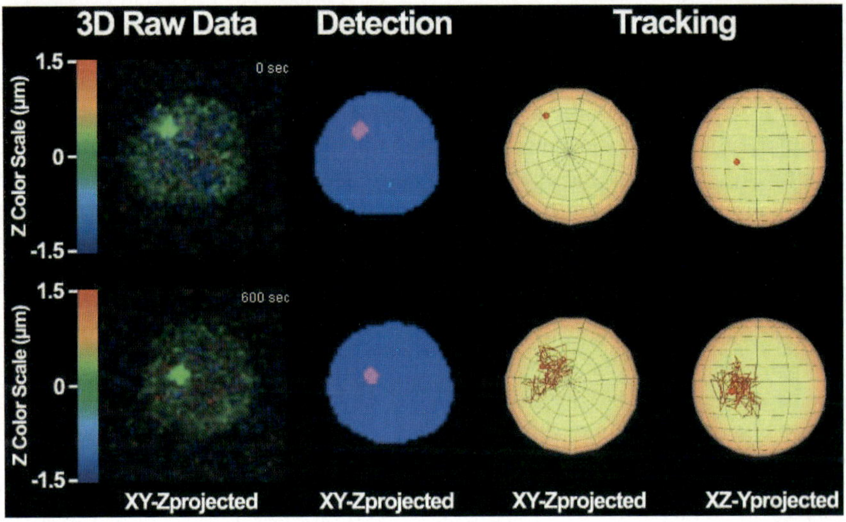

Fig. 8.1 Detection, localization, and tracking of a single chromosomal locus with QUIA. A *Saccharomyces cerevisiae* nucleus with a single chromosomal locus labeled with the *tet* operator (TetO)/*tet* repressor (TetR)–green fluorescent protein (GFP) system is imaged each 4 s for 10 min. Only the first and the last times from this time-lapse sequence are shown. The *panel on the left* shows a projection along the *Z* axis of the raw data acquired by 3D time-lapse confocal microscopy. Each slide was given a different color according to the *Z*-position of the labeled locus as indicated on the *left*. The projection along the *Z* axis of the recorded sequence after image processing with the detection plug-in is shown in the *middle panel*. The tracking sequence after removal of the general nucleus movement is shown in the *two rightmost panels*. One of these panels shows a projection along the *Z* axis and the other one a projection along the *Y* axis

In this chapter we will provide the key steps to tag single chromosomal loci for in vivo analysis. We have chosen to describe the protocols used in the yeast *Saccharomyces cerevisiae* because the labeling of single loci was first developed in this organism (see before), and we are currently using this model organism. This approach is quite similar in other model organisms (bacteria, Webb et al. 1997; *Drosophila*, Vazquez et al. 2001; *Caenorhabditis elegans*, Gonzalez-Serricchio and Sternberg 2006); and human cells, Chubb et al. 2002). We will then describe how to process (Box 8.1), analyze, and interpret (Box 8.2) images of chromosomal loci, in order to obtain the maximum information about the physical properties of the chromatin fiber motion.

8.2.3 The Design of Strains Containing TetO Repeats and Expressing TetR–GFP

First, one has to construct a plasmid containing tandem repeats of TetO. This plasmid can be obtained by multimerizing a head-to-tail 350-bp PCR fragment containing

Box 8.1 Implementation of single molecule tracking algorithms in biological systems

To overcome barriers to single-molecule tracking of multiple biological objects we have implemented a dedicated image analysis algorithm for the three 3D plus time detection and tracking of single or multiple fluorescent objects in biological microscopic images with the assistance of the Quantitative Image Analysis Group at Institut Pasteur. Essentially four approaches are combined to provide spatial and temporal information on the behavior of single particles in multiple-particle environments. A key advance of this approach is the decoupling of the detection and the tracking process in a two-step procedure. The method has the advantage of being able to track multiple objects in 3 dimensions, in complex situations where they cross paths or temporarily split. A model was first developed using synthesized data to confirm that this algorithm enables a robust tracking of a high density of particles. The method enables the extraction and analysis of information such as the number, position, speed, movement and diffusion phases of single fluorescently labeled molecules (Example of tracking of single chromosomal locus is shown in Fig. 8.1). First the particles (or particle as in the case of single chromosomal loci) are detected in 3 dimensional image stacks using a shift-invariant or un-decimated 3-D wavelet transformation for the detection of the fluorescent objects. This type of detection gives the advantage of not being subject to variations in image noise and contrast. In a second step the tracking is performed within a Bayesian framework wherein each particle is represented by a state vector evolving according to a biologically realistic dynamic model. Such a framework predicts the spatial position of a particle based on knowledge of its prior position and increases the reliability of the data associations. The Interacting Multiple Model (IMM) estimator algorithm is used with different transition models for the prediction and estimation of the state of the biological particle since it can readily self-adapt to transitions (Genovesio et al. 2006). The IMM has been adapted in our case to include several models corresponding to different biologically realistic types of movement (Genovesio et al. 2004). The tracks are then constructed by a data association algorithm based on the maximization of the likelihood of each IMM (Genovesio and Olivo-Marin 2004).

These models are able to model both Brownian motion and directed movement (motor-dependent) with constant speed or acceleration. A further assumption is made that during motion of the biological object abrupt switching between the three models is possible. It is hypothesized that the three models can be represented by a linear application. The random-walk model makes the assumption that the next state is defined by the previous state and an additive Gaussian noise. The first-order linear extrapolation model assumes the next state is defined by the linear extrapolation of the

(continued)

Box 8.1 (continued)

previous two 3D locations while retaining similar volume and intensity. The final model, second-order linear extrapolation, makes the assumption that the subsequent state is defined by the linear extrapolation of the last 3D locations while keeping the same intensity and volume.

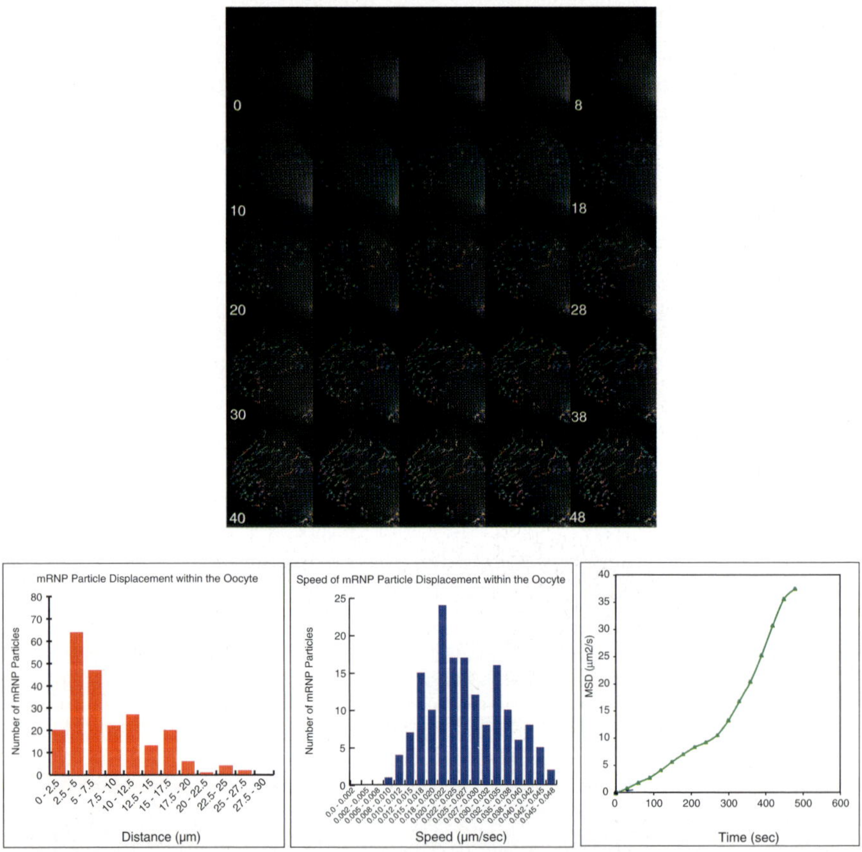

Fig. 8.2 Detection, localization, and tracking of *oskar* messenger RNA (mRNA) in drosophila oocytes. Molecular beacons complementary to *oskar* mRNA were injected in stage 8 drosophila oocytes at time 0 whilst imaging on a Nipkow disc confocal microscope as shown. The acquisition was done in 3D plus time over 50 minutes in continuous acquisition mode. Here the acquisition is shown as a Z-projection. The acquired images are then loaded into the tracking and detection module of the QUIA software. The detection and tracking is performed in 3D plus time. The result of this detection and tracking of the mRNA particles is shown here and is most notable after 10 minutes. Though it is Z-projected the tracking is in 3D and can be visualized as such. Other data from the entire population or each tracked particle can be obtained (as indicated) such as total displacement, mean squared displacement and diffusion kinetics. mRNP = mRNA-Ribonucleic protein complex)

Box 8.2 Brownian motion analysis

The use of Brownian motion analysis is applicable to all diffusive movement such as that of DNA, RNA, and proteins. For example, to access the laws underlying chromatin dynamics, random movement of the tagged loci can be analyzed by computing the mean-squared displacements (MSD) in the 3D position of the locus as a function of the time interval: $<\Delta d^2>=<[X(t)-X(t+\Delta t)]^2>+<[Y(t)-Y(t+\Delta t)]^2>+<[Z(t)-Z(t+\Delta t)]^2>$, where Δd is in microns and Δt is in seconds, currently denoted as $<\Delta d^2>=<[\mathbf{d}(t)-\mathbf{d}(t+\Delta t)]^2>$ in the literature, where $\mathbf{d}(t)$ is a vector (Fig. 8.3). The average is performed over all overlapping time intervals (Fig. 8.3). The *X*, *Y*, and *Z* coordinates of the tagged locus obtained with the detection algorithm (Box 8.1) are then normalized relative to a nuclear reference of choice. Since the cell nucleus is moving over time, we subtract the position vector of the nucleus centroid from the position vector of the locus to get the movement of the locus relative to the nucleus.

Previous studies have described the dynamics of genetic loci as confined random walks exhibiting a free diffusion motion within the region of confinement (Bystricky et al. 2005; Heun et al. 2001; Marshall et al. 1997). In this case, the expected MSD increases linearly with Δt at small time scales [i.e., $<\Delta d^2>=c(\Delta t)^\alpha$, with $\alpha=1$ and a coefficient of diffusion $D=c/6$] and reaches a plateau at longer times (Fig. 8.4a). Most recently the Nehrbass group has described a markedly different behavior, with MSD curves being well fitted at small time scales by a power law: $<\Delta d^2>=c(\Delta t)^\alpha$, with $\alpha\neq1$ (Fig. 8.4a, b). In this case gene movement cannot be described as a "confined random walk." In fact, either $\alpha>1$, indicating overdiffusion, or $\alpha<1$, indicating subdiffusion movement (Fig. 8.4c; Bouchaud and Georges 1990; Havlin and Ben-Avraham 2002).

In the case of Brownian motion, the plateau reached by the MSD at longer time scales indicates that the locus motility is spatially constrained within a certain volume (Fig. 8.4a). Note that this spatial confinement is independent of the type of diffusion previously determined at small time scale. One can make the approximation that this volume is spherical and estimate that at a long time interval $<\Delta d^2>\approx1.2r_c^{\,2}$, where r_c is the radius of confinement.

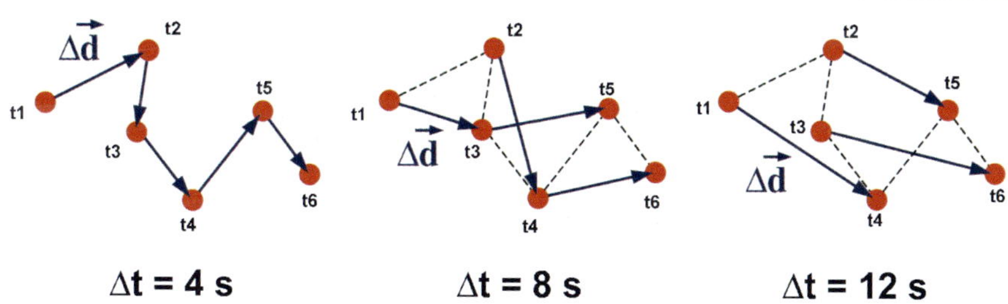

Fig. 8.3 MSD computing method. MSD analysis is performed by computing the mean-squared change in distance Δd^2 (microns squared) as a function of the time interval Δt (seconds). The Δd vectors are calculated using the *X*, *Y*, and *Z* coordinates given by the detection software. Here is shown as an example the determination of the Δd vectors for the three first Δt on the six first time positions. This should be repeated for each Δt of the sequence and with all the positions at each Δt.

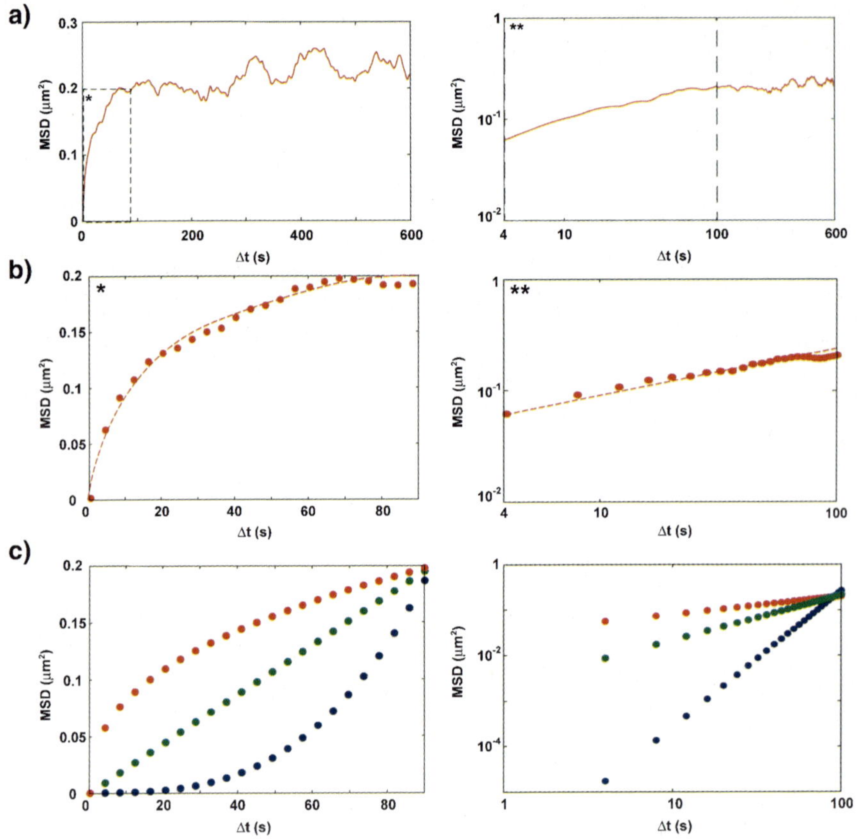

Fig. 8.4 MSD analysis method. **a** MSD of the tracking sequence from datasets of Fig. 8.1 (*left*). The same set of data is also plotted with log–log axes (*right*). The entire 900s sequence is used for the calculation but only Δt from 0 to 600 s is shown. The MSD reaches a plateau at time scale exceeding 100 s. **b** Short time scale zoom in on the MSD plotted in **a**. The entire 900s sequence is used for the calculation but only Δt from 0 to 90 s is shown. This MSD curve is well fitted by the power law $y=0.04x^{0.39}$ (*dashed curve*). **c** Theoretical MSD curves at small time scale for subdiffusion (*red*, $\alpha< 1$), free diffusion (*green*, $\alpha=1$) and overdiffusion (*blue*, $\alpha>1$) movements

seven TetO sequences (GAGTTTACCACTCCCTATCAGTGATAGAGAAAAGTG AAAGTC, for details see Michaelis et al. 1997). The number of repeats generally used is 112 (insert size 5.6 kb) or 224 (11.2 kb) depending on the expression rate of TetR–GFP or the signal-to-noise ratio required for the image processing. The stability of such repeats during cloning and amplification is largely improved by using either SURE2 (Stratagene), STBL2 (Invitrogen), or EC100 (Epicentre Technologies) Escherichia coli strains which minimize recombination.

Integration of the TetO repeats into the genome is achieved in yeast by homologous recombination (Fig. 8.5a). The crucial step lies in the choice of where to integrate the repeats. It is important to consider that the integration of an approximately 15 kb plasmid containing tandem repeats is not neutral. For example, we have observed that integration of TetO repeats in very close proximity to a subtelomeric marker largely impairs its repression by the silencing machinery. As a consequence, the integration site has to be appropriately chosen in order not to damage the chromatin environment and metabolism of the target locus. Promoters, terminators, or other known regulatory sequences have to be preserved. Conversely, one has to be careful to not insert the TetO repeats too far from the target gene and thus uncouple the behavior of the labeled site from the locus of interest. During interphase in yeast the chromatin fiber exhibits a compaction level of 10–15kb per 100 nm (Bystricky et al. 2004). Considering the precision of detection of the majority of tracking algorithms to be around 50–100 nm, we strongly recommend integrating the TetO repeats at a maximum of 5 kb away from the target locus. In any case, the expression level of the surrounded genes should be checked by reverse transcription PCR or northern blot after integration of the TetO repeats. Instability of the repeats has been observed during yeast transformation, but once integrated, the TetO array is quite stable. TetR fixation stabilizes the repeats and using a strain already expressing the TetR–GFP construct could enhance the TetO repeats stability during transformation. A Southern-blot verification should be performed to check for either multiple or unspecific insertion, as well as to check the number of remaining repeats.

Visualization of the TetO array requires the binding of the TetR protein fused to a fluorescent protein. We are currently using the integrative plasmid, expressing the TetR protein fused via the N-terminus to the SV40 nuclear localization signal and to the SuperGlow GFP at the C-terminus (Michaelis et al. 1997). This construct is expressed under the control of the *URA3* promoter, and termination is ensured via the *ADH1* terminator. In our hands this plasmid used with 112 TetO repeats gives a good signal to noise ratio and allows a wide range of applications (Fig. 8.5b).

Studying nuclear architecture requires, in most cases, a structural reference to be labeled together with the single chromosomal locus. The nuclear envelope, the nucleolus, and the spindle pole body are all good candidates (Heun et al. 2001b; Bystricky et al. 2005; Cabal et al. 2006). Intuitively, one would use spectrally well separated fluorescent proteins. Though it is easiest to use this approach for the processing/analysis of images, this strategy is laborious during image acquisition and results in poorer temporal resolution if one multiplies the exposure time by the number of different fluorescent proteins excited. Using the same fluorophore can be a good solution to overcome this restriction. If a single wavelength is used, it is possible to discriminate between two GFP signals if both the intensity and the expected spatial localization remain sufficiently different (compare the TetR–GFP signal with GFP–Nup49 in Fig. 8.5b).

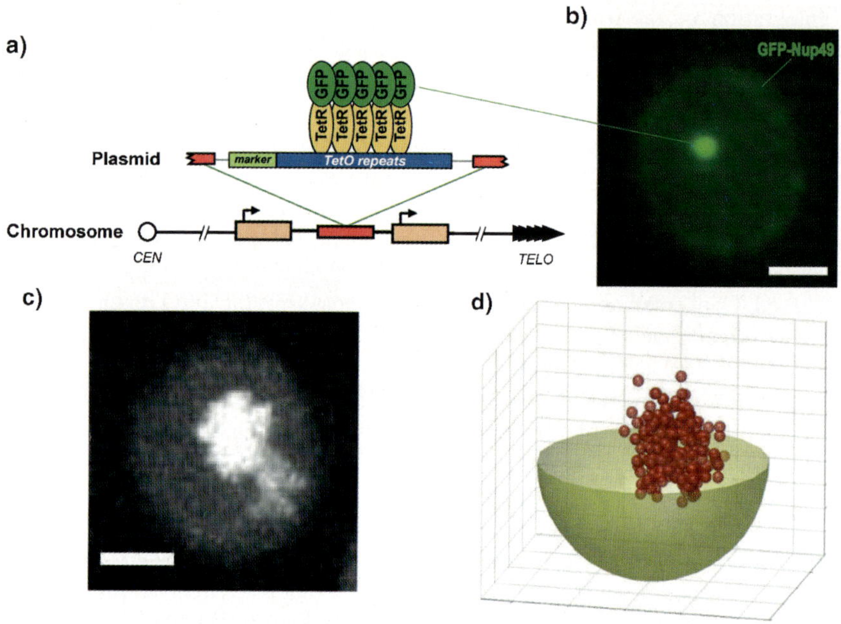

Fig. 8.5 In vivo single chromosomal locus tracking. **a** Integration of the TetO repeats into the genome is performed in the yeast *Saccharomyces cerevisiae* by homologous recombination allowed by linearizing the TetO-containing plasmid into a previously cloned PCR fragment (*red box*), of at least 500 bp and homologous to the chosen chromosomal integration site. Note that after this recombination step, the homologous sequence is duplicated. Following this insertion, the visualization of the TetO array is achieved by the binding of the TetR protein fused to a fluorescent protein. **b** Under fluorescent microscopy the TetR–GFP signal is visualized simultaneously with the nuclear envelope stained with GFP–Nup49. *Scale bar* 1 μm. **c** Time-lapse confocal microscopy in three dimensions was performed to track the single labeled locus. A 25-image *Z*-stack (*Z*-step 250 nm) was taken every 4 s over 15 min (226 time points). *Z* and time projection of the whole 3D raw data sequence is shown. *Scale bar* 1 μm. **d** 3D plot of the tracking sequence obtained after image processing. *CEN* centromere, *TELO* telomere

8.2.4 In Vivo Microscopy for Visualization of Single Tagged Chromosomal Loci

A mounting protocol for yeast is provided in Sect. 8.7.1. The fluorescence efficiency and the signal-to-noise ratio obtained by the use of the TetO/TetR–GFP labeling system are good compared with those for the majority of endogenous GFP-tagged yeast proteins (Fig. 8.5b). This allows acquisition with exposure times below 500 ms with the majority of available confocal or widefield microscopes. We advise the use of a Nipkow disc confocal microscope as it allows rapid acquisition with minor phototoxicity and photobleaching effects. We also recommend the use of a highly sensitive CDD or preferably an electron-multiplying CCD (EMCCD) camera to improve the sensitivity when used with yeast owing to its small size and the weak protein expression compared with those of metazoan cells.

To date, gene loci have been followed at a single confocal plane over time (Heun et al. 2001) mostly owing to technical limitations and because the small size of the interphase yeast nucleus (about 2 μm diameter) and the inherent limited resolution of optical microscopy (around 0.25 μm in *X*–*Y* and 0.7 μm in *Z* for GFP signal with a ×100 1.4 numerical aperture objective) make distance measurements in two dimensions potentially highly inaccurate. Moreover, since this kind of 2D approach implies a stark sampling bias in favor of nuclei for which the locus lies in the focal plane, a large proportion of nuclei in the population are excluded, leaving the assay insufficiently exhaustive for accurate statistical analysis. To overcome these limitations, we suggest detecting and localizing single chromosomal loci in three dimensions, which is largely possible with modern confocal microscopy (Fig. 8.5c).

The overall setup for acquisition (exposure time, excitation intensity, *Z*-step size and number, delay between time points, etc.) will strongly depend on the application. Two major experimental protocols can be followed. First, the spatial organization of chromatin structures can be described by statistical quantities such as the probability distribution of positions within a population, which does not require any time-lapse approach but only static acquisitions at a single time point. Second, to access the laws underlying chromatin dynamics, gene motion can be described by parameters such as diffusion coefficients and confinement radii. Here we will focus on the second category, which requires time-lapse acquisition.

For 3D plus time acquisition (4D), the initial steps involve adjusting the 3D acquisition for single time points. First, a *Z*-scan over 6–7 μm is enough to capture the entire yeast nucleus. Second, the optimum axial (*Z*) sampling rate is derived via the Nyquist rate, the calculation of which will depend on the type of microscope used. However, particle tracking requires specific adaptation of this rate and oversampling must be avoided (Ober et al. 2004). Third, one has to keep in mind that the coefficients of diffusion of chromatin fiber computed in yeast are around 10^{-11}–10^{-12} cm^2/s (Bystricky et al. 2005; Marshall et al. 1997). Therefore, considering that the detection precision of the large majority of tracking algorithms is around 50–100 nm, the total acquisition time for one *Z*-stack should not exceed 10 s to avoid detectable movement of the locus over the acquisition.

Adding the time dimension to the 3D acquisition process diminishes the signal-to-noise ratio and forces a compromise between temporal resolution and spatial resolution on the *Z* axis. Furthermore, the investigator has to devote attention to both getting the best spatiotemporal resolution and avoiding phototoxicity and photobleaching. To lower phototoxicity and photobleaching, the number of *Z*-steps per *Z*-stack and both the exposure time and the intensity of the excitation source should be decreased to the limit of detection by the tracking algorithm used. This will also allow a reduction in acquisition time per *Z*-stack and by consequence will enhance temporal resolution. To date, the best time resolution we have been able to achieve by capturing the entire nucleus in three dimensions (25 slices per *Z*-stack with 250-nm *z*-step) is a 4s delay between two time points over a total acquisition time of 15 min (Fig. 8.5c, d; Cabal et al. 2006). But the use of an EMCCD camera, which is more sensitive than classic CDD cameras, will certainly shorten this interval by lowering the exposure time. However, depending on the application (see later), one can either decrease the time delay by reducing the spatial resolution in the *Z* direction or increase the total acquisition time by augmenting the time delay and/or diminishing the *Z*-step number per *Z*-stack. Lastly, acquisition conditions should be compatible with life; thus, it is advisable to check the cells after acquisition to determine if laser irradiation has heavily damaged them.

8.2.5 Limits and Extension of Operator/Repressor Single Loci Tagging System

Even if the operator/repressor system employed to tag a single chromosomal locus is powerful and allows in vivo analyses to give access to the dynamics of the chromatin fiber, there are limitations. At present it cannot be accurately predicted to what extent the TetO repeats and the ensuing TetR–GFP binding affect surrounding chromatin compaction and/or folding. This applies to the properties of motion as well. Some studies report some mislocalization of chromatin fiber by using this tagging system (Tsukamoto et al. 2000; Fuchs et al. 2002). Those artifacts appear to be locus-dependent; thus, the investigator must be prudent in the choice of locus and perform adequate control experiments (see earlier) (Chubb et al. 2002).

Nonetheless, the method presented allows significant advances in our comprehension of the organization and properties of chromatin fiber. This approach can be enhanced by its combination with other modern chromatin engineering tools. Tagging multiple loci in the same cell can be achieved by combining TetO insertion with other operator systems such as LacO rather than using multiple TetO insertions, which can cause mislocation (Bressan et al. 2004; Bystricky et al. 2005; Fuchs et al. 2002). One can also artificially tether the tagged locus to nuclear subcompartments of interest or excise it on a freely diffusing episome (Feuerbach et al. 2002; Taddei et al. 2004; Gartenberg et al. 2004). Finally, it is possible to follow in situ the expression of the tagged gene either by performing mRNA fluorescent in situ hybridization (FISH) (Cabal et al. 2006) or by combining the DNA tagging approach with an in vivo RNA labeling method (reviewed in this chapter and in Janicki et al. 2004). Combining the tracking of individual chromosomal loci with measurements on the activity of transcription and the motion of RNA particles will yield tremendous improvements of our understanding of eukaryotic gene expression.

8.3 Single-Molecule Tracking of mRNA

8.3.1 Overview

The intracellular transport of mRNAs from sites of transcription of active chromosomal loci in the nucleus to specific destinations in the cytoplasm is highly regulated and conserved among eukaryotic species (Jansen 2001; St Johnston 2005). The localization of mature mRNAs acts as an important posttranscriptional mechanism to spatially restrict the synthesis of proteins. Such mechanisms are found throughout the plant and animal phyla, playing diverse and important functions in development, stem cell fate, memory formation, and cell division among others (Jansen 2001). Much progress has been made in understanding the various *trans*-acting proteins that interact with these mRNAs. The transport of mRNAs in eukaryotic cells from sites of transcription in the nucleus to specific destinations in the cytoplasm occurs via heterogeneous nuclear ribonucleoproteins, (hnRNPs) or mRNA–protein complexes (mRNPs) (Dreyfuss et al. 2002). Understanding interactions between mRNPs and these *trans*-acting proteins, which range from molecular motors to proteins implicated in RNA interference, necessitates approaches permitting the direct observation of such dynamic events in vivo (Tekotte and Davis 2002). Approaches that facilitate the tracking and covisualization of *trans*-acting factors with individual mRNP particles in real time permit the precise description of transport events and how they are influenced by *trans*-acting factors or cellular structures. Such descriptions range from the speed and motion characteristics of a given mRNA allowing the determination of whether it is energy-dependent or energy-independent. By simultaneous tracking of the motion of a putative *trans*-acting protein and a mRNA, one can determine when and where the protein is implicated in energetic components of mRNA transport. Further the control of RNA transport has emerged as an important regulatory mechanism in gene expression (Gorski et al. 2006). Thus, the ability to spatially and temporally resolve single molecules of mRNPs and proteins in living cells will assist in unlocking the biological mechanisms at work in the dynamics of mRNA transport and metabolism.

Earlier we described how the intelligent use of fluorescent fusion proteins that bind to specific chromosomal sites has permitted cell biologists to gain a spatial and temporal understanding of chromosomal loci dynamics. Similar approaches have been developed to track the movements of mRNPs. A number of technical approaches have been developed to fluorescently tag mRNAs in living cells (Dirks et al. 2001). Here we will cover two systems which we feel are best suited to single-molecule tracking of mRNA in living cells (Bertrand et al. 1998, Tyagi et al. 1996). They use distinct approaches in fluorescently labeling mRNAs to enable single-molecule tracking of mRNA.

8.3.2 The MS2–GFP System

The MS2–GFP system was first used to visualize mRNA in yeast cells using a novel two-plasmid approach (Bertrand et al. 1998; Fig. 8.6a). One plasmid encoded a GFP fused to the coding sequence for the single-stranded phage RNA phage capsid protein, MS2. The MS2 capsid protein has a high affinity (K_d=39 nM in vitro) for a specific RNA sequence with a short stem–loop structure (MS2 binding sequence) encoded by the phage. The second plasmid contains the MS2 binding sequence multimerized in six, 12, or 24 copies. To reduce the background signal in the cytoplasm from unbound GFP–MS2 protein a nuclear localization sequence is engineered into the sequence, thereby restricting it to the nucleus. Both plasmids are cotransfected into cells or alternatively cell lines, or GAL4/UAS systems can be created for the inducible expression of GFP-MS2. The combination of the highly specific binding of GFP–MS2 to the MS2 binding sequence and the ability to engineer the MS2 binding sequence into an RNA of choice makes this a powerful system to fluorescently tag mRNAs in living cells.

Tracking mRNPs, one has to consider that the biological activity of some mRNPs resides in their 3′ untranslated region (3′UTR), or is intrinsic to their secondary structure, and any disruptions affecting these regions can have negative impacts. Therefore, an approach that is based on mRNPs containing multiple repeat sequence motifs that are recognized by fluorescent fusion proteins may interfere with the endogenous behavior of the individual mRNPs. Thus, we will present an alternative approach using specific biosensor molecules, so-called molecular beacons, as a different means to track individual mRNPs in the subsequent paragraphs.

8.3.3 The Molecular Beacon System

Molecular beacons are oligonucleotide-based probes that fluoresce only upon hybridizing specifically to complementary mRNA sequences (Tyagi and Kramer 1996; Fig. 8.6b). Unbound molecular beacons are nonfluorescent, and it is not

Fig. 8.6 (continued) proteins on a single mRNA can be detected via epifluorescent or confocal microscopy. **b** Molecular beacons are stem–loop oligonucleotide probes that fluoresce upon hybridization to DNA or RNA targets. They consist of a stem portion with as few as three and as many as eight bases paired to each other, with a fluorophore and quencher at the 5′ and 3′ extremities. The loop portion can posses as few as ten and as many as 50 nucleotides. In the absence of a target, the fluorophore and quencher remain in close proximity to each other and the fluorescence is quenched. When the loop portion of the molecular beacon comes into contact with a perfectly complementary DNA or RNA target, the entire molecule undergoes a spontaneous change in conformation, resulting in the loop hybridizing to the target and the 5′ and 3′ extremities of the stem being well separated from each other. As a result the fluorophore is no longer quenched and the resulting fluorescence is detectable via epifluorescent or confocal microscopy. **c** RNA secondary structure is the greatest impediment to targeting molecular beacons to mRNA owing to the inaccessibility of regions of mRNA to the binding of the probes. Secondary structure predictions, similar to the Zuker fold prediction shown, can provide a guideline as to which regions are accessible for hybridization. Bases in "cooler" colors (e.g., *blue*, *black*) are likely to be double-stranded, whereas those in "warmer" colors (e.g., *orange*, *red*) are likely to be single-stranded. Molecular beacons can be designed and tested in vitro for their ability to bind these sites prior to using them in vivo (see Sect. 8.7)

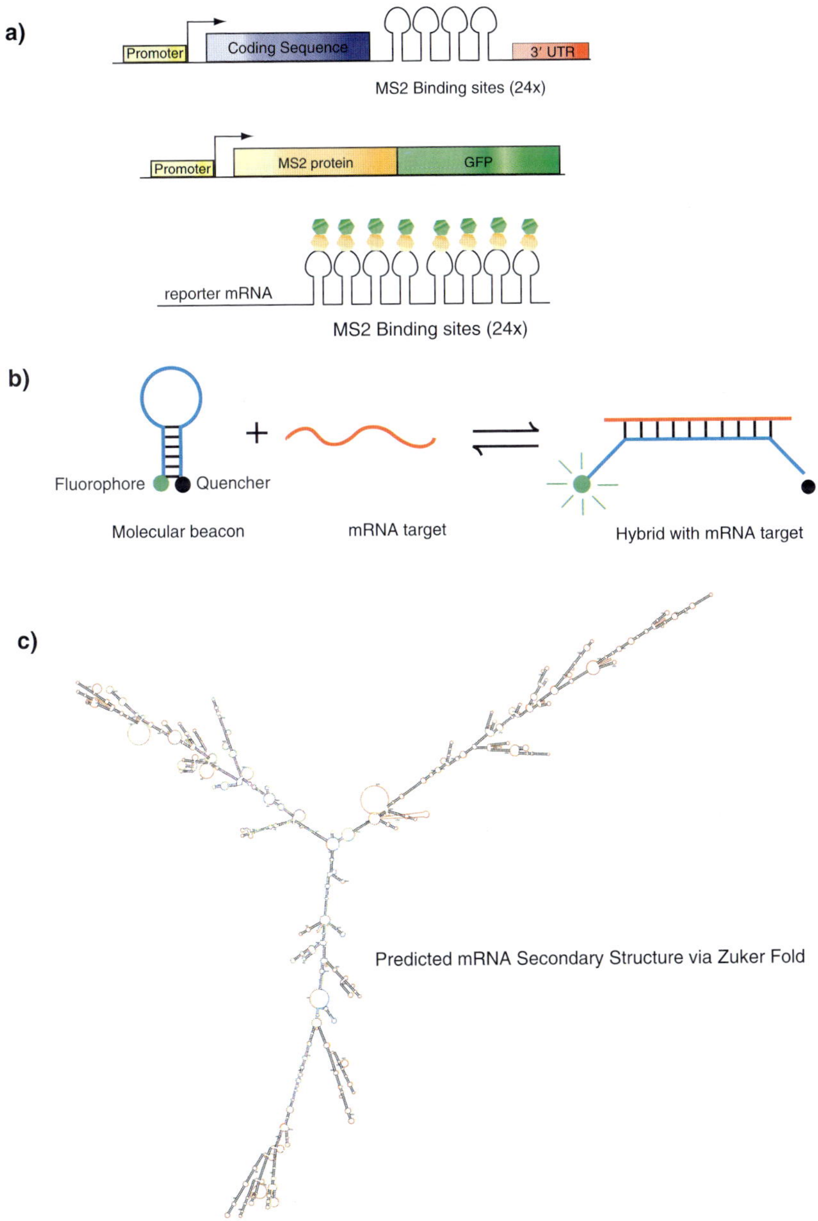

Fig. 8.6 Imaging mRNA in vivo. **a** The MS2–GFP system relies on the use of two DNA constructs. This first encodes the mRNA of interest and the MS2 binding sites which form stem–loop repeats. The MS2 binding site is usually inserted between the 3′ untranslated region (*3′UTR*) of the mRNA of interest and the coding sequence. The second construct encodes the MS2 protein fused to GFP. When both are expressed in a cell, the MS2–GFP fusion will bind with very high affinity to the MS2 binding sites on the reporter mRNA. The accumulation of several MS2– GFP

necessary to remove excess probes to detect the hybrids. Molecular beacons hybridize spontaneously to their targets at physiological temperatures; hence, their introduction into cells is sufficient to fluorescently illuminate their target mRNAs. Hybridization of complementary nucleic acids is one of the most specific biological interactions known, occurring with extremely high specificity. For this reason molecular beacons will "fish out" their targets in a "sea" of noncomplementary RNA targets. Thus, like the MS2 system, molecular beacons enable single molecule spatiotemporally resolved studies to understand the orchestrated relationship between various proteins involved in mRNA transport, and allow the precise determination of the time points at which association or disassociation with the mRNP complex occurs.

Subsequently, we will focus on using the molecular beacon technique for single-molecule tracking of mRNA. We will describe how to set up the system to perform single-particle tracking to be used in mammalian cell lines and in *Drosophila melanogaster* (for examples of biological contexts where this has been used, refer to Bratu et al. 2003). Molecular beacons have been used for in vivo tracking of mRNA in the *Xenopus* system as well as in human cell lines (Mhlanga et al. 2005; Vargas et al. 2005). We will then present an overview of how to analyze data from experiments using molecular beacons to understand the behavior of mRNA in differing biological contexts. For further information on the use of the MS2 system, see Bertrand et al. (1998) and Goldman and Spector (2005).

8.3.4 Setting Up the Molecular Beacon System for the Detection of mRNA

In theory, the entire length of an mRNA molecule can be targeted with molecular beacons; however, in practice several constraints exist over the length of the sequence, precluding the use of the entire length. Primary amongst these constraints are regions of the mRNA with known biological function, such as regions where exon junction complexes form, or where other proteins necessary for the activity of the mRNA are known to bind. The secondary constraint is the existence of complex secondary and tertiary intramolecular structures within the mRNA (Fig. 8.6c). These structures, often difficult to predict with currently available software, mask regions of the mRNA and render them inaccessible to the probes. Several in vitro assays and theoretical algorithms are available to help identify putative target sites within mRNA sequences, as well as probes with high affinity for binding (Mathews et al. 1999; Zuker 2003). Approaches to find accessible probe-binding target sites within an RNA sequence and to design efficient molecular beacons for RNA detection in vivo are described in Bratu 2006 and Bratu et al. 2003.

8.3.5 Ensuring the Observed Fluorescent Particles in Vivo Consist of Single Molecules of mRNA

To be able to detect single mRNA molecules in living cells it is important to relate the quantum fluorescence yield of a molecular beacon to the number of mRNA molecules. To perform highly resolved studies with single-molecule quantification of mRNA in vivo, it is important to ensure that the number of probes bound to the mRNA is detectable by instruments currently available. One approach is to target several molecular beacons to different regions of the mRNA to generate sufficient signals (Mhlanga et al. 2005). Another is to engineer a series of repeated binding sites for molecular beacons at the 3′UTR of the mRNA sequence (Vargas et al. 2005). The idea of this is similar to the tagging of the genome loci described earlier in this chapter; however, it requires molecular beacons instead of fluorescent fusion proteins to bind to the repeat sequences. Using this approach, one can introduce a defined number of binding sites, in our case 96, downstream of the gene of interest. In our model gene system the sites are engineered downstream of GFP and an inducible promoter (Fig. 8.7). Previous studies have indicated that the quantum yield of fluorescence from 48 GFP molecules or approximately 70 Cy3 moieties is detectable via available detection methods (this excludes the use of EMCCD cameras, which are more sensitive) (Babcock et al. 2004; Shav-Tal et al. 2004). Therefore, it may be possible to obtain robust signals using fewer than 96 binding sites.

It is important to check for any aggregation effects that may occur in vivo when introducing tandem array constructs into cells as described earlier. To check for this, two approaches can be used. In the first approach, the tandem array construct, along with its GFP fusion, can be transcribed in vitro by varying the number of binding sites, 16, 32, 64, 96, for example (Fig. 8.7a). These in vitro transcribed mRNAs can then be prehybridized to molecular beacons that hybridize to the binding sites and then are injected into the cell type where the studies will eventually be conducted. The intensity of the particles produced should vary in direct proportion to the number of binding sites present in the in vitro transcribed RNA, thus indicating that there is no formation of aggregates of the particles in the repeated arrays. The second approach is to prehybridize each in vitro transcribed mRNA to molecular beacons labeled with two differently colored fluorophores (Fig. 8.7b). The differently colored molecular beacons are separately hybridized to the mRNA. The resulting hybrids are then mixed together and coinjected. The observed fluorescent mRNA particles in vivo should contain a single color if aggregation of the mRNA particles does not occur.

Statistical analyses of RNA particle intensity in vivo can then be used to determine that the observed fluorescent particles are the product of the interaction of the molecular beacons with one mRNA molecule and not the result of aggregates (Vargas et al. 2005). If the observed fluorescent particles consist of aggregates of many molecules, then differently sized complexes will be observed, resulting in a multimodal distribution of particle intensities (Fig. 8.7c). Thus, a unimodal distribution of fluorescent particle intensity is indicative of the imaging of a single molecule of mRNA. Each of these analyses should be performed to ensure that

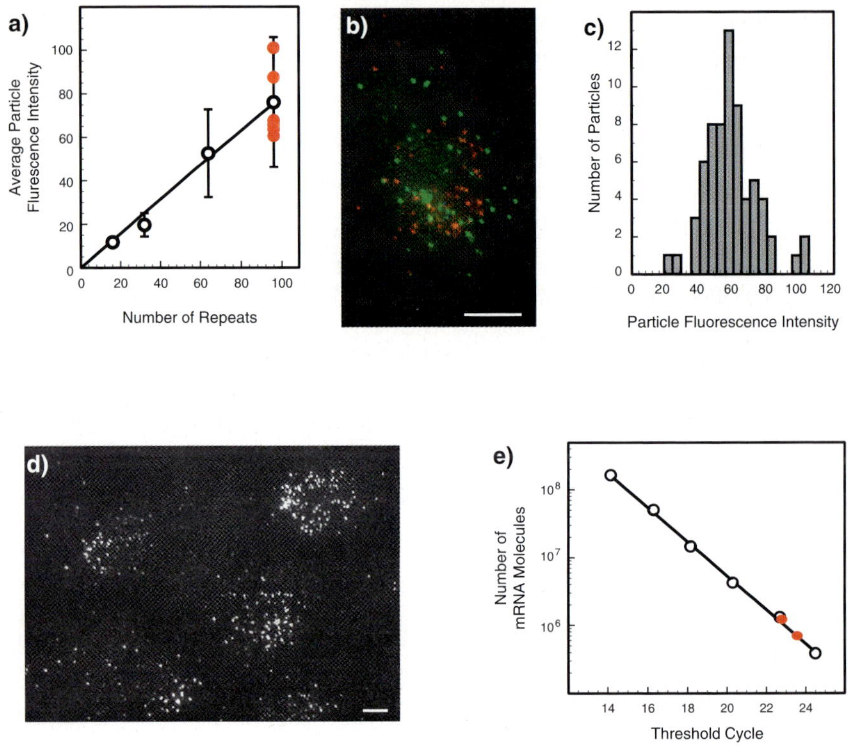

Fig. 8.7 Setting up the molecular beacon system for RNA detection. **a** To be sure that particle intensity reflects the number of molecular beacons bound, the fluorescence intensity is measured and correlated to the number of binding sites. Thus, synthetic hybrids containing 96 binding sites produced particles with intensities roughly equal to the intensities displayed by the particles containing endogenous mRNA (**d**), indicating that the fluorescence of both types of particles arose from an equal number of molecular beacons. By measuring the fluorescence intensity resulting from the injection of synthetic hybrids containing 16, 32, and 64 binding sites, one can directly correlate their intensity to particle intensity. **b** Since it is conceivable that the particles observed were produced by multimerization of mRNAs or by the association of multiple mRNAs to structures present within the cell, synthetic hybrids are prepared as described in the text and are coinjected into CHO cells with differently colored molecular beacons. **c** If the mRNA molecules have a tendency to aggregate in the cell, complexes of different sizes should occur, resulting in a multimodal distribution of particle intensities. Measurements of the intensities of a large number of particles from the same nucleus found their intensity distribution was unimodal, as shown here. **d** In situ hybridization on fixed cells using probes specific for the repeated sequence, rather than using molecular beacons in live cells, to ensure that all cytoplasmic particles are counted. Direct counting of particles indicated that there were, on average, 65 GFP–mRNA–96-mer molecules per cell. **e** RNA extracted from 10,000 cells was used to initiate a real-time PCR. The resulting threshold cycle indicates that 80 molecules of GFP–mRNA–96-mer were initially present per cell (red dots). The close agreement between this measurement and that found in **d**, in combination with the results of the previous experiments, strongly indicates that the endogenous mRNP particles observed in the cells each contain a single mRNA molecule *Image Courtesy of Sanjay Tyagi*

the observed fluorescent particle represents a single molecule of mRNA and not the aggregation of several mRNA molecules (Fig. 8.7).

The Kramer/Tyagi laboratory has also developed an approach utilizing a comparison between the average number of particles observed per cell using FISH and comparing this result by quantitative real-time PCR with the average number of tandem array 96-mer molecules per cell in an identical cell preparation. Using this approach, they were able to determine that on average 65 of the 96 positions were bound per cell using the FISH method, which uses the direct counting of cells. In contrast, quantitative PCR revealed that in a population of 10,000 cells approximately 80 of the 96 positions were bound (Fig. 8.7e). The close correlation between these two measurements is strongly indicative that each observed fluorescent particle consists of a single molecule of mRNA. Once these parameters have been defined, it is possible to track the observed fluorescent particles in vivo with very high confidence that one is tracking single molecules of mRNA as shown in Fig. 8.7d (Vargas et al. 2005).

The use of molecular beacons is not without its pitfalls and limitations. Primary among these is the problem of target selection. RNA is a highly folded structure which masks the accessibility of certain sequences to hybridization by molecular beacons. As previously mentioned, several software solutions have been designed to assist in probe design. However, in most cases there is no substitute for the empirical testing in vitro of different molecular beacons for their ability to bind a given RNA. This is usually done using a spectrofluorometer and is described in Bratu et al. 2003. The delivery of molecular beacons into different cellular contexts can also pose an impediment to their use if these cells are contained in deep tissue or are inaccessible to standard oligotransfection or microinjection approaches.

The use of imaging techniques to visualize native mRNAs is still in its infancy compared with imaging of proteins. The field has begun to address many biological questions in the characterization of single particles of mRNA to establish the basic principles of mRNA dynamics. The mobility of mRNA in the nucleus is the subject of intense study and a good deal has been determined (Gorski et al. 2006; Shav-Tal et al. 2004; Vargas et al. 2005). Major questions remain as to how these mRNAs then form aggregates of many mRNAs and the role that nuclear history (the proteins that interact with mRNA in the nucleus) plays in the cytoplasmic fate of mRNA (Palacios 2002). The techniques described in this chapter provide an important basis for addressing several of these questions.

8.4 Single-Particle Tracking for Membrane Proteins

8.4.1 Overview

Proteins are relatively small in comparison to the size of oligomeric RNA or DNA molecules (Waggoner 2006). So far, it remains very challenging to detect individual proteins or multimeric protein complexes in living cells. The reason for this is that single fluorophores, such as GFPs or small organic fluorescent molecules coupled to proteins of interest, do not exhibit a signal-to-noise ratio that is above the autofluorescence background of living cells. Several approaches have been developed to tackle this problem, and here we would like to present one of these: small inorganic semiconductors, generally named quantum dots or qdots, that can be coupled to proteins of interest (Bruchez 2005). The fluorescence properties of quantum dots are advantageous in tracking single particles as outlined in the following parts of this chapter.

Other microscopic techniques, such as fluorescent correlation spectroscopy (FCS) and fluorescence cross-correlation spectroscopy (FCCS), have been developed to detect the movements of individual proteins and to measure other physiological parameters (e.g., protein–protein interaction) inside living cells. It is important to mention that FCS and FCCS do not represent approaches for tracking individual molecules in the strict sense. FCS and FCCS count photons in a small sample volume with a high resolution on the time scale, and they use mathematical algorithms to deduce how fast fluorescently labeled molecules are moving in and out the sample volume. Therefore, FCS and FCCS are discussed in this chapter and are presented Chap. 7 by Wachsmut and Weisshart.

8.4.2 Quantum Dots As Fluorescent Labels for Biological Samples

Quantum dots were developed as biological fluorescent sensors almost a decade ago as a powerful alternative to render molecules of interest fluorescent (Chan and Nie 1998). These molecules are single crystals made from semiconducting materials, and are characterized by the differences in their bulk band-gap energy (Michalet et al. 2005). Spherical quantum dots have a typical diameter of about 2.5–10 nm (Fig. 8.8a, c); this size and the chemical composition of a specific quantum dot determine its spectral fluorescence properties (Fig. 8.8b). There are several advantages of quantum dots in comparison to generic fluorescent labels, offering the possibility to use quantum dots for the tracking of individual particles. Firstly, quantum dots are 2–10 times brighter than other fluorophores, with a quantum yield that is higher than 50%. Secondly, quantum dots have a much higher photostability than other fluorescent markers commonly used to label biological macromolecules; continuous illumination does not bleach quantum dots. This allows continuous image acquisition of quantum dot labeled biological samples up to several hours. Thirdly, quantum dots have distinct spectral properties (Fig. 8.8b). Their emission spectra are typically 30–50-nm wide (at the half maximum), and symmetric, and quantum dots can be synthesized with emission peaks throughout the entire visible spectrum. With use of the appropriate filters, this allows the measurement of multiple quantum dots in a biological sample.

For tracking individual particles in living cells, the two major advantages are the brightness and the photostability of quantum dots. Of course, quantum dots can also be used to label DNAs or RNAs. However, here we would like to focus on using

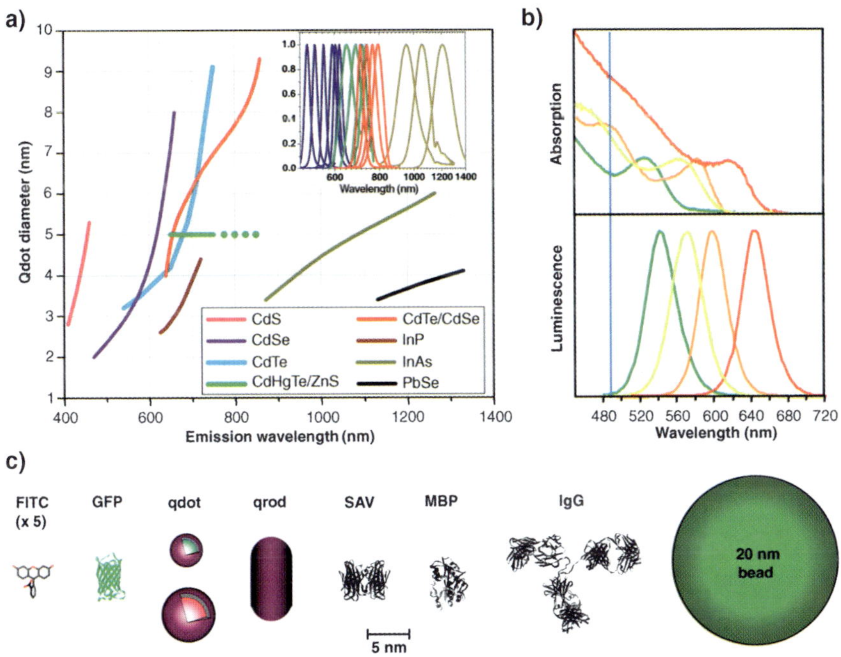

Fig. 8.8 Size and Spectral Properties of quantum dots (qdots). **a** Emission maxima and sizes of quantum dots of different composition. Qdots can be synthesized from various types of semiconductor materials (II–VI: CdS, CdSe, CdTe, etc.; III–V: InP, InAs, etc.; IV–VI: PbSe, etc.). The *curves* represent experimental data from the literature for the dependence of peak emission wavelength on qdot diameter. The range of emission wavelengths is 400–1,350 nm, with the size varying from 2 to 9.5 nm. All spectra are typically around 30–50 nm (full width at half maximum). *Inset*: Representative emission spectra for some materials. **b** Absorption (*upper curves*) and emission (*lower curves*) spectra of four CdSe/ZnS qdot samples. The *vertical blue line* indicates the 488-nm line of an argon-ion laser, which can be used to efficiently excite all four types of qdots simultaneously. **c** Size comparison of qdots and comparable objects. The qdot at the *top* is 4 nm CdSe/ZnS (*green*) and that at the *bottom* is 6.5 nm CdSe/ZnS (*red*). Three proteins – streptavidin (*SAV*), maltose-binding protein (*MBP*), and immunoglobulin G (*IgG*) – have been used for further functionalization of qdots. *FITC* fluorescein isothiocyanate, *qrod* rod-shaped qdot. (From Michalet et al. 2005, printed with permission from AAAS)

quantum dots to track proteins, and we will discuss an example of how the quantum dot properties have been exploited to track individual glycin receptors at the synaptic cleft.

8.4.3 Functionalizing Quantum Dots To Label Specific Proteins

One major challenge using quantum dots in conjunction with biological samples, and particularly for labeling proteins, is the need to solubilize them and to functionalize their surface. Since quantum dots are generally synthesized in nonpolar organic solutions, their hydrophobic surface ligands have to be replaced with amphiphilic ligands. Several solubilization strategies have been developed (Michalet et al. 2005); however, they will not be covered here because hydrophilic quantum dots with functionalized surfaces are now commercially available. Functionalizing of the quantum dot surface depends on the biological application (Fig. 8.9). For tracking of individual particles, particularly proteins, it is useful to use quantum dots with a surface that has been coated either with streptavidin or with a secondary antibody. These surfaces can then be detected either by a biotinylated first antibody or by the first antibody against the protein of choice. One drawback of this labeling technique is the size of the resulting complex between the protein of interest, the detecting antibody, and the functionalized quantum dot. Therefore, such approaches require stringent controls to test the functionality of the resulting protein–antibody–quantum dot complexes. It is also possible to make customized surface-coated quantum dots with antibody FAb against the

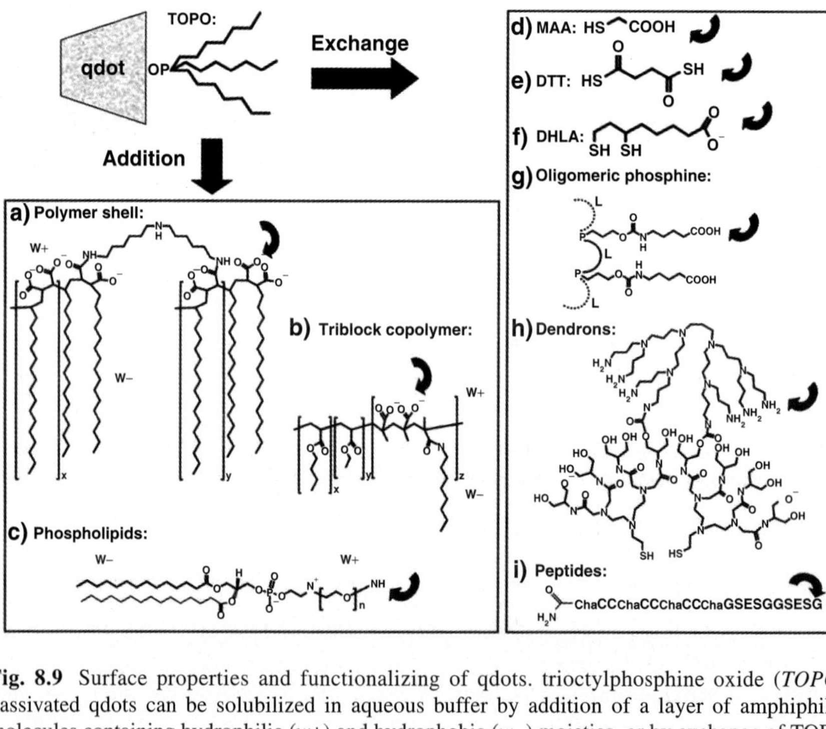

Fig. 8.9 Surface properties and functionalizing of qdots. trioctylphosphine oxide (*TOPO*) passivated qdots can be solubilized in aqueous buffer by addition of a layer of amphiphilic molecules containing hydrophilic (*w*+) and hydrophobic (*w*–) moieties, or by exchange of TOPO with molecules that have a Zn-coordinating end (usually a thiol group, SH) and a hydrophilic end. Examples of addition include **a** formation of a cross-linked polymer shell, **b** coating with a layer of amphiphilic triblock copolymer, and **c** encapsulation in phospholipid micelles. Examples of exchange include **d** mercaptoacetic acid (*MAA*), **e** dithiothreitol (*DTT*), **f** dihydrolipoic acid (*DHLA*), **g** oligomeric phosphines, **h** cross-linked dendrons, and **i** peptides. The *curved arrow* indicates sites available for further functionalization. (From Michalet et al. 2005, printed with permission from AAAS)

protein of choice. This alternative approach reduces the size of the forming quantum dot–protein conjugate.

8.4.4 Tracking the Glycin Receptor 1 at the Synaptic Cleft Using Quantum Dots

One impressive example of how the tracking of individual proteins inside living cells led to new insights into the functionality of cells was the investigation of the lateral dynamics of individual glycin receptors at neuronal membranes. Antoine Triller and colleagues labeled the glycin receptors, GlyR1 α, of spinal cultured neurons with quantum dots, and compared the lateral movements at the plasma membrane in the synaptic, perisynaptic, and extrasynaptic regions (Dahan et al. 2003; Fig. 8.10). The labeling was achieved via detection of GlyR1 with a primary antibody that was recognized by a biotinylated antimouse FAb fragment. This complex bound to streptavidin-coated quantum dots, and allowed tracking of the fluorescently labeled receptors in real time. The quantum dots were photostable, and allowed imaging of individual receptors up to 20 min. Diffusion coefficients of individual receptors were determined for the various membrane regions. This revealed that diffusion coefficients that had been measured previously using large 500-nm-diameter beads were about 4 times lower than those measured using the quantum dot–receptor complex. This indicated that larger beads impede proper receptor diffusion. The researchers found that the diffusion coefficients decreased 4–7 times in the synaptic cleft, and they also found that individual receptors were able to enter the synaptic cleft, and could subsequently exit this site again. Finally, the quantum dot–receptor complexes could be visualized by electron microscopy, allowing the correlation between live-cell imaging and high-resolution analysis.

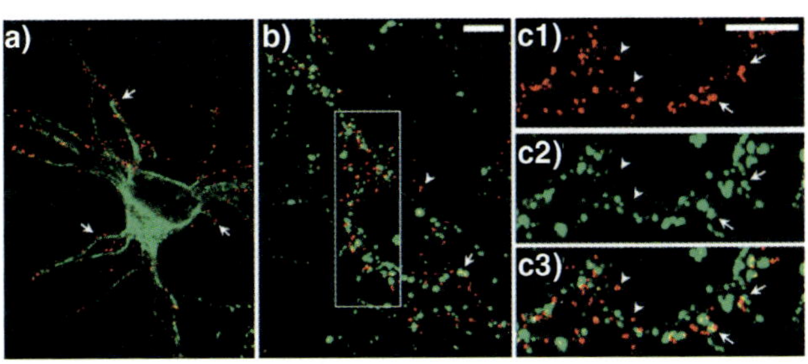

Fig. 8.10 Qdots as a marker for GlyR localization in neurons. **a** Qdot–GlyRs (*red*) detected over the somatodendritic compartment identified by microtubule-associated protein 2 (*green*). *Arrows* mark clusters of qdot–GlyRs located on dendrites. **b**, **c** Relation of qdot–GlyRs (*red*) with inhibitory synaptic boutons labeled for vesicular inhibitory amino acid transporter (*green*). Qdots are either in front of (*arrows*) or adjacent to (*arrowheads*) inhibitory synaptic boutons. The boxed region in **b** is shown as enlarged single-channel images in **c1**–**c3**. Images are projections of confocal sections. *Scale bars* 10 μm. (From Dahan et al. 2003, printed with permission from AAAS)

8.5 Tracking Analysis and Image Processing of Data from Particle Tracking in Living Cells

As illustrated by the examples in the previous paragraphs, single-molecule or single-particle tracking of mRNA, DNA, or proteins requires the rapid acquisition of microscopic images using rapid microscope systems, such as a Nipkow disc confocal microscope. This particular system produces images in three dimensions plus time with high spatial and temporal resolution. The trajectories followed by the individual molecules need to be extracted from the image sequences acquired. This process can be done manually when the number of molecules in a given image is very low. For example, in the case of single-chromosome tracking there is only a single locus to monitor. However, in many biological contexts there are dozens if not hundreds of molecules to track as is the case with mRNA. The tracking of mRNA bound by molecular beacons represents a unique case. Though the molecular beacons only fluoresce upon hybridizing to an mRNA, there is a certain amount of background fluorescence and thus the signal-to-noise ratio can be low. Conventional methods for particle tracking are based on simple intensity thresholding to detect individual particles and nearest-neighbor association to perform tracking. Such methods can function well when the number of particles is limited and of very high intensity in a uniform background. In most biological contexts the intensity is usually nonmodal and the images are noisy with a very high density of spots. We strongly recommend that scientists who desire to perform single-molecule tracking studies work closely with research groups experienced in design and implementation of quantitative image analysis algorithms. This permits the nuances of the biological question to be integrated in the image acquisition and image analysis algorithms. By working with a group composed of mathematicians, physicists, and biologists, all the necessary "skill sets" to perform quantitative imaging are present and enable highly sensitive and reproducible approaches to be used. We provide an example of such a "working group" in Box 8.1.

8.6 Conclusion

In this chapter we covered various approaches for how to track biological macromolecules in live cells that are currently used by cell biology laboratories. Obviously, all the methods presented are still very limited, mainly owing to the size of the fluorescent label required to detect the macromolecule of choice, and owing to the technical limitations of the currently available microscopes. Tremendous progress has been made to overcome these limitations. During the writing of this chapter, we started to test novel EMCCD cameras that are several times more sensitive than the CCD cameras that have been in use during the last few years. These novel cameras allow highly accelerated image acquisition, or the imaging of samples that are less fluorescent. Therefore, imaging of chromosomal loci or mRNP particles could possibly be achieved using fewer repeats than are currently used. This would diminish the possibility of artifacts due to the large size of the fluorescent reporter. Therefore, we expect that technical improvements during the coming years will facilitate single-particle analysis in live samples.

8.7 Protocols for Laboratory Use

8.7.1 Protocol: Single-Molecule Tracking of Chromosomal Loci in Yeast

One of the main challenges working with yeast cells is be to keep them alive and growing as well as to prevent them from moving around during the microscopic acquisition process.

First of all, cells from an exponentially growing culture should be used (optical density at 600 nm of less than 1). Furthermore, if possible for the specific experimental approach, we encourage the use of rich media in order to obtain healthy yeast for the imaging procedure. Rich media are often autofluorescent; therefore, cells growing in such media have to be washed several times with a synthetic medium prior to mounting them on slides for microscopic observation. Also note that yeast cells with the *ade*$^-$ genetic background accumulate an intermediate that is fluorescent when exited with blue light.

To prevent yeast cells from floating, they should be spread on slides coated with a synthetic medium patch containing 1.5% agarose. Subsequently, the slide containing the specimen should be sealed with VaLaP (one third vaseline, one third lanoline, one third paraffin). This mounting protocol has been shown to prevent both rotation and other movements of the entire yeast nuclei during image acquisition (Bystricky et al. 2005; Cabal et al. 2006).

8.7.2 Protocol: Single-Molecule Tracking of mRNA – Experiment Using Molecular Beacons

2′-*O*-Methyl molecular beacons (Mhlanga and Tyagi 2006) are designed and synthesized using standard protocols described previously (Bratu et al. 2003) and can be ordered from most suppliers of primers or oligonucleotides.

8.7.2.1 Determination of Quenching Efficiency

The signal-to-background ratio of all molecular beacons constructed with a fluorophore and a quencher is measured with a spectrofluorometer to ensure it can hybridize to its target in a specific manner and elicit a spontaneous increase in fluorescence upon hybridization:

1. Determine the baseline fluorescence of the solution at 25°C. Transfer an aliquot a 200-μl solution containing 30 nM molecular beacons in 5 mM $MgCl_2$ and 20 mM tris(hydroxymethyl)aminomethane hydrochloride (pH 8.0) into a quartz cuvette used with a spectrofluorometer.
2. Using maximal excitation and emission wavelengths add a twofold molar excess of the mRNA transcript of interest that has been in vitro transcribed (complementary to the loop portion of the molecular beacon) and monitor the rise in fluorescence until a stable level is reached. The rise in fluorescence over the signal of the molecular beacon alone (without the addition of oligonucleotide target) should be calculated to determine the signal-to-background ratio and quenching efficiency of the molecular beacon.

8.7.2.2 Visualizing and Tracking Single Particles of mRNA in Living Cells

1. Culture CHO/HELA cells in the alpha modification of Dubelcco's medium, supplemented with 10% fetal bovine serum on T4 culture dishes with a 0.17-mm cover glass with a coating of conductive material at the bottom to permit controlled heating. Ensure that the temperature of the T4 culture dish and the microscope objective are maintained at 37°C, preferably by using two Bioptech controllers. Just prior to imaging, exchange the Dubelcco medium (supplemented with 10% fetal bovine serum) with Leibovitz's L-15 medium (free of phenol red).
2. Using a Femtojet microinjection apparatus, microinject molecular beacons targeting your desired sequence in the mRNA. Collect images via epifluorescent or confocal microscopy. Your expected result is shown in Fig. 8.7d.
3. Alternatively you can introduce the molecular beacons into CHO/HELA cells via transfection. Culture the cells to 70% confluency in T4 culture dishes as in step 1.
4. Wash the CHO/HELA cells with serum-free Opti-MEM1. Incubate the Opti-MEM1 in transfection reagent, oligofectamine, for 5 min at a ratio of 1 μl oligofectamine to 9 μl oOpti-MEM1.
5. Combine the premixed oligofectamine and Opti-MEM1 with (1 ng/μl diluted in Opti-MEM1) the molecular beacon.
6. Incubate at 25°C for 20 min and then dilute the complex with 200 μl of serum-free medium. Then gently add this entire dilution to the CHO/HELA cells.
7. Incubate for 3 h and wash with Leibovitz medium supplemented with serum just prior to imaging. Collect images via epifluorescent or confocal microscopy. This should be preferably with a Nipkow disc confocal microscope as it provides the speed of acquisition required to acquire fast-moving mRNA particles.
8. Analyze the image stacks with QUIA to obtain tracking data as shown in Fig. 8.2.

The single molecule tracking algorithm, QUIA, can be implemented to track a few hundred particles of mRNA or single labeled locus in three dimensions plus time on image stacks acquired from rapid Nipkow disc microscopy. In these experiments key information must be predetermined prior to performing the acquisition to ensure that tracking analysis with QUIA yields accurate results. Such information includes the verification of chromatic aberration and the determination of the pixel size as determined by the camera and objective used. For mRNA tracking our experimental setup was a Zeiss Axiovert 200 with ×63 objective (PlanNeofluar, 1.4 numerical aperture oil immersion) with a Hamamatsu Orca II cooled CCD camera objective pixel size of 65.84 nm in the *X* and *Y* axes. The *Z* axis is determined by the piezo step size in the experiment e.g., 500 nm. The *X*, *Y*, and *Z* axis information can then be entered into QUIA to determine the voxel size. QUIA also requires the input of the exposure time so as to give the correct spatial, temporal, and kinetic information for single particles. In general, QUIA requires 8-bit TIFF files for tracking in three dimensions plus time. Most cameras produce 12-bit images and these must be converted within QUIA prior to performing 3D plus time single-molecule tracking. The data produced from such 3D plus time tracking experiments are shown in Fig. 8.2.

References

Babcock HP, Chen C, Zhuang X (2004) Using single-particle tracking to study nuclear trafficking of viral genes. Biophys J 87:2749–2758

Bertrand E, Chartrand P, Schaefer M, Shenoy SM, Singer RH, Long RM (1998) Localization of ASH1 mRNA particles in living yeast. Mol Cell 2:437–445

Bouchaud JP, Georges A (1990) Anomalous diffusion in disordered media: statistical mechanics, models and physical applications. Phys Rep 195:127

Bratu D (2006) Molecular beacons: fluorescent probes for detection of endogenous mRNAs in living cells. Methods Mol Biol 319:1–14

Bratu DP, Cha BJ, Mhlanga MM, Kramer FR, Tyagi S (2003) Visualizing the distribution and transport of mRNAs in living cells. Proc Natl Acad Sci USA 100:13308–13313

Bressan DA, Vazquez J, Haber JE (2004) Mating type-dependent constraints on the mobility of the left arm of yeast chromosome III. J Cell Biol 164:361–371

Bruchez MP (2005) Turning all the lights on: quantum dots in cellular assays. Curr Opin Chem Biol 9:533–537

Bystricky K, Heun P, Gehlen L, Langowski J, Gasser SM (2004) Long-range compaction and flexibility of interphase chromatin in budding yeast analyzed by high-resolution imaging techniques. Proc Natl Acad Sci USA 101:16495–16500

Bystricky K, Laroche T, van Houwe G, Blaszczyk M, Gasser SM (2005) Chromosome looping in yeast: telomere pairing and coordinated movement reflect anchoring efficiency and territorial organization. J Cell Biol 168:375–387

Cabal GG, Genovesio A, Rodriguez-Navarro S, Zimmer C, Gadal O, Lesne A, Buc H, Feuerbach-Fournier F, Olivo-Marin JC, Hurt EC, Nehrbass U (2006) SAGA interacting factors confine sub-diffusion of transcribed genes to the nuclear envelope. Nature 441:770–773

Chan WC, Nie S (1998) Quantum dot bioconjugates for ultrasensitive nonisotopic detection. Science 281:2016–2018

Chubb JR, Boyle S, Perry P, Bickmore WA (2002) Chromatin motion is constrained by association with nuclear compartments in human cells. Curr Biol 12:439–445

Cremer T, Cremer C (2001) Chromosome territories, nuclear architecture and gene regulation in mammalian cells. Nat Rev Genet 2:292–301

Dahan M, Levi S, Luccardini C, Rostaing P, Riveau B, Triller A (2003) Diffusion dynamics of glycine receptors revealed by single-quantum dot tracking. Science 302:442–445

Dirks RW, Molenaar C, Tanke HJ (2001) Methods for visualizing RNA processing and transport pathways in living cells. Histochem Cell Biol 115:3–11

Dreyfuss G, Kim VN, Kataoka N (2002) Messenger-RNA-binding proteins and the messages they carry. Nat Rev Mol Cell Biol 3:195–205

Feuerbach F, Galy V, Trelles-Sticken E, Fromont-Racine M, Jacquier A, Gilson E, Olivo-Marin JC, Scherthan H, Nehrbass U (2002) Nuclear architecture and spatial positioning help establish transcriptional states of telomeres in yeast. Nat Cell Biol 4:214–221

Fuchs J, Lorenz A, Loidl J (2002) Chromosome associations in budding yeast caused by integrated tandemly repeated transgenes. J Cell Sci 115:1213–1220

Gartenberg MR, Neumann FR, Laroche T, Blaszczyk M, Gasser SM (2004) Sir-mediated repression can occur independently of chromosomal and subnuclear contexts. Cell 119:955–967

Genovesio A, Olivo-Marin J (2004) Split and merge data association filter for dense multi-target tracking. In: Proceedings of the 17th international conference on pattern recognition, vol 4, pp 677–680

Genovesio A, Belhassine Z, Olivo-Marin J (2004) Adaptive gating in Gaussian Bayesian multi-target tracking. In: Proceedings of the international conference on image processing, vol 1, pp 147–150

Genovesio A, Liedl T, Emiliani V, Parak WJ, Coppey-Moisan M, Olivo-Marin JC (2006) Multiple particle tracking in 3-D+t microscopy: method and application to the tracking of endocytosed quantum dots. IEEE Trans Image Process 15:1062–1070

Goldman R, Spector D (2005) Live cell imaging, a laboratory manual. Cold Spring Harbor Press, Cold Spring Harbor

Gonzalez-Serricchio AS, Sternberg PW (2006) Visualization of C. elegans transgenic arrays by GFP. BMC Genet 7:36

Gorski SA, Dundr M, Misteli T (2006) The road much traveled: trafficking in the cell nucleus. Curr Opin Cell Biol 18:284–290

Havlin S, Ben-Avraham D (2002) Diffusion in disordered media. Adv Physics 51:187–292

Heun P, Laroche T, Shimada K, Furrer P, Gasser SM (2001) Chromosome dynamics in the yeast interphase nucleus. Science 294:2181–2186

Janicki SM, Tsukamoto T, Salghetti SE, Tansey WP, Sachidanandam R, Prasanth KV, Ried T, Shav-Tal Y, Bertrand E, Singer RH, Spector DL (2004) From silencing to gene expression: real-time analysis in single cells. Cell 116:683–698

Jansen RP (2001) mRNA localization: message on the move. Nat Rev Mol Cell Biol 2:247–256

Marshall WF, Straight A, Marko JF, Swedlow J, Dernburg A, Belmont A, Murray AW, Agard DA, Sedat JW (1997) Interphase chromosomes undergo constrained diffusional motion in living cells. Curr Biol 7:930–939

Mathews DH, Sabina J, Zuker M, Turner DH (1999) Expanded sequence dependence of thermodynamic parameters improves prediction of RNA secondary structure. J Mol Biol 288:911–940

Mhlanga M, Tyagi S (2006) Using tRNA-linked molecular beacons to image cytoplasmic mRNAs in live cells. Nat Protocols 1:1392–1398

Mhlanga MM, Vargas DY, Fung CW, Kramer FR, Tyagi S (2005) tRNA-linked molecular beacons for imaging mRNAs in the cytoplasm of living cells. Nucleic Acids Res 33:1902–1912

Michaelis C, Ciosk R, Nasmyth K (1997) Cohesins: chromosomal proteins that prevent premature separation of sister chromatids. Cell 91:35–45

Michalet X, Pinaud FF, Bentolila LA, Tsay JM, Doose S, Li JJ, Sundaresan G, Wu AM, Gambhir SS, Weiss S (2005) Quantum dots for live cells, in vivo imaging, and diagnostics. Science 307:538–544

Misteli T (2004) Spatial positioning; a new dimension in genome function. Cell 119:153–156

Ober RJ, Ram S, Ward ES (2004) Localization accuracy in single-molecule microscopy. Biophys J 86:1185–1200

Palacios IM (2002) RNA processing: splicing and the cytoplasmic localisation of mRNA. Curr Biol 12:R50–52

Robinett CC, Straight A, Li G, Willhelm C, Sudlow G, Murray A, Belmont AS (1996) In vivo localization of DNA sequences and visualization of large-scale chromatin organization using lac operator/repressor recognition. J Cell Biol 135:1685–1700

Shav-Tal Y, Darzacq X, Shenoy SM, Fusco D, Janicki SM, Spector DL, Singer RH (2004) Dynamics of single mRNPs in nuclei of living cells. Science 304:1797–1800

St Johnston D (2005) Moving messages: the intracellular localization of mRNAs. Nat Rev Mol Cell Biol 6:363–375

Straight AF, Belmont AS, Robinett CC, Murray AW (1996) GFP tagging of budding yeast chromosomes reveals that protein-protein interactions can mediate sister chromatid cohesion. Curr Biol 6:1599–1608

Taddei A, Hediger F, Neumann FR, Bauer C, Gasser SM (2004) Separation of silencing from perinuclear anchoring functions in yeast Ku80, Sir4 and Esc1 proteins. EMBO J 23: 1301–1312

Taddei A, Van Houwe G, Hediger F, Kalck V, Cubizolles F, Schober H, Gasser SM (2006) Nuclear pore association confers optimal expression levels for an inducible yeast gene. Nature 441: 774–8.

Tekotte H, Davis I (2002) Intracellular mRNA localization: motors move messages. Trends Genet 18:636–642

Tsukamoto T, Hashiguchi N, Janicki SM, Tumbar T, Belmont AS, Spector DL (2000) Visualization of gene activity in living cells. Nat Cell Biol 2:871–878

Tyagi S, Kramer FR (1996) Molecular beacons: probes that fluoresce upon hybridization. Nat Biotechnol 14:303–308

Vargas DY, Raj A, Marras SA, Kramer FR, Tyagi S (2005) Mechanism of mRNA transport in the nucleus. Proc Natl Acad Sci USA 102:17008–17013

Vazquez J, Belmont AS, Sedat JW (2001) Multiple regimes of constrained chromosome motion are regulated in the interphase Drosophila nucleus. Curr Biol 11:1227–1239

Waggoner A (2006) Fluorescent labels for proteomics and genomics. Curr Opin Chem Biol 10:62–66

Webb CD, Teleman A, Gordon S, Straight A, Belmont A, Lin DC, Grossman AD, Wright A, Losick R (1997) Bipolar localization of the replication origin regions of chromosomes in vegetative and sporulating cells of B. subtilis. Cell 88:667–674

Zuker M (2003) Mfold web server for nucleic acid folding and hybridization prediction. Nucleic Acids Res 31:3406–3415

9
From Live-Cell Microscopy to Molecular Mechanisms: Deciphering the Functions of Kinetochore Proteins

Khuloud Jaqaman, Jonas F. Dorn, and Gaudenz Danuser

Abstract The goal of cell biology research is to explain cell behavior as a function of the dynamics of subcellular molecular assemblies. Live-cell light microscopy has emerged as the method of choice for probing molecular function in a near-physiological environment. However, light-microscopy data are on the cellular scale, while data interpretation occurs on the molecular scale. To bridge the gap between these two scales, empirical mathematical models of the relationship between molecular action and cellular behavior must be devised and calibrated using the experimental data. In this chapter we discuss several necessary steps to achieve this task. First, experiments should be designed such that the molecular action of interest is probed with sufficient spatial and temporal resolution and such that the resulting imagery is amenable to computational analysis. Second, automated image analysis tools must be developed to extract from the experiments reliable and reproducible quantitative data necessary for model calibration. Third, since molecular action is generally stochastic, experimental data and model simulation results cannot be compared directly. Rather, they have to be analyzed to obtain a set of descriptors that allows their indirect comparison for the purpose of model calibration. These descriptors should be complete, unique and sensitive. Throughout the chapter, we illustrate these steps using the regulation of microtubule dynamics by kinetochore proteins during chromosome segregation as an illustrative example.

9.1 Introduction

The goal of cell biology research is to explain normal and aberrant cellular-scale behavior, such as in mitosis, cell migration and endocytosis, in terms of the underlying molecules and their interactions. In vitro methods, such as messenger RNA expression profiling, protein affinity chromatography and coimmunoprecipitation, reveal which proteins bind to each other and which play a role in a certain cellular phenomenon. Such information can be used to group proteins into larger complexes

S.L. Shorte and F. Frischknecht (eds.), *Imaging Cellular and Molecular Biological Functions.*

that act as functional units (De Wulf et al. 2003). However, the understanding of where and when these proteins interact in their native cellular environment and what their contributions are to the cellular function of interest requires the study of protein dynamics in situ.

The one method that allows the study of protein function in an environment close to the natural milieu of proteins is live-cell light microscopy. In particular, light microscopy allows us to monitor the dynamics of fluorescently labeled molecules (or macromolecules), such as proteins and chromosomes. Thus, techniques such as colocalization, fluorescence resonance energy transfer and fluorescence correlation spectroscopy have been used to devise models of the functional relationships between proteins in space and time.

However, the models constructed are generally qualitative. They are limited to small and relatively simple interaction networks. The agreement between model predictions and experimental data is only qualitative. Furthermore, they cannot be tested extensively because it is difficult to predict the cellular-scale consequences of their molecular-level manipulation.

These problems can be overcome when quantitative models of the molecular interactions are constructed instead. Mathematical formulae are a convenient method for representing interactions and dependencies between arbitrarily many proteins. The predictions of mathematical models are readily obtained by solving the set of equations representing the model, no matter its complexity. Mathematical models can be also manipulated in a straightforward manner to mimic perturbations of the system, and thus they can be comprehensively tested by comparing the predictions of the perturbed model with the corresponding experimental data.

However, biological systems are complex, prohibiting the construction of quantitative models from first principles, where, ideally, the quantum mechanical equations describing the system are solved to obtain its configuration as a function of time. Even the structure of a single protein cannot be obtained from first principles! Thus, quantitative models of biological systems have to be empirical. In contrast to first-principles models, empirical models contain a set of parameters whose values must be determined from experimental data. But live-cell microscopy data are on the cellular scale, while model parameters pertain to interactions on the molecular scale. Hence, these parameters cannot be directly derived from the experimental data; instead, cellular-scale data must be generated from the model and compared with experimental data. In this approach, the set of parameters that reproduces the experimentally observed dynamics is considered to be the correct set of parameters.

To achieve quantitative accuracy in a model, the matching between experimental data and model predictions must be done quantitatively and not only qualitatively (such as by visual inspection, as is usually done). This is not straightforward, however, owing to the stochastic nature of the dynamics of molecules. This stochasticity results from a combination of the inherent probabilistic nature of molecular interactions (intrinsic source) and the information loss between observed states due to undersampling (extrinsic source) (Jaqaman et al. 2006). By definition, the state at time t of a dynamic system that is driven by a stochastic process only determines the probabilities of its possible states at time $t+1$, and not the exact state that it will transition to. Consequently, it is meaningless to compare the stochastic dynamics of molecules and macromolecules time point by time point. What is meaningful is to compare the processes that have generated the observed dynamics. But these processes, namely, the underlying molecular-level interactions, are not available. In fact, the whole purpose of this chapter is to provide a method to obtain them!

A practical step that facilitates the comparison of simulation results with experimental data is analysis of the dynamics with relatively simple models that describe them on the cellular scale. These models have the advantage that their parameters can be obtained directly from the data, since both the model and the data are on the same scale. If a model is appropriate, it will require different parameters for dynamics with different characteristics; thus, model parameters can be used as descriptors of the dynamics. Dynamics under different conditions can be indirectly compared by comparing their descriptors. Furthermore, descriptors can be used as intermediate statistics for matching simulation with experiment for the purpose of model calibration (Gourieroux et al. 1993; Smith 1993).

With the above issues in mind, our strategy for elucidating the molecular interactions that underlie a certain cellular function is presented in Fig. 9.1. Experimentally, the cellular system of interest is imaged, potentially after some molecular perturbation, and the dynamics of the labeled molecules are obtained via image analysis. Then the dynamics are analyzed and their descriptors are determined, a task generally referred to as data mining. In parallel, simulated dynamics of the labeled molecules are generated using a model of the known relevant molecular interactions. The descriptors of simulated and experimental dynamics are compared, and model parameters are iteratively adjusted until simulated and experimental descriptors are statistically equivalent. Owing to functional redundancy in complex protein interaction networks, the perturbation of certain molecules might not have an effect on the cellular-scale dynamics. For this reason it is sometimes necessary to perform experiments with multiple perturbations to identify the functions of such components.

In this chapter, we will elaborate on several of the tasks presented in Fig. 9.1, particularly image acquisition (Sect. 9.3), image analysis (Sect. 9.4) and data mining (Sect. 9.5). The iterative model calibration procedure is one of the most challenging

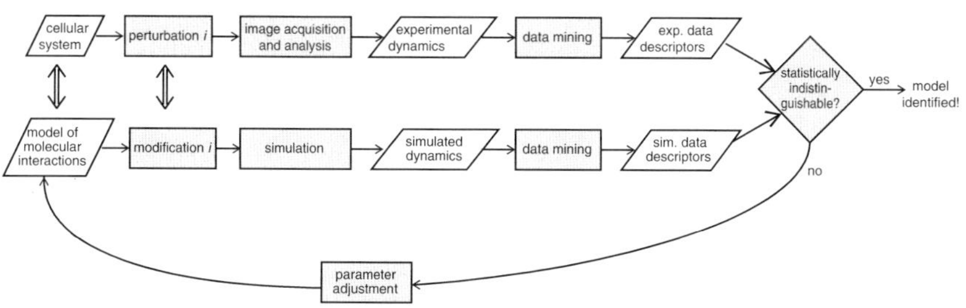

Fig. 9.1 The integration of live-cell microscopy with molecular-scale mathematical models of protein interactions, allowing the study of protein function in situ. *Double arrows* indicate equivalence between the system to be studied and the model representing it

tasks within this framework, but we will not address it further because of space limitations. Throughout the chapter, we will use a specific biological question namely, elucidating kinetochore protein function (introduced in Sect. 9.2), as an illustrative example. We present some of our biological results, obtained via the methods discussed throughout the chapter, in Sect. 9.6. Finally, Sect. 9.7 includes some concluding remarks.

9.2 Biological Problem: Deciphering the Functions of Kinetochore Proteins

One of the most central questions in cell biology is how dividing cells ensure that replicated chromosomes are correctly transferred to the two daughter cells. The machinery responsible for chromosome segregation is the mitotic spindle, which is principally composed of microtubules (MTs) that emanate from two oppositely located spindle poles (Alberts et al. 2002). During mitosis, MTs grow and shrink and switch between the two states in a process called dynamic instability (Mitchison and Kirschner 1984) in order to capture chromosomes and achieve bipolar attachment (Alberts et al. 2002). Once bipolar attachment is achieved, the MTs jointly shrink and pull sister chromatids apart.

MT–chromosome attachment takes place at a specific site on the chromosome, termed the centromere (CEN), onto which a protein complex, called the kinetochore, assembles. The kinetochore acts as an interface between chromosomes and MTs, and it is highly likely that kinetochore proteins regulate kinetochore–MT (k-MT) dynamics. However, little is known about the specific functions of kinetochore proteins in terms of how they regulate k-MT dynamics, what chemical or mechanical signals they process, and in what hierarchy they transmit these signals to k-MTs.

We intend to elucidate the interactions between kinetochore proteins and build a mathematical model of their mechanochemical regulation of k-MT dynamics using the framework of Fig. 9.1. We have chosen the budding yeast *Saccharomyces cerevisiae* as our model system because (1) its kinetochore is composed of a relatively small number (about 70) of known proteins (Cheeseman et al. 2002; De Wulf et al. 2003), (2) in budding yeast we can thoroughly manipulate the kinetochore and dissect the functional interactions between kinetochore proteins and (3) in contrast to mammalian spindles where there are around 20 k-MTs per sister chromatid, *S. cerevisiae* has only one k-MT per sister chromatid (O'Toole et al. 1999) that establishes attachment to the spindle pole body (SPB). *S. cerevisiae* k-MTs do not seem to treadmill (Maddox et al. 2000) and their minus-ends are fixed at the SPB: thus, the motion of a chromatid in budding yeast is the direct result of assembly and disassembly at the plus-end of one k-MT.

However, these many advantages come at a price. The small size of yeast poses considerable challenges for imaging and image analysis. The only way to observe the dynamics of a single k-MT is to label the SPB and a CEN proximal region on a chromosome (Robinett et al. 1996; Straight et al. 1996; Fig. 9.2a). Because the

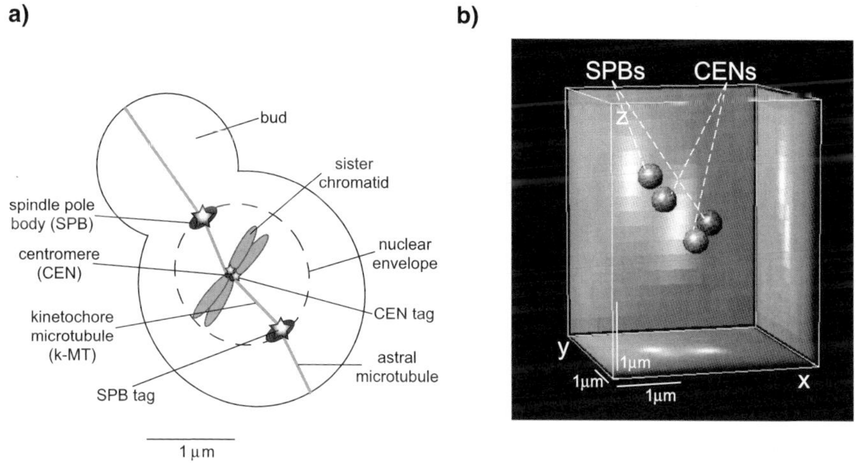

Fig. 9.2 **a** The relatively simple bipolar attachment of sister chromatids in budding yeast in metaphase. **b** Typical 3D image of two spindle pole body (*SPB*) tags and two centromere (*CEN*) tags in metaphase (image filtered for clarity)

distances between tags are at the resolution limit and their images are highly overlapping (Fig. 9.2b), advanced image analysis techniques are needed to extract the dynamics accurately. Furthermore, the small size of the yeast nucleus implies that a k-MT switches very frequently between growth and shrinkage, and hence very fast temporal sampling is needed. But, as discussed in Sect. 9.3, we can only sample at a rate of one frame per second, so the observed dynamics are most likely undersampled and aliased, posing a challenge for data analysis. To get a rough estimate of the timescale of events, let us assume that k-MTs have the same shrinkage rate as MTs in vitro (0.45 μm/s; Walker et al. 1988) and that a k-MT in metaphase spans half of the nucleus before switching from shrinkage to growth (a distance of around 0.75 μm). Thus, the time spent in shrinkage before switching to growth is approximately $(0.75\,\mu\text{m})/(0.45\,\mu\text{m/s}) \approx 1.5\,\text{s}$.

9.3 Experimental Design

Probing a dynamic process involves gathering both spatial and temporal information. Accurate spatial information requires images with a high signal-to-noise ratio (SNR). In terms of sampling, the pixel size and temporal sampling interval should be at most one third the extent of the point-spread function (PSF) and the timescale of the fastest process of interest, respectively (note that Nyquist's theorem gives a ratio of one half as the upper limit, but that applies to noise-free data and is not sufficient for noisy images; Stelzer 2000). Moreover, in the case of stochastic processes, the measurement window has to be long enough to capture all states of the underlying process.

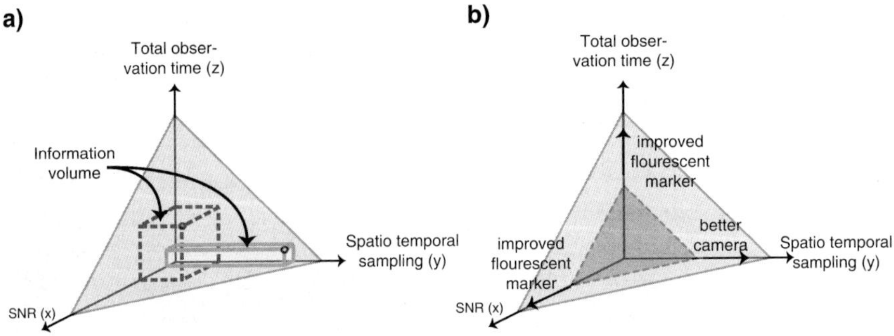

Fig. 9.3 Conflict between spatiotemporal sampling, total observation time and signal-to-noise ratio (*SNR*) in experimental design. **a** An improvement in one experimental aspect leads to a deterioration in another. **b** An increase in maximal information volume can be achieved only by improving experimental conditions

However, there are conflicts between these requirements (depicted graphically in Fig. 9.3):

1. SNR vs. temporal sampling (*x*,*y*-axes in Fig. 9.3). The acquisition of an image with high SNR requires a long enough exposure time to collect enough light. This minimum exposure time sets an upper limit for the sampling frequency that might be lower than what is needed to fulfill the required sampling criterion. Thus, one improves SNR at the expense of temporal resolution, and vice versa.
2. SNR vs. spatial sampling (*x*,*y*-axes in Fig. 9.3). As the pixel size gets smaller, the fluorescence signal from one tag gets distributed over a larger number of pixels. Thus, the observed intensity becomes lower, reducing with it the SNR in the image. This leads to a conflict between spatial sampling and SNR.
3. SNR vs. total observation time (*x*,*z*-axes in Fig. 9.3). There is a conflict between SNR and total observation time, since the acquisition of images with higher SNR at fixed spatiotemporal sampling requires an increase in the amount of excitation light, leading to faster photobleaching and sample inviability owing to phototoxicity.
4. Temporal sampling vs. total observation time (*y*,*z*-axes in Fig. 9.3). Faster sampling implies a higher exposure of the sample to light in a shorter period of time, leading to a shorter overall observation time owing to photobleaching and phototoxicity.
5. Spatial sampling vs. total observation time (*y*,*z*-axes in Fig. 9.3). In order to increase spatial sampling while retaining the same SNR, brighter sample illumination is needed. This leads to faster photobleaching and expedites sample inviability, reducing the total observation time. Thus, there is a conflict between spatial sampling and total observation time.

Note that the information output of an experiment is maximal for a certain combination of spatiotemporal sampling rate, SNR and total observation time (Fig. 9.3a, parallelepiped with dashed edges), and an improvement in one of these aspects not only leads to a deterioration of the other aspects, but also to a loss in the total amount of information obtained (Fig. 9.3a, parallelepiped with solid edges). The maximal total amount of information yielded by an experiment can be increased only by improving the experimental conditions, such as by using a better fluorescent marker or a better camera (Fig. 9.3b).

Another issue to consider when designing an experiment is whether two-dimensional (2D) imaging is sufficient, or whether three-dimensional (3D) imaging is needed. 3D imaging involves taking a stack of 2D images, which makes it 1–2 orders of magnitude slower than 2D imaging. It also increases the speed of photobleaching due to out-of-focus light, decreasing the total observation time. If the system studied is relatively flat, 2D imaging is sufficient and one spatial degree of freedom can be sacrificed in order to gain temporal resolution and observation time. Otherwise, 3D imaging is necessary, even if that means less temporal sampling and a shorter observation window.

Given these conflicts, prior knowledge about the system to be studied, as well as characteristics of the available image and data analysis techniques, should be used to optimally design an experiment that efficiently yields the necessary information. For example, if the objective of an experiment is to probe the diffusion of a presumed Brownian particle, then high temporal resolution is not necessary; on the contrary, long measurement times and good spatial resolution are needed. On the other hand, if the mechanism of state transitions in a dynamic process is of interest, then prior knowledge about the time scales of the dynamics can be used to determine the required temporal sampling frequency. Furthermore, an analysis of the data characterization methods employed informs us of the number of observations needed to fully sample the process of interest (see, for example, the study of the convergence of descriptor estimation shown in Fig. 9.4). Given the total observation time of one experiment, knowledge of the necessary number of observations helps us determine the number of times an experiment must be repeated to get a good sample of the dynamics. Notice that this repetition of experiments and collection of data are not needed to increase the sample size for the sake of improving statistics, but are essential to get a complete picture of the dynamics.

When there is little prior knowledge about the system studied, one can use the experimental data themselves to determine whether the current experimental conditions allow the accurate probing of system dynamics. Analysis of experimental data will reveal the reliability of an experiment, its limitations and which of its aspects need improvement. For example, given the high switching frequency of k-MTs in budding yeast between growth and shrinkage, and the fact that 3D imaging is much slower than 2D imaging, we have investigated the possibility of limiting the image acquisition to two dimensions to increase temporal sampling. In order to get data similar to what would be obtained via 2D image acquisition, we projected tag coordinates from our 3D data sets onto the imaging plane and retained only those frames where both tags were visible in the imaging plane. The resulting SPB–CEN distance trajectories (i.e., distance as a function of time) were significantly distorted and suffered from substantial data loss (compare the in-focus trajectory with the 3D data trajectory in Fig. 9.5a). Note that retaining all of the time points, which is

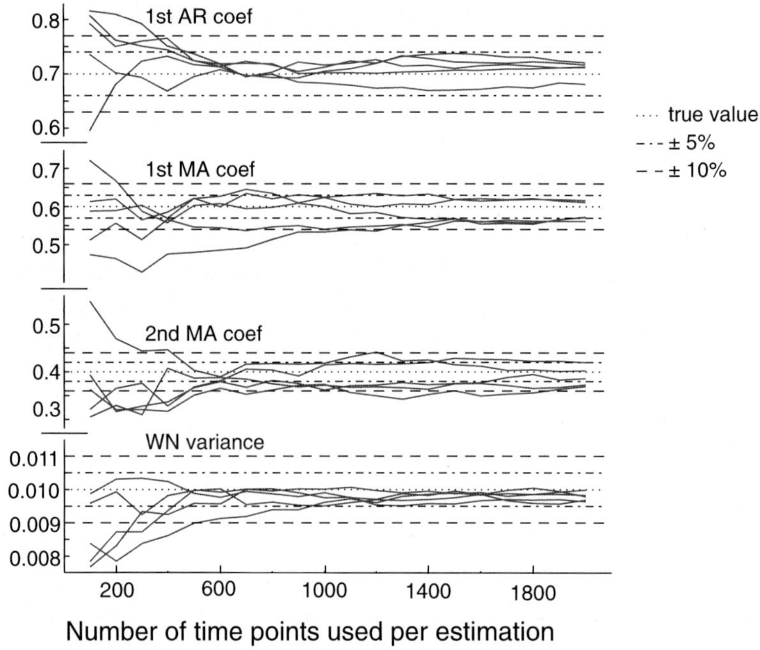

Fig. 9.4 Convergence of estimated autoregressive moving average (ARMA) descriptors (see Sect. 9.5.3 for a discussion of ARMA descriptors) toward their true values as sample size increases. A total of 1,500–2,000 data points are needed to get estimates of ARMA descriptors that are within 5–10% of their true values. *AR* autoregressive, *MA* moving average, *WN* white noise. (Reproduced from Jaqaman et al. 2006 with permission from *Biophysical Journal*)

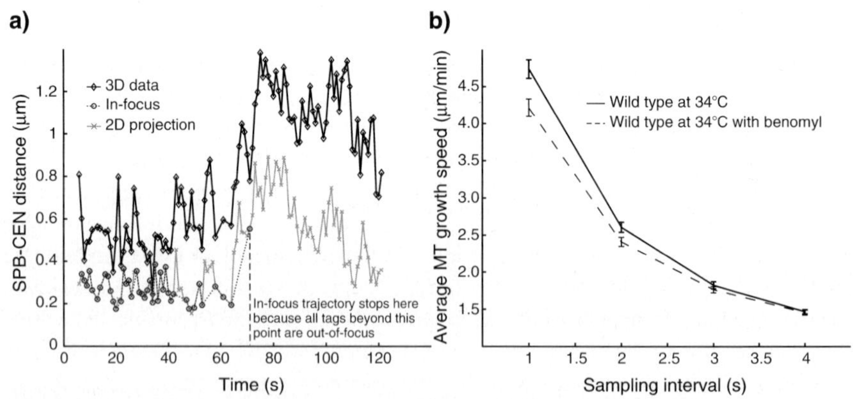

Fig. 9.5 Analysis of experimental data to reveal the reliability and limitations of experiments. **a** Budding yeast requires 3D imaging; 2D imaging distorts trajectories and leads to substantial data loss. **b** A sampling rate of at least one frame per second is needed to capture kinetochore–MT (k-MT) dynamics in budding yeast. (Reproduced from Dorn et al. 2005 with permission from *Biophysical Journal*)

equivalent to 2D projection that is done sometimes to simplify image analysis, significantly distorts the trajectories obtained (2D projection trajectory in Fig. 9.5a). Thus, in our case, 3D image acquisition is essential to get accurate SPB and CEN tag coordinates and thus accurate MT length trajectories.

Another aspect of probing a system's dynamics that must be investigated is the effect of temporal sampling on the dynamics and their descriptors. In our case, temporal sampling is limited to one frame per second, and so we must check whether it is sufficient to capture the essential dynamics or whether there are processes that are too fast to be captured. One way to tackle this issue is by artificially downsampling the experimental data and investigating the resulting changes in the calculated descriptors. If the descriptors do not change, then even the slower sampling rate is sufficient to probe the dynamics. Otherwise, the original higher sampling rate must be used, and processes that take place at frequencies more than half the sampling frequency are not observable. To examine the sampling rate in our experiments, we have downsampled 1-s data from wild-type (WT) yeast at 34°C with and without the MT drug benomyl and analyzed the effect of 2-, 3- and 4-s sampling on the average growth speed in those two conditions (Fig. 9.5b). The growth speed at 2-s sampling drops to about 55% of its value at 1-s sampling in both cases, indicating a significant information loss; thus we should not sample any slower than one frame per second. Notice that the ability to distinguish between the two conditions is also lost with slower sampling. In fact, at 3-s sampling, the growth speeds of the two conditions are no longer distinguishable.

In summary, given the substantial effect of sampling rate on the observed dynamics and their descriptors, experiments have to be designed such that they fully capture the dynamics. If this is not possible, data that are to be compared with each other must be obtained from experiments with identical experimental setting to avoid artifacts.

9.4 Extraction of Dynamics from Images

To allow the comparison of model predictions with experimental data, quantitative information has to be extracted from the raw images. The data should be reliable, reproducible and consistent, which is ideally achieved by fully automated computer vision methods that rely on a rigorous hypothesis testing framework. Computer vision methods are particularly superior to manual image analysis in a case like ours where tag images highly overlap (owing to the small size of yeast) and where the SNR is low (owing to the necessity of both 3D spatial sampling and fast temporal sampling). Under such conditions, it is very difficult for the human eye to locate tag images, let alone their centers, to determine tag positions. Computer vision algorithms, on the other hand, can use prior knowledge about the shape of tag images to search for them (since a tag is a subresolution feature, its image is the PSF of the microscope). This use of prior knowledge leads to both subpixel localization of all detected tags and the superresolution of tags. Superresolution refers to the ability to resolve tags that are closer than the Rayleigh limit. Note that, even if image conditions are ideal (with low overlap and high SNR), it is very difficult to accurately locate tag positions in three dimensions by visual inspection. In contrast, computer vision methods can be used to locate tag positions in any number of dimensions.

In the following subsections we describe an image processing scheme for the fully automated extraction of the CEN and SPB tags in the yeast spindle. Our algorithm involves three steps: (1) tag detection and localization via mixture-model fitting (MMF), (2) tag tracking, i.e., tag linking between frames, and (3) enhancement of localization and resolution via multitemplate matching (MTM).

9.4.1 Mixture-Model Fitting

The MMF algorithm addresses the problem of tag localization when tag images overlap. As we have discussed elsewhere (Thomann et al. 2002, 2003; Dorn et al. 2005), it exploits the prior knowledge that tag images are PSFs and that the images contain a finite number of discrete tags. While in our yeast system the number of tags is particularly low, MMF approaches can be applied to images with several thousand discrete tags.

To initialize mixture models, the images are first treated with a blob detector that segments each image into regions containing one or more tags each. These blobs are then fitted with a model of the form

$$M(\mathbf{x},\mathbf{a},b,\mathbf{c},n)=\sum_{i=1}^{n} a_i \cdot \mathrm{PSF}(\mathbf{x}-\mathbf{c}_i)+b. \qquad (9.1)$$

The free parameters in this model are the number n of PSF kernels (i.e., tag images contributing to a blob), the positions $\mathbf{c}_i=(x_i,y_i,z_i)$ of their centers, their signal magnitudes a_i, and a common background intensity b. The vector $\mathbf{x}$ denotes the coordinates of any voxel inside the volume enclosing the spot analyzed.

The main challenge in the fitting is the determination of n. In view of the low number of tags in our images, we apply a bottom-up strategy to identify the optimal balance between the number of degrees of freedom of the model and the χ^2 statistics of the residuals from the fit (i.e., the actual intensity minus the intensity from the fit, at all voxels considered). Bottom-up strategies begin with an order n=1 and increase the number of kernels until adding another kernel is no longer statistically justified. This is in contrast to top-down strategies, which begin with a very large number of kernels and reduce it until it is no longer statistically justified. From the intensity residuals at n=1, we estimate the uncertainties in signal magnitude and position by means of Gaussian error propagation (Koch 1988). If the signal magnitude is not significantly above the noise magnitude, the blob is rejected. Otherwise, the procedure is repeated for mixture models of increasing order until the improvement in the residuals of the model of order n+1 relative to the residuals of the model of order n is not significant, or until the distance between any two kernels is not significant compared with their combined positional uncertainties. Kernels for which any of these tests fails are rejected. The output of the MMF module is a list of tags, where each tag is assigned a position and brightness, and their uncertainties.

9.4.2 *Tag Tracking*

After detecting tags in each frame, we need to track them by linking corresponding tags in consecutive frames. When the image data are sparse, simple nearest-neighbor assignment can be used. Our algorithm uses a modified nearest-neighbor approach where we jointly minimize the displacement of tags between frames (corrected for stage drift) and the change in tag intensity (corrected for photobleaching), while conserving the number of tags. When the MMF algorithm has failed to separate all tags in a frame, the linking module assigns multiple tags from the previous frame to the same position, thereby creating a fusion blob.

When there are many tags and their images are overlapping, nearest-neighbor assignment does not work. Tag tracking in this case is a nontrivial problem that is beyond the scope of this chapter. The interested reader can refer to Blackman and Popoli (1999) for more details.

9.4.3 *Multitemplate Matching*

In order to further enhance the resolution and precision in tag localization, we have devised an algorithm that performs a relative matching of linked tag signals, exploiting the prior knowledge that tag brightness (corrected for photobleaching) is constant over time and that tags in our images do not disappear. This algorithm is based on the fundamental principle that relative measurements are more accurate than absolute measurements, since systematic errors, such as optical aberrations, cancel out in a relative measurement. Thus, tag signals in a source frame where tags have been resolved by MMF are taken as templates that are matched to the corresponding signals in a target frame that contains several unresolved tags (in fusion blobs); thus, the name multitemplate matching (MTM). The displacements of tags between source and target frames are taken to be those which minimize the difference between the combined image of all source tags, each shifted by an unknown displacement, and the target image (Thomann et al. 2002, 2003; Dorn et al. 2005).

As with MMF, the least-squares optimization used in MTM allows the rigorous propagation of the effect of image noise and tag overlap on the precision of the displacement estimates (Thomann et al. 2002, 2003; Dorn et al. 2005). Thus, the positional uncertainty of every tag localized in the target frame is calculated as the sum of the positional uncertainty of the tag in the source frame and the uncertainty of the MTM displacement estimate.

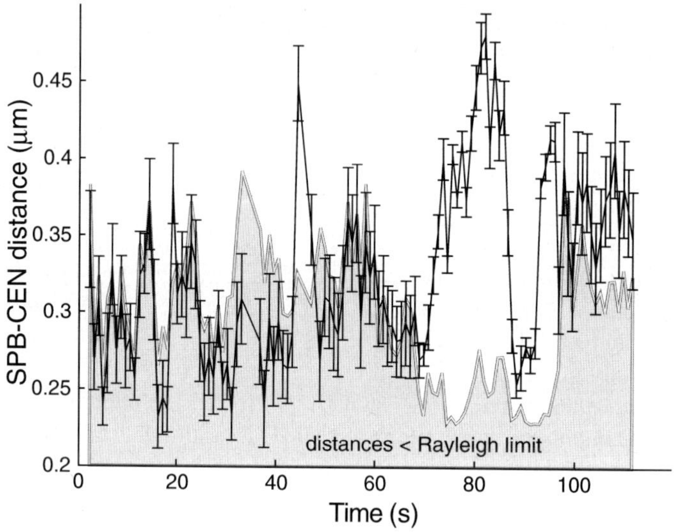

Fig. 9.6 A SPB–CEN distance trajectory and its uncertainty, displaying superresolution of tags. (Reproduced from Dorn et al. 2005 with permission from *Biophysical Journal*)

Figure 9.6 shows a SPB–CEN distance trajectory (approximating a k-MT length trajectory in the case of chromosome attachment to SPB) extracted via the three steps discussed above. The error bars on the graph indicate the distance uncertainty at each time point, as derived by yet another propagation of the positional uncertainties of the two tags. Note that the SPB–CEN distance between 20 and 40 s is smaller than the limit of optical resolution as expressed here by the Rayleigh limit. These results indicate the power of computer vision methods not only in automating detection and tracking, but also in determining tag coordinates beyond the visual perception of a human observer.

9.5 Characterization of Dynamics

As mentioned Sect. 9.1, the dynamics of molecules are stochastic; thus, for the purpose of model calibration, one cannot directly compare the dynamics but must indirectly compare them via a set of descriptors that are extracted from the data. As with image analysis, data analysis should be reliable, reproducible and consistent. Again, this is best achieved by automated data analysis algorithms. In the following, we discuss three models that we have used to characterize SPB–CEN distance/k-MT length trajectories. These models are the confined Brownian motion model, the simple MT dynamic instability (MTDI) model and the autoregressive moving average (ARMA) model.

9.5.1 Confined Brownian Motion Model

One of the simplest ways to characterize motion that has a significant random element is by considering it to be Brownian motion (diffusion) that is possibly confined within a certain volume. The descriptors are thus the diffusion constant and confinement radius of the labeled molecule. Figure 9.7 shows the mean-square SPB–CEN distance changes (MSQDC) for many budding yeast mutants, calculated as described in Dorn et al. (2005). To get reliable values of the MSQDC, we have averaged both over time in each trajectory and over several trajectories of the same strain, giving each distance change a weight based on its uncertainty.

As expected for random diffusion, the MSQDC is initially linear with time. At long times, however, trajectories sampled at one frame every 5 s reach a plateau, indicating that the motion is confined. Assuming a spherical confinement region, the plateau, MSQDC($\tau \rightarrow \infty$), is related to the confinement radius, R_c, by the equation (Dorn et al. 2005)

$$R_C^{\ 2} = \frac{25}{8}\text{MSQDC}(\tau \rightarrow \infty). \tag{9.2}$$

Trajectories sampled at one frame per second seem to be heading toward a plateau, but they do not reach it because they are too short owing to photobleaching. In this case, one can use the relationship

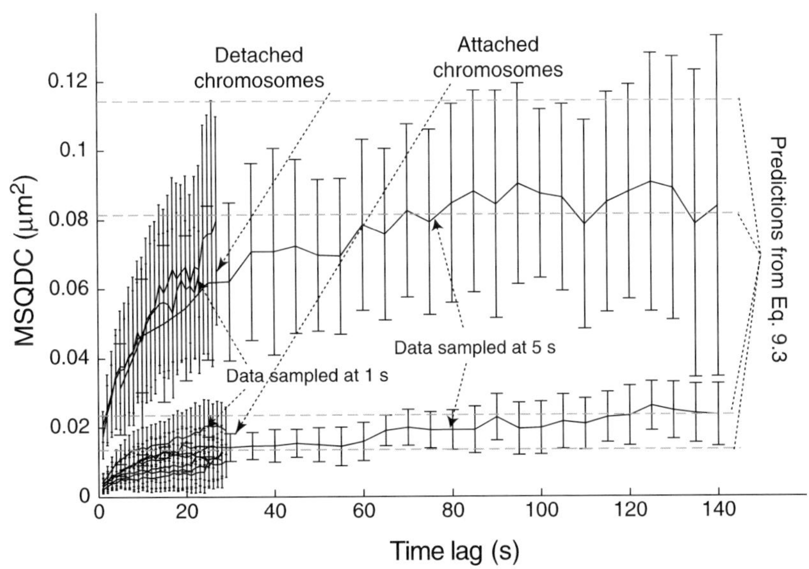

Fig. 9.7 Brownian motion analysis of SPB–CEN distance trajectories (see text for details). *MSQDC* mean square SPB–CEN distance change. (Reproduced from Dorn et al. 2005 with permission from *Biophysical Journal*)

$$\mathrm{MSQDC}(\tau \to \infty) = \frac{8}{25} R_{\mathrm{C}}^{\,2} = 2\sigma^2, \tag{9.3}$$

where σ^2 is the variance of the SPB–CEN distance trajectory, to calculate the plateau [MSQDC($\tau\to\infty$)] and confinement radius (Dorn et al. 2005).

Note that Brownian motion analysis is a relatively crude trajectory characterization that does not look at the details of transitions from one time point to another, but simply characterizes the overall averaged behavior. Thus, it can reveal dramatic differences in behavior between chromosome motion with and without attachment to k-MTs. However, it cannot detect subtler differences – which are the interesting differences – resulting from most kinetochore protein mutations, where attachment is maintained but k-MTs are differentially regulated.

9.5.2 *Simple Microtubule Dynamic Instability Model*

Owing to the insensitivity of the previous set of descriptors to differences between most of the mutants studied, we have looked for a better set of descriptors that captures the details of state transitions in a dynamic process, and not only its averaged behavior. The MTDI model used here is MT-specific, although in principle it can be used to analyze any motion that switches between forward and backward movement along one dimension. This model is quite intuitive: it characterizes an MT by its growth and shrinkage speeds, and frequencies of switching from growth to shrinkage (catastrophe) and from shrinkage to growth (rescue).

To characterize k-MT behavior within the context of a simple MTDI model, we have designed a statistical classification scheme that identifies the most probable mode of motion between consecutive time points while accounting for the uncertainty of SPB–CEN distances (Dorn et al. 2005; Fig. 9.8a). In brief, intervals between consecutive time points in a trajectory are classified in three steps:

1. Each interval is tested for a statistically significant change in distance. If the distance change is significant, the motion in that interval is classified as either antipoleward (AP) or poleward (TP), depending on whether the CEN tag moves away from or toward the SPB tag, respectively.
2. Consecutive intervals that could not be classified in step 1 are collectively tested for pause. A pause has to consist of at least two intervals.
3. Consecutive intervals that could not be classified in either step 1 or step 2 are collectively tested for long-term AP or TP motion. Also, intervals that were classified in step 1 are combined with neighboring undetermined intervals and are collectively tested for long-term AP or TP motion.

This classification allows for heterogeneity in speeds and frequencies. Thus, in contrast to most MT dynamics analyses where only the average speeds and frequencies (Walker et al. 1988; Dhamodharan and Wadsworth 1995) are measured, we obtain a spectrum of growth and shrinkage speeds and rescue and catastrophe frequencies.

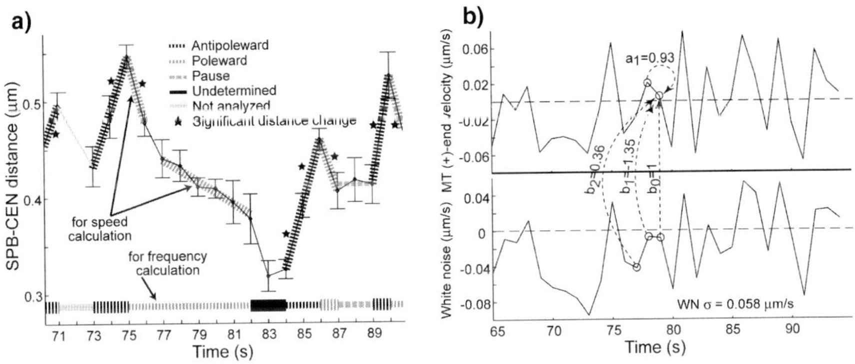

Fig. 9.8 **a** Simple MT dynamic instability (MTDI) classification of a SPB–CEN distance trajectory. **b** Illustration of an ARMA(1,2) model. The MT plus-end velocity at time *t* is the sum of $a_1 \times$(velocity at time *t*–1), WN at time *t*, $b_1 \times$(WN at time *t*–1) and $b_2 \times$(WN at time *t*–2). (Reproduced from Dorn et al. 2005 and from Jaqaman et al. 2006 with permission from *Biophysical Journal*)

Within this MTDI analysis of SPB–CEN distance trajectories, we use the mean speeds and frequencies (taking into account their uncertainties), as well as the mean-corrected speed and frequency distributions, as descriptors.

9.5.3 Autoregressive Moving Average Model

The last set of descriptors that we will discuss in this section comes from the field of time series analysis. In time series analysis, a stochastic trajectory is characterized by a detailed model of its transitions from one state to another, taking its randomness into account. Such models have been employed in various fields, from economics to ecology, to characterize stochastic series in order to predict their future values (Brockwell and Davis 2002). Here we use the parameters of one such model, the ARMA model, as descriptors that we use to compare trajectories with each other. ARMA models are the simplest of many parametric analysis models and are thus a good point to start and to illustrate the general method.

An ARMA model relates the value of an observed variable to its values at previous time points (the autoregressive, AR, component of the model) as well as to the present and past values of an associated white noise (WN) variable that renders the series stochastic (the moving average, MA, component). An ARMA(*p*,*q*) process is defined as

$$x_i = a_1 x_{i-1} + \ldots + a_p x_{i-p} + \varepsilon_i + b_1 \varepsilon_{i-1} + \ldots + b_q \varepsilon_{i-q}, \quad \varepsilon \sim N\left(0,\sigma^2\right), \tag{9.4}$$

where x_i (i=1,2,…,n) is the series being analyzed, ε the WN (assumed to be normally distributed with mean zero and variance σ^2), p the AR order, $\{a_1,\ldots,a_p\}$ the AR coefficients, q the MA order and $\{b_1,\ldots,b_q\}$ the MA coefficients. An ARMA(1,2) model is depicted graphically in Fig. 9.8b.

Time series to be characterized by parametric models must satisfy certain conditions. For example, trajectories to be described by ARMA models must be nonperiodic and stationary with zero mean (Brockwell and Davis 2002). The MT length trajectories studied here are nonperiodic, but not stationary; hence, Eq. 9.4 cannot be applied to them. The time series that we alternatively analyze are the instantaneous MT plus-end velocity series, defined as $v_i^+=(l_{i+1}-l_i)/(t_{i+1}-t_i)$ (l is the MT length, t is time and i is the time point). Calculating v^+ is equivalent to taking the first difference of MT length trajectories, removing linear trends and rendering the series stationary with zero mean.

The characterization of time series with ARMA models involves not only estimating ARMA coefficient values and WN variance, but also the AR and MA orders (in a sense, this is similar to MMF described in Sect. 9.4.1, where not only do we need to estimate the position and amplitude of each kernel, but also the number of kernels to be used). Thus, fitting is done in two steps (Jaqaman et al. 2006): (1) several models with various AR and MA orders are fitted to the data and their parameters are estimated and then (2) the most appropriate model is determined by seeking a balance between improvement of fit on the one hand, and decrease in parameter reliability on the other hand, when model order is increased. The parameters of the most appropriate model, $\{a_1,\ldots,a_p,b_1,\ldots,b_q,\sigma^2\}$, are thus used as trajectory descriptors. Note that the algorithm computes the uncertainties in these parameters as well.

9.5.4 Descriptor Sensitivity and Completeness

For the proper characterization of dynamics, descriptors must be sensitive and reliable, such that they distinguish between dynamics that are different and detect similarity in dynamics that are the same. There are two ways to investigate the sensitivity of a set of descriptors: (1) by constructing a molecular-scale model, varying its parameters and checking how the descriptors respond and (2) by characterizing experimental trajectories and verifying the descriptors' proper detection of differences and similarities.

Whichever approach is taken, the comparison of descriptors must be quantitative. In particular, the statistical properties of the descriptors must be employed to compare them within a hypothesis-testing framework (Koch 1988; Papoulis 1991; Sheskin 2004). For MTDI descriptors, we use Student's t test to compare mean speeds and frequencies, taking into account their uncertainties, and the Kolmogorov–Smirnov test to compare speed and frequency distributions. For ARMA descriptors, we use the Fisher test to compare the coefficients $\{a_1,\ldots,a_p,b_1,\ldots,b_p\}$, taking into account their uncertainties and interdependencies, and a second Fisher test to compare the WN variances σ^2.

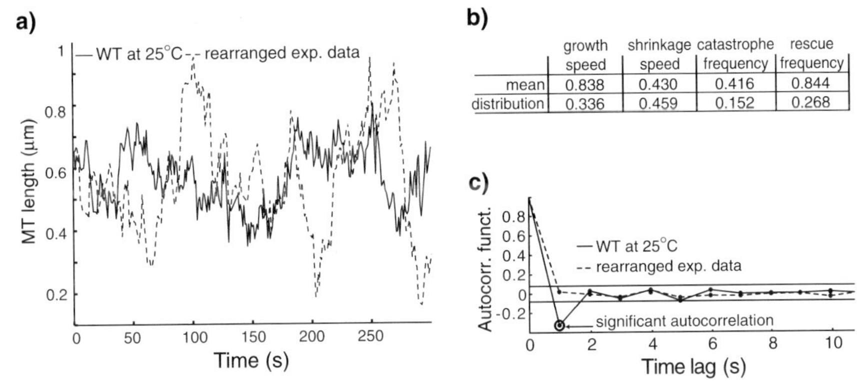

Fig. 9.9 **a** k-MT length trajectories from the wild type (*WT*) at 25°C and resulting from a random rearrangement of the sequence of experimental MT velocities in the WT at 25°C. **b** *p*-values for comparing the simple MTDI descriptors of the two trajectories. See Papoulis (1991) and Sheskin (2004) for a discussion of hypothesis testing. **c** Autocorrelation functions of the plus-end velocities derived from the two trajectories. The *two horizontal black lines* indicate the 99% confidence range for significant correlation values. (Reproduced from Jaqaman et al. 2006 with permission from *Biophysical Journal*)

To test the sensitivity of MTDI descriptors, we first applied them to experimental trajectories. There we found that they distinguished properly between mutants and conditions (Dorn et al. 2005). However, a deeper analysis of these descriptors showed that they did not extract from k-MT trajectories the correlation between an MT's state at time t and its state at time some later time t' (Jaqaman et al. 2006). This shortcoming is illustrated in Fig. 9.9. In Fig. 9.9a, we show original trajectories from WT at 25°C and a synthetic series generated by randomly rearranging the MT states from the original trajectories over time. In Fig. 9.9b, we show the *p*-values for comparing the MTDI descriptors, and it is seen that they do not distinguish between the original trajectories and the synthetic series. In contrast, in Fig. 9.9c we show a plot of the autocorrelation function of these two series, which clearly indicates a difference between them. Thus, MTDI descriptors are relatively sensitive, but they are not complete and cannot be trusted when they indicate that there is no difference between dynamics under different experimental conditions.

The correlation that is ignored by simple MTDI descriptors is captured by ARMA models, which are specifically designed for that purpose (Fig. 9.10a). Furthermore, ARMA models implicitly contain the information captured by the traditional MTDI descriptors (see Fig. 9.10b for a comparison of AP/growth speeds). Thus, from this analysis, we can conclude that ARMA descriptors are a more complete set of descriptors of k-MT dynamics, and are expected to be more sensitive. This was found to be indeed the case (Jaqaman et al. 2006).

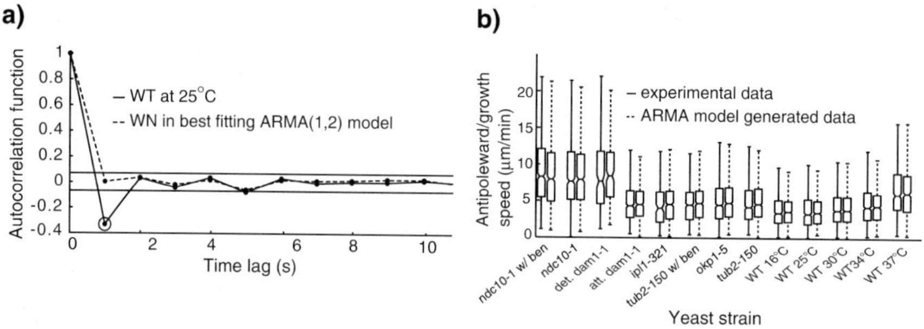

Fig. 9.10 **a** Autocorrelation function of the residuals from fitting plus-end velocities in the WT at 25°C with an ARMA(1,2) model. **b** Antipoleward/growth speed distributions from experimental trajectories and from synthetic trajectories generated using the corresponding ARMA descriptors. *Box lines* indicate the 25th percentile, the median and the 75th percentile of a distribution, while *whiskers* show its extent. *Boxes whose notches do not overlap* represent distributions whose medians differ at the 5% significance level. (Reproduced from Jaqaman et al. 2006 with permission from *Biophysical Journal*)

9.6 Quantitative Genetics of the Yeast Kinetochore

The ultimate purpose of estimated descriptors within the framework of Fig. 9.1 is to allow the comparison of experimental data with simulations of stochastic, mechanistic models and thus model calibration. However, descriptors can be also used to identify phenotypes associated with molecular and genetic interventions in a system, without reference to mechanistic models. In this section, we illustrate the use of the descriptors of k-MT dynamics in the quantitative genetics of the yeast kinetochore in the G1 phase of the cell cycle. Budding yeast in G1 provides us with a yet simpler system in which to study the regulation of k-MTs, since there are no sister chromatids and hence no tension affecting k-MT dynamics.

The following is a brief discussion of k-MT dynamics in several mutants, based on the quantitative comparison of the descriptors of those mutants to the descriptors of the WT (Fig. 9.11):

- *ndc10-1*: This mutant fails to form a kinetochore (Goh and Kilmartin 1993; Hyman and Sorger 1995), and its chromosomes do not get attached to MTs; therefore, chromosome motion in *ndc10-1* is expected to be different from motion in the WT. Furthermore, motion in *ndc10-1* does not depend on MTs, and hence should not be affected by drugs or mutations that affect MT dynamics. This was found to be indeed the case: ARMA descriptors of *ndc10-1* were statistically different from those of the WT, but statistically indistinguishable from those of *ndc10-1* with 40 μg/ml of the tubulin-binding drug benomyl (Gupta et al. 2004). In contrast, the descriptors of the WT changed significantly in the presence of benomyl. This example illustrates the ability of ARMA descriptors to properly detect differences and similarities between mutants and conditions.

var \ coef	WT 37°C	WT 37°C w/ ben	*ndc10-1*	*ndc10-1* w/ ben	*ipl1-321*	att. *dam1-1*	det. *dam1-1*
WT 37°C		7.8e-16	0	0	1.1e-06	0	3.3e-14
WT 37°C w/ ben	4.9e-24		2.2e-02	2.3e-04	4.0e-02	7.6e-04	3.1e-01
ndc10-1	0	0		3.8e-01	7.2e-05	3.9e-01	5.5e-02
ndc10-1 w/ ben	0	0	3.7e-01		1.5e-06	3.2e-01	2.8e-01
ipl1-321	1.2e-12	4.3e-02	1.4e-207	1.3e-192		3.9e-07	3.1e-06
att. *dam1-1*	6.6e-06	1.7e-07	7.9e-191	2.5e-175	2.3e-03		2.6e-01
det. *dam1-1*	0	0	1.9e-01	1.4e-01	0	0	

coef: ▒ p-value<1e-3, significantly different ARMA coefficients
var: ▒ p-value<1e-10, significantly different WN variance

Fig. 9.11 *p* values for the comparison of chromosome/k-MT dynamics (*above diagonal* ARMA coefficients, *below diagonal* WN variance) in various strains at 37°C. See Papoulis (1991) and Sheskin (2004) for a discussion of hypothesis testing. (Reproduced from Jaqaman et al. 2006 with permission from *Biophysical Journal*)

- *ipl1-321*: Ipl1p (yeast homologue of Aurora kinase) is a key regulator in the mitotic spindle. In metaphase, it induces k-MTs of sister chromatids with improper attachment to depolymerize, facilitating their detachment and consequent proper attachment to two SPBs. Once bipolar attachment is achieved, tension is thought to inhibit Ipl1p, stabilizing k-MTs (McAinsh et al. 2003; Dewar et al. 2004). Our analysis indicates that Ipl1p has a regulatory role in G1 as well: both ARMA coefficients and WN variance were significantly altered in the *ipl1-321* mutant. Interestingly, the WN variance of the *ipl1-321* mutant was significantly smaller than that of the WT, indicating that k-MTs are generally less dynamic when the function of Ipl1p is hindered.
- *dam1-1*: Dam1p is an MT-binding protein. It is part of the DASH complex that forms rings around k-MTs, facilitating k-MT attachment to kinetochores (Miranda et al. 2005; Westermann et al. 2006). Mutations in Dam1p are thus expected to disrupt the attachment of MTs to chromosomes, and possibly the transmission of regulatory signals from inner kinetochore proteins to the k-MT. Interestingly, we found two classes of chromosome dynamics in this mutant. Some *dam1-1* cells had ARMA descriptors that were statistically indistinguishable from those in *ndc10-1*, indicating that their chromosomes were detached from MTs. Other *dam1-1* cells had a WN variance similar to that of the WT, indicating that they were attached, but ARMA coefficients that were significantly different, suggesting that k-MT regulation in *dam1-1* was different from regulation in the WT. These observations reveal the essential role played by Dam1p in both chromosome–MT attachment and signal transduction.

Comparing descriptors of experimentally observed dynamics, as done above, reveals which proteins play a role in the process of interest and which do not. Proteins whose mutations result in dynamics with different descriptors from those of dynamics in the WT definitely play a role in the mechanism of interest. Mutations that do not change the descriptors indicate that either the corresponding protein is not involved in the mechanism of interest, or that the protein might play a role that is not reflected in the dynamics. In either case, the interesting mutations are those which change the observed dynamics. However, a comparison of descriptors alone cannot reveal protein functions. To elucidate function, we need mathematical models of the underlying molecular processes that are calibrated using the experimental data. Differences between model parameters that yield dynamics matching those of a certain mutant and model parameters that yield dynamics similar to those of the WT reveal changes in the underlying molecular processes due to a certain protein mutation. This indirectly reflects the function of the protein in normal physiology. An account of this kind of mechanistic data modeling, however, is outside the scope of this chapter.

9.7 Conclusion

In summary, we have addressed several issues that aid our use of live-cell imaging to test hypotheses about the molecular-level interactions that lead to the observed cellular-scale phenomena. First, experiments must be designed such that they yield data that are amenable to quantitative analysis. Second, the images must be analyzed in an objective manner to yield quantitative data. This can only be done in an automated fashion using computer vision algorithms. Third, the stochastic data obtained via image analysis must be further analyzed to yield descriptors that characterize them. These descriptors can be used for either phenotype comparison, in which case we get a survey of proteins that play a role in the mechanism of interest, or model calibration, in which case we quantitatively dissect the molecular interactions that underlie the cellular-scale dynamics of interest.

References

Alberts B, Johnson A, Lewis J, Raff M, Roberts K, Walter P (2002) Molecular biology of the cell. Garland, New York

Blackman SS, Popoli R (1999) Design and analysis of modern tracking systems. Artech House, Norwood

Brockwell PJ, Davis RA (2002) Introduction to time series and forecasting. Springer, New York

Cheeseman IM, Drubin DG, Barnes G (2002) Simple centromere, complex kinetochore: linking spindle microtubules and centromeric DNA in budding yeast. J Cell Biol 157:199–203

De Wulf P, McAinsh AD, Sorger PK (2003) Hierarchical assembly of the budding yeast kinetochore from multiple subcomplexes. Genes Dev 17:2902–2921

Dewar H, Tanaka K, Nasmyth K, Tanaka TU (2004) Tension between two kinetochores suffices for their bi-orientation on the mitotic spindle. Nature 428:93–97

Dhamodharan R, Wadsworth P (1995) Modulation of microtubule dynamic instability in vivo by brain microtubule associated proteins. J Cell Sci 108:1679–1689

Dorn JF, Jaqaman K, Rines DR, Jelson GS, Sorger PK, Danuser G (2005) Interphase kinetochore microtubule dynamics in yeast analyzed by high-resolution microscopy. Biophys J 89:2835–2854

Goh PY, Kilmartin JV (1993) Ndc10 – a gene involved in chromosome segregation in Saccharomyces cerevisiae. J Cell Biol 121:503–512

Gourieroux C, Monfort A, Renault E (1993) Indirect inference. J Appl Econ 8:S85–S118

Gupta K, Bishop J, Peck A, Brown J, Wilson L, Panda D (2004) Antimitotic antifungal compound benomyl inhibits brain microtubule polymerization and dynamics and cancer cell proliferation at mitosis, by binding to a novel site in tubulin. Biochemistry 43:6645–6655

Hyman AA, Sorger PK (1995) Structure and function of kinetochores in budding yeast. Annu Rev Cell Dev Biol 11:471–495

Jaqaman K, Dorn GF, Jelson GS, Tytell JD, Sorger PK, Danuser G (2006) Comparative autoregressive moving average analysis of kinetochore microtubule dynamics in yeast. Biophys J 91:2312–2325

Koch K-R (1988) Parameter estimation and hypothesis testing in linear models. Springer, Berlin

Maddox PS, Bloom KS, Salmon ED (2000) The polarity and dynamics of microtubule assembly in the budding yeast Saccharomyces cerevisiae. Nat Cell Biol 2:36–41

McAinsh AD, Tytell JD, Sorger PK (2003) Structure, function, and regulation of budding yeast kinetochores. Annu Rev Cell Dev Biol 19:519–539

Miranda JJ, De Wulf P, Sorger PK, Harrison SC (2005) The yeast DASH complex forms closed rings on microtubules. Nat Struct Mol Biol 12:138–143

Mitchison T, Kirschner M (1984) Dynamic instability of microtubule growth. Nature 312:237-242

O'Toole ET, Winey M, McIntosh JR (1999) High-voltage electron tomography of spindle pole bodies and early mitotic spindles in the yeast Saccharomyces cerevisiae. Mol Biol Cell 10:2017–2031

Papoulis A (1991) Probability, random variables, and stochastic processes, 3rd edn. McGraw-Hill, New York

Robinett CC, Straight A, Li G, Willhelm C, Sudlow G, Murray A, Belmont AS (1996) In vivo localization of DNA sequences and visualization of large-scale chromatin organization using lac operator/repressor recognition. J Cell Biol 135:1685–1700

Sheskin DJ (2004) Handbook of parametric and nonparametric statistical procedures. Chapman and Hall, Boca Raton

Smith AA Jr (1993) Estimating nonlinear time-series models using simulated vector autoregression. J Appl Econ 8:S63–S84

Stelzer EHK (2000) Practical limits to resolution in fluorescence light microscopy. In:Yuste R, Lanni F, Konnerth A (eds) Imaging neurons. Cold Spring Harbor Press, Cold Spring Harbor, pp 12.1–12.9

Straight AF, Belmont AS, Robinett CC, Murray AW (1996) GFP tagging of budding yeast chromosomes reveals that protein-protein interactions can mediate sister chromatid cohesion. Curr Biol 6:1599–1608

Thomann D, Rines DR, Sorger PK, Danuser G (2002) Automatic fluorescent tag detection in 3D with super- resolution: application to the analysis of chromosome movement. J Microsc 208:49–64

Thomann D, Dorn J, Sorger PK, Danuser G (2003) Automatic fluorescent tag localization II: improvement in super-resolution by relative tracking. J Microsc 211:230–248

Walker RA, O'Brien ET, Pryer NK, Soboeira MF, Voter WA, Erickson HP, Salmon ED (1988) Dynamic instability of individual microtubules analyzed by video light-microscopy – rate constants and transition frequencies. J Cell Biol 107:1437–1448

Westermann S, Wang HW, Avila-Sakar A, Drubin DG, Nogales E, Barnes G (2006) The Dam1 kinetochore ring complex moves processively on depolymerizing microtubule ends. Nature 440:565–569

III
Cutting Edge Applications & Utilities

10
Towards Imaging the Dynamics of Protein Signalling

Lars Kaestner and Peter Lipp

Abstract In living cells protein signalling is a complex dynamic process. We discuss the spatiotemporal aspects of such signalling and how the variety of different optical imaging techniques can deal with the necessary temporal and high optical resolution. We consider techniques like confocal laser scanning microscopy, kilobeam scanning, nonlinear microscopy, selective plane imaging, structured illumination, total internal reflection fluorescence microscopy, near-field optical microscopy and fluorescence lifetime imaging. Complementary to the technological aspects we discuss different approaches of how to make specific proteins visible, including bioluminescence, autofluorescence, immunohistochemistry, fluorescence-fusion proteins and specific covalent labelling of proteins. Finally we introduce concepts to image protein signalling. Protein dynamics are considered on the basis of translocations of fluorescent protein tagged conventional protein kinase C. Protein–protein interactions can best be investigated in living cells by molecular interaction of fluorescence-protein tags, namely bimolecular fluorescence complementation or Förster resonance energy transfer (FRET). Today, modern imaging techniques even allow the investigation of biochemical reactions such as kinase-mediated phosphorylation in living cells. Such genetically engineered biosensors make use of conformational changes that induce changes of intramolecular FRET.

10.1 Spatiotemporal Aspects of Protein Signalling Dynamics

Proteins are the workforce in living cells and, as such, interactions between proteins are vital for the survival of cells. Protein interactions are historically a domain of biochemists traditionally investigating protein properties *in vitro*. Nowadays this has changed into a rather interdisciplinary approach significantly depending on fluorescence imaging. As a consequence, the test tube has been at least partially replaced by the living cell (live-cell imaging). As proteins are small molecules or molecular complexes (size on the order of 10^{-10} m) one would spontaneously opt for an imaging method that is able to resolve such small domains, e.g. electron microscopy (resolution on the order of 10^{-9} m). Unfortunately, optical microscopy using visible light has a much lower resolution (on the order of 10^{-7} m) and is thus not able to directly picture individual proteins or their complexes. This said, later in this chapter we will introduce methods that, albeit indirectly, allow visualisation of protein–protein interaction. Furthermore, one should abandon notions that image-based methods just produce a *single* high-resolution representation of the microscopic specimen.

To emphasise that, let us consider the following analogy. You may not have heard of football at all and look at a number of snapshots from a football game, including the spectators, the excitement of the players, referees discussing, etc. It will be nearly impossible to understand the rules of the game without any temporal connection between those snapshots. In a sense the same holds true for studies of cell signalling: the full understanding of cellular signalling cascades, including protein–protein interactions, can only be appreciated when visualising such processes with spatial *and* temporal resolution. In this, time domains of biological processes are extremely diverse: developmental mechanisms can last for days, weeks or even months, cell division takes place within many hours, apoptotic processes take several hours, while hormone signal transmission can take minutes and mitochondrial motility is in the domain of seconds. In contrast, cardiac contraction only lasts a fraction of a second and synaptic transmission takes place within milliseconds.

Since we do not want to solely detect such processes but intend to foster their understanding by following their time course, the acquisition speed needs to be faster than the lifetime of the event of interest. At the same time, a superior optical resolution is required for visualisation of these processes in living cells, tissues or even living organisms. While classic light microscopy usually produces 2D images, modern fluorescence microscopes are able to generate stacks of 2D images over time, yielding added dimensionalities such as simple 3D reconstructions or 3D views over time (4D) and/or in different colours (5D). Such data objects can be generated either by so-called deconvolution microscopy or directly by one of the incarnations of laser scanning confocal microscopes.

10.2 How to Be Fast While Maintaining the Resolution

There are a number of highly developed so-called diffraction-limited imaging methods to achieve optical sectioning (necessary for high-resolution image acquisition), such as two-photon imaging, selective plane imaging (SPIM) and structured illumination (for more details, see later). However, the most popular and most versatile imaging method offering high resolution used in living cells is laser scanning confocal microscopy. Since high-speed acquisition is essential in imaging cell/protein signalling we would like to discuss technological and methodological approaches available today to achieve that goal.

The basic principle of laser scanning confocal microscopy is an optical sectioning of the specimen along the optical axis. In single-photon excitation this is achieved by excluding light originating from above or below the plane of focus. This can be obtained by fixed pinholes or variable irises depending on the construction of the confocal microscope. For an illustration of the confocal principle see Fig. 7.2.

The price one has to pay for the optical sectioning capabilities offered with this approach is that measurements can only be obtained at a single point at a time (but see the discussion of the multipoint scanner below). From this it becomes obvious that a true confocal image cannot be seen by the human eye; instead a computer is required to construct the resulting image from consecutive points recorded during the scanning process. One can envisage that scanning an entire high-resolution image by such a mechanism is a time-consuming process and thus the resulting image acquisition rates have traditionally been rather low.

Below we briefly enumerate the two basic scanning processes still available today in various incarnations:

1. Scanning the image by moving the specimen (microscope stage). This is the historical approach of the first confocal device patented by Marvin Minsky (1957). This technology is still used, e.g. in single-molecule applications, commercially available from PicoQuant (Berlin, Germany).
2. Scanning a single laser beam across the specimen. This method is much faster than stage scanning and is still the standard approach. Single-beam scanners are usually equipped with galvanometer-based scanning mirrors for both the x- and the y-direction. While this keeps the scanning electronics and the acquisition algorithms reasonably simple since recording is usually performed during the linear part of the mirror movements, this method is inherently slow because the mirrors have to be moved physically. Image acquisition rates with reasonable resolutions are usually limited to single-figure frames per second. This rate can be increased significantly when switching from "normal" mirrors to resonating mirrors (at least in the more demanding x-direction). When doing this, frame rates can easily reach video or double video rate. One of the major difficulties with such an approach is the fact that the scanning process does not display a linear part; thus, sinusoidal scanning has to be accounted for with rather complex optical and/or electronics designs for compensation. Another approach to further gain 1 order of magnitude of scanning speed is to substitute the x-scanning mirror by an acousto-optical deflector (AOD) crystal. The operating principle of an AOD is illustrated in Fig. 10.1. This mass-free scanning pushes the frame rate up to several hundred Hertz. The price for this speed is a reduced spatial resolution in both the x- and the z-direction since the AOD operates in a wavelength-dependent fashion and can therefore not be used for the descanning of the image (Stokes shift). As a result, the fluorescence is not a stationary point but a linearly moving point at the level of the detector and thus the pinhole needs to be replaced by a slit. Such a scanner is commercially available from VisiTech International (Sunderland, UK). For this kind of device, the restriction in scanning speed in experiments is no longer the scanning mechanism itself but most often the amount of fluorescence available and the viability of the specimen.

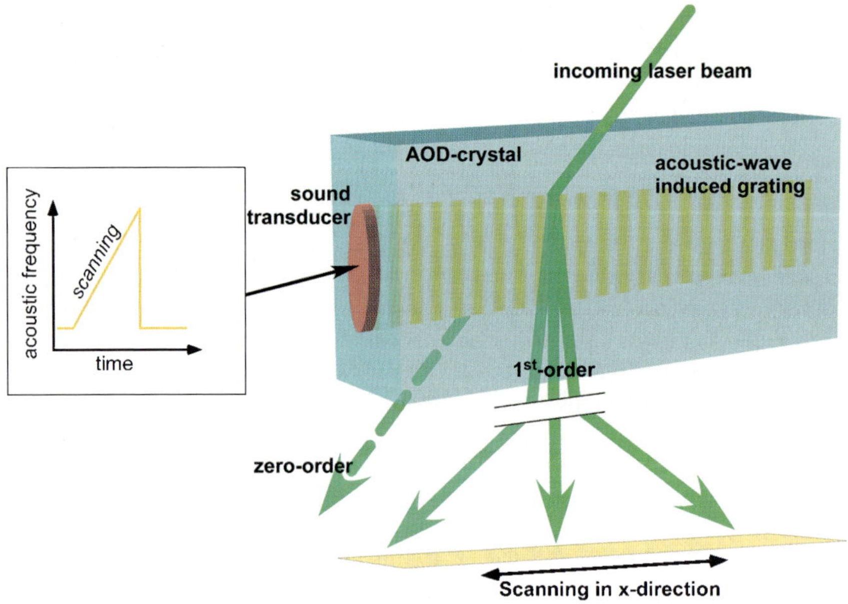

Fig. 10.1 Operating principle of an acousto-optical deflector (AOD). The AOD is an optical crystal through which an acoustic wave is transmitted that induces a diffraction grating. The first-order diffracted laser beam is used for scanning. Since the lattice parameter can be changed almost instantaneously by modulating the acoustic frequency, the degree of diffraction and thereby the movement (scanning) of the laser beam are inherently fast

In addition, specialised devices are commercially available (Molecular Devices, Sunnyvale, USA) where stage scanning and beam scanning are combined: The *x*-scan is performed with a mirror, while for the *y*-scan the microscope stage is moved.

A totally different approach of increasing the frame rate of confocal scanning is the idea of exciting with more than one beam simultaneously. This can be realised either by a linear array of points (a line) such as in the swept-field microscope from Prarie Technologies (Madison, USA) or by a 2D array of points. Since the incarnations of the latter method use several thousands of parallel scanning beams, they are referred to as kilobeam scanners. They have many advantageous properties that are essential in imaging protein–protein signalling dynamics. These advantages comprise high acquisition speed (as fast as the attached camera can capture images), high efficiency in terms of simultaneous imaging of thousands of beams, low bleaching and low phototoxicity (Lipp and Kaestner 2006). These kilobeam scanners are available in two versions:

1. The Nipkow-disc system is based on a rotating disc with a specific pattern of pinholes originally invented, designed and built to code and transmit TV images (Nipkow 1884). This scanning principle was made popular some 10 years ago

by Yokogawa Electric (Tokyo, Japan) in their confocal scanning unit (CSU-10). They overcame the original drawback of single-disc scanners, namely the extremely low excitation throughput by including a second disc with matching microlenses that served as light collectors (i.e. focussing the excitation light into each of the pinholes). This approach massively increased the excitation throughput and represents one of the major advancements in confocal microscopy in the last decade, since it allowed long-term live-cell imaging with high acquisition rates. The operating principle is displayed in Fig. 10.2.

2. 2D-array scanners actively move the array of parallel laser beams generated by a microlens array across the specimen. Here, in contrast to the Nipkow-disc system, the microlens and pinhole arrays are stationary. The only moving part is a single mirror responsible for scanning and descanning on the front surface as well as

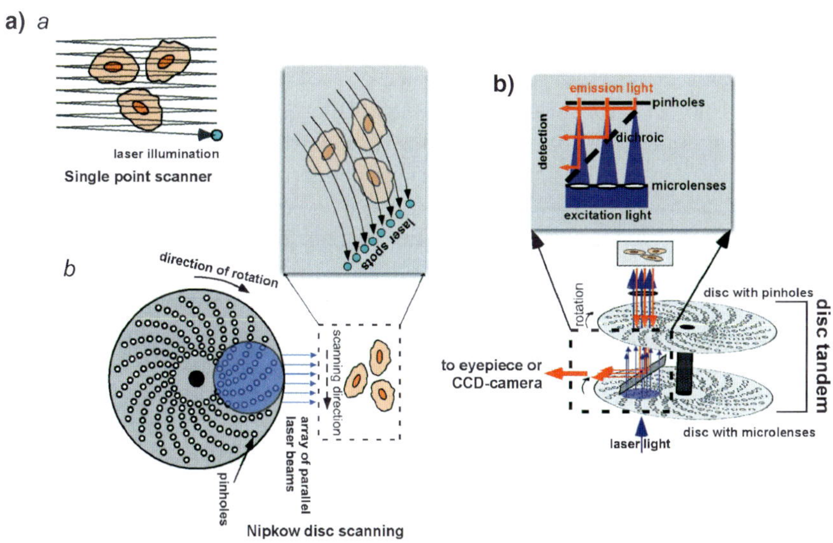

Fig. 10.2 The Nipkow-disc scanner. **a** Comparison of the scanning process for a single-point scanner (*a*) and a multibeam scanner employing a Nipkow disc (*b*). While the single-beam scanner needs to address all pixels in a specimen in succession, the Nipkow disc generates a 2D array of thousands of parallel laser beams that functionally scan across the specimen simultaneously (*upper right*). **b** Optical principle of a Nipkow-disc scanner. Laser light (*blue lines*) for excitation is beam-shaped and directed onto the microlens disc of the Nipkow-disc scanner. The lenses effectively collect the excitation light and focus it down into the pinholes of the pinhole-disc. The microlens disc ensures an excitation light throughput of more than 80%. The parallel beam of laser excitation scans across the specimen when the discs rotate in synchrony. Fluorescence from the sample is separated from excitation light by a dichroic mirror sandwiched between the two Nipkow discs and directed to the detection unit. Here the user can inspect a confocal image, or alternatively the emitted light can be multiplexed by a series of dichroic mirrors and filter wheels in front of up to four CCD cameras (Evotec Technologies, Hamburg, Germany, or VisiTech International, Sunderland, UK)

rescanning the image across the detection camera on its back surface. This approach ensures absolute synchrony between the scanning and the detection side of the confocal head. The novel concept of the 2D-array scanner from VisiTech International (Sunderland, UK) also includes changeable pinholes for variable resolutions in the direction of the optical axis, a feature not found in current versions of Nipkow-disc-based systems. For high-speed imaging (video rate and above), the linear movement of the scanning mirror allows easier synchronisation between the image generation of the confocal head and the image detection by the attached CCD camera, although this has been solved to a certain extent also for Nipkow-disc-based systems.

One of the most frequently used preconceptions about kilobeam scanners is the possible crosstalk between the pinholes, most prominent in thicker specimens. This will result in a diminished sectioning ability when recording 3D stacks for reconstructions. Despite earlier reports (Egner et al. 2002), in our hands the sectioning quality in thicker specimens (here we used the same kind of pollen grain as in Egner et al. 2002) appears artefact-free as illustrated in Fig. 10.5c. We made an additional attempt to probe the sectioning ability of kilobeam scanners in thicker specimens by acquiring data for a 3D reconstruction from a transplanted beta-islet. This beta-islet from the pancreas of hamster donors was transplanted into a dorsal skinfold chamber of a living animal (hamster; Menger et al. 1990; Fig. 10.3). It is clearly visible that even at depths approaching 100 μm, single capillary vessels and beta-islet tissue can readily be reconstructed and distinguished. In this example, the capillary bed has been stained green and the beta-islet cells are shown in red. In our opinion there is little doubt that single-beam scanners might be more appropriate when deep penetration with single-photon confocal microscopy is the prime goal of the experiment, but in situations where one could compromise on the penetration depth (less than 100 μm) such experiments still benefit from the higher quantum yield of state-of-the-art electron-multiplying CCD cameras and the resulting minimised impact on tissue viability. The situation is strictly different when considering multiphoton applications as outlined further below.

Besides the confocal point and multipoint scanners there are a number of further optical sectioning techniques:

1. In contrast to single-point scanners, slit scanners do not use an individual point, but use an entire line for excitation. Consequently the pinhole is replaced by a slit. The idea is to gain acquisition speed (matching the AOD-driven scanner) by sacrificing some of the resolution (see later). This technique recently underwent a revival in the latest developments by Carl Zeiss Jena (Germany) with the introduction of the LSM 5 Live.
2. Multiphoton microscopy is based on nonlinear effects during the sample excitation. By condensing the laser energy in time (femtosecond pulses) and space (focus), the energy density in the focus becomes so high that a molecule of the sample can absorb two (or multiple) photons simultaneously. Since the underlying

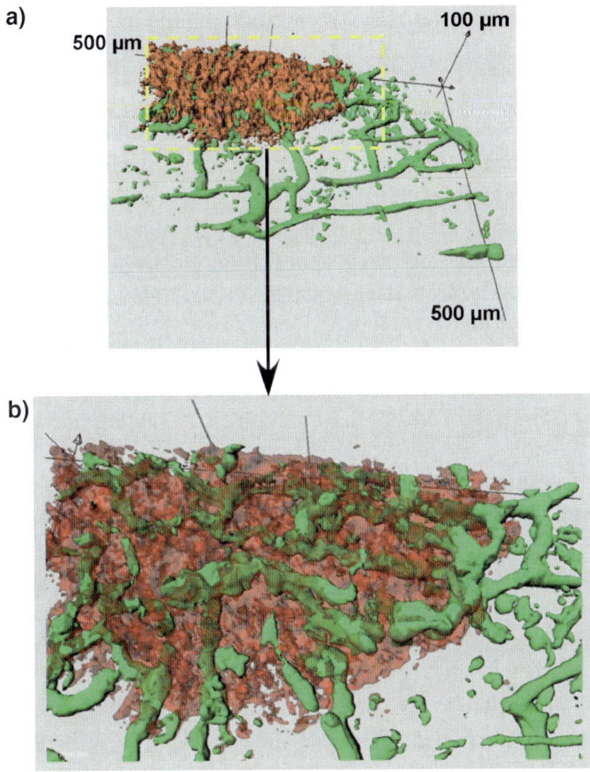

Fig. 10.3 Langerhans islet: 3D reconstruction of the surface part of the skin of a hamster model of Langerhans islet transplantation *in vivo*. **a** A reconstruction of the total acquired volume in the skin of the living hamster. We used a Nipkow-based single-photon real-time confocal microscope. The vasculature as well as the islet were stained by intravenous injection with fluorescein dextran (*green*) and rhodamine (*red*), respectively. The rhodamine accumulates in the islet. **b** Part of **a** (*yellow dashed box*) in a magnified view. For clarity we have rendered the islet cells with 50% transparency. (Kaestner, Laschke, Menger and Lipp 2006, unpublished data)

molecular excitation process remains the same, both photons have to deliver approximately half (or other fractions depending on the multiplicity of excitation) of the energy. Half of the energy translates into a doubling of the wavelength, which explains why far-red and infrared light emitting titanium:sapphire lasers are used. Thus, if the chromophore requires, e.g., single-photon excitation at 450 nm, the equivalent two-photon excitation wavelength would be 900 nm. For other multiphoton processes this shift is multiplied by higher factors. One should be aware that the absorption cross-section of single-photon and multiphoton excitation can be quite different. Compared with single photon confocal imaging the basic advantage of using multiphoton excitation is reduced photobleaching in out-of-focus planes, since multiphoton excitation is restricted to the focal plane. For a thorough discussion of multiphoton microscopy see Denk et al. (1995). Here, we solely mention two additional significant advantages of multiphoton excitation:

(a) Deeper penetration depth. Since excitation can be performed with red, far-red or even infrared light, the penetration of the excitation light in living tissue is considerably increased in comparison with that of shorter wavelengths. It should be mentioned here that the maximal penetration depth is tissue-dependent. One still has to consider that emission light scattering becomes more prominent with increasing penetration depth.

(b) Intrinsic sectioning.
As described above, the "multiphoton effect" is restricted to the core of the excitation light focus in the specimen owing to the nonlinear excitation probability. From this it follows that in contrast to single-photon confocal scanners that employ optical sectioning on the "emission" side (spatial filter, the pinhole or pinholes), scanners using multiphoton excitation already generate sectioning on the excitation side. We thus refer to this as "excitation sectioning". For single-point scanners this translates into their liberation from descanning; all light emitted only originates from the excitation volume that is diffraction-limited and thus light collection does not require the ability for spatial discrimination. As a result, single-point multi-photon scanners do not need pinholes on the detection side. Similar to single-beam scanners, in multi-photon microscopy the construction of a 2D image is also realised by scanning processes. In the easiest of all cases, scanning is performed by means of two galvanometer-controlled mirrors with all the limitations discussed above. The application of fast AOD crystals for *x*-scanning is certainly more complex when using pulsed femtosecond light sources, because diffraction of AODs is wavelength-dependent and femtosecond pulsed lasers produce a spectral band (e.g. with 75-fs pulses the bandwidth of the resulting spectrum can reach around 10 nm). This means that the degree of diffraction of the excitation light will be different for the "red" and the "blue" components of the excitation spectrum. Nevertheless, multiple approaches exist that offer possible solutions (Bouzid and Lechleiter 2002; Roorda and Miesenbock 2002). In addition, LaVision BioTec (Bielefeld, Germany) has introduced a commercial multiphoton multipoint multiplexed scanner (up to 64 parallel excitation points arranged in a line), which, in principle, allows up to 64 times faster image acquisition.

3. Selective plane imaging (SPIM) relies on the illumination of the sample by a light sheet perpendicular to the optical axis of the microscope. The recent revival of that technique (Huisken et al. 2004) is based on a sample holder with a rotational axis parallel to the gravitation field that enables rotation while keeping the sample itself distortion-free. The serial optical sectioning is realised by moving the sample through the light sheet and for every position an image is collected. Although the lateral resolution is lower than with confocal techniques, the biggest advantage of SPIM is its ability to come close to the diffraction limit in all

three axes. Nevertheless it has to be mentioned that acquisition speed is limited and the specimen setup is much more complex when manipulation of the cells is required for the recording (solution changes, patch-clamp, etc.). Quite some computational work has to be invested before a 3D reconstruction can be visualised. Similarly but not identical to multiphoton excitation "SPIM excitation" happens in the focal plane only ("light sheet"). SPIM appears extremely promising and it can be expected to be commercially available soon.

4. Structured illumination is another technique that allows optical sectioning at the level of axial resolutions offered by a classic pinhole-based confocal arrangement. The principle of structured illumination is an optical grid that is placed in the illumination beam in such a way that the image of the grid is projected exactly into the focal plane of the microscope objective. For the data acquisition the representation of the grid in the focal plane needs to be moved laterally and images are acquired in each position. Lateral movement of the grid image can be obtained by a piezo (if the grid itself is moved as introduced in the initial setup; Neil et al. 1997) or by moving a glass block in the excitation light path (as incorporated in the commercialised version Apotome from Carl Zeiss Jena, Germany). Exposures are taken from three defined grid image positions. From such three consecutive exposures, an image of the optical section is calculated. In contrast to confocal laser scanning, the structured illumination imposes fewer requirements on the illumination source (no laser is necessary) and it is therefore less expensive. In terms of acquisition speed, structured illumination recordings can be faster than classic confocal recordings (see above). However, the acquisition speed is limited by the readout of the camera used and since three exposures are necessary for the calculation of one image it is inherently slower than other camera-based confocal systems like Nipkow-disc systems or slit scanners (see above). This "delay" can partially be overcome by using the principle of the "running average", i.e. exposures 1–3 give the first image, exposures 2–4 the second and so on. In any case, if the process one is looking at is very fast, there is not only a smear in the *x/y*-plane but also a resolution decrease in the *z*-axis since the algorithm relies on the assumption that there are no sample changes within the three exposures needed to calculate the image. When comparing structured illumination with confocal imaging one has to take into account that the structured illumination image is already a processed image and especially when taking *z*-stacks for 3D reconstruction, the deconvolution of confocal images leads to a resolution increase.

Speaking of (diffraction-limited) imaging technologies, we need to discuss two additional subjects:

1. Time resolution can always be increased by sacrificing spatial resolution. The first step in this direction is the reduction of the number of pixels per image. This can be realised by limiting the overall size of the image or by binning adjacent pixels or at its extreme by reducing image sizes to individual lines (line scans). Furthermore if the entire fluorescence signal of the microscope is detected in a point detector we speak about confocal spot measurements (Parker and Ivorra 1993). In this case sampling rates can be as high as several Megahertz. Nonetheless, even without any

optical spatial resolution, the functional spatial resolution (even beyond the diffraction limit) can originate from the experimental design (see below).
2. Spatial resolution can be increased to an almost spherical point-spread function by illuminating and detecting the specimen from more than one angle. In practice, all of these techniques are time-consuming and far away from video-rate or even faster recordings. One of these incarnations is the SPIM (see above). Another one is the 4π technology, i.e. simultaneous recording with two objectives from opposing directions, commercialised by Leica Microsystems (Wetzlar, Germany).

At this point we would like to briefly mention technologies that allow spatially resolved recordings that even break the diffraction limit by various optical designs:

1. Total internal reflection fluorescence (TIRF) microscopy based on evanescent wave imaging is now offered by basically all of the major microscope manufacturers as an add-on (Wazawa and Ueda 2005). At the angle of total internal reflection, excitation around that phase boundary is induced by the energy transfer of the evanescent wave. In most cases these are cells growing on a glass substrate (e.g. glass cover slip). This evanescent wave travels parallel to the cover slip illuminating a layer less than 100 nm above the cover slip; therefore, this is the method of choice for membrane-related imaging.
2. Another technique is stimulated emission depletion (STED). This method requires a high technical complexity and is so far limited to academic applications. In brief, an already excited fluorophore is illuminated with bright light of the same wavelength as the emission of the fluorophore. This will induce depletion of the excited state. The point-spread function of the excitation spot is partly "switched off" by a donut-shaped depletion structure that "cancels out" (i.e. depletes) fluorescence and generates an effective emission spot with sub-diffraction-limited properties. This principle is realised by providing an excitation light pulse that is directly followed by a 20–40-ps depletion pulse (Hell and Wichmann 1994; Egner and Hell 2005).
3. A totally different approach is based on scanning near-field optical microscopy (SNOM). Here we do not find an optical image plane. Instead, a probe with a diameter of 100 nm or less is used. This sensor has to be mechanically scanned across the specimen, under the condition that the probe head needs to stay within a 20-nm distance to the surface of interest (Wabuyele et al. 2005).
4. Fluorescence, once generated, reveals an exponential decay, the so-called fluorescence lifetime, which is characteristic for every fluorophore. Since the fluorescence lifetime is a material constant and thus concentration-independent it has been proven to be a potentially useful parameter for microscopic imaging, fluorescence lifetime microscopy (FLIM). Interestingly, the lifetime of the fluorophores can be extremely sensitive to the surrounding in its very proximity (nano environment). This interaction between the lifetime and its environment can indeed be used for, e.g., ion concentration imaging with FLIM (Lakowicz et al. 1992). Another extremely interesting approach is the combination of FLIM and Förster resonance energy transfer (FRET; discussed in Sect. 10.5 and in Chap. 6 by Hoppe), the so-called FRET-induced fluorescence lifetime alterations (FIFA). This combination will allow, e.g., for the quantitative measurement of protein–protein interactions.

The last three approaches are inherently slow and so far they are far away from video-rate imaging. This technical limitation restricts their significance predominantly to localisation studies or to experiments that do not tend to be time-critical.

10.3 How To Make Proteins Visible

This question needs to be specified further since we do not want to make proteins visible in general, instead we intend to visualise individual proteins. Basically we know of two different light–protein interactions: absorption and emission. Owing to the properties of many aromatic side chains of amino acids, proteins inherently display both properties, they can absorb light of specific wavelengths and many proteins also show fluorescence, so called autofluorescence, since we do not have to stain the protein. Despite these properties we are not in the position to discriminate between proteins of interest and others. The absorption of light by proteins is well known and is used in a molecular biology laboratory for spectrometric measurements of the protein concentration of a protein solution on a daily basis. However, to resolve a protein entity by absorption, at least 1% attenuation of light is necessary. According to the Lambert–Beer law this would need a concentration of at least 10^{-7} M, equivalent to 10^{16} molecules. In contrast, in fluorescence recordings a single molecule can be detected, especially when the background fluorescence is very low. We thus use fluorescence techniques on the detection side, but this method still lacks specificity as will be discussed shortly. For an illustration of the differences between absorption and fluorescence microscopy, see Fig. 10.4.

Besides the methods that will be described later to link proteins to fluorescent tags, there is a small number of proteins that do produce light by themselves (bioluminescence; e.g. no excitation light necessary). An example is the aquorin–green fluorescent protein (GFP) complex that naturally occurs in the jellyfish *Aequorea victoria*. Since the application of such fluorescent proteins has revolutionised live-cell imaging we have dedicated a longer section to this technique.

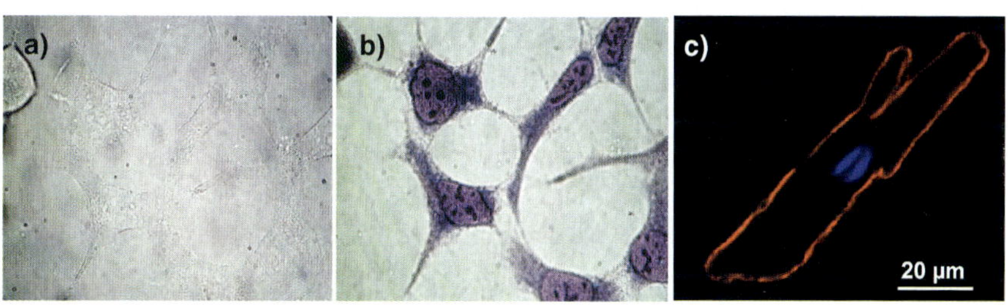

Fig. 10.4 Comparison of transmission and fluorescence microscopy for stained and unstained cells. **a**, **b** White-light transmission images of the same cells: **a** unstained, **b** stained using a panoptic staining procedure according to Pappenheim (1908); nucleus and mitochondria are visible. **c** A fluorescence image of a human cardiac myocyte upon illumination with 488- and 635-nm light, stained with di-8-anepps for the cell membrane (*red*) and with DRAQ-5 for the nuclei (*blue*). The fluorescence of an unstained cell would yield a basically black image (apart from some very low autofluorescence).

We can summarise four additional approaches for the visualisation of proteins:

1. As mentioned above many proteins do show so-called autofluorescence, typically upon illumination with ultra violet (UV) light. These properties can be attributed either to an individual amino acid, such as tryptophan[1] or to cofactors bound to the proteins such as reduced nicotinamide adenine dinucleotide (NADH)[2]. The latter is also popular for directly imaging mitochondrial activity (Piston et al. 1995; Fig. 10.5a). However, there is also autofluorescence upon illumination with visible or even red light, e.g. protoporphyrin IX[3] in the erythrocytes of patients suffering from Erythropoitic protoporphyria (Kaestner et al. 2004).

An additional interaction of proteins with light is light scattering. If light scattering involves nonlinear effects induced by femtosecond pulsed lasers, certain proteins can produce light exactly at half the wavelength of the incident light. This effect is called second-harmonic generation and can, e.g., be used in living tissue to visualise the extracellular matrix (i.e. collagen fibres) without the need to stain the preparation at all (Sun 2005; Fig. 10.5d).

Another nonlinear technique is Raman microscopy, but this method produces weak signals that require data acquisition times too long to follow protein signalling in living cells. However, there is a related technique that is inherently fast, reaching video-rate (30-Hz) acquisition speed: coherent anti-Stokes scattering (CARS) microscopy (Dudovich et al. 2002; Evans et al. 2005), a detailed explanation of which would go beyond the scope of this chapter.

For a selection of cells imaged without any staining, see Fig. 10.5.

2. Identifying proteins by specific antibodies is called immunohistochemistry. The primary antibody may already contain a fluorescent tag or it may contain binding sites for secondary antibodies that are tagged with fluorophores. The latter method has the advantage of containing an amplification step but is inherently nonquantitative since the ratio of primary to secondary antibodies is difficult to determine quantitatively. Antibodies are available in basically all colours throughout the entire range of the visible spectrum. Today, there are even kits that allow simple labelling of your own antibodies. The chromophore is usually an inorganic group. However, recently another approach to label proteins was established with some interesting properties in fluorescence imaging, using so-called quantum dots (Q-dots). These nanostructures (5–15 nm in diameter) comprise a crystalline core (usually cadmium mixed with selenium or tellurium). This core is the "heart" of the Q-dots in that it provides the structure and mechanism for the generation of fluorescence. The other coatings protect the core, make it biocompatible

---

[1] Tryptophan shows a broad UV absorption with a local peak at 280 nm and an emission peak at about 360 nm, but this extends into the visible range.

[2] NADH has absorption and emission peaks at 340 and 465 nm, respectively.

[3] Protoporphorin IX is the direct precurser of haemoglobin – it has a broad absorption almost over the entire visible range, but the most prominent Soret-band peak is at about 410 nm. The emission has a double-peak at both 635 and 705 nm.

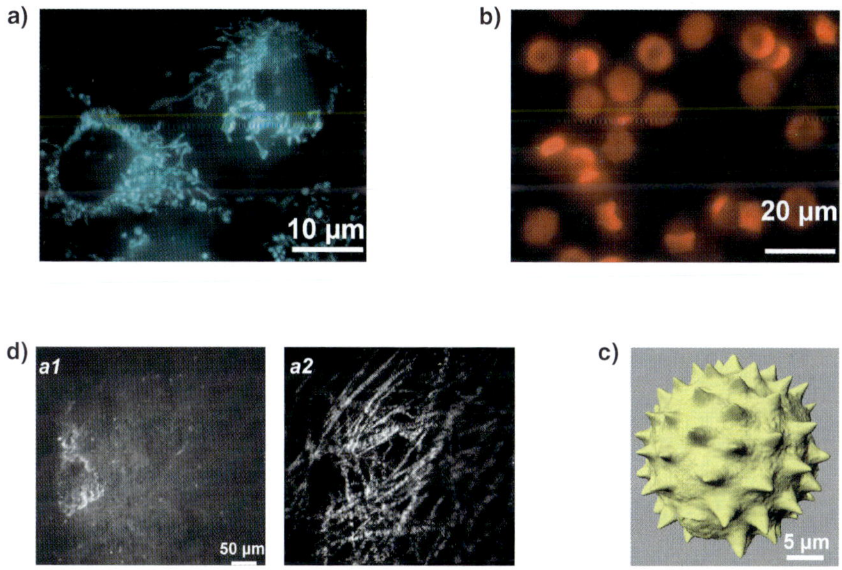

Fig. 10.5 Images of native (unstained) cells. **a**, **b** Autofluorescent cells recorded by video imaging. **a** COS-1 cells excited at a wavelength of 340 nm and emission was collected at wavelengths above 440 nm, revealing mitochondria. **b** Autofluorescence of human erythrocytes excited at a wavelength of 480 nm, emission was collected at wavelengths above 500 nm. **c** A 3D reconstruction of a pollen grain. Images were recorded using a Nipkow-disc confocal scanner with excitation at a wavelength of 561 nm. The reconstruction was performed on raw data without any further deconvolution. **d** Simultaneous fluorescence (*a1*) and second-harmonic generation (SHG) imaging (*a2*) of the surface of a living mouse heart. The sample was excited with a femtosecond laser at 830 nm and fluorescence was collected at 500 nm, while the SHG image was recorded simultaneously at 415 nm. The SHG image illustrates the distribution of collagen fibres in an unstained heart

and provide binding sites for attachment of the Q-dot to the target protein. Q-dots display the advantage of having an extremely broad excitation spectrum while maintaining a symmetrical and narrow emission band. Furthermore, Q-dots only exhibit extremely low rates of photobleaching, especially when compared with inorganic dyes, but also have drawbacks such as an unpredictable fluorescence blinking.

Nevertheless, it is usually difficult to use fluorescently labelled antibodies in live-cell imaging owing to their membrane impermeability. Thus, applications of labelled antibodies are usually limited to micro-injection (or diffusion through a patch-clamp pipette) or to labelling epitopes on the outside of living cells.

3. An entirely different approach has been taken with genetically encoded fluorescence. Here, the target cell itself produces its own fluorescent entities. The most popular of such fluorescent proteins is the GFP and its derivatives blue fluorescent protein (BFP), cyan fluorescent protein (CFP), yellow fluorescent protein (YFP), etc. (Zimmer 2005). GFP was the first fluorescent protein used in biological research applications. However, nowadays the selection of colours of the

fluorescent proteins is almost unlimited, from the BFP over the entire visible range to the far-red HcRed (derived from the reef coral *Heteractis crispa*). Of the three techniques discussed here, the fluorescent protein approach is certainly amongst the most widely used methods when considering live-cell imaging. It appears fair to say that the availability of fluorescent fusion proteins has clearly revolutionised live-cell work, it has made many experimental approaches much more feasible and has, in addition, opened new ways for studying the morphology and physiology of living cells in situ. Furthermore, approaches using fluorescent protein techniques in vivo are also growing and will certainly reveal details of cellular functions in the context of the living organism that are obscured in isolated cells. One of the major properties of fluorescent proteins permitting their use in living cells is their almost ideal biological inertia, i.e. the fluorescent protein molecules do not interact with other partners in living cells. Thus, when they are coupled to other proteins they allow visualisation of that protein, partners of that protein and even biochemical reactions in situ (see below) without affecting the underlying structure or function. It has to be mentioned here that, of course, this paradigm has to be verified for each fluorescent fusion protein constructed. In addition to that biological inertia, they are more stable towards photobleaching than most other inorganic chromophores and since they are produced by the cells themselves imaging can be performed over a rather extended period of time. The "only" prerequisite the experimenter has to ensure is a sufficient transfer of genetic material into the target cell. This can be a rather tricky business. Although transfection of cell lines and most other cells can be achieved by means of specialised transfectants, this can prove more difficult in certain cell types such as T cells, in which electroporation has proven a much more versatile tool. For end-differentiated nonproliferating cells such as cardiac muscle cells and neurones, gene transfer with the help of transfectants typically results in low transfection rates, sometimes in single-figure percentages, which very often is not sufficient. Here, virus-mediated gene transfer (e.g. adenovirus or lentivirus) has proven easy and efficient, with transfection rates often approaching 100%. Viral gene transfer even allows the transfection of cells in the living animal. Here, especially the lentivirus-mediated gene transfer has been proven to ensure stable expression for more than 6 months after initial virus delivery (Sugiyama 2005).

4. An approach to overcome the disadvantages of the significant size of fluorescent proteins (see below) is the specific covalent labelling of proteins in living cells by the FlAsH or ReAsH (Gaietta et al. 2006; Griffin et al. 1998). To achieve labelling of the target protein several steps are necessary. Firstly one has to engineer a target sequence into the protein of interest. This sequence has to incorporate four cysteines at the specific positions x, $x+1$, $x+4$ and $x+5$ (Griffin et al. 1998). It is the final binding of the fluorophore to these four cysteines that forms the fluorescent label inside the living cell. Although originally it was thought that these structures would form an α-helix, evidence is growing for a 3D hairpin structure for this binding motif (Adams et al. 2002). After the protein of interest incorporating the binding motif has been expressed

in the living cell, a multistep staining process follows comprising the application of biarsenical coupled fluorophores such as FlAsH–EDT_2 or ReAsH–EDT_2 (where EDT is 1,2-ethanedithiol). These biarsenical dyes are membrane-permeable, nonfluorescent in solution and only gain fluorescence when binding to their four cysteine-residues target motif. After incubation, the nonbound or only weakly bound dye molecules have to be removed by thorough washing with EDT. This is a very important step to achieve proper staining. By means of this procedure, proteins can be labelled with a minimum amount of change in the protein itself (only 12 amino acids have to be added) and also the chromophore itself is rather small (around 700 Da). Although it was shown that the physiology of cell lines is not altered by addition of the organic compounds, we recommend checking this for each particular cell type, especially when using primary cultures or more sensitive cells. Recently, this technique has been expanded in combination with GFP to allow a combination of light and electron microscopy on the level of small vesicles of the Golgi apparatus. The authors initially performed live-cell imaging using the GFP tag and in a second step after photoconversion of the FlAsH tag they implemented correlated electron microscopy (Gaietta et al. 2006). Other application areas of these chromophores are labelling of protein domains (owing to their small size) and FRET approaches, including intramolecular and intermolecular FRET.

10.4 Concepts To Image Protein Dynamics

While the static location of proteins can be determined by methods such as immunohistochemistry, for which cells usually have to be killed and fixed to get the antibodies into the cells, following protein dynamics in living cells requires techniques that (1) specifically only label the protein(s) of interest and (2) do not interfere with the biological function of the protein(s). Here, two similar but not equal approaches have proven very successful and influential in the field of live-cell imaging. The technique of fluorescent protein tagging was described in Sect. 10.3. In brief, a new fusion protein is created comprising the protein of interest fused (N- or C-terminally) to a fluorescent protein of which basically a whole array of proteins can be applied spanning the entire visible spectrum from deep blue to far red (for a novel class of longer-wavelength fluorescent proteins, see Shaner et al. 2004). They are transferred into living cells by means of gene transfer (described in Sect. 10.3) and the target cells finally express the fluorescent fusion protein. Within a couple of days, for some viral expression system even quicker, the fluorescence in the cells is sufficiently high to start imaging experiments. The technique of fusion proteins has reached such a sophisticated level that in combination with fast imaging techniques translocation of, e.g., 110-kDa fusion proteins can be readily detected as depicted in Fig. 10.6. Today, these fusion proteins have even reached the level of routine tools in cell research and they are on the verge of also allowing studies on the level of tissues or even entire animals. For the latter, transgenic

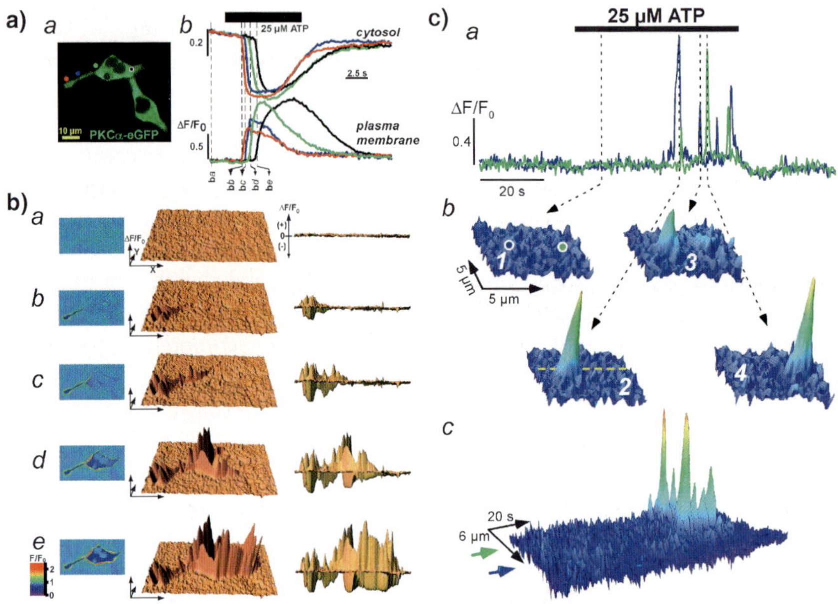

Fig. 10.6 Imaging of PKCα–eGFP fusion proteins reveals highly complex translocation patterns in living cells. **a** PKCα translocation waves. *a* HEK-293 cells expressing the PKCα construct. The *coloured circles* depict cellular locations from which fluorescence was averaged for the fluorescence traces in *b*. For those time points marked with *dashed lines* individual confocal images have been analysed in **b**. The labels in **a**, panel *b* correspond to panels *a–e* in **b**. For the *left column* in **b**, we calculated the relative fluorescence and colour-coded the ratio according to the colour wedge in panel *e*. 3D surface plots of the relative fluorescence have been constructed for the *middle and right columns*, which represent two different viewing angles onto the same surface. Upward deflections indicate relative increases and downward deflections relative decreases in the local PKCα–eGFP concentration. **c** Spatiotemporal properties of local translocation events (LTEs) in COS-1 cells. *a* The time course of the relative fluorescence signals at two LTE sites plotted in *blue* and *green*, respectively. For selected time points (marked by the *vertical dashed lines* in *a*) 3D surface plots of the fluorescence distribution have been redrawn in *b*. *b1* The positioning of the two regions of interests from which the fluorescence was averaged (traces in *a*). The *dashed yellow line* in *b2* depicts the line across the two LTEs used for constructing the 3D surface pseudo-line-scan image displayed in *c*. Here time runs from left to right and the preserved spatial dimension goes into the paper plane, the relative fluorescence has been coded in both the colour and the height of the 3D surface. *PKC* protein kinase C, *GFP* green fluorescent protein. (Reproduced from Reither et al. 2006. *The Journal of Cell Biology*, 2006, 174:521–533. Copyright 2006 The Rockefeller University Press)

animals are one possibility to perform such studies; in particular gene-substitution experiments are of major interest since they indeed substitute the native protein and can thus even be used to study gene expression under "physiological" conditions. A further interesting approach is the use of virus-mediated gene transfer (briefly described in Sect. 10.3). Although targeted expression (restricted to a certain type of tissue or cells) can readily be realised through specific promoters in the genetic construct, the delivery of the virus can impose difficulties. Such difficulties can be overcome by using special infusion techniques that result in a preferential adherence of the virus to the tissue of interest. A subsequent problem is posed by the particular virus used. In this respect the lentiviris system has proven to induce very little immunological response by the acceptor animal. As a retrovirus it integrates its own genome into the host genome. The result of this is a slower onset of the expression but an extremely long lasting and constant protein level (Sugiyama et al. 2005).

In all of this, one has to consider the following: tagging proteins with proteins (here fluorescent proteins) increases the molecular mass in many instances significantly. In the case of, e.g., a protein kinase C (PKC) molecule, the molecular mass is increased from around 70–80 kDa to almost 110 kDa, by roughly 30%. When considering dynamic protein translocation (as shown in Fig. 10.6) such increases in molecular mass will unavoidably impinge on the diffusion characteristics of the molecule. These effects are by far not linear but depend much more on the 3D arrangement of the molecule itself. In the case of PKC fusion proteins, it was reported that the PKCα–eYFP construct displayed only a modest decrease in apparent diffusion within the cytosol (Schaefer et al. 2001).

At this point we would like to extend our discussion of fluorescent fusion proteins to fusion between functional groups and fluorescent proteins rather than complete host proteins. By means of such constructs, researchers are able to custom-design their own bioindicators, e.g. for lipids such as PIP3 or PI3P (Anderson et al. 2000; Ellson et al. 2001). These bioindicators comprise a functional domain of a well-known protein with appropriate binding characteristics (dissociation constant and specificity). When imaging the distribution and in particular changes of this distribution, we can learn a lot about the spatiotemporal properties of, e.g., second messenger synthesis and NADPH oxidase activation as shown for PI3P in Fig. 10.7. Similar concepts have been used in detection systems of inositol-1,4,5-trisphosphate production with pleckstrin homology (PH) domains coupled to fluorescent proteins, e.g. (Bartlett et al. 2005).

In addition to the translocation experiments depicted above the methods of fluorescence recovery after photobleaching und fluorescence correlation spectroscopy as described in Chap. 7 by Wachsmuth and Weisshart can be applied to fluorescent proteins in order to characterise protein dynamics.

10.5 Concepts To Image Protein–Protein Interactions

Up to a few years ago, investigating protein–protein interactions was a major domain of biochemists working with in vitro systems and applying techniques such as yeast two-hybrid screens, immunoprecipitation or chemical cross-linking. Not solely but massively supported by the readily available fluorescent protein approaches, this has changed and we are now able to explore such interactions on the level of single molecules *in vivo*. One would assume that simply tagging possible protein interaction partners with different tags (e.g. red and green) would do the job, assuming an interaction when colocalisation (in that particular example a yellow

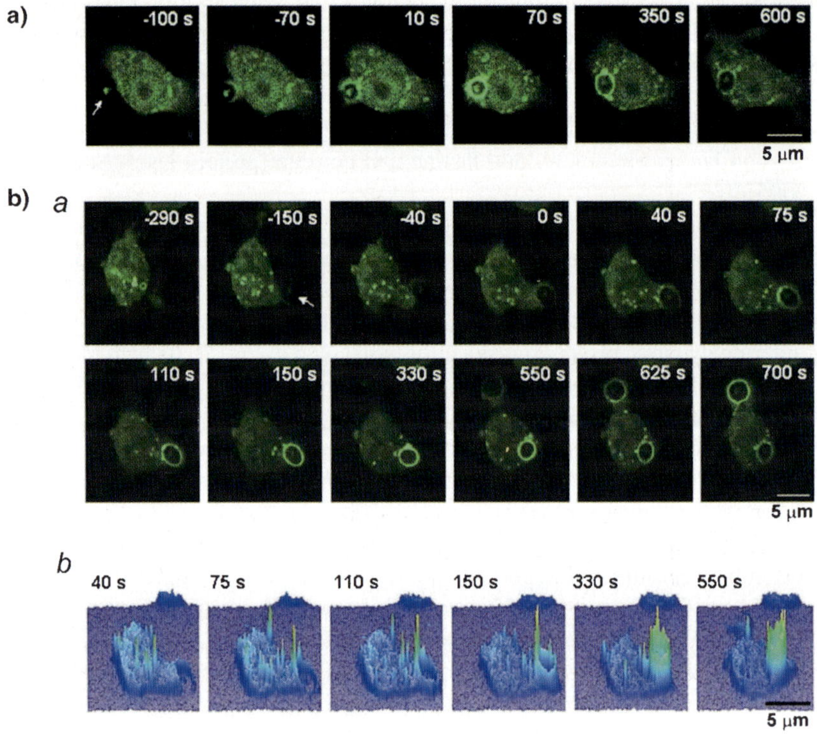

Fig. 10.7 PX-domain, Time-lapse confocal images of the effects of phagocytosis on the distribution of GFP–FYVE and GFP–PX in RAW 264.7 cells. **a** Images collected using a GFP–FYVE expressing RAW mouse cell line. Opsonised zymosan particles were added to the cells (approximately 1×10^6/ml) and images (0.5-s exposure) were captured repetitively; phagosomal ring closure has been defined as time zero. The images depict the attachment and uptake of a single particle; the *arrow* indicates the position of the particle at the point of attachment. **b** Images collected using a GFP–PX expressing RAW cell line. Images (0.5-s exposure) were collected using continual *z*-sectioning (15 slices per stack, 1-μm step size) during particle uptake. *a* The attachment and uptake of two successive particles visualised in a single confocal plane; the attachment of the first particle is indicated by an *arrow*. *b* Redrawing of individual confocal planes at the given time point from the confocal recording depicted in *a*. For this, the distribution of the GFP fluorescence has been coded into the height and the colour ("warmer colours" for high fluorescence) of the 3D surface. The asymmetric rise in intensity of GFP fluorescence around the phagosome, the cytosolic drop in fluorescence and the maximal intensity achieved on the phagosome can be seen. (Reprinted from Ellson et al. 2001, copyright 2001, with permission from Elsevier)

pixel) is detected. Colocalisation on the light-microscope level can be read as protein populations that share the same *voxel* (volume element) of our image. In the best of all cases this voxel will be diffraction-limited and thus display an ellipsoid size of around 250 nm × 250 nm × 750 nm. The resulting volume is extremely large in comparison to the dimensions of a single protein molecule; thus, the two proteins of interest might have come closer to one another but, frankly, cannot really be regarded as interacting (see also Chap. 5 by Oheim and Li). For the observation of protein–protein interactions we need methods that allow identification of such an interaction on the level of a few nanometres. The most promising image-based methods to detect protein–protein interactions will be discussed in detail: bimolecular fluorescence complementation (BiFC) and FRET:

1. BiFC based on the idea that proteins can be split into two halves which when recombined are able to recreate the fully functional parent protein. Tom Kerppola and co-workers have transferred this basic idea to the concept of BiFC by using fragments of fluorescent proteins to study interactions between molecules (for a recent review see Kerppola 2006). At first glance BiFC appears as a rather simple and ideal technique to study protein–protein interactions. In the resting state, i.e. no interactions, there is simply no aggregation between the two halves of the fluorescent protein; thus, there is no fluorescence. Upon interaction, the two fluorescent protein parts connect and finally form a fully functional fluorescent protein whose spectral properties are basically indistinguishable from those of their native counterparts; thus, we see fluorescence. This inherent simplicity is the strongest advantage of the BiFC approach. Although in the case when you see fluorescence there was an interaction between the two partners, it does not necessarily tell you *when* and *where* and *how much*. Here one faces the possible limitation of the technique: as pointed out by Kerppola in his recent review, in most cases fluorescent protein formation has been proven to be basically irreversible. This notion has been supported by independent data provided by Lynne Regan and co-workers (Magliery et al. 2005). They demonstrated irreversible formation of GFP. Once the fluorescent protein complex has been secured by the two interacting proteins, the formation of the fully functional fluorescent protein appears as a multistep process, involving irreversible binding between the two halves (yet not fluorescent) and final steps of the maturation of the functional fluorescent protein which exhibits its "original" fluorescence properties (Kerppola 2006). This irreversibility of the functional fluorescent protein poses a potential problem or limitation when attempting to visualise transient interactions. Nevertheless, some reports indeed describe apparent reversible fluorescence, the mechanisms of which are still not understood (Kerppola 2006). In conclusion, BiFC appears to be an extremely sensitive and versatile approach when trying to detect principal interactions or, in other words, when addressing the question of whether two proteins dointeract at all with restricted information about the dynamics and kinetics of that interaction process. BiFC can probably be seen as an integrating, accumulating method. Interacting proteins will be "collected" and stored in the form of fully functional fluorescent protein constructs, even at times when they lose interaction again (Magliery et al. 2005). When using this powerful technique, one has to keep these characteristics in mind and consider them when interpreting the imaging results obtained with BiFC. From this it follows that when BiFC is used, numerous control experiments are absolutely essential to rule out false-positive and also false-negative imaging results.

2. An entirely different approach for monitoring protein–protein interactions uses energy transfer between two fluorophores with matching fluorescent properties, the so-called FRET. We will not discuss the mechanistic principles of FRET; for this we refer to Chap. 6 by Hoppe. Nevertheless, we would like to discuss FRET in light of the discussion we provided earlier. In comparison with BiFC, FRET only uses and imposes weak interactions between the fluorophores. Thus, interactions between the fluorophores usually do not grossly modulate the overall binding properties between the protein partners and do not influence their binding kinetics. This, of course, has to be tested and we want this statement to be interpreted in light of the most likely irreversible binding of the fluorescent protein halves discussed above. For fast interactions, FRET partners, especially when these are fluorescent proteins, will impose kinetic differences on the reaction and the imaging results, but most likely these are on a different order of magnitude from the possible restrictions discussed for BiFC. For FRET no such measurable maturation of the FRET complex has to occur and FRET signals dissipate rapidly upon unbinding of the two protein partners (steep, nonlinear relationship of FRET efficiency and distance between FRET partners); therefore, kinetic recordings of dynamic protein–protein interactions are possible. For FRET to occur both partners (the donor and the acceptor) not only have to have some matching spectral properties but they have to approach each other very closely. The efficacy of energy transfer in FRET is extremely dependent on the distance between the chromophores (i.e. proportional to the sixth power of the distance). Apart from what we discussed above and what is discussed in Chap. 6 by Hoppe we would like to highlight a combination of methods that might offer important advantages over solely recording fluorescence intensities. When fluorophores are excited by absorbing photons they usually undergo a transition from the energy ground state (S_0) on an excited state (S_1). The molecules reside in this S_1 state until they "fall" back to the S_0 state (referred to as relaxation). The energy gained by this process is at least partially released in the form of a photon (emission). The time spent in the excited state for an ensemble of molecules of a certain fluorophore is referred to as the fluorescence lifetime. This fluorescence decay can be measured, e.g. when delivering excitation pulses in the picosecond time domain (or shorter) which excite an ensemble of fluorophore molecules simultaneously. From the fluorescence decay one can derive the fluorescence lifetime, which can show mono exponential or multiexponential behaviour. Interestingly, the lifetime is not only characteristic for a given fluorophore, but very often is modulated by environmental parameters such as the composition of the surrounding medium or the temperature. Another process that can alter the fluorescence lifetime is FRET. Since the S_1 state of the donor is partly depleted by FRET, the fluorescence decay becomes faster, i.e. the fluorescence lifetime shortens. Considering the FRET partners, with increasing proximity (within the Förster radius – typically a few nanometres), FRET between donor and acceptor increases and the lifetime of the donor fluorescence decreases. Such FRET-induced fluorescence lifetime alterations (FIFA) can be used to relatively easily record quantitative FRET data. Moreover, such measurements can

be performed with a solely absorbing acceptor, since only the donor fluorescence lifetime is required (Ganesan et al. 2006).

In principle there are four major approaches for how to measure fluorescence lifetimes: (1) frequency-domain imaging, (2) time-correlated single-photon counting, (3) imaging with time gates and (4) streak camera based imaging.

In the area of frequency-domain imaging there is a system commercially available from Lambert Instruments (Leutingewolde, The Netherlands). The time domain has traditionally been covered by time-correlated single-photon counting systems commercially available, e.g., from PicoQuant (Berlin, Germany) and Becker & Hickl (Berlin, Germany). More recent approaches are the time gating for photomultiplier tubes that has been commercialised by Nikon Instruments Europe (Badhoevedorp, The Netherlands) or the time gating for cameras (e.g. Kentech Instruments, South Moreton, UK) or LaVision BioTec, Bielefeld, Germany). Very promising is also the streak camera based approach of Hamamatsu Photonics (Hamamatsu City, Japan).

Besides the FLIM modality itself, the achievable overall acquisition speed of such measurements depends very much on the brightness of the sample and (since most FLIM detectors are add-on modules to existing fluorescence measurement setups) on the configuration of the basic equipment, like laser power and scanning efficiency. Nevertheless, the approaches described above are still far away from "real-time" data provided by standard fluorescence measurements for FRET. Whatever method one chooses for FRET measurements (fluorescence intensity or FLIM) the resulting data allow for a direct analysis of protein–protein interactions.

10.6 Concepts To Image Biochemistry with Fluorescent Proteins

While tagging of proteins or functional binding domains allows visualisation of protein translocation, generation of second messengers and protein–protein interactions, there is another field of active research that aims at constructing probes that allow the direct visualisation of biochemical processes in living cells, such as phosphorylation. We would like to discuss two different approaches as general examples of how to potentially tackle similar problems. Both approaches use intramolecular FRET whereby the fluorescent protein partners have been N- and C-terminally fused to the central re cognition protein. The groups of Alexandra Newton and Roger Tsien have used the concept of intramolecular binding between a phosphorylation motif and a binding domain recognising the successful phosphorylation for probing two kinase families, PKC (Violin et al. 2003) and protein kinase A (PKA) (Zhang et al. 2001). In the nonphosphorylated state, the binding domain (phosphoamino acid binding domain, 14-3-3) has very little affinity to the PKA-specific phosphorylatable peptide sequence; thus, the protein is in its open conformation, the distance between the FRET partners CFP and YFP is too large and as a result the FRET efficiency is low. Upon phosphorylation of the appropriate motif, the 14-3-3 domain binds to the phosphorylated amino acid(s) and the conformational change of the protein brings CFP and YFP closer together. This increases the FRET efficiency. A construct applying this concept was used to probe oscillatory PKC-mediated phosphorylation (Violin et al. 2003) and recently also in combination with subcellular targeted expression of such a sensor to probe spatial aspects of PKC activity in living cells (Gallegos et al. 2006). In particular, the latter example very nicely combines various aspects of fluorescence protein work as described in this chapter, namely probing biochemical reactions and protein targeting. An inherent problem of such constructs is the possibility of unwanted interactions of the target or the binding domains with endogenous proteins resulting in FRET reduction by intramolecular reactions. In order to address that problem Karsten Schultz and co-workers have gone down a different avenue. They have constructed probes for PKC and PKA activity reporters that rely on altered intramolecular interactions of two neighbouring domains whereby this interaction is only indirectly dependent on the phosphorylation of a target sequence. This target domain is flanked by the two interaction domains [pleckstrin comprising a DEP (Dishevelled, Egl-10, pleckstrin) and PH with a central PKC-consensus phosphorylation domain]. The DEP and the PH domains are themselves fused to a GFP variant and CFP, respectively. Thus, it is not the binding affinities of the interacting domains that change the FRET efficiency between donor and acceptor; instead, the phosphorylation-dependent conformational change of the entire molecule modulates the distance between the two interacting domains and hence the CFP–GFP distance (KCP-1) (Schleifenbaum et al. 2004). More recently the Schultz group has extended that concept when introducing a dual probe that is able to detect PKA-dependent and PKC-dependent phosphorylation (Brumbaugh et al. 2006). Under conditions of a lack of phosphorylation the GFP and CFP partners display an intermediate FRET distance. In this construct the group introduced a second phosphorylation site into the linker between the DEP and the GFP domains that is a consensus site for PKA-dependent phosphorylation. PKA activity at this site introduces additional negative charges to the molecule, leading to an additional conformational change of the DEP–GFP domain that increases the distance between the FRET partners. Thus, while PKA-dependent phosphorylation results in a FRET decrease, PKC-mediated phosphorylation brings both FRET partners close together and thus leads to an increase in the FRET efficiency (Brumbaugh et al. 2006). Although such dual probes might enable researchers to detect two biochemical activities with a single probe, as of yet it appears a principal problem of such a probe that it is not able to directly distinguish between PKC and PKA activity in the situation where both kinase activities are changing *in vivo*.

One of the major drawbacks of approaches translating such small conformational changes (e.g. compared with calmodulin upon Ca^{2+} binding) into fluorescence recordings is very often a low signal-to-noise ratio. Nevertheless, we expect more probes like those described here to appear in the future, enabling researchers to fully explore the biochemistry of living cells *in vivo*.

References

Adams SR, Campbell RE, Gross LA, Martin BR, Walkup GK, Yao Y, Llopis J, Tsien RY (2002) New biarsenical ligands and tetracysteine motifs for protein labeling in vitro and in vivo: synthesis and biological applications. J Am Chem Soc 124:6063–6076

Anderson KE, Lipp P, Bootman M, Ridley SH, Coadwell J, Ronnstrand L, Lennartsson J, Holmes AB, Painter GF, Thuring J, Lim Z, Erdjument-Bromage H, Grewal A, Tempst P, Stephens LR, Hawkins PT (2000) DAPP1 undergoes a PI 3-kinase-dependent cycle of plasma-membrane recruitment and endocytosis upon cell stimulation. Curr Biol 10:1403–1412

Bartlett PJ, Young KW, Nahorski SR, Challiss RA (2005) Single cell analysis and temporal profiling of agonist-mediated inositol 1,4,5-trisphosphate, Ca^{2+}, diacylglycerol, and protein kinase C signaling using fluorescent biosensors. J Biol Chem 280:21837–21846

Bouzid A, Lechleiter J (2002) Laser scanning fluorescence microscopy with compensation for spatial dispersion of fast laser pulses. US Patent Appl 2002/0176076

Brumbaugh J, Schleifenbaum A, Gasch A, Sattler M, Schultz C (2006) A dual parameter FRET probe for measuring pK_c and pK_a activity in living cells. J Am Chem Soc 128:24–25

Denk W, Piston DW, Webb WW (1995) Two-photon molecular excitation in laser-scanning microscopy. In: Pawley JB (ed) Handbook of biological confocal microscopy. Plenum, New York, pp 445–458

Dudovich N, Oron D, Silberberg Y (2002) Single-pulse coherently controlled nonlinear Raman spectroscopy and microscopy. Nature 418:512–514

Egner A, Andresen V, Hell SW (2002) Comparison of the axial resolution of practical Nipkow-disc confocal fluorescence microscopy with that of multifocal multiphoton microscopy: theory and experiment. J Microsc 206:24–32

Egner A, Hell S (2005) Fluorescence microscopy with super-resolved optical sections. Trends Cell Biol 15:207–15

Ellson CD, Anderson KE, Morgan G, Chilvers ER, Lipp P, Stephens LR, Hawkins PT (2001) Phosphatidylinositol 3-phosphate is generated in phagosomal membranes. Curr Biol 11:1631–1635

Evans CL, Potma EO, Puoris'haag M, Cote D, Lin CP, Xie XS (2005) Chemical imaging of tissue in vivo with video-rate coherent anti-Stokes Raman scattering microscopy. Proc Natl Acad Sci USA 102:16807–16812

Gaietta GM, Giepmans BN, Deerinck TJ, Smith WB, Ngan L, Llopis J, Adams SR, Tsien RY, Ellisman MH (2006) Golgi twins in late mitosis revealed by genetically encoded tags for live cell imaging and correlated electron microscopy. Proc Natl Acad Sci USA 103:17777–17782

Gallegos LL, Kunkel MT, Newton AC (2006) Targeting protein kinase C activity reporter to discrete intracellular regions reveals spatiotemporal differences in agonist-dependent signaling. J Biol Chem 281:30947–30956

Ganesan S, Ameer-Beg SM, Ng TT, Vojnovic B, Wouters FS (2006) A dark yellow fluorescent protein (YFP)-based resonance energy-accepting chromoprotein (reach) for Forster resonance energy transfer with GFP. Proc Natl Acad Sci USA 103:4089–4094

Griffin BA, Adams SR, Tsien RY (1998) Specific covalent labeling of recombinant protein molecules inside live cells. Science 281:269–272

Hell SW, Wichmann J (1994) Breaking the diffraction resolution limit by stimulated emission: stimulated emission depletion microscopy. Opt Lett 19:780–782

Huisken J, Swoger J, Del Bene F, Wittbrodt J, Stelzer EH (2004) Optical sectioning deep inside live embryos by selective plane illumination microscopy. Science 305:1007–1009

Kaestner L, Juzeniene A, Moan J (2004) Erythrocytes-the 'house elves' of photodynamic therapy. Photochem Photobiol Sci 3:981–989

Kerppola TK (2006) Visualization of molecular interactions by fluorescence complementation. Nat Rev Mol Cell Biol 7:449–456

Lakowicz JR, Szmacinski H, Nowaczyk K, Johnson ML (1992) Fluorescence lifetime imaging of calcium using Quin-2. Cell Calcium 13:131–147

Lipp P, Kaestner L (2006) Image based high content screening – a view from basic science. In: Hüser J (ed) High-throughput screening in drug discovery. Wiley-VCH, Weinheim, pp 129–149

Magliery TJ, Wilson CG, Pan W, Mishler D, Ghosh I, Hamilton AD, Regan L (2005) Detecting protein-protein interactions with a green fluorescent protein fragment reassembly trap: scope and mechanism. J Am Chem Soc 127:146–157

Menger MD, Jager S, Walter P, Hammersen F, Messmer K (1990) A novel technique for studies on the microvasculature of transplanted islets of Langerhans in vivo. Int J Microcirc Clin Exp 9:103–117

Minsky M. (1957) Microscopy apparatus. US Patent 3,013,467

Neil MA, Juskaitis R, Wilson T (1997) Method of obtaining optical sectioning by using structured light in a conventional microscope. Opt Lett 22:1905–1907

Nipkow P (1884) Elektrisches teleskop. Kaiserliches Patentamt, Germany

Pappenheim A (1908) Panoptische Universalfärbung für Blutpräparate. Med Klin 32:1244.

Parker I, Ivorra I (1993) Confocal microfluorimetry of Ca^{2+} signals evoked in xenopus oocytes by photoreleased inositol trisphosphate. J Physiol 461:133–165

Piston DW, Masters BR, Webb WW (1995) Three-dimensionally resolved NAD(P)H cellular metabolic redox imaging of the in situ cornea with two-photon excitation laser scanning microscopy. J Microsc 178:20–27

Reither G, Schaefer M, Lipp P (2006) PKCα: a versatile key for decoding the cellular calcium toolkit. J Cell Biol 174:521–533

Roorda RD, Miesenbock G (2002) Beam-steering of multi-chromatic light using acousto-optical deflectors and dispersion-compensatory optics. US Patent Appl 2002/0149769

Schaefer M, Albrecht N, Hofmann T, Gudermann T, Schultz G (2001) Diffusion-limited translocation mechanism of protein kinase C isotypes. FASEB J 15:1634–1636

Schleifenbaum A, Stier G, Gasch A, Sattler M, Schultz C (2004) Genetically encoded fret probe for pK_c activity based on pleckstrin. J Am Chem Soc 126:11786–11787

Shaner NC, Campbell RE, Steinbach PA, Giepmans BN, Palmer AE, Tsien RY (2004) Improved monomeric red, orange and yellow fluorescent proteins derived from Discosoma sp. red fluorescent protein. Nat Biotechnol 22:1567–1572

Sugiyama O, An DS, Kung SP, Feeley BT, Gamradt S, Liu NQ, Chen IS, Lieberman JR (2005) Lentivirus-mediated gene transfer induces long-term transgene expression of BMP-2 in vitro and new bone formation in vivo. Mol Ther 11:390–398

Sun CK (2005) Higher harmonic generation microscopy. Adv Biochem Eng Biotechnol 95:17–56

Violin JD, Zhang J, Tsien RY, Newton AC (2003) A genetically encoded fluorescent reporter reveals oscillatory phosphorylation by protein kinase C. J Cell Biol 161:899-909

Wabuyele MB Culha M, Griffin GD, Viallet PM, Vo-Dinh T (2005) Near-field scanning optical microscopy for bioanalysis at nanometer resolution. Methods Mol Biol 300:437–452

Wazawa T, Ueda M (2005) Total internal reflection fluorescence microscopy in single molecule nanobioscience. Adv Biochem Eng Biotechnol 95:77–106

Zhang J, Ma Y, Taylor SS, Tsien RY (2001) Genetically encoded reporters of protein kinase A activity reveal impact of substrate tethering. Proc Natl Acad Sci USA 98:14997–15002

Zimmer M (2005) Glowing genes: a revolution in biotechnology. Prometheus, Amherst

11
New Technologies for Imaging and Analysis of Individual Microbial Cells

Byron F. Brehm-Stecher

Abstract Despite their size, microbes manage to exert profound effects on the human macroworld. Traditionally, the field of microbiology has been focused at the population level, limited primarily by the resolution of its workhorse technologies – growth after inoculation into liquid or solid media. However, such "bulk-phase" or population-scale measurements are composed of data from thousands or millions of individual cellular inputs. Averaging the measured parameter across the population may "smooth over" and obscure information on discrete cellular phenomena that may otherwise provide key insights into how microbial cells function or interact with their environments. As alternatives to population-based methods, new tools are needed to visualize discrete microbial structures or phenomena that are inaccessible via traditional means. Classically defined, "imaging" refers chiefly to methods that use visible light to generate a likeness of an object. However, approaches using other regions of the electromagnetic spectrum or nonphotonic processes can also be used to generate graphic representations of microbial cells or provide quantitative measurements, often capturing details not available via other means. Examples include Raman microspectroscopy, microbeam analyses and atomic force microscopy. This chapter seeks to provide a cross-sectional sampling of the diversity of new technologies for imaging and analysis of individual microbial cells and how they have been applied to the problems of single-cell microbiology. Special attention is paid to commercially available technologies allowing dynamic observation of discrete phenomena occurring within individual living microbial cells.

11.1 Introduction

Despite their size, microbes manage to exert profound effects on the human macroworld – on individuals as the causative agents of disease or through production of vital nutritional cofactors within our bodies; on human culture as cellular catalysts in processes fundamental to civilization, such as baking and brewing (Samuel 1996); and on our history as agents of societal change (Doyle and Lee 1986; Achtman et al.

S.L. Shorte and F. Frischknecht (eds.), *Imaging Cellular and Molecular Biological Functions.*

2004) or as a force affecting the preservation of our artwork and monuments (Ciferri 1999; Gonzalez and Saiz-Jimenez 2005). Traditionally, the field of microbiology has been focused at the population level, limited primarily by the resolution of its workhorse technologies – growth after inoculation into liquid or solid media. However, such "bulk phase" or population-scale measurements are composed of data from thousands or millions of individual cellular inputs. Averaging the measured parameter across the population (even a remarkably small population – see Le et al. 2005) may "smooth over" and obscure information on discrete cellular phenomena that may otherwise provide key insights into how microbial cells function or interact with their environments. As alternatives to population-based methods, new tools are needed to visualize the discrete microbial structures or phenomena that are inaccessible via traditional means.

Classically defined, "imaging" refers chiefly to methods that use visible light to generate a likeness of an object. However, approaches using other regions of the electromagnetic spectrum or nonphotonic processes can also be used to generate graphic representations of microbial cells, often capturing details not available via other means. For example, Raman microspectroscopy has provided dynamic information on the metabolic activities of single living cells (Huang et al. 2004), microbeam techniques have been used to study the dynamics of biochemical and physiological changes in living algal cells or elemental distributions in individual diatoms (Heraud et al. 2005; Twining et al. 2003) and atomic force microscopy (AFM) can directly probe the nanomechanical properties (aggregation forces, elasticity, discrete oscillatory motions, etc.) of single microbial cells at subcellular resolution (Ma et al. 2005; Pelling et al. 2004; Touhami et al. 2003, Zhao et al. 2005).

This chapter does not aim to provide a comprehensive review of all the imaging methods that have been developed for the characterization and analysis of individual microbial cells, but seeks instead to provide a cross-sectional sampling of the diversity of approaches that are available and how they have been applied to the problems of single-cell microbiology. Special attention is paid to commercially available technologies allowing dynamic observation of discrete phenomena occurring within individual living microbial cells.

11.2 Live-Cell Imaging

Since the dawning of modern microbiology more than 100 years ago, and with the exception of a few remarkable early examples (Roux et al. 2004), much has been learned from the study of nonliving cells – cells that have been fixed to glass with a flame and subjected to a multiplicity of toxic dyes, or cells dipped in glutaraldehyde, serially dehydrated with ethanol, coated with electron-dense heavy metals and placed under high vacuum. Such techniques allowed the discrimination of various classes of microbes based on morphology and dye uptake, facilitated the detection of these microbes within diseased tissues and enabled exquisite studies of individual cellular and microbial community ultrastructure. However, apart from advances in magnification, our abilities to view *living* microbial cells as they went about their daily routines remained essentially unchanged from van Leeuwenhoek's time (van Leeuwenhoek 1684). Today, several lines of steadily advancing technology are converging to create unprecedented opportunities for imaging and characterizing single, living microbial cells. These include advances in molecular biology, chemistry, optics, microfabrication techniques and nanotechnology as well as the ready commercial availability of fast, inexpensive computing power (Andersson and van den Berg 2004; Brehm-Stecher and Johnson 2004; Maiti et al. 1997; Skakun et al. 2005). These advances have given rise to new methods for fast, accurate and minimally invasive observation or measurement of dynamic microbial phenomena at both the molecular and the cellular scales. Such methods still typically involve exposure of cells to external electromagnetic or physical forces and/or exposure to chemical probes interacting with key cellular components, but basic problems limiting long-term live-cell observations, such as photoxicity, can now be minimized by using noncontinuous, low-intensity or low-energy (e.g., far-red) illumination sources (Frischknecht et al. 2006; Lizundia et al. 2005; Roux et al. 2004), new nonlinear optical imaging modalities (Campagnola et al. 1999; Débarre et al. 2006), label-free vibrational contrast microscopies (Nan et al. 2006; Potma et al. 2001) or new chemical probes having decreased phototoxicities (Q. Li et al. 2006). Multidimensional imaging techniques enable visualization of cells and their processes in three dimensions (3D, or volumetric imaging), through time (4D, or time-lapse imaging) and at multiple wavelengths (5D, or multicolor imaging) (Presley 2005; Roux et al. 2004).

11.3 Imaging Infection

A key beneficiary of new developments in minimally invasive, live-cell imaging techniques is the area of pathogen–host interactions (see also Chap. 12 by Amino et al.). The ability to follow the cycle of infection from first contact to host cell death, without perturbing the process, allows new insights into and more thorough characterization of infectious processes (Frischknecht et al. 2006; Lizundia et al. 2005; Roux et al. 2004). In some instances, such approaches enable "full-picture" correlation of molecular, physiological or morphological events occurring in the system under observation. An example is the study of Lizundia et al. (2005), which used microrotation (MR) imaging to identify and follow discrete subcellular phenomena associated with infection of individual B lymphocytes by the protozoan parasite *Theileria parva*. Cotransfection of lymphocytes with JNK-APF, a dominant-negative mutant of c-JUN N-terminal kinase, enabled these authors to probe the role that modulation of the expression of this enzyme may play in the survival and proliferation of lymphocytes infected with *T. parva*. The microrotation imaging configuration used by these authors enabled high-content, time-resolved visualization of individual lymphocytes in three dimensions. Trapping of lymphocytes within a dielectric cage allowed static positioning of cells, while a constant supply of reagents (annexin V-phycoerythrin, calcium) was provided via microfluidic flow. The dielectric cage was also used to provide precise control over cell rotation (0.2–1 revolutions per second), and images (five to eight frames per second) were recorded periodically with a Zeiss Axiovert 200M microscope using minimally invasive epifluorescent illumination lasting only a few seconds. This design facilitated viewing of infected cells from multiple angles, not just a single focal plane, and allowed these authors to follow discrete changes in annexin uptake or subcellular compartmentation that could not have been captured using standard microscopic techniques. Additional use of green fluorescent protein (GFP) labeled histone H2B facilitated concomitant visualization of chromatin reorganization. Combined analyses enabled the identification of five distinct phenotypes in lymphocytes infected with *T. parva*. Phenotypes characteristic of various stages of apoptosis occurred exclusively in infected cells downregulated for JNK expression, demonstrating a role for this enzyme in the proliferation of cells infected with *T. parva*, a hallmark of infection with this pathogen.

Using microrotation imaging, Lizundia and colleagues were able to capture phenotypic subtleties that were not apparent via conventional flow cytometry. New flow cytometry technologies can now be used to provide high-resolution, multicolor images of individual apoptotic cells within larger populations (George et al. 2004; Sect. 11.9). While such methods represent a major advance in data collection capabilities, the microrotation imaging approach described by Lizundia et al. (2005) is unique in that it can track and image morphological, biochemical or genetic events occurring within a *single* living cell as a function of time.

Whether they are externally applied or expressed under genetic control, fluorescent labels provide a sensitive means for following specific events occurring within individual cells. As a fundamental strategy, however, dependence on extrinsic staining may suffer from several potentially critical limitations, including labeling inefficiencies, interference with the cell's delicate biochemical balance or overt chemical or phototoxicities. Genetically expressed labels may also impose a metabolic burden on expressor cells, leading to reduced competitiveness of labeled vis-á-vis wild-type cells (Füchslin et al. 2003). For some studies, the use of such extrinsic labeling approaches is a must – there are simply no existing alternatives for visualizing the desired cellular properties. However, for studies focused on the dynamic and biochemically delicate interplay between pathogens and their hosts, it is arguably more desirable to employ a system having fewer "moving parts" that could potentially cloud the observational window. These are the types of studies that may benefit from the use of "label-free" approaches – technologies capable of resolving cellular features without the need for extrinsic tags or stains. An example of a label-free technology that has been used successfully in the study of host–pathogen interactions is atomic force microscopy (AFM) (Haberle et al. 1992; Horber et al. 1992; A. Li et al. 2006; Ohnesorge et al. 1997). First described by Binnig et al. (1986), AFM produces topographical images through z-plane deflection of a microscopic stylus rastering in the x–y plane. The stylus can be positioned directly above a sample (and therefore influenced by "atomic forces" – electrostatic, electrosteric and van der Waals), dragged across the sample's surface or used to intermittently "tap" the sample (Brehm-Stecher and Johnson 2004). An important advantage of AFM for the study of infection is that imaging can be carried out in a liquid environment, with little or no sample preparation, and can therefore be used to image living cells. Other benefits include excellent height resolution for cell-surface features (A. Li et al. 2006) and the ability to quantitatively examine multiple biomechanical parameters, including microrheology, elasticity and cellular or molecular binding forces, all of which may vary during the course of an infection (Haberle et al. 1992). AFM can provide valuable multidimensional data when an individual cell is imaged at different times during an infection cycle – from first contact with a pathogen to its ultimate death (Haberle et al. 1992; A. Li et al. 2006). Although AFM is a versatile tool for label-free imaging and surface characterization, many events in host–pathogen interactions are likely short-lived and may not be captured at the relatively slow scan rates of most current atomic force microscopes – tens of seconds for a typical 256×256 pixel image (Hobbs et al. 2005). However, a recent innovation in AFM technology, video-rate AFM, is able to provide full-resolution nanometer-scale AFM images at up to 1,000 times this rate – 25 frames per second (Hobbs et al. 2005). This is achieved by mounting the sample on a microresonant (approximately 20 kHz) tuning fork that is scanned in the *x*–*y* plane underneath the stylus. Video-rate AFM (VideoAFM, Infinitesima, Oxford, UK; Infinitesima 2007) expands the power of AFM by allowing relatively large areas to be quickly imaged at the nanometer scale and is available as a modular attachment to existing instruments. Although it has primarily been applied to imaging of relatively planar systems (DNA adsorbed to a solid surface, collagen fibers, polymer crystals), this approach, once it is applied to cells, could potentially provide further insight into temporally ephemeral cellular phenomena occurring during the infection process. Interestingly, a relatively standard technique for nanoscale imaging, scanning electron microscopy (SEM), has also recently been adapted to allow visualization of unstained, fully hydrated cells ("wet SEM"; Thiberge et al. 2004).

Another exciting modular retrofit device has shown promise for increasing the resolution and label-free imaging capabilities of light microscopy. At the heart of the CytoViva™ system (CytoViva 2007) is a cardioid annular condenser, coupled with high-aperture microscope objective containing an iris. The condenser forms an oblique hollow cone of light with a numerical aperture (NA) of 1.2–1.4. When the objective's iris is closed (darkfield illumination), only refracted, scattered or diffracted light is admitted into the objective (Vainrub et al. 2006). The system combines Köhler illumination with high-efficiency transfer of light from the source to the specimen, so that very high light intensities at the sample can be achieved using a relatively low power light source. At such high light intensities and with an exceptionally high signal-to-noise ratio, small cellular features begin to scatter enough light to become visible, enabling high-definition visualization of cells without the need for staining. Because a low-power light source is used, undesirable sample heating is minimized, facilitating live-cell work. According to company literature, the unique configuration of the system also gives rise to non-diffraction-limited phenomena, including "…standing evanescent waves, plasmon resonance, and fluorescence…" and the system allows resolution of features as small as 90 nm (Vainrub et al. 2006). Combined, these remarkable features provide an imaging platform that is uniquely

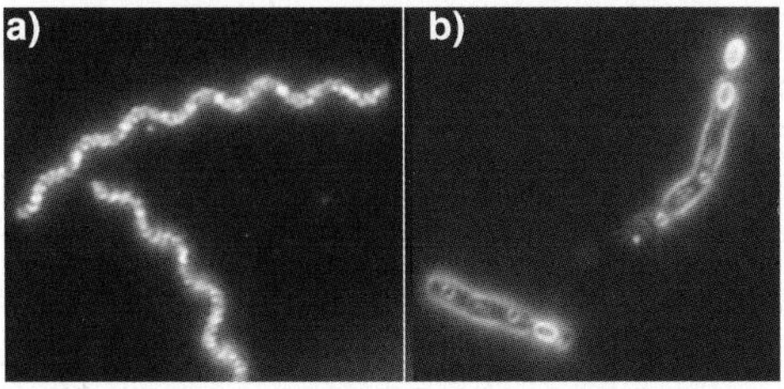

Fig. 11.1 Video micrographs of live, unstained bacterial cells taken with the CytoViva™ imaging system. Both photographs highlight the capacity for this imaging system to resolve fine-scale cellular features in live, unstained cells. **a** Smaller, periodic coil-like structures can clearly be seen within the major helices of these cells. **b** Release of a highly refractive endospore from a cell of *Bacillus anthracis*. Both micrographs were taken using a standard research microscope (×100, oil, 1.4 numerical aperture objective) retrofitted with the CytoViva™ high-resolution adaptor, which replaces the microscope's condenser. (Data courtesy of CytoViva, Auburn, AL, USA)

suited for noninvasive monitoring of infectious processes. Video photomicrographs of free-swimming spirochetes made using a standard research microscope (×100, oil, 1.4NA) show excellent detail, demonstrating a fine-scale periodicity of 12 smaller coils for each major coil of the cell (Fig. 11.1). With its fluorescence capabilities, the system can be used to monitor labeled intracellular pathogens within their host cells, with contrast and visualization of host cell structures provided by the enhanced light scattering effects noted above. Adjustment of the iris allows the user to "dial in" the brightness of the host cell context in which the pathogen resides, an approach that has been used to monitor GFP-labeled *Plasmodium falciparum* within cultured red blood cells. Microbial and host cell features are also readily distinguished without labeling of the pathogen, opening the possibilities for monitoring infection by a wide range of intracellular pathogens, including *Mycobacterium tuberculosis*, *Chlamydia*, *Ehrlichia*, *Legionella*, *Rickettsia* and *Shigella* spp., and *Listeria monocytogenes*. Access to cutting-edge-resolution capabilities typically requires a high degree of technical sophistication and the financial wherewithal to afford the latest "toys." However, the development and commercial availability of versatile off-the-shelf imaging systems such as the CytoViva™ and the increasing number of institutes affording imaging centers promise to place sophisticated imaging capabilities in the hands of nonspecialist users as well. The ability to noninvasively image cellular processes over time (with or without extrinsic labels) will be a boon to both basic cell biology and to the study of host–pathogen dynamics.

11.4 Imaging Single Molecules (Within Single Cells)

Advances in imaging technologies and techniques are now leading us beyond the mere imaging of individual living cells or their subcellular compartments and into an exciting realm in which investigators have been able to image single molecules *within* single cells! Viollier et al. (2004) used a fluorescent repressor-operator reporter system in *Caulobacter crescentus* based on the binding of a LacI–cyan fluorescent protein chimera to tandem arrays of *lac* operator sites inserted at defined regions around the chromosome. Using this approach, these authors investigated the behavior of 112 chromosomal loci in living cells undergoing replication. This approach revealed that each locus occupied a unique "subcellular address" and was transported, in chronological order after replication, to its final destination in the newly formed half of the as yet undivided cell (Viollier et al. 2004). This remarkably ordered process hints at a complex molecular machinery responsible for "on the fly" organization of the replicating chromosome (Viollier et al. 2004).

Golding and Cox (2004) used a similar nucleic acid binding fluorescent protein reporter approach to observe the dynamic life cycle of individual messenger RNA (mRNA) molecules inside single cells of *Escherichia coli*. These authors used time-lapse photography to image both static/localized mRNAs (still tethered to DNA during transcription) and free mRNAs as they diffused throughout the cell (Golding and Cox 2004; Fig. 11.2). Using an enhanced yellow fluorescent protein fusion, Deich et al. (2004) obtained similar images for single histidine kinase molecules in living cells of *Caulobacter crescentus*.

Further development of these tools will open new vistas into the now still largely invisible cellular mechanics governing the addressing, compartmentation, modification and turnover of individual biomolecules within living microbial cells.

11.5 Measuring Discrete Cell Properties and Processes

What if we could shrink ourselves down to the microscale, put on our boots and lug a rheometer through the cytoplasm of our favorite cell type? Or devise a pH probe small enough to fit in a subcellular compartment and accurately report the pH of this microenvironment? What if we had a thermometer that could measure the temperature in the local vicinity of a cell to within 0.1°C and with micrometer precision? And wouldn't it be nice to be able to watch the uptake and compartmentation of a particular sugar by yeast cells, simply by looking at them? These and other small technological wonders have already come to pass (minus the shrinking and boots, of course) – the literature is filled with many inventive approaches for either the direct or indirect measurement of such discrete cell properties and processes.

Fluorescence ratio imaging microscopy (FRIM) is a general imaging approach that can be used for such measurements. The FRIM technique is a microscopy-based approach used to monitor the ratio between two different fluorescence wavelengths and assay specificity is provided by the probe system used. Because the local chemical or physical environment of a fluorophore can affect its behavior, this property can be used to measure those environmental variables a particular probe is responsive to. The fluorescence of other probes may be modulated through the binding of ions. With probes free in solution throughout the cytosol, these FRIM approaches are capable of resolving regions of subcellular microheterogeneity for

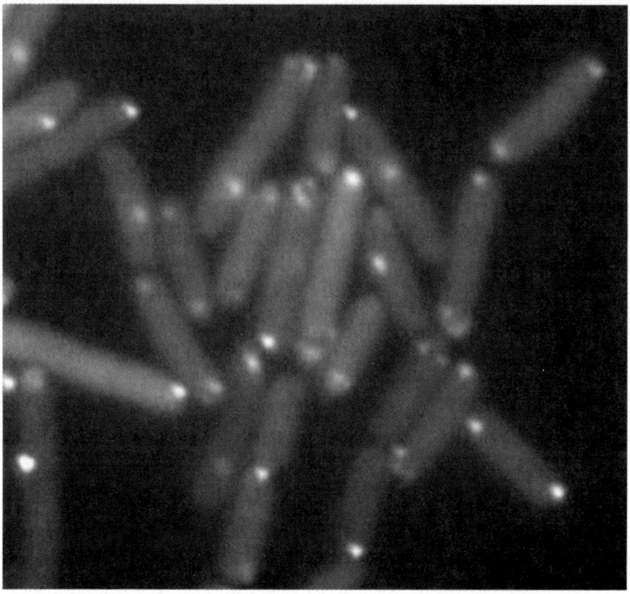

Fig. 11.2 Visualization of individual green fluorescent protein (GFP)-labeled RNA molecules in single, living bacterial cells. A two-component messenger RNA detection system comprising a plasmid-borne GFP–bacteriophage MS2 coat protein fusion (GFP–MS2) and a reporter RNA (on a bacterial artificial chromosome) containing multiple MS2 binding sites was introduced into *Escherichia coli*. When coexpressed with the reporter RNA, multiple copies of GFP–MS2 bind to the tandem repeat in the reporter, enabling it to be visualized within living cells for observational periods extending over several hours. Three distinct types of molecular behavior were observed using this system: localized motion, where the fluorescent particle behaved as if it were tethered by a spring to one spot; movement of individual molecules throughout the entirety of the cell; and "stretching and writhing" of single RNA molecules in behaviors that may represent direct observation of DNA transcription. This novel labeling approach places otherwise invisible genetic phenomena in frame with its cellular context, providing unique insights into dynamic molecular processes occurring within individual living cells over relatively long observational periods. (I. Golding and E. Cox, unpublished data)

the measured parameter. Using viscosity-responsive styrylpyridinium dyes, Wandelt et al. (2005) were able to map the viscosity of DB1X rat embryonic thoracic aorta smooth muscle cells via FRIM, and FRIM-based measurements of internal pH are commonplace in microbial analyses (Imai and Ohno 1995; Olsen et al. 2002). In contrast to the use of fortuitously sensitive chemical compounds to report on local environmental conditions, substrate-specific nanosensor probes have also be designed and used for FRIM studies of properties such as substrate concentration or enzymatic activity. Examples include fluorescence resonance energy transfer (FRET) based maltose or phosphorylation-sensitive probes (Fehr et al. 2002; Sato and Umezawa 2004). The use of such nanosensor approaches has enabled researchers to study discrete phenomena such as subcellular compartmentation of metabolic reactions and spatiotemporal dynamics of phosphorylation-based signal transduction systems in living cells (Fehr et al. 2002; Sato and Umezawa 2004).

Approaches that bridge the use of chemical probes with application-specific microhardware or nanohardware include a microscale thermometer formed from a micropipette tip loaded with a thermosensitive dye (Zeeb et al. 2004) and an optical pH meter based on the use of gold nanoshell particles coated with a pH-responsive molecular adsorbate (Bishnoi et al. 2006). The microscale thermometer used a closed-tip patch pipette having a tip diameter between 1 and 2 μm in size. The tip was filled with an europium compound exhibiting a thermosensitive (thermobleachable) phosphorescence after brief (nonheating) laser excitation. The whole system, mounted on a microscope stage, was capable of measuring the temperature in the immediate locality of the tip with a precision of 0.1°C, providing a tool that may have important applications for measuring microenvironmental temperature fluctuations occurring at the cellular scale.

For the optical pH meter, the size of the nanoshell particle was controlled to yield a surface-enhanced Raman scattering (SERS) effect when an excitation wavelength of 785 nm was used. The SERS spectrum of the molecular adsorbate (*para*-mercaptobenzoic acid) varied in a predictable fashion in response to the pH of the nanoshell's local environment, and the resulting system was capable of reporting microenvironmental pH over a range from 5.80 to 7.60. The coupling of infrared excitation with Raman spectral analysis was chosen to enable this system to operate away from regions of cellular autofluorescence and through tissues or blood (Bishnoi et al. 2006). Ultimately, this type of system may find applications in probing phenomena of direct interest to microbiologists, such as measuring the pH regime to which these particles (and by extension, bacterial pathogens) are exposed following phagocytosis.

Other studies have focused on various aspects of microbial nanobiomechanics – characterizing the process of DNA ejection in bacteriophage T5 through the combined use of hydrodynamic flow and fluorescent DNA stains (Mangenot et al. 2005; Fig. 11.3), or measuring the binding strength of the holdfast of *Caulobacter crescentus* via micromanipulation (Tsang et al. 2006). As they are of central importance to the life cycle or ecology of these organisms, the study of these mechanisms at ultrahigh resolution is expected to provide fundamental insights into both phage and bacterial biology.

11.6 "Wetware"

Although advances in optics, computing and related technological hardware have been important drivers for the development and commercial availability of new single cell imaging methods, this burgeoning field would not be as accessible as it is today without parallel advances in the "wetware," or fluorescent labeling chemistries, that have been developed or adapted for this work. Fluorescent probe molecules interacting with all major classes of macromolecules within microbial cells (proteins, lipids, nucleic acids, carbohydrates) are now commercially available, as are fluorescent enzyme substrates, indicators of pH or metallic ion concentration and

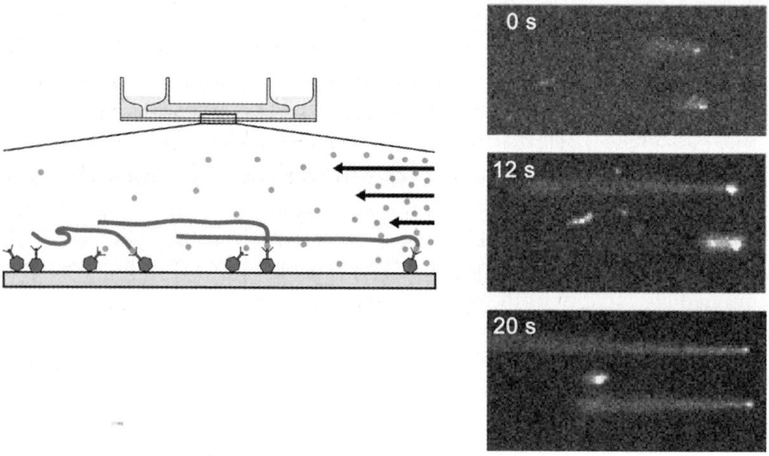

Fig. 11.3 Real-time visualization of receptor-mediated DNA ejection from individual phage particles. Bacteriophage infection is characterized by receptor-mediated attachment, followed by injection of viral DNA into host cells. It is has been estimated that packaging forces required for DNA encapsidation approach 50 pN, corresponding to an internal pressure of approximately 6 MPa (Mangenot et al. 2005). However, little is known about the mechanics of DNA release from phage particles. Mangenot et al. (2005) investigated this process in greater detail for bacteriophage T5 using a microfluidic system comprising a microscopic flow chamber, an intercalating DNA stain (YO-PRO-1) and the bacterial receptor to which the phage selectively binds (the outer-membrane receptor FhuA, from *Escherichia coli*). The cartoon on the *left* illustrates the experimental setup used. Phages (*surface-bound hexagons*) were adsorbed to the surface of the flow-chamber slide and DNA ejection was triggered by addition of purified FhuA (*small circular particles*). A peristaltic pump provided a hydrodynamic flow (direction given by *arrows*), stretching the DNA as it was ejected (*long strings* extending from the phage to which FhuA has bound). Adsorption of phages to the flat surface of the microslide allowed in-focus visualization of phage DNA as it was ejected and subsequently stained by YO-PRO-1 present in the buffer. The photographic series on the *right* follows ejection of DNA from two phage particles. The time points shown immediately after (0 s) or 12 and 20 s after addition of FhuA and YO-PRO-1. Data from these single-virion experiments reveal that DNA ejection from bacteriophage T5 is an unexpectedly complex process. Some phages ejected their entire genome in a single step shortly after receptor binding, while ejection of DNA from other phages occurred in multiple steps, punctuated by potentially sequence specific pauses. The biological significance of these observations is still not fully understood. (Reprinted from Mangenot et al. 2005 with permission)

kit-based dye combinations for estimating cell viability (Invitrogen 2007). Ingenious new chemistries are being applied to "old" problems, with exciting results. For example, Silverman and Kool (2005) have described "quenched autoligation" (QUAL) probes capable of hybridizing with single-nucleotide sequence discrimination to *living* Gram-negative bacteria via fluorescence in situ hybridization (FISH). Although various in vivo hybridization strategies have been described for mammalian cells (Dirks et al. 2003), FISH-based detection of bacteria typically requires fixation of cells prior to hybridization. The ability to hybridize nucleic acid probes to living bacterial cells opens new possibilities for postdetection growth of strains of interest for further study. The development of GFP and related autofluorescent protein technology has been of particular importance to our abilities to sensitively visualize specific molecules, mechanisms and events within living microbial cells (Cluzel et al. 2000; Le et al. 2005; Lippincott-Schwartz and Patterson 2003; Southward and Surette 2002), the activities and interactions of microbes in the environment, or even the subcellular activities of microbial toxins (Dong et al. 2004; Larrainzar et al. 2005). The combination of such “smart” fluorescent probes and the hardware needed to resolve and analyze the resulting signals provides a flexible and expandable palette from which new and powerful single cell imaging strategies may be blended.

11.7 Hardware and Applications

11.7.1 Nonphotonic Microscopies

AFM is the major nonphotonic microscopy in use for biological analyses. AFM can provide exquisite topographic imaging, coupled with detailed microphysical and nanophysical probing and characterization of cell surfaces. Figure 11.4 provides an example of the resolution possible with this approach. An array of related scanning

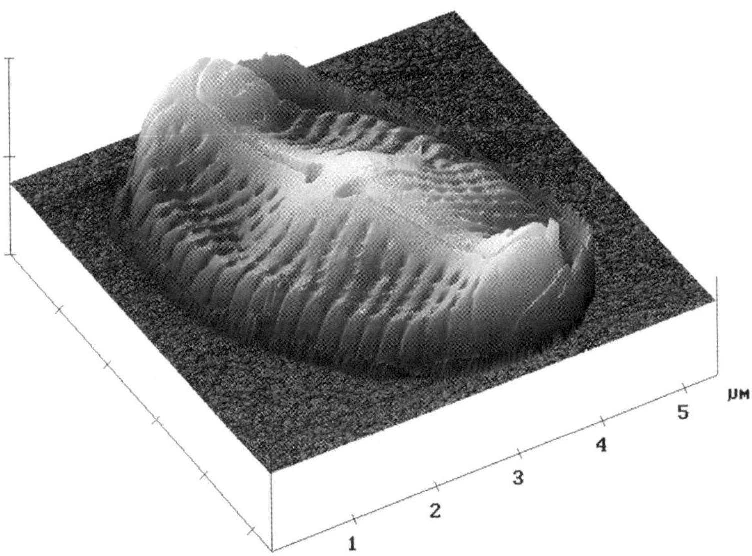

Fig. 11.4 Example atomic force microscope image. This image of the freshwater diatom *Navicula pelliculosa* provides an example of the exquisite topographical detail possible using the atomic force microscope. This diatom was grown on a mica surface, washed with ethanol and dried prior to imaging. Apart from qualitative data (images), atomic force microscopy may also be used to collect quantitative data on such micromechanical and nanomechanical properties such as hardness and elasticity. (Reprinted from Almqvist et al. 2001 with permission)

probe microscopy (SPM) technologies, including magnetic force microscopy (MFM), magnetic resonance force microscopy (MRFM), scanning electrochemical microscopy (SECM), scanning ion conductance microscopy (SICM) and dielectrophoretic force microscopy (DEPFM) further expand the diagnostic capabilities of this family of nonphotonic imaging technologies. As with other imaging techniques, multimode correlation approaches also enable parallel characterization of individual cells using both light or fluorescence microscopy and AFM, expanding analytical capability and depth. Outside the SPM family of instruments, additional nonphotonic microscopies capable of providing unique insights into the mechanics and structure of individual microbial cells include NMR microscopy and acoustic microscopy.

11.7.2 Image Analysis

Unlike conventional, chemical emulsion-based photography, the informational content of digital images is deep, yet readily accessible. Digital image data may be enhanced, extracted, exploded or exported with ease (see Chap. 2 by Meijering and van Cappellen). The marriage of digital imaging technologies with computer analyses has enabled the development of systems capable of tasks ranging from automated detection of bacterial cell division times (Fig. 11.5), to cell lineage tracing in *Caenorhabditis elegans* (Bao et al. 2006) and high-throughput, multidimensional drug profiling via combined automated microscopy and multivariate single-cell analysis (Perlman et al. 2004). Microrobot technology, combined with multiple layers of antibody binding, imaging and subsequent fluorescence bleaching (multi-epitope-ligand cartography, or MELC) has also been used to mine data on the topology of protein expression, including hierarchical organization of protein networks (the "toponome") within single cells (Schubert et al. 2006). Developments in other high-throughput data collection hardware could conceivably outpace our capacity for processing and making sense of the data generated using these platforms (Baatz et al. 2006; Chap. 3 by Swedlow). Such concerns have fueled the development of robust, flexible and customizable software approaches for object-oriented image analysis. One example is the Cellenger image analysis package (Definiens, Munich, Germany), which preserves the topological and hierarchical relationships between subcellular features in the original data, is able to extract data from less-than-optimal, "real-world" samples (e.g., poorly stained or illuminated or confluently growing cells) and is capable of handling data sourced from a variety of platforms (Baatz et al. 2006). In applied clinical markets, microfluidic cassette-based automated image analysis systems are now available for culture-independent concentration, identification and antibiotyping of clinical isolates (BACcelr8r diagnostic system; Accelr8 Technology 2007). The deep informational content available with digital imaging approaches invites the continued development of methods capable of full extraction and utilization of these data, which promises to further the field of single cell microbial analysis.

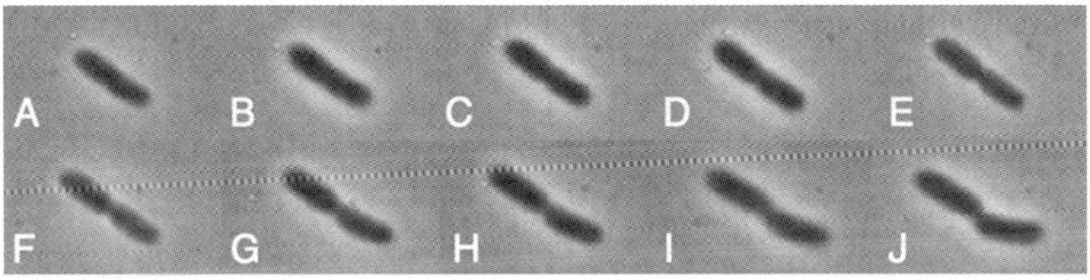

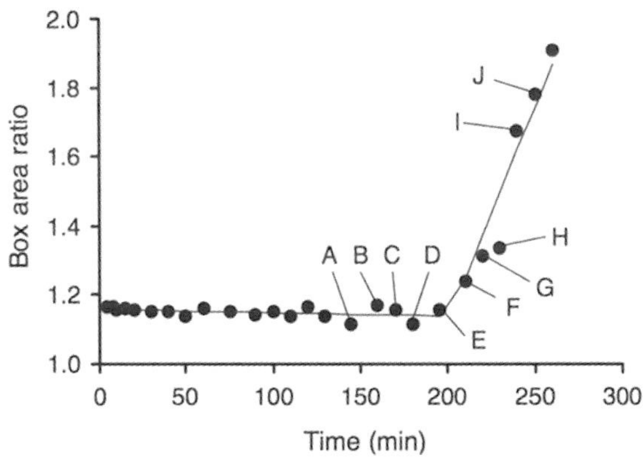

Fig. 11.5 Estimation of cell division times using digital image analysis. Knowledge of the distribution of lag times for individual bacterial cells within a population may contribute to a greater understanding of poststress cell physiology and growth, with important implications for the fields of predictive microbiology and risk assessment (Niven et al. 2006). Manual estimation of division times from digital images is labor-intensive, nonobjective and impractical for the study of larger populations. Cells (*Escherichia coli*) were grown at room temperature on the surface of agar-coated microscope slides and digital images were taken at 10–20-min intervals. A series of programs were then used to analyze cell images (measurements of length, width, thresholded areas), track individual cells within image sequences and calculate the "box area ratio" – a simple parameter whose sudden increase was found to correlate well with cell division times, as determined through visual inspection of the photographic series. This work highlights how the informational content of even simple digital images can be harvested using relatively simple and automatable processes. (Reprinted from Niven et al. 2006 with permission)

11.7.3 Spectroscopic Methods

In general terms, spectroscopy is the study of how light interacts with matter – through absorption, scattering or emission. Analogous processes involving other portions of the electromagnetic spectrum or other forms of applied energy (sound, for example, or even mechanical force) are also termed "spectroscopies." Spectroscopic approaches can be used to provide detailed information on the elemental or biochemical composition of biological samples, and therefore may be used as noninvasive alternatives to traditional "wet chemistry" analytical techniques (Brehm-Stecher and Johnson 2004). Spectroscopic technologies designed to characterize micron-scale samples are referred to aptly as microspectroscopic methods. A number of ingenious and powerful approaches for microspectroscopic analysis of individual microbes at both cellular and subcellular scales have been developed and have been reviewed in detail elsewhere (Brehm-Stecher and Johnson 2004). These include Raman, Fourier transform infrared, dielectric and microbeam spectroscopies. An excellent example of a microbeam approach for elemental mapping of an individual, living algal cell is provided in Heraud et al. (2005). Another member of the microspectroscopic family of tools is fluorescence correlation spectroscopy (FCS). As an emerging and versatile new spectroscopic tool for microbiological analysis, FCS is introduced and discussed further in the next section (see also Chap. 7 by Wachsmuth and Weisshart).

11.8 Fluorescence Correlation Spectroscopy

FCS provides a noninvasive means for real-time analysis of dynamic molecular interactions, including those of short-lived species or intermediates present in low concentrations within a sample (Le et al. 2005; Vukojevic et al. 2005). FCS provides quantitative information about the spatial and temporal behaviors of target species and can be used to follow their diffusion in solution or inside a cell. The technique is based on measurement of the fluctuations in fluorescence intensity of a small number of molecules as they pass through a microvolumetric observation "window" (Maiti et al. 1997; Vukojevic et al. 2005). These fluctuations exhibit randomness in both their intensities and their timescales. Statistical analysis (autocorrelation) of the resulting molecular "noise" as a function of time can yield important information about the behavior of the molecules or particles giving rise to these signals (Skakun et al. 2005; Van Craenenbroeck and Engleborghs 2000; Vukojevic et al. 2005). FCS studies can yield data on a variety of molecular properties, including the average number of target molecules present in a sample, their apparent hydrodynamic volumes, molecular binding and dissociation constants, coefficients of diffusion and reaction rates or rate constants for chemical reactions (Chen et al. 1999; Vukojevic et al. 2005).

A typical FCS instrument is based on confocal microscopy architecture. Precision focusing of a laser by a microscope objective produces a minute volume element within the sample. Fluorescence from molecules transiting the volume element is passed through a pinhole, or confocal aperture, eliminating background signals from molecules outside the focal plane and yielding an effective sampling volume of about 2×10^{-16} l (Maiti et al. 1997; Vukojevic et al. 2005). Fluorescence signals are detected by photodiodes and the resulting electronic signals are statistically analyzed (autocorrelated) to detect informational patterns in the data (Vukojevic et al. 2005).

Although the basic principles behind FCS have been recognized for more than 30 years, the complexity of custom FCS instrumentation has not made this technique amenable for use in nonspecialist laboratories. However, parallel developments in both fast and affordable computing technology and in new fluorescent reagents for use in live-cell applications (Larrainzar et al. 2005; Lippincott-Schwartz and Patterson 2003; Maiti et al. 1997) have facilitated both the development and demand for commercial systems. Commercial FCS instruments are now available from a number of manufacturers. These include the ConfoCor 3 system from Carl Zeiss/EvoTec, the Alba™ FCS from ISS, Leica's FCS2, an instrument from Hamamatsu and the Olympus MF20 system.

As a general technique, FCS has many potential applications in microbiology. Interest in the use of FCS for the analysis of bacterial flagellar motors dates to the early days of this technique (Hoshikawa and Asai 1985). However it was not until the advent of suitable methods for the introduction of inducible fluorescent reporters that FCS was used to correlate biochemical phenomena with associated cellular behaviors. Cluzel et al. (2000) used a combination of FCS and video microscopy to correlate intracellular levels of a chemotactic signaling protein (CheY-P, fused to GFP) with rotational bias in an individual flagellar motor. The ability to visualize the dynamic interplay between cellular biochemistry and observed behavior in a single bacterium provides an unprecedented window into the inner mechanics of living cells (Cluzel et al. 2000). In a related study, FCS was used to monitor induced RNA transcription in real time within individual bacterial cells. Cellular division was monitored via microscopy, allowing RNA transcriptional dynamics to be observed within the context of the cell cycle. In contrast with population-scale measurements, in which an apparent steady level of transcriptional activity was observed over time, individual cells exhibited bursts or "pulses" of transcriptional activity following cell division. It is interesting to note that when the values for just 14 single-cell observations were later averaged, the resulting data resembled the expected population-scale profile of steady transcriptional activity, with the discrete pulses in RNA levels seen for single cells no longer apparent (Le et al. 2005). These experiments show again that population-based determinations of cellular activities may not accurately portray the processes occurring within single cells, and discrete phenomena that may be key to understanding cellular processes may be lost within the milieu of even relatively small populations.

Chemotaxis to cyclic AMP plays an important role in amoebal aggregation and transition from a single cell to a multicellular lifestyle in the slime mold *Dictyostelium discoideum*. Plekstrin homology (PH) protein domains, homologous regions common to a diverse set of cellular proteins, are known to be actively involved in the organization of these proteins during the chemotaxis process (Ruchira et al. 2004). However, little is known about the positional dynamics of PH domain-containing proteins within living cells undergoing chemotaxis (Ruchira et al. 2004). Ruchira et al. (2004) followed the fate of proteins containing PH domains in *Dictyostelium discoideum* cells during chemotaxis using GFP-labeled PH domain proteins and FCS analysis. With this approach, they were able to observe the subcellular distribution of three different PH–GFP protein fusions prior to and following the onset of chemotaxis, thereby helping to further define the roles of these proteins within actively chemotaxing cells (Ruchira et al. 2004).

Apart from its applications in studying dynamic intracellular processes, FCS has also been used for detection of whole bacterial cells and other colloidal particles (Kunst et al. 2002; Qing et al. 2003). These approaches differ from other FCS applications in that they focus on cellular, rather than molecular targets. Also unique about these reports is that detection was accomplished in flowing (vs. stationary) liquids within glass capillaries (Kunst et al. 2002; Qing et al. 2003). Special problems associated with FCS-based detection of whole bacteria include the similarity in size of bacteria to the confocal volume element and the potential for cell clumping. FCS theory has been validated for the case of very small molecules passing through a relatively large volume element. It is not clear how the results may be affected when the size of the target species approaches the size of the volume element. For example, if only a portion of a cell passes through the detection window, this would likely result in measurement of a shorter apparent diffusion time (Kunst et al. 2002). In the study by Qing et al. (2003), individual dye-stained cells of *Escherichia coli* were calculated to be 4×10^8 times brighter than individual dye molecules. Therefore, clumping of stained cells would be expected to produce very bright spikes in fluorescence that would likely impact the autocorrelation curve and interfere with analyses (Qing et al. 2003).

Hink et al. (2000) used FCS to study the interactions of both molecular *and* cellular-scale ligands and coupled this work with time-resolved fluorescence anisotropy to help explain the high success rate of GFP (see Sect. 11.6) as a reporter molecule in fusion proteins. Unlike other reporters, which may misfold to form nonfunctional aggregates, fusion proteins containing GFP are typically more successful, yielding partnerships where GFP retains its fluorescence and labeled proteins retain their biological activities (Hink et al. 2000). Focusing on the capacity of FCS to measure the diffusional characteristics of different species (scFv-GFP and whole bacterial cells) in solution or suspension, these authors were able to show that a lipopolysaccharide-targeted single chain antibody fusion (scFv-GFP) retains its ability to bind to native lipopolysaccharide on the surface of the Gram-negative bacterium *Ralstonia solanacearum* (Fig. 11.6). Coupling these observations with the anisotropy data, these authors posited that the fusion partners in scFv-GFP are bound by a highly flexible bridge that provides each partner with enough freedom to behave as if it were in the native conformation (Hink et al. 2000).

Due in part to the difficulties associated with both the resolution and the propagation of viruses, fewer studies have been done in the area of "single-cell" (single virion) microbiology. However, as a special application of colloidal particle tracking, FCS has been used to study the diffusion of fluorescently labeled bacteriophages inside microbial biofilms (Lacroix-Gueu et al. 2005). The ability of phages to penetrate and navigate the solvent channels within bacterial biofilms suggests a possible role for biofilm structure in protecting phages from external stresses or inactivation. These results may have practical implications in the persistence of phage infection in cheesemaking and other industrial fermentations (Lacroix-Gueu et al. 2005).

An interesting variant of this technology is high-pressure FCS, introduced by Müller and Gratton (2003). Pressure-induced denaturation of biological molecules has been an important tool for biomolecular characterization, facilitating studies of macromolecular assembly (protein folding, subunit association, etc.) or enzyme structure–activity relationships (Müller and Gratton 2003; Northrop 2002). However, because single-molecule FCS requires high-NA objectives, which have short working distances (less than 1 mm), and traditional pressure cells have observation

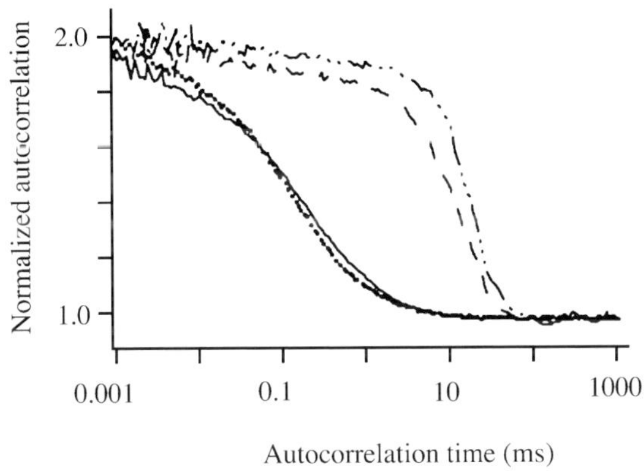

Fig. 11.6 Example fluorescence correlation spectroscopy (FCS) autocorrelation curve. Among its applications, FCS can be used to explore binding interactions between molecular partners, based on differences in the diffusional properties of free partners vs. those of their complexes. These analyses can also be extended to interactions between molecular/cellular binding pairs, as this figure illustrates. A single-chain antibody–GFP fusion (scFv–GFP) targeted to the lipopolysaccharide of Gram-negative bacteria was examined via FCS (Zeiss-Evotec ConfoCor® system) for its ability to bind its target. Binding of this small antibody fragment to its cellular target is expected to result in a large (cellular-scale) complex. Given its size, this scFv-cell complex will display a correspondingly longer diffusion time through the microvolumetric observational window of the instrument vis-à-vis the free antibody fragment. The autocorrelation curves shown here support these expectations: the *solid line* depicts the diffusional profile of the scFv–GFP fusion alone, and the *dotted-and-dashed line*, with its substantial increase in diffusion time, is the profile for the scFv–GFP fusion bound to the surfaces of Gram-negative (target) bacteria. Autofluorescent bacteria examined by themselves and nontarget (Gram-positive) bacteria incubated with the scFv–GFP fusion were included as positive and negative controls (*dashed line* and *dotted line*, respectively). This work demonstrates that although FCS is typically used to characterize molecular-scale phenomena, it can also be used to study binding interactions between whole cells and their ligands. (Reprinted from Hink et al. 2000 with permission)

windows in excess of these dimensions, FCS analyses of systems under high pressure has not been previously possible. The instrument developed by Müller and Gratton (2003) uses a cylindrical fused-silica capillary in which the body of the capillary serves as both a pressure-containment vessel and, at approximately 150 μm in thickness (comparable to the glass cover slips typically used in FCS), as an optical window. To mitigate problems associated with bending of incoming light by the curved surface of the capillary, glycerol (a liquid with the same refractive index as quartz) was used as a coupling medium. The system was found to be capable of single-molecule spectroscopy at pressures up to 350 MPa, with a theoretical load of 700 MPa, and was sensitive enough to detect small pressure-induced effects on the density and viscosity of the aqueous medium (Müller and Gratton 2003). Although pressure-mediated inactivation of pathogenic bacteria was first suggested more than 100 years ago (Hite 1899; Gould 2006), it is now only becoming more widespread as a food processing tool, as an alternative to thermal methods (Gould 2006; San Martin et al. 2002). High-pressure FCS could therefore serve as a key link between the basic science of molecular behavior at high pressures and the organismal-level physiological effects needed for pressure-mediated inactivation of pathogens in foods.

Outside applications focused on detection of target molecules within microbial cells or the detection of the cells themselves, FCS has also found use in microbiology for the sensitive detection of pathogen-specific PCR products in complex molecular mixtures (Walter et al. 1996). In this application, a 30-s post-PCR FCS measurement enabled these authors to monitor the incorporation of an amplicon-specific fluorescently labeled probe into a double-stranded product as a function of the probe's diffusional mobility, enabling the detection of as few as ten target molecules against a background of 2.5 μg nonspecific DNA (Walter et al. 1996). This wide variety of microbiological problems to which FCS has been applied highlights the power and versatility of this approach for the characterization of molecular and/or colloidal interactions, both in vivo and in vitro.

11.9 A Picture is Worth a Thousand Dots – New Developments in Flow Cytometry

An often-cited limitation of flow cytometry surrounds the fact that fluorescence and scatter data are not image-based, but rather represent photon-generated electrical pulses of varying intensity. In complex, or "noisy" samples, it may therefore be difficult to distinguish between "real" signals due to positive labeling of target cells, and spurious signals arising from the presence of intrinsically fluorescent background particles or nonspecific adsorption of probe molecules to components of the sample matrix. With good labeling and at relatively high target cell concentrations, positive populations in such samples may often be readily distinguished from background visually by combining both scatter and fluorescence data (Fig. 11.7). However, when dealing with rare target cells, the distinction between "real" and spurious signals may not be as clear and may require more sophisticated approaches, including special sample preparation procedures and multicolor analyses (Gross et al. 1993; Radbruch and Recktenwald 1995).

A key strength of whole-cell staining approaches such as "phylogenetic staining" using ribosomal RNA targeted nucleic acid probes (DeLong et al. 1989) is that such methods preserve potentially vital diagnostic information on cell morphology (Brehm-Stecher et al. 2005). Cytometric light scatter measurements convey some morphological information on cell size (forward scatter) and internal content or "granularity" (side scatter), but the relationship between light scatter properties and cell shape, size or grouping (e.g., chains, tetrads, heterogeneous cell clumps or even cell-particle associations) is not always clear or direct. Moreover, many of the event types that may prove problematic to cytometric analysis are readily identifiable by eye using a microscope.

These problems have spurred the development of "hybrid" flow cytometry systems capable of both scatter-based or Coulter-volume detection and simultaneous

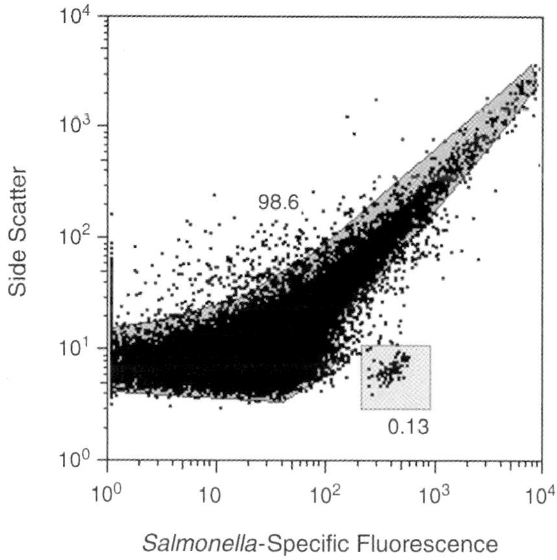

Fig. 11.7 Combination of fluorescence in situ hybridization (FISH) and flow cytometry for detection of *Salmonella typhimurium* in contaminated alfalfa sprouts. Sprout samples contaminated with 10^5 CFU/g *S. typhimurium* (10^4 CFU/ml sprout wash) were concentrated via tangential flow filtration and hybridized with a Cy5-labeled DNA FISH probe. A discrete subpopulation of *Salmonella* was easily detected at this level of contamination, despite the high background of nontarget microbial flora and particulate matter present in these samples. The *Salmonella* subpopulation comprised only 0.13% of the total population (82 events out of about 62,000 total events collected), highlighting the capacity of this method for detection of relatively low levels of specific bacteria in complex, "noisy" food matrices. *CFU* colony forming units. (B. Bisha and B.F. Brehm-Stecher, unpublished data)

imaging of individual events (George et al. 2004; Kubota et al. 1995; Wietzorrek et al. 1999). Because flow cytometry is a dynamic analytical approach with cells passing rapidly (on a millisecond scale) through the observation window, successful imaging, matching and storage of collected images with their correlating electronic signals can be a technically demanding and memory-intensive process (George et al. 2004). Although the technology behind imaging cytometry has been in development for almost 30 years (Hüller et al. 1991, 1994; Kachel et al. 1979; Kay et al. 1979; Kubota et al. 1995; Sieracki et al. 1998; Wietzorrek et al. 1999) it has only been recently, with the widespread availability of digital imaging and sufficient computing power, that such systems have been made commercially available. Existing commercial systems include the ImageStream® system from the Amnis® Corporation (Amnis 2007), the FlowCAM® from Fluid Imaging Technologies (2007) and the Sysmex FPIA-3000, available from Malvern Instruments (2007). The most versatile of these systems is the ImageStream®, which is capable of providing six separate images of each cell analyzed (brightfield, darkfield and up to four fluorescence colors) at throughput rates of up to 300 cells per second. According to company literature, over 35 morphometric and signal intensity features are calculated for each image, yielding over 200 measurements per cell, with high enough resolution and sensitivity (50 molecules of equivalent soluble fluorochrome) to enable subcellular localization of fluorescence signals. The resulting data-rich files are correspondingly larger than typical cytometry datasets, with a 10,000-event file containing both raw and processed data requiring approximately 500 MB of disc space. Although biological applications have been reported for the Sysmex machine (Kubota et al. 1995; Kubota 2003), it is currently being marketed for particle analysis in the pharmaceutical and related industries. The FlowCAM® has also been used for particle characterization (Sterling et al. 2004), but its chief application has been for automated in situ cytometry and imaging of marine microplankton (Sieracki et al. 1998). In this application, the FlowCAM® may be useful as a remote sensor and early-warning system for the presence of harmful, toxigenic microalgal blooms (Sieracki et al. 1998). The FlowCAM® instrument is capable of imaging particles or cells ranging from 3 to 100 μm at sampling volumes between 1 and 12 ml/min, and is available in benchtop and portable models, as well as a version that is submersible to 100 ft (Sieracki et al. 1998; Fig. 11.8).

In order to solve the special problems associated with video capture of individual cells in motion, imaging cytometers may incorporate gas discharge flash or stroboscopic illumination (Kubota 2003; Wietzorrek et al. 1998) or, in the case of the Amnis instrument, "virtual panning" of the CCD detector in synchrony with the velocity of a moving cell (George et al. 2004). This latter approach enables observational windows of up to 10 ms, without image blurring. Strategies for ensuring that cells in flow remain in the focal plane include hydrodynamic focusing to create a flat pane of sheath fluid in the flow cell (Sysmex FPIA-3000, product literature)

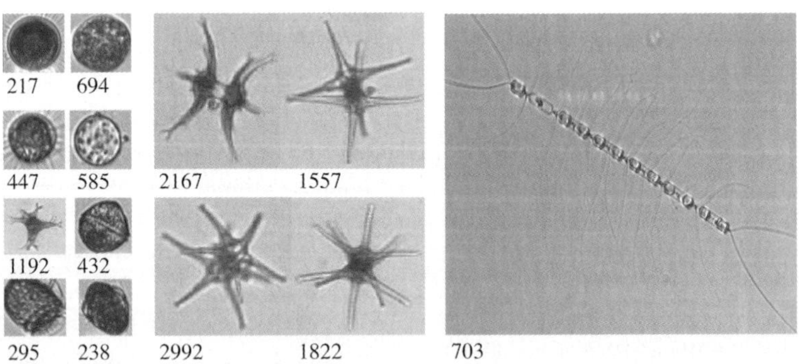

Fig. 11.8 Composite of typical FlowCAM® image data. This figure demonstrates the capacity of the FlowCAM® instrument to provide high-quality, flow-through imaging of individual diatoms in natural waters. Selection of individual events on a scattergram allows automated recall of a stored digital image of the cell or particle associated with each event. The ability to correlate a cytometric scatter event with its corresponding image provides an additional layer of single-cell resolution, enabling fuller characterization of sample composition than is possible with cytometry alone. The most recent version of the FlowCAM® instrument provides natural color images, further blurring the distinction between cytometry and microscopy. (Courtesy of Fluid Imaging Technologies, Edgecomb, ME, USA)

or use of a concentrically etched optical element that introduces a controlled spherical aberration, providing a continuous focal depth of about 300 µm (FlowCAM®; Sieracki et al. 1998). Scatter or fluorescence data from these imaging cytometers are available as dot plots (scattergrams), with captured images provided as interactive photomontages (Sieracki et al. 1998). The ability to key in on select events on a dot plot and subsequently call up stored images of the cells corresponding to these events combines the best aspects of both flow cytometry and digital imaging technologies, allowing both rapid analysis of large cell populations and digital-image-based cell recognition and morphometry (George et al. 2004).

Although it does not incorporate optical imaging, another "hybrid" cytometry system that bears mentioning here is the on-chip dielectric spectroscopy device described by Cheung et al. (2005). This instrument takes advantage of the fact that impedance measurements made over a range of frequencies can provide key information on cellular physiology (Cheung et al. 2005). Biological cells are composed of adjacent layers of materials having widely variant dielectric properties (Pethig and Markx 1997). As an example, an individual cell of the Gram-negative bacterium *Escherichia coli* is composed of the following layered structures: an outer membrane, a thin cell wall and protein-rich periplasmic space, a cytoplasmic membrane and a gel-like, solute-rich cytoplasm. Field-induced interfacial polarizations at the boundaries of these layers can give rise to dielectric signals that can provide information on a number of physiological parameters, including cell size, surface charge, structure and organization, membrane integrity and cytoplasmic conductivity (Pethig and Markx 1997). Typically, dielectric spectroscopy is performed either at the population scale (Asami et al. 1999) or on single cells held in a dielectrophoretic trap (Hölzel 1998, 1999). However, the device described by Cheung et al. (2005) enables flow-through dielectric spectroscopy of microbial populations on a cell-by-cell basis. Because dielectric spectroscopy derives its resolving power from intrinsic (structural) cellular properties, this approach is label-free, allowing the direct analysis of cell mixtures without the need for extensive sample processing. Figure 11.9 demonstrates the ability of this approach to distinguish between various cell types or particles (*Bacillus cereus* spores or vegetative cells, 4-µm latex beads, erythrocytes and *Mucor* spores). The device, now being developed commercially by Axetris Microsystems (2007), provides an additional "flavor" to the spectrum of novel cytometry-based products now available for single-cell analysis. Lastly, the COPAS™ Biosorter from Union Biometrica (2007) provides a means for flow cytometry analysis and sorting of relatively large objects (20–1,500 µm in diameter), including multicellular animals (*Caenorhabditis elegans*) or embryos (*Drosophila melanogaster*, zebra fish and *Xenopus*). Because of the increasingly important roles that such nonmammalian animal hosts play as models for bacterial and fungal infection processes (Fuchs and Mylonakis 2006; Sifri et al. 2005), the COPAS™ system provides a potentially valuable cytometric tool for following host–pathogen interactions (see Sect. 11.3). Although the first flow cytometer was described almost 60 years ago (Davey and Kell 1996; Gucker et al. 1947), the continual appearance of these and other new variant forms highlights that this is still a viable and evolving technology platform for single-cell analysis.

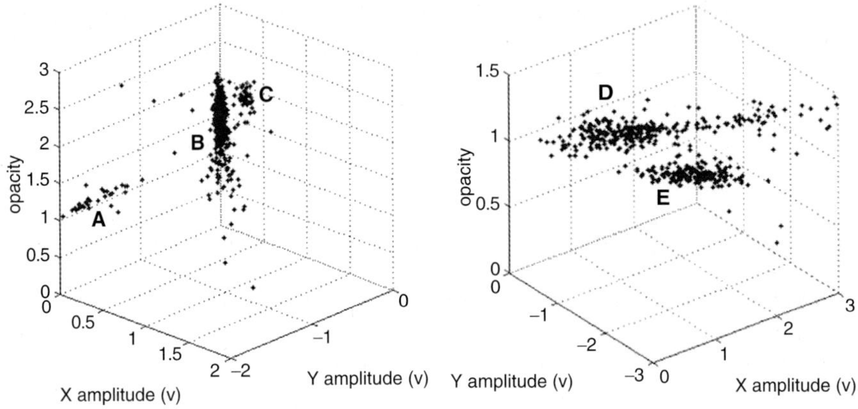

Fig. 11.9 Label-free differentiation of distinct cell types via flow-through dielectric spectroscopy. Mixtures of different microbial or eukaryotic cell types or latex beads (4 μm) were analyzed at frequencies between 8 and 14 MHz using the on-chip dielectric flow cytometer manufactured by Axetris Microsystems (2007). Distinct cell types or latex beads were clearly distinguished via on-chip spectroscopy as follows: *A Bacillus cereus* (vegetative cells), *B Bacillus cereus* (spores), *C* latex beads, *D* erythrocytes and *E* fungal spores (*Mucor* spp.). These results highlight the capacity of flow-through dielectric spectroscopy to provide accurate and label-free differentiation of cells or particles based on their dielectric response to an applied field. (Courtesy of Axetris Microsystems, Sarnen, Switzerland)

11.10 Strength in Numbers – Highly Parallel Analysis Using Cellular Arrays

Given the fact that microbes typically live and grow in nature as multispecies assemblages, not as individual cells (Donlan 2002), the ability to relate observations made at the single-cell level to larger-scale (e.g., populational) phenomena would add another, powerful layer of resolution. However, the choice of one imaging technology over another often involves compromise – techniques that allow high-resolution observations of individual cells over time may not permit easy comparison of such observations for large numbers of cells. Conversely, technologies designed for collection of single-cell parameters across microbial populations may provide only a limited "snapshot" of how the cells appeared the moment they passed in front of the lens or photomultiplier tube. Recent advances in cellular array technology may help bridge this gap, allowing collection of multidimensional data on single cells, with subsequent comparison of these data across larger populations. In an early example of this approach, Ericsson et al. (2000) used a manual laser tweezer/optical trapping technique to form ordered arrays of individual cells of a nonflagellated strain of *Escherichia coli* (500 cells per array) on the surfaces of gridded counting chambers. Prior to the cells being addressed within the array, they were stained with a commercially available dye-based viability kit (LIVE/DEAD

*Bac*Light kit; Invitrogen 2007). Position on the array and viability staining results were recorded for each cell. With the arrayed cells held in place by the optical trap, fresh nutrients were added and cells were scored for their ability to divide. In this way, these authors were ultimately able to screen 10,000 cells and correlate dye-conferred viability reporting with the observed ability of these cells to divide and form daughter cells. While this was a pioneering experiment, each experiment required at least 4 h, not including manual arraying of cells and data analysis. As 20 individual experiments were needed to collect data on 10,000 cells, this approach was fairly cumbersome. As single-cell microbiology is a rapidly maturing field, subsequent developments in technology have forwarded the capabilities of such cellular array approaches, automating many aspects and increasing throughput.

As an example, Chiou et al. (2006) have described an approach using optically generated dielectrophoretic traps formed by projecting concentric patterns of light on a chemically doped photoconductive surface. With these optically controlled "virtual" dielectrophoretic traps, which, at $10\,nW/\mu m^2$, use 1/100,000th of the optical intensity required for typical laser tweezers, these authors demonstrated massively parallel operation of 15,000 optical dielectrophoretic traps on a $1.3 \times 1.0\,mm^2$ surface (Chiou et al. 2006). Another approach to creating high-occupancy arrays is to chemically etch the end of a fiberoptic bundle (50,000 individual fibers) to form microwells (3.1-µm wide) into which cells can be loaded for analysis (Kuang et al. 2004). As each well provides an independent light path, this approach provides up to 50,000 individually readable output channels, with each occupied well dedicated to collecting information on a single living cell (Kuang et al. 2004). This method has been demonstrated for use with both bacteria and yeast (Biran and Walt 2002). Additional strategies have been described for the creation of high-content (about 30,000 cells) lymphocyte arrays (Yamamura et al. 2005). In this work, fluorescence analysis of microscopic images was done using existing microarray software and cells of interest could be removed via aspiration with a micromanipulator for further analysis. It is likely that similar strategies could be readily adopted for high-throughput, parallel analysis of sessile microbial cells (e.g., yeasts or non-flagellated bacteria). Such approaches would fill the gap between methods designed to examine a single cell over time and those that collect data on large numbers of cells in flow-through systems, yet lack the capability of requerying these cells at a later time point.

11.11 Nontactile Manipulation of Individual Cells and "Wall-less Test Tubes"

Single-cell studies can be frustrated by the dynamic nature of their subjects – in order to study a cell in detail, you must first get it to hold still. Cells may be independently motile, subject to the vagaries of Brownian motion and bulk fluid effects, or simply nonadherent. One approach may be to anchor such cells to a surface (such as a microscope slide) via chemical, electrostatic or physical (thin gelatin or agar overlay) means. Although these approaches may be effective in restricting cell movement to the field of view during time-course studies, they are artificial, contact-based methods and do not allow the study of cells in a planktonic state. Nontactile approaches, on the other hand, may be less invasive and allow selective repositioning or rotation of cells, flow-through addition and removal of reagents, and 3D visualization (Roux et al. 2004). Dielectrophoretic and optical trapping are examples of nontactile methods now widely used for manipulation of individual cells (Brehm-Stecher and Johnson 2004). Additional nontactile approaches are now being developed that allow stable trapping and/or manipulation of individual cells for optical, physiological or biochemical analyses – essentially "wall-less test tubes". These include biphasic physical (water-in-oil) systems and acoustic levitation ("acoustic force field") approaches (He et al. 2005; Santesson et al. 2000, 2003, 2004; Wood et al. 2005). In the first approach, described by He et al. (2005), a microfluidic system was used to generate aqueous droplets within an immiscible (soybean oil) phase. Optical trapping was used to position ("park") beads, cells or organelles at an aqueous–oil interface within adjoining microfluidic channels, followed by a pressure pulse to shear off individual aqueous droplets and encapsulate the desired objects (He et al. 2005). Once loaded with their cargo, these droplets could be further manipulated and positioned using optical trapping. Cells or beads remained stably entrapped and were free to move around within such droplets, but could not be moved across the water–oil interface. A single, rapid (about 5-ns) pulse from a 355-nm YAG laser enabled photolysis of encapsulated cells, converting these droplets into miniature biochemical reaction vessels having near-cellular-scale volumes. The ability to nearly instantaneously decompartmentalize the contents of individual cells into such optically transparent and diffusionally confined reaction vessels opens up new possibilities for biochemical analyses ("chemical cytometry") at the level of the single cell (He et al. 2005; Wu et al. 2004).

Acoustic levitation involves the suspension of a liquid droplet in a nodal point within a standing wave of ultrasound energy generated between a radiator and a reflector (Santesson et al. 2000; Fig. 11.10). Suspended droplets can range from 20 to 500 nl in volume and are applied to the node with custom dispensers or a syringe. Additional liquids (water, to counter evaporative effects and biochemical reagents) are dispensed as needed (Santesson et al. 2000). Individual cells may be suspended within such droplets and observed via optical or spectroscopic means (Santesson et al. 2000; Wood et al. 2005). Advantages of this type of approach include the fact that individual cells may be confined in space for analysis, analytes or reagents are not lost through nonspecific adhesion to the walls of the "test tube," and apart from the reflectivity of the droplet's surface, there are no additional barriers to optical transmission at either excitation or emission wavelengths, resulting in improved signal-to-noise ratios for certain analytical approaches (Wood et al. 2005). Acoustic levitation has been used successfully for the study of human adipocytes via fluorescence analysis (Santesson et al. 2000) and for Raman spectral profiling of suspended algal cells (Wood et al. 2005). In an excellent review, Santesson and Nilsson (2004) detail bioanalytical chemistry applications for acoustic levitation technology and describe additional, related approaches including optical,

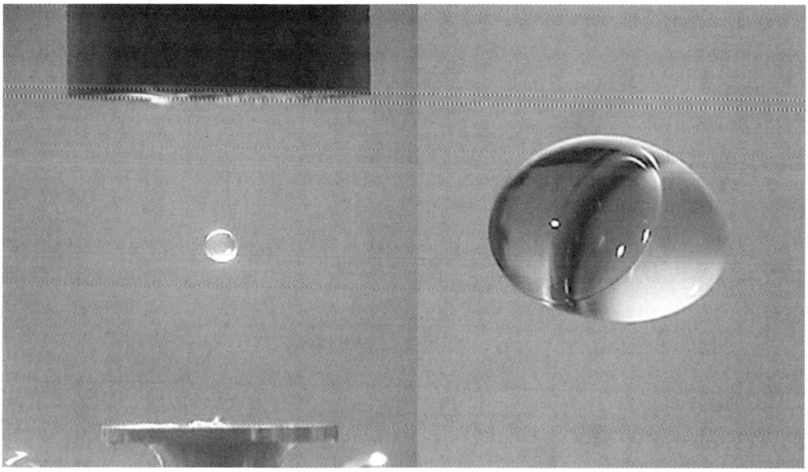

Fig. 11.10 Acoustic levitation for nontactile manipulation of nanovolume liquids. Nontactile methods for liquid handling such as acoustic levitation can be used to create "wall-less test tubes" – containerless reaction chambers whose contents can be monitored using optical imaging or spectroscopic techniques. The lack of reflective or refractive surfaces (other than the liquid–air interface itself) can dramatically improve results obtained from such investigations. For example, Wood et al. (2005) combined acoustic levitation with Raman spectroscopy to collect spectra from live microalgal cells (about 10^4 cells per levitated droplet). These authors observed an approximate 15-fold increase in the signal-to-noise ratio for spectra collected from levitated samples vs. spectra collected using a quartz microcuvette. Although these data were collected on cell populations, acoustic levitation has also been used for fluorescence or spectroscopic analysis of individual cells and for biphasic chemical reactions (Santesson et al. 2000, 2003, 2004; Santesson and Nilsson 2004). The *left-hand panel* shows a 500-nl drop of water suspended within a node in a standing wave generated between an ultrasonic transducer (*bottom*) and a reflector (*top*). The *right-hand panel* shows a biphasic organic–aqueous system contained within a single levitated droplet. (Reprinted from Santesson and Nilsson 2004 with permission)

diamagnetic, electrostatic and aerodynamic levitation techniques. The availability of acoustic levitation and related techniques adds another level from which to formulate a nontactile analytical approach for the study of individual microbial cells.

11.12 Conclusions

Continual advances in our understanding of microbial cell structure and organization are gradually shifting our view of microbes from "amorphous bags of enzymes" to complex cells containing discrete subcellular structures (e.g., cytoskeletal elements, organelles) and domains in which distinct genetic or biochemical processes occur and are regulated over space and time (Brehm-Stecher and Johnson 2004; Harold 2005; Kerfeld et al. 2005; Losick and Shapiro 1999; Löwe et al. 2004; Viollier et al.

2004). Technologies capable of high-resolution, real-time imaging of individual living microbes will facilitate further exploration of this emerging complexity within microbial cells.

References

Accelr8 Technology (2007) Rapid identification of antibiotic resistance. http://www.accelr8.com. Cited 02 May 2007

Achtman M, Morelli G, Zhu P, Wirth T, Diehl I, Kusecek B, Vogler AJ, Wagner DM, Allender CJ, Easterday WR, Chenal-Francisque V, Worsham P, Thomson NR, Parkhill J, Lindler LE, Carniel E, Keim P (2004) Microevolution and history of the plague bacillus, *Yersinia pestis*. Proc Natl Acad Sci USA 101:17837–17842

Almqvist N, Delamo Y, Smith BL, Thomson NH, Bartholdson Å, Lal R, Brzezinski M, Hansma, PK (2001) Micromechanical and structural properties of a pinnate diatom investigated by atomic force microscopy, J Microsc 202:518–532

Amnis (2007) Amnis Corporation home. http://www.amnis.com. Cited 2 May 2007

Andersson H, van den Berg A (2004) Microtechnologies and nanotechnologies for single-cell analysis. Curr Opin Biotechnol 15:44–49

Asami K, Gheorghiu E, Yonezawa T (1999) Real-time monitoring of yeast cell division by dielectric spectroscopy. Biophys J 76:3345–3348

Axetris Microsystems (2007) Alextris Microsystems. http://www.axetris.com. Cited 2 May 2007

Baatz M, Arini N, Schäpe A, Binnig G, Linssen B (2006) Object-oriented image analysis for high content screening: detailed quantification of cells and sub cellular structures with the Cellenger software. Cytometry A 69:652–658

Bao Z, Murray JI, Boyle T, Ooi SL, Sandel MJ, Waterston RH (2006) Automated cell lineage tracing in *Caenorhabditis elegans*. Proc Natl Acad Sci USA 103:2702–2712

Binnig G, Quate CF, Gerber C (1986) Atomic force microscope. Phys Rev Lett 56:930–933

Biran I, Walt DI (2002) Optical imaging fiber-based single live cell arrays: a high-density cell assay platform. Anal Chem 74:3046–3054

Bishnoi SW, Rozell CJ, Levin CS, Gheith MK, Johnson BR, Johnson DH, Halas NJ (2006) All-optical nanoscale pH meter. Nano Lett 6:1687–1692

Brehm-Stecher BF, Johnson EA (2004) Single-cell microbiology: tools, technologies, and applications. Microbiol Mol Biol Rev 68:538–559

Brehm-Stecher BF, Hydlig-Nielsen JJ, Johnson EA (2005) Design and evaluation of 16S rRNA-targeted peptide nucleic acid probes for whole-cell detection of members of the genus *Listeria*. Appl Environ Microbiol 71:5451–5457

Campagnola PJ, Wei M, Lewis A, Loew LM (1999) High-resolution nonlinear optical imaging of live cells by second harmonic generation. Biophys J 77:3341–3349

Chen Y, Müller JD, Berland KM, Gratton E (1999) Fluorescence fluctuation spectroscopy. Methods 19:234–252

Cheung K, Gawad S, Renaud P (2005) Impedance spectroscopy flow cytometry: on-chip label-free cell differentiation. Cytometry A 65:124–132

Chiou PY, Ohta AT, Wu MC (2005) Massively parallel manipulation of single cells and microparticles using optical images. Nature 436:370–372

Ciferri O (1999) Microbial degradation of paintings. Appl Environ Microbiol 65:879–885

Cluzel P, Surette M, Leibler S (2000) An ultrasensitive bacterial motor revealed by monitoring signaling proteins in single cells. Science 287:1652–1655

CytoViva (2007) CytoViva – see things you've never seen before – live cell imaging. http://www.cytoviva.com. Cited 2 May 2007

Davey HM, Kell DB (1996) Flow cytometry and cell sorting of heterogeneous microbial populations: the importance of single-cell analyses. Microbiol Rev 60:641–696

Débarre D, Supatto W, Pena A-M, Fabre A, Tordjmann T, Combettes L, Schanne-Klein, M-C, Beaurepaire E (2006) Imaging lipid bodies in cells and tissues using third-harmonic generation microscopy. Nat Methods 3:47–53

Deich J, Judd EM, McAdams, HH, Moerner, WE (2004) Visualization of the movement of single histidine kinase molecules in live *Caulobacter* cells. Proc Natl Acad Sci USA 101:15921–15926

DeLong EF, Wickham GS, Pace NR (1989) Phylogenetic stains: ribosomal RNA-based probes for the identification of single microbial cells. Science 243:1360–1363

Dirks RW, Molenaar C, Tanke HJ (2003) Visualizing RNA molecules inside the nucleus of living cells. Methods 29:51–57

Dong M, Tepp WH, Johnson EA, Chapman ER (2004) Using fluorescent sensors to detect botulinum neurotoxin activity in vitro and in living cells. Proc Natl Acad Sci USA 101:14701–14706

Donlan RM (2002) Biofilms: microbial life on surfaces. Emerg Inf Dis 9:881–890

Doyle RJ, Lee NC (1986) Microbes, warfare, religion, and human institutions. Can J Microbiol 32:193–200

Ericsson M, Hanstorp D, Hagberg P, Enger J Nyström T (2000) Sorting out bacterial viability with optical tweezers. J Bacteriol 182:5551–5555

Fehr M, Frommer WB, Lalonde S (2002) Visualization of maltose uptake in living yeast cells by fluorescent nanosensors. Proc Natl Acad Sci USA 99:9846–9851

Fluid Imaging Technologies (2007) Particle analysis in a fluid stream – Fluid Imaging Technologies. http://www.fluidimaging.com. Cited 2 May 2007

Frischknecht F, Renaud O, Shorte SL (2006) Imaging today's infectious animalcules. Curr Opin Microbiol 9:297–306

Fuchs BB, Mylonakis E (2006) Using non-mammalian hosts to study fungal virulence and host defense. Curr Opin Microbiol 4:346–351

Füchslin HP, Rüegg I, van der Meer JR, Egli T (2003) Effect of integration of a GFP reporter gene on fitness of *Ralstonia eutropha* during growth with 2,4-dichlorophenoxyacetic acid. Environ Microbiol 5:878–887

George TC, Basiji DA, Hall BE, Lynch DH, Ortyn WE, Perry DJ, Seo MJ, Zimmerman CA, Morrissey PJ (2004) Distinguishing modes of cell death using the ImageStream® Multispectral imaging flow cytometer. Cytometry A 59:237–245

Golding I, Cox EC (2004) RNA dynamics in live *Escherichia coli* cells. Proc Natl Acad Sci USA 101:11310–11315

Gonzalez JM, Saiz-Jimenez C (2005) Application of molecular nucleic acid-based techniques for the study of microbial communities in monuments and artworks. Int Microbiol 8:189–194

Gould GW (2006) History of science – spores. J Appl Microbiol 101:507–513

Gross HJ, Verwer B, Houck D, Recktenwald D (1993) Detection of rare cells at a frequency of one per million by flow cytometry. Cytometry 14:519–526

Gucker FT, O'Konski C, Pickard HB, Pitts JN (1947) A photoelectric counter for colloidal particles. Am J Chem 69:2422–2431

Haberle W, Horber JK, Ohnesorge F, Smith DP, Binnig G (1992) In situ investigations of single living cells infected by viruses. Ultramicroscopy 42–44:1161–1167

Harold FM (2005) Molecules into cells: specifying spatial architechture. Microbiol Mol Biol Rev 69:544–564

He M, Edgar JS, Jeffries GDM, Lorenz RM, Shelby JP, Chiu DT (2005) Selective encapsulation of single cells and cellular organelles into picoliter-and femtoliter-volume droplets. Anal Chem 77:1539–1544

Heraud P, Wood B, Tobin MJ, Beardall J, McNaughton D (2005) Mapping of nutrient-induced biochemical changes in living algal cells using synchrotron infrared microspectroscopy. FEMS Microbiol Lett 249:219–225

Hink MA, Griep RA, Borst JW, van Hoek A, Eppink MHM, Schots A, Visser AJWG (2000) Structural dynamics of green fluorescent protein alone and fused with a single chain Fv protein. J Biol Chem 275:17556–17560

Hite, B (1899) The effect of pressure in the preservation of milk. Bull W Va Univ Agric Exp Stn 58:15–35

Hobbs JK, Vasilev C, Humphris ADL (2005) VideoAFM – a new tool for high speed surface analysis. Analyst 131:251–256

Hölzel R (1998) Nystatin-induced changes in yeast monitored by time-resolved automated single cell electrorotation. Biochim Biophys Acta 1425:311–318

Hölzel R (1999) Non-invasive determination of bacterial single cell properties by electrorotation. Biochim Biophys Acta 1450:53–60

Horber JKH, Haberle W, Ohnesorge F, Binnig G, Liebich HG, Czerny H, Mahnel H, Mayr A (1992) Investigation of living cells in the nanometer regime with the scanning force microscope. Scanning Microsc 6:919–930

Hoshikawa H, Asai H (1985) On the rotational Brownian motion of a bacterial idle motor. II. Theory of fluorescence correlation spectroscopy. Biophys Chem 22:167–172

Huang WE, Griffiths RI, Thompson IP, Bailey MJ, Whiteley AS (2004) Raman microscopic analysis of single microbial cells. Anal Chem 76:4452–4458

Hüller R, Paffgen W, Glossner E, Hummel P, Kachel V (1991) A PC-at based video device for flow cytometrically triggered cell imaging in flow. Comput Med Imaging Graph 15:85–91

Hüller R, Glossner E, Schaub S, Weingartner J, Kachel V (1994) The macro flow planktometer: a new device for volume and fluorescence analysis of macro plankton including triggered video imaging in flow. Cytometry 17:109–118

Imai T, Ohno T (1995) The relationship between viability and intracellular pH in the yeast Saccharomyces cerevisiae. Appl Environ Microbiol 61:3604–3608

Ininitesima (2007) Microscopy products from Ininitesima. http://www.infinitesima.com. Cited 2 May 2007

Invitrogen (2007) Invitrogen | Country select. http://www.probes.invitrogen.com. Cited 2 May 2007

Kachel V, Benker G, Lichtnau K, Valet G, Glossner E (1979) Fast imaging in flow: a means of combining flow-cyometry and image analysis. J Histochem Cytochem 27:335–341

Kay DB, Cambier JL, Wheeless LL Jr (1979) Imaging in flow. J Histochem Cytochem 27:329–334

Kerfeld CA, Sawaya MR, Tanaka S, Nguyen CV, Phillips M, Beeby M, Yeates TO (2005) Protein structures forming the shell of primitive bacterial organelles. Science 309:9365–938

Kuang Y, Biran I, Walt DR (2004) Living bacterial cell array for genotoxin monitoring. Anal Chem 76:2902–2909

Kubota F (2003) Analysis of red cell and platelet morphology using an imaging-combined flow cytometer. Clin Lab Haematol 25:71–76

Kubota F, Kusuzawa H, Kosaka T, Nakamoto H (1995) Flow cytometer and imaging device used in combination. Cytometry 21:129–132

Kunst BH, Schots A, Visser AJWG (2002) Detection of flowing fluorescent particles in a microcapillary using fluorescence correlation spectroscopy. Anal Chem 74:5350–5357

Lacroix-Gueu P, Briandet R, Lévêque-Fort S, Bellon-Fontaine, M, Fontaine-Aupart M (2005) In situ measurements of viral particle diffusion inside mucoid biofilms. C R Biol 328:1065–1072

Larrainzar E, O'Gara F, Morrissey JP (2005) Applications of autofluorescent proteins for in situ studies in microbial ecology. Annu Rev Microbiol 59:257–277

Le TT, Harlepp S, Guet C, Dittmar K, Emonet T, Pan T, Cluzel P (2005) Real-time RNA profiling within a single bacterium. Proc Natl Acad Sci USA 102:9160–9164

Li A, Mansoor AH, Tan KSW, Lim CT (2006) Observations on the internal and surface morphology of malaria infected blood cells using optical and atomic force microscopy. J Microbiol Methods 66:434–439

Li Q, Kim Y, Namm J, Kulkarni A, Rosania GR, Ahn Y-H, Chang Y-T (2006) RNA-selective, live cell imaging probes for studying nuclear structure and function. Chem Biol 13:615–623

Lippincott-Schwartz J, Patterson GH (2003) Development and use of fluorescent protein markers in living cells. Science 300:87–97

Lizundia R, Sengmanivong L, Guergnon J, Müller T, Schnelle T, Langsley G, Shorte SL (2005) Use of micro-rotation imaging to study JNK-mediated cell survival in *Theileria parva*-infected B-lymphocytes. Parasitology 130:629–635

Losick R, Shapiro L (1999) Changing views on the nature of the bacterial cell: from biochemistry to cytology. J Bacteriol 181:4143–4145

Löwe J, van den Ent F, Amos LA (2004) Molecules of the bacterial cytoskeleton. Annu Rev Biophys Biomol Struct 33:177–198

Ma H, Snook LA, Kaminskyj SGW, Dahms TES (2005) Surface ultrastructure and elasticity in growing tips and mature regions of *Aspergillus* hyphae describe wall maturation. Microbiology 151:3679–3688

Maiti S, Haupts U, Webb WW (1997) Fluorescence correlation spectroscopy: diagnostics for sparse molecules. Proc Natl Acad Sci USA 94:11753–11757

Malvern Instruments (2007) Brochure: Sysmex FPIA 3000 flow particle image analyzer. http://www.malvern.com/common/downloads/MRK652-02_FPIA-3000_brochure_lr.pdf. Cited 2 May 2007

Mangenot S, Hochrein M, Rädler J, Letellier L (2005) Real-time imaging of DNA ejection from single phage particles. Curr Biol 15:430–435

Müller JD, Gratton E (2003) High-pressure fluorescence correlation spectroscopy. Biophys J 85:2711–2719

Nan X, Potma EO, Xie XS (2006) Nonpertubative chemical imaging of organelle transport in living cells with coherent anti-Stokes Raman scattering microscopy. Biophys J 91:728–735

Niven GW, Fuks T, Morton JS, Rua SACG, Mackey BM (2006) A novel method for measuring lag times in division of individual bacterial cells using image analysis. J Microbiol Methods 65:311–317

Northrop DB (2002) Effects of high pressure on enzymatic activity. Biochim Biophys Acta 1595:71–79

Ohnesorge FM, Horber JK, Haberle W, Czerny CP, Smith DP, Binnig G (1997) AFM review study on pox viruses and living cells. Biophys J 73:2183–2194

Olsen KN, Budde BB, Sigumfeldt H, Rechinger KB, Jakobsen M, Ingmer H (2002) Noninvasive measurement of bacterial intracellular pH on a single-cell level with green fluorescent protein and fluorescence ratio imaging microscopy Appl Environ Microbiol 68:4145–4147

Pelling AE, Sehati S, Gralla EB, Valentine JS, Gimzewski JK (2004) Local nanomechanical motion of the cell wall of *Saccharomyces cerevisiae*. Science 305:1147–1150

Perlman ZE, Slack MD, Feng Y, Mitchison TJ, Wu LF, Altschuler SJ (2004) Multidimensional drug profiling by automated microscopy. Science 306:1194–1198

Pethig R, Markx GH (1997) Applications of dielectrophoresis in biotechnology. Trends Biotechnol 15:426–432

Potma EO, de Boeij WP, van Haastert PJM, Wiersma DA (2001) Real-time visualization of intracellular hydrodynamics in single living cells. Proc Natl Acad Sci USA 98:1577–1582

Presley JF (2005) Imaging the secretory pathway: the past and future impact of live cell optical techniques. Biochim Biophys Acta 1744:259–272

Qing D-K, Mengüç MP, Payne FA, Danao M-GC (2003) Convective-diffusion-based fluorescence correlation spectroscopy for detection of a trace amount of *E. coli* in water. Appl Opt 42:2987–2994

Radbruch A, Recktenwald D (1995) Detection and isolation of rare cells. Curr Opin Immunol 7:270–273

Roux P, Münter S, Frischknecht F, Herbomel P, Shorte SL (2004) Focusing light on infection in four dimensions. Cell Microbiol 6:333–343

Ruchira, Hink MA, Bosgraaf L, van Haastert PJM, Visser AJWG (2004) Plekstrin homology domain diffusion in *Dictyostelium* cytoplasm studied using fluorescence correlation spectroscopy. J Biol Chem 279:10013–10019

Skakun VV, Hink MA, Digris AV, Engel R, Novikov EG, Apanasovich VV, Visser AJWG (2005) Global analysis of fluorescence fluctuation data. Eur Biophys J 34:323–334

Samuel D (1996) Investigation of ancient Egyptian baking and brewing methods by correlative microscopy. Science 273:488–490
San Martin MF, Barbosa-Canovas GV, Swanson BG (2002) Food processing by high hydrostatic pressure. Crit Rev Food Sci Nutr 42:627–645
Santesson S, Nilsson S (2004) Airborne chemistry: acoustic levitation in chemical analysis. Anal. Bioanal. Chem. 378:1704–1709
Santesson S, Andersson M, Degerman E, Johansson T, Nilsson J, Nilsson S (2000) Airborne cell analysis. Anal Chem 72:3412–3418
Santesson S, Johansson J, Taylor LS, Levander I, Fox S, Speaniak M, Nilsson S (2003) Airborne chemistry coupled to Raman spectroscopy. Anal Chem 75:2177–2180
Santesson S, Barinaga-Rementeria Ramírez I, Viberg P, Jergil B, Nilsson S (2004) Affinity two-phase partitioning in acoustically levitated drops. Anal Chem 76:303–308
Sato M, Umezawa Y (2004) Imaging protein phosphorylation by fluorescence in single living cells. Methods 32:451–455
Schubert W, Bonnekoh B., Pommer AJ, Philpsen L, Böckelmann R, Malykh Y, Gollnick H, Friedenberger M, Bode M, Dress AWM (2006) Analyzing proteome topology and function by automated multidimensional fluorescence microscopy. Nat Biotechnol 24:1270–1278
Sieracki CK, Sieracki ME, Yentsch CS (1998) An imaging-in-flow system for automated analysis of marine microplankton. Mar Ecol Prog Ser 168:285–296
Sifri CD, Begun J, Ausubel FM (2005) The worm has turned – microbial virulence modeled in *Caenorhabditis elegans*. Trends Microbiol 13:119–127
Silverman AP, Kool ET (2005) Quenched autoligation probes allow discrimination of live bacterial species by single nucleotide differences in rRNA. Nucleic Acids Res 33:4978–7986
Southward CM, Surette MG (2002) The dynamic microbe: green fluorescent protein brings bacteria to light. Mol Microbiol 45:1191–1196
Sterling MC Jr, Bonner JS, Ernest ANS, Page CA, Autenrieth RL (2004) Characterizing aquatic sediment-oil aggregates using in situ instruments. Mar Pollut Bull 48:533–542
Thiberge S, Nechushtan A, Sprinzak D, Gileadi O, Behar V, Zik O, Chowers Y, Michaeli S, Schlessinger J, Moses E (2004) Scanning electron microscopy of cells and tissues under fully hydrated conditions. Proc Natl Acad Sci USA 101:3345–3351
Touhami A, Hoffmann B, Vesella A, Denis FA, Dufrêne YF (2003) Aggregation of yeast cells: direct measurement of discrete lectin-carbohydrate interactions. Microbiology 149:2873–2878
Tsang PH, Li G, Brun YV, Freund LB, Tang JX (2006) Adhesion of single bacterial cells in the micronewton range. Proc Natl Acad Sci USA 103:5764–5768
Twining BS, Bains SB, Fisher NS, Maser J, Vogt S, Jacobsen C, Tovar-Sanchez A, Sañudo-Wilhelmy SA (2003) Quantifying trace elements in individual aquatic protist cells with a synchrotron X-ray fluorescence microprobe. Anal Chem 75:3806–3816
Union Biometrica (2007) Union Biometrica. http://www.unionbio.com. Cited 2 May 2007
Vainrub A, Pustovyy O, Vodyanoy V (2006) Resolution of 90 nm (λ/5) in an optical transmission microscope with an annular condenser. Opt Lett 31:2855–2857
Van Craenenbroeck EV, Engelborghs Y (2000) Fluorescence correlation spectroscopy: molecular recognition at the single molecule level. J Mol Recognit 13:93–100
van Leeuwenhoek A (1684) Microscopical observations, about animals in the scurf of the teeth, the substance call'd worms in the nose, the cuticula consisting of scales. Philos Trans R Soc Lond 14:568–574
Viollier PH, Thanbichler M, McGrath PT, West L, Meewan M, McAdams HH, Shapiro L (2004) Rapid and sequential movement of individual chromosomal loci to specific subcellular locations during bacterial DNA replication. Proc Natl Acad Sci USA 101:9257–9262
Vukojevic V, Pramanik A, Yakovleva T, Rigler R, Terenius L, Bakalkin G (2005) Study of molecular events in cells by fluorescence correlation spectroscopy. Cell Mol Life Sci 62:535–550
Walter NG, Schwille P, Eigen M (1996) Fluorescence correlation analysis of probe diffusion simplifies quantitative pathogen detection by PCR. Proc Natl Acad Sci USA 93:12805–12810

Wandelt B, Cywinkski P, Darling GD, Stranix BR (2005) Single cell measurement of microviscosity by ratio imaging of fluorescence of styrylpyridinium probe. Biosens Bioelectron 20:1728–1736

Wietzorrek J, Plesnila N, Baethmann A, Kachel V (1999) A new multiparameter flow cytometer: optical and electrical cell analysis in combination with video microscopy in flow. Cytometry 35:291–301

Wood, BR, Heraud P, Stojkovic S, Morrison D, Beardall J, McNaughton D (2005) A portable Raman acoustic levitation spectroscopic system for the identification and environmental monitoring of algal cells. Anal Chem 77:4955–4961

Wu H, Wheeler A, Zare RN (2004) Chemical cytometry on a picoliter-scale integrated microfluidic chip. Proc Natl Acad Sci USA 101:12809–12813

Yamamura S, Kishi H, Tokimitsu Y, Kondo S, Honda R, Rao SR, Omori M, Tamiya E, Muraguchi A (2005) Single-cell microarray for analyzing cellular response. Anal Chem 77:8050–8056

Zeeb V, Suzuki M, Ishiwata S (2004) A novel method of thermal activation and temperature measurement in the microscopic region around single living cells. J Neurosci Methods 139:69–77

Zhao L, Schaefer D, Xu H, Modi SJ, LaCourse WR, Marten MR (2005) Elastic properties of the cell wall of *Aspergillus nidulans* studied with atomic force microscopy. Biotechnol Prog 21:292–299

12
Imaging Parasites in Vivo

Rogerio Amino, Blandine Franke-Fayard, Chris Janse, Andrew Waters, Robert Ménard, and Freddy Frischknecht

Abstract Pathogens interact with their hosts and host cells in many different ways to achieve successful entry, replication and dissemination. Recent progress in the development of molecular tools and imaging techniques have allowed to gain exciting new insights into these interactions on the cellular and molecular level. Often these studies have led to the discoveries of basic cellular mechanisms that the pathogen, be it virus, bacterium, fungus or parasite usurps for its own advantage. In this chapter we present and discuss recent in vivo imaging studies of unicellular parasites, emphasizing mainly on malaria parasites. The aim of the chapter is to highlight how simple fluorescent and bioluminescent imaging techniques can provide interesting insights in medically important interactions of these pathogens with their hosts.

12.1 Introduction

Imaging the interactions of pathogens with their respective hosts or host cells faces challenges that often lie beyond mere image acquisition technology. While studying tissue culture cells usually faces few challenges from sample preparation, the study of pathogens might fail when they cannot be isolated in the first place. Frequently, they have to be handled in high-containment areas where access to microscopes might be limited. Also many pathogens cannot be transfected, poorly express green fluorescent protein (GFP) fusion proteins, or simply do not interact with cultured cells. Nevertheless, several studies have now been performed using a number of imaging approaches on a wide variety of viral and bacterial pathogens at different levels of complexity (Enninga et al. 2005; Larson et al. 2003, 2005; Pelkmans et al. 2005; Rietdorf et al. 2001; Rust et al. 2004; Schlumberger et al. 2005). These and other reports have proven that the study of host–pathogen interaction is a fruitful endeavor that can yield interesting insights into the cell biology of host cells as well as uncover biologically interesting and medically important details of the infectious process (Frischknecht et al. 2006b; Greber and Way 2006; Münter et al. 2006; Srinivasan and McSorley 2004).

S.L. Shorte and F. Frischknecht (eds.), *Imaging Cellular and Molecular Biological Functions.*

Here, we review recent studies of unicellular parasites and the interactions with their living hosts, whether they be insects or small mammals, using bioluminescence and fluorescence microscopy. As a paradigm we use the malaria parasite to highlight the problems that had to be overcome and point to the challenges that lie ahead. This chapter also serves to illustrate how comparatively simple imaging approaches can still yield interesting insights into medically important biological processes.

> "...parasites live invisibly, and parasitologists usually can see what they are doing only by killing their hosts and dissecting them. These grisly snapshots slowly add up to a natural history."
> — Carl Zimmer (2000) in *Parasite Rex*

> "Watch ultra-high-speed processes down into the time domain of just a few microseconds."
> — Recent advertisement for a commercially available microscope

12.2 The Life Cycle of Malaria Parasites

Malaria parasites are transmitted by the bite of an infected mosquito, which introduces saliva containing *Plasmodium* sporozoites into the host skin. The mosquito penetrates the host skin with its proboscis and salivates in search for a suitable blood meal (Fig. 12.1). The mosquito can suck up blood either from a punctured blood vessel or from a pool of blood that leaked from damaged blood vessels into the skin tissue. In contrast to what is often believed, *Plasmodium* sporozoites are not injected directly into the bloodstream but in the dermis, along with the saliva (Sidjanski and Vanderberg 1997). Sporozoites differentiate into thousands of merozoites within liver cells, but how they reach hepatocytes from the skin has, in the absence of in vivo imaging approaches, remained largely elusive (Amino et al. 2005; Baldacci and Menard 2004).

Merozoites in turn infect red blood cells and it has only recently been shown how they leave the infected liver cell and enter the bloodstream (Sturm et al. 2006). Merozoites invade red blood cells by first binding to their surface and then use their own actin- and myosin-based motor to push themselves into a vacuole. Within this parasitophorous vacuole the parasites differentiate and multiply. They also profoundly modify the infected red blood cell (iRBC). The remarkable structures known as knobs, which are foci of several parasite proteins on the iRBC surface, allow the iRBC to bind to the endothelium. This is thought to enable the parasites to circumvent the clearance of the infected host cells in the spleen, where macrophages either ingest the iRBC or pinch the parasite from the iRBC, thus returning still functioning red blood cells to the circulation. Intravital microscopy has shown how parasites can interact with the endothelium of blood vessels. To this end, either rodent or human parasite (*Plasmodium falciparum*) iRBCs were filmed in the exposed cremaster muscle (Kaul et al. 1998) or in human skin grafted on SCID mice (Ho et al. 2000), respectively.

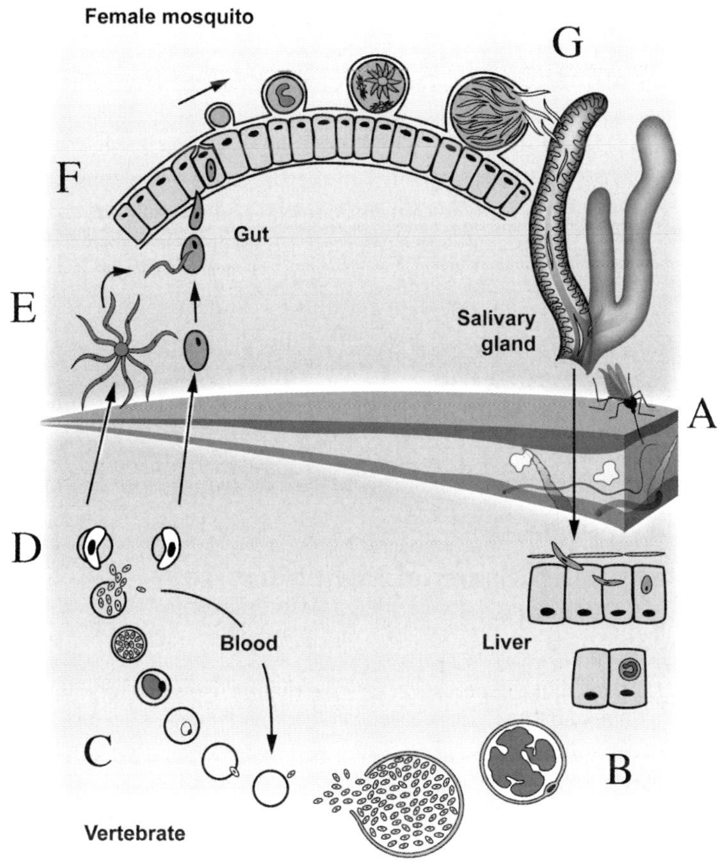

Fig. 12.1 Life cycle of mammalian malaria species. *Plasmodium* sporozoites enter a host during the bite of a mosquito (**A**). Deposited in the skin, they reach the liver via the blood circulation. They enter into hepatocytes and differentiate within them into thousands of merozoites (**B**). Merozoites reenter the blood circulation and invade red blood cells. During the cycles of replication within and reinvasion of red blood cells (**C**) the symptoms of malaria are generated. Eventually some merozoites can give rise to male or female gametocytes (**D**). These can further develop into gametes after being taken up by a mosquito (**E**). After fertilization a zygote develops, which in turn develops into an ookinete. Ookinetes can penetrate the midgut epithelium (**F**) and develop into oocysts. Within oocysts the sporozoites are generated that eventually invade the salivary glands (**G**), from where they can be injected into a vertebrate host with the saliva

Merozoites eventually differentiate into male or female gametocytes. Once a mosquito takes up blood during a blood meal the gametocytes sense both a drop in temperature and mosquito-specific chemicals and differentiate into gametes (Billker et al. 1998). After fertilization a zygote forms, from which a so-called ookinete develops. This motile form of the parasite can escape the digesting blood meal and penetrate the epithelial cell layer of the mosquito's gut (Baton and Ranford-Cartwright 2005). During this journey, the ookinete encounters a number of recently discovered defense mechanisms of the mosquito and only in some mosquito species the parasite manages to overcome them (Blandin and Levashina 2004; Osta et al. 2004). At the far side of the gut epithelium the ookinete comes to a halt and develops into an oocyst. Over 1–2 weeks, hundreds to thousands of sporozoites form within the oocyst. The sporozoites then use proteases to break the oocyst capsule (Aly and Matuschewski 2005; Wang et al. 2005), and once they are in the hemolymph (the circulatory liquid of insects) they float until they are arrested at the salivary glands, which they invade (Akaki and Dvorak 2005; Frischknecht et al. 2006a). Within the salivary glands the sporozoites can survive for several weeks before being transmitted. It is still largely enigmatic how some mosquitoes can harbor thousands of sporozoites in their glands but only transmit a few to a few dozen during a single bite (Beier 1998; Frischknecht et al. 2004; Medica and Sinnis 2005).

12.3 A Very Brief History of Light Microscopy and Malaria Parasites

The discovery of the malaria parasite was, like that of many pathogens at the time, crucially dependent on light microscopy. In 1880 Alphone Laveran, working in Algeria, observed the blood smear of a sick patient and found what he considered could be the causative agent of the disease. A careful study showed that 148 out of 192 patients had similar crescent-shaped bodies in their blood. Important for the rapid verification of Laveran's observation was the development of the oil immersion lens in 1878 by Ernst Abbe from Zeiss Optical Works.

It still remained a puzzle how the parasite could enter into our bodies. In 1897–1898 Ronald Ross, who first studied human malaria in India, but then had to confine himself to the study of bird malaria, demonstrated that mosquitoes transmit the parasite. This was not only a major breakthrough in tropical medicine but also demonstrated the successful use of an animal model system in malaria research based mainly on experimentation with a simple light microscope. Ross dissected mosquitoes which fed on infected humans and found cystlike structures at the mosquito gut wall. Working on malaria parasites of birds, he finally observed the sporozoites in the salivary glands. He managed to transmit malaria to 21 out of 28 birds by using mosquitoes fed on infected birds. Shortly afterwards his studies were confirmed in human malaria by Battista Grassi.

Over 100 years later we can still make major discoveries using new imaging techniques in combination with model systems that can reliably advance our knowledge about the human parasite. In the following, we briefly describe some imaging approaches to investigate parts of the malaria life cycle and other parasites as well as the interactions with their respective hosts.

12.4 In Vivo Imaging of Luminescent Parasites

A dream of many microscopists working on host–pathogen interactions must certainly be to follow an infection from the time when a few individual pathogens enter into the body to the final stages of the disease in real time using intact, living animals. While it will still take some time to see this dream come true, some steps can already be made. Today, it is possible to detect and follow for some time individual fluorescent pathogens deposited in the outermost parts of the body, mainly the skin (see below). Also, the use of bioluminescent pathogens can reveal interesting insights into where the pathogens replicate and cause disease. As this technology relies on the detection of larger numbers of individuals, it provides mainly insight into the final stages of a natural infection. This technique has allowed remarkable progress in uncovering unexpected bacteria–host interactions (Doyle et al. 2004).

To image bioluminescence in whole animals, the anesthetized mouse is usually placed in a dark chamber with an attached charge coupled device camera some 15–20 cm from the specimen. Photons are collected and integrated for 10–180 s and a pseudocolor image representing the bioluminescence intensity is generated. This allows the detection of faint signals representing ideally just a few dozen bioluminescent cells. This commercially available technique has been used to investigate several host–pathogen interactions (Doyle et al. 2004). These included among others the study of interferons on vaccinia and herpes simplex virus systemic spread (Luker et al. 2003, 2005), and gave new insights into the replication sites and efficiencies of several bacteria, including *Listeria monocytogenes* (Hardy et al. 2001) and *Staphylococcus aureus* (Yu et al. 2005). The parasites *Leishmania* (Lang et al. 2005) and the malaria-related *Toxoplasma* (Hitziger et al. 2005; Saeij et al. 2005) have also been investigated using bioluminescence imaging. The work by Saeij et al. (2005) on *Toxoplasma* showed that bioluminescence imaging can be used to reveal the kinetics of parasite spread in strains with different virulence and the reemergence of parasites after induced immunosuppression. Hitziger et al. (2005) additionally investigated the role of Toll-like receptors in host resistance and visualized a full-blown infection of nonvirulent parasites in mice lacking the adaptor protein MyD88.

In the case of malaria the combination of GFP–luciferase expressing parasites with whole-body imaging of intact mice revealed the role for the CD36 host receptor in anchoring of iRBCs to the lung and adipose tissue but not the spleen (Franke-Fayard et al. 2005; Fig. 12.2). This study also confirmed that in mice models of cerebral malaria no parasites accumulated in the brain and no CD36 was needed to cause adverse effects in the brain (Hunt and Grau 2003; Turner et al. 1994). A combination of parasites that were expressing GFP–luciferase from different promoters allowed the demonstration that gametocytes do not adhere to any organ, while schizonts (late-stage iRBCs) accumulated specifically in a CD36-dependent way in the lung and adipose tissue. Taken together, these pioneering studies thus demonstrate that bioluminescence imaging of host–pathogen interactions has come

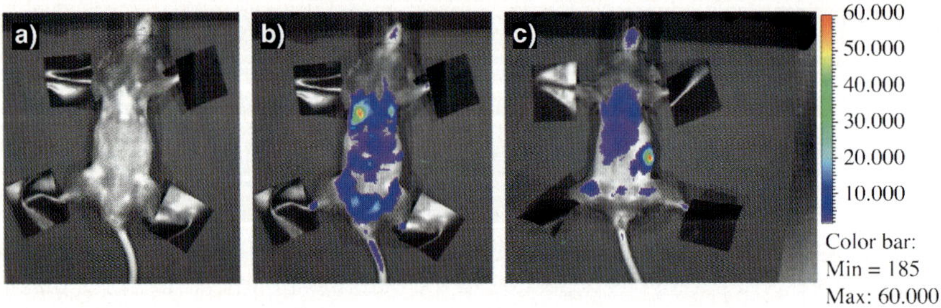

Fig. 12.2 Luminescent malaria parasites in the wild-type and CD36 knockout mouse. Visualization of the sequestration of *Plasmodium berghei* luciferase-expressing parasites (p*Pbgfp-luc*$_{SCH}$; Franke-Fayard et al. 2005) in synchronized infections in wild-type and CD36-deficient mice. **a** Bright-field image of a mouse. **b, c** Luminescent image of an infected wild-type mouse (**b**) and an infected mouse lacking the CD36 gene (**c**). Rainbow images show the relative level of luciferase activity ranging from low (*blue*), to medium (*green*), to high (*red*). Measurements of anesthetized mice injected with D-luciferin were performed 22h after infection of purified merozoites when sequestration of schizonts occurs (15-cm Field of view, medium binning, and 30-s exposure time). The images show the change in distribution of the parasites in the CD36 knockout host compared with the wild type. The major organs that support sequestration are the lungs and fat tissue in this system. Note the increased concentration of parasites in the spleen in the CD36 knockout mouse which is most likely due to increased clearance of circulating schizonts

of age and while having already provided a wealth of interesting insights promises a bright future.

12.5 In Vivo Imaging of Fluorescent Parasites

P. berghei, a rodent malaria parasite, has been the model of choice to combine microscopy with fluorescence-based observations. In order to image malaria parasites at different stages, a number of recombinant *P. berghei* parasites were generated expressing the GFP from a variety of promoters. Some of these parasites were constructed to yield fluorescence at one particular stage of the life cycle (de Koning-Ward et al. 1998; Khan et al. 2005; Natarajan et al. 2001; Sultan et al. 1999; Vlachou et al. 2004), while others expressed the GFP at all stages (Franke-Fayard et al. 2004). Depending on the promoters used, a strong fluorescence signal revealed parasites in the mammalian host or the insect vector. This allowed the follow-up of earlier in vitro studies that had shown the motile behavior of *Plasmodium* ookinetes and sporozoites (Freyvogel 1966; Vanderberg 1974). Both ookinetes and sporozoites use a specific mode of substrate-dependent translocation termed "gliding motility" that relies on an actin–myosin motor in the parasite (Baum et al. 2006; Kappe et al. 2004; Schüler and Matuschewski 2006; Yuda et al. 1999). In vitro the parasites were shown to glide on a substrate at different speeds (about 5–15 μm/min for ookinetes, about 60–80 μm/min for sporozoites). However, the role of gliding in parasite transmission in vivo remained to be uncovered. The generation of knockout parasites lacking the ookinete-specific CTRP and sporozoite-specific TRAP gene demonstrated that these proteins are important for motility and the invasive capacity of the respective parasite stages (Dessens et al. 1999; Sultan et al. 1997; Templeton et al. 2000; Yuda et al. 1999).

12.6 Imaging Malaria Parasites in the Mosquito

The first attempts to image motile parasites in their natural environment used explanted mosquito guts. Video sequences from differential interference contrast recordings showed that ookinetes actively penetrate the epithelial cells when crossing from their apical to their basolateral sides (Zieler and Dvorak 2000). Studies using fluorescent parasites also imaged explanted guts with confocal microscopes. This showed how the ookinete interacts with epithelial cells, and also provided insights into how the mosquito reacts to the presence of parasites by mounting an innate immune response (Danielli et al. 2005; Vlachou and Kafatos 2005; Vlachou et al. 2004). Acquisition of 4D datasets with spinning disc confocal microscopy revealed that the ookinete displays various modes of motility within intact mosquitoes, which are all used in seemingly random order during its journey across the gut epithelium (Vlachou et al. 2004). For this study infected mosquitoes were glued to a glass slide and viewed with a ×20 (0.75 numerical aperture and long working distance) immersion objective, while dissected midguts were kept in the medium on glass-bottomed well dishes and investigated with a ×40 (1.25 numerical aperture) oil immersion lens. Both approaches, albeit simple to set up, suffered from the intrinsic movement of the observed mosquitoes and the peristaltic movement of the isolated guts, which required a lot of patience to collect useful datasets. Nevertheless, continuous recording periods of almost 2 h were achieved for a few mosquitoes and allowed the detailed quantitative analysis of 58 ookinetes. This revealed that stationary rotation, during which the parasites turn in circles but do not migrate, was the predominant form of motility. Stationary rotation was observed on average for 53 min per parasite and accounted for 49% of the total observation time. The other two modes of forward motility, termed "straight segment" and "spiraling" made up for an additional 38% of the observation time, with the ookinetes spending the remaining time immobile (Vlachou et al. 2004). Observations on isolated midguts complemented and expanded these findings. Using conventional confocal scanning microscopes, Vlachou and colleagues showed that ookinetes penetrate through a number of individual midgut epithelial cells before coming to rest at the basal lamina. Once the ookinetes passed a cell, a purse-string-type mechanism caused the epithelium to push the damaged cells out of the monolayer, a mechanism that might play a role in protecting the mosquito against physical damage as well as from infections (Baton and Ranford-Cartwright 2005; Danielli et al. 2005; Vlachou et al. 2004).

The release of sporozoites from oocysts has not yet been investigated by real-time imaging. However, once the sporozoites are released from the oocysts, simple wide-field epifluorescence microscopy with long working distance (×10 to ×40) objectives was used to detect and follow them within the hemolymph of immobilized mosquitoes (Akaki and Dvorak 2005; Frischknecht et al. 2006a). Sporozoites can thus be visualized in any tissue bathed by the hemolymph, including the legs, the palps and the wings of the mosquito. Akaki and Dvorak (2005) showed that sporozoites collected from mature oocysts and placed in matrigel can move towards homogenates of salivary glands, suggesting that they might use chemotactic clues to move the last few microns towards the salivary gland. This, however, was not verified by in vivo imaging. Interestingly, sporozoites isolated from inside the salivary glands do not show any such chemotactic behavior towards salivary gland extracts, and it remains to be shown whether the parasites use cues from the mammalian host for directed migration, e.g., towards blood or lymph vessels in the dermis. After adhesion to the basal lamina of the glands, sporozoites cross the acinar cells and aggregate in the salivary cavities (Pimenta et al. 1994; Sterling et al. 1973). In explanted salivary glands, it was shown that these aggregated sporozoites move less than the surrounding parasites (Frischknecht et al. 2004). Unlike in vitro, sporozoites moved within explanted salivary glands in the absence of serum, which was believed to be essential for sporozoite motility (Fig. 12.3a). However, intrasalivary sporozoites typically displayed a back-and-forth type of locomotion, which produced little translocation. Only a small proportion of intrasalivary parasites moved forward, usually at very slow average speed; however, they could occasionally reach maximum velocities similar to those of sporozoites gliding in vitro. Incubation of salivary glands in various media that either support or suppress migration showed that over time sporozoites migrate from the distal part of the salivary glands into the ducts (Frischknecht et al. 2004).

Transmission of sporozoites occurs when the mosquito ejects saliva into the skin, probing for a blood vessel. Despite some mosquitoes harboring thousands of sporozoites in their salivary glands it is known that only very few (ten to 20) are injected during a bite. This suggests that the parasite can somehow regulate the numbers that are ejected and it might be that only those sporozoites get transmitted that migrated into the narrow salivary ducts (Beier 1993; Frischknecht et al. 2004). During ejection, the parasites float in the saliva that is ejected by the mosquito into the skin of the mammal. In order to follow sporozoites during ejection by mosquitoes, the insects were glued onto glass slides and the stylets of the proboscis that normally penetrate the skin were exposed by micromanipulation. Mosquitoes, initially kept on ice, started to salivate upon transfer to the microscope stage (Fig. 12.3b). Sporozoite recording necessitated an image acquisition system that allowed a frame rate of at least four images per second, although counting the number of ejected sporozoites can be performed on any fluorescence microscope. Observations of over 100 salivating mosquitoes showed that most parasites are ejected early during

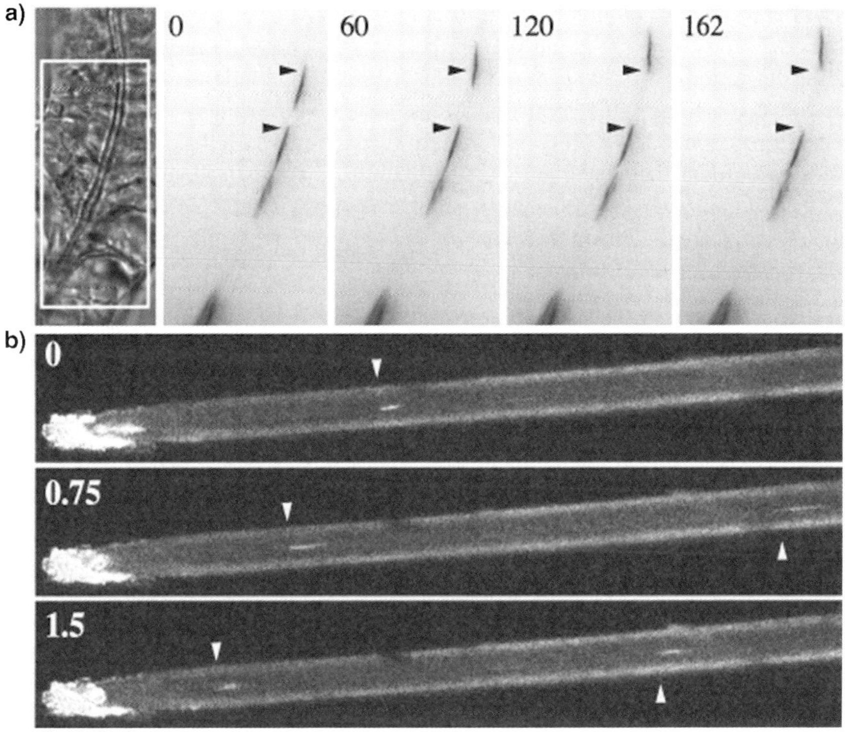

Fig. 12.3 Fluorescent sporozoites in mosquitoes. **a** Sporozoites within a salivary duct move only rarely and at low speed. The panel on the *left* shows a transmission image of an isolated salivary gland with the *boxed region* indicating the area that was filmed. Inverted images of sporozoites in a salivary duct show that only one sporozoite moves slowly down a duct. *Numbers* indicate time in seconds. *Arrowheads* indicate the front end of two sporozoites at the beginning of the sequence. **b** Sporozoites are ejected through the proboscis by an immobilized mosquito. *Arrowheads* indicate the front end of two sporozoites. Note the various length of the fluorescence signal due to varying speed of saliva flow. Numbers indicate time in seconds

salivation. Interestingly, some mosquitoes, despite having thousands of sporozoites in their salivary glands, did not eject any or ejected just a few sporozoites during salivation (Frischknecht et al. 2004; Medica and Sinnis 2005). These and other studies also revealed the large variability in sporozoite numbers at any stage and during any type of process, which necessitated the quantification from many individual observations. Indeed, mosquitoes with similar numbers of sporozoites in their salivary glands could eject either no or over 1.000 sporozoites during a 10-min artificial salivation period. During the first minute of salivation a mean of 20 (but median of only 1!) sporozoites were ejected from mosquitoes fed 18 days prior to the experiment. After observation of hundreds of sporozoites from differently aged or treated mosquitoes, it was suggested that the slow motility of sporozoites within the salivary glands could contribute to the creation of a releasable pool of sporozoites, i.e., those present in the salivary ducts, which could in principle be ejected during salivation. (Frischknecht et al. 2004). These observations raise a number of questions, including ones about the fitness of the sporozoites. For example, are ejected sporozoites more efficient in transmitting the disease than sporozoites isolated from infected mosquitoes? Are sporozoites ejected in large numbers more infectious than when ejected in low numbers? What is the role of the mosquito saliva? As malaria is transmitted by a number of different mosquito species, is there a difference from vector to vector?

12.7 Imaging Malaria Parasites in the Mammalian Host

In order to observe sporozoites in the dermis of the host, an anesthetized mouse bitten by infected mosquitoes can be examined directly on a microscope stage (Vanderberg and Frevert 2004), or the mouse can be placed on the stage after examination of the bite site has revealed the presence of a hematome (Amino et al. 2006). An ideal site where sporozoites can be observed in living mice is the ear, as it is easily accessible and owing to the thinness of ear tissues allows the observation of circulating blood cells using transmitted light. Once the sporozoites have been injected into the dermis of the host, their gliding capacity is increased and they switch from the sluggish, back-and-forth (in the mosquito salivary ducts) into a "full steam" and forward (in the mammalian dermis) mode of locomotion. Sporozoites can move over long distances for several tens of minutes, covering up to about 2 mm in 30 min. However, they follow winding paths rather than a straight trajectory, often returning close to their starting point after several minutes of movement over hundreds of micrometers (Amino et al. 2005, 2006; Fig. 12.4). Also, quantitative analysis of sporozoite motility in the skin showed that parasite speed decreases over time, and that only a few sporozoites leave the bite site as outlined by the area observed with a ×10 long-distance objective. This raises a number of questions: Are sporozoites able to penetrate into blood vessels only at the bite site? Could a self-attracting random walk whereby the sporozoites follow cues that they laid down themselves during gliding (or other modes of chemotaxis) determine that parasites stay at the bite site? Could the modification of the tissue by mosquito saliva modulate gliding so as to ensure optimal migration only at the bite site? What happens to the parasites over time apart from their movement getting slower? Surprisingly, it was found that sporozoites enter both blood (Fig. 12.5) and lymphatic vessels (Amino et al. 2006; Fig. 12.6). Again, the observation of large numbers of individual parasites revealed that although more parasites enter into blood capillaries, some 20% of the parasites deposited in the skin enter lymphatics and accumulate in the draining lymph node. Investigations of parasites in explanted lymph nodes over time showed that sporozoites rarely pass the first lymph node and progressively associate with dendritic cells in the node. While the

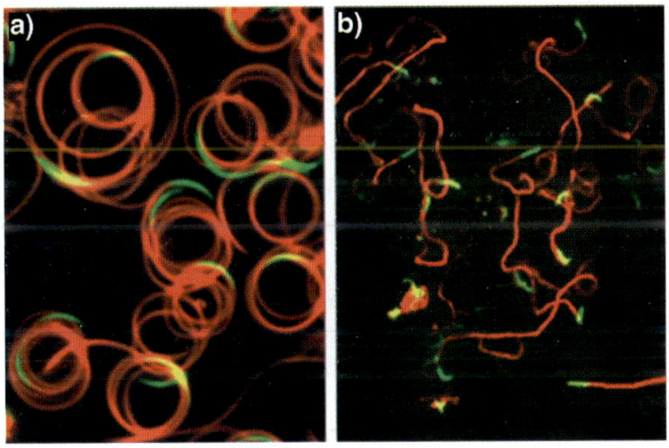

Fig. 12.4 Motility of fluorescent sporozoites. **a** Isolated sporozoites (*green*) move on a glass surface in a circular path as indicated by the maximum fluorescence intensity projection (*red*). **b** Sporozoites filmed in the skin of a living mouse after a natural bite display a seemingly random movement pattern

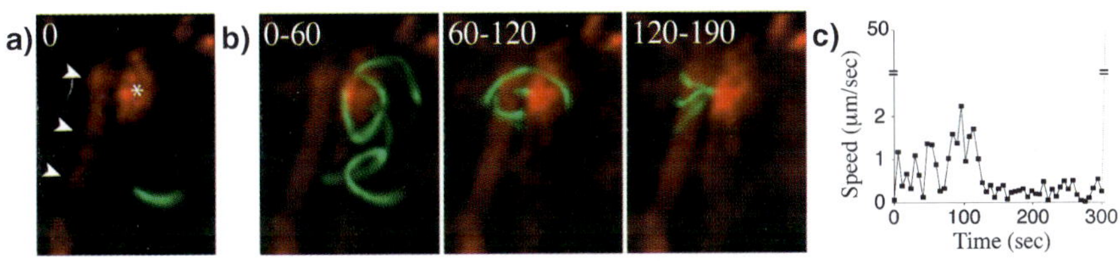

Fig. 12.5 Entry of sporozoites into blood vessels. **a** A sporozoite (*green*) is shown in the skin of a mouse ear. The mouse was previously injected intravenously with fluorescently conjugated bovine serum albumin, which highlights the blood vessel (*arrowheads*) as well as cells that take up the dye (*asterisk*). **b** Maximum intensity projections of three subsequent recordings showing that the sporozoite moves through the skin (0–60), associates with and enters the blood vessel. Numbers indicate the time frames used for the maximum projections in seconds. **c** The decrease of speed after the sporozoite associates with the blood vessel and the sudden increase after invasion. (Modified from movie 2 of Amino et al. 2006)

most of the sporozoites in the lymph nodes are destroyed, a small number can develop into early extraerythrocytic stages, which were thought to develop in vivo only in the liver. These early extraerythrocytic stages are found mainly associated with endothelial cells (Amino et al. 2006).

Once the sporozoites in the blood arrive in the liver, they must adhere to the walls of the sinusoids and find their way through the endothelial cell layer to the underlying hepatocytes. To do so, they are thought to enter into liver resident macrophages (Kupffer cells), which they leave again to then penetrate hepatocytes (Baer et al. 2007; Baldacci and Menard 2004; Frevert 2004; Frevert et al. 2006). Curiously, the sporozoite, once in the liver parenchyma, does not infect the first hepatocyte. Instead, it traverses several hepatocytes before invading the final one in

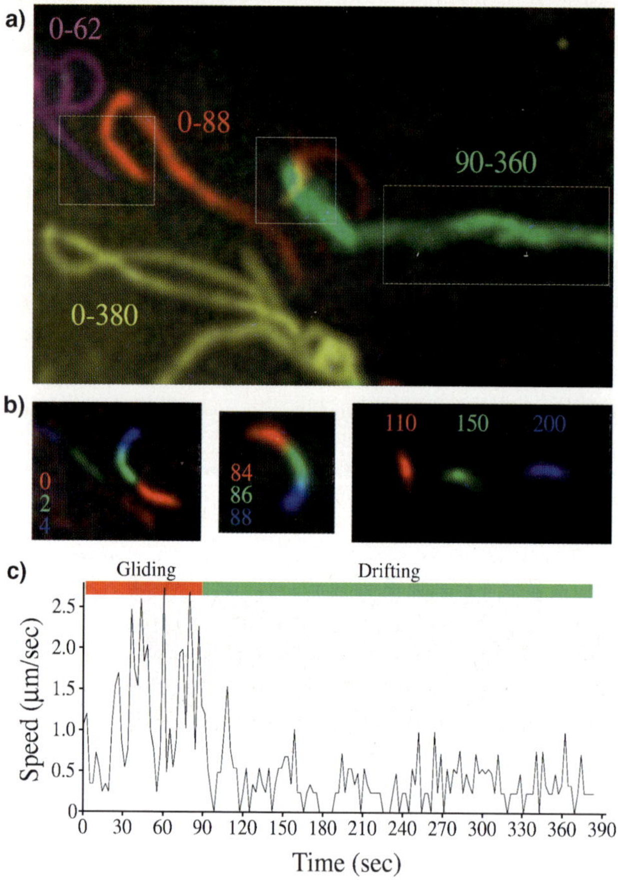

Fig. 12.6 Drifting movement of sporozoites in the skin. **a** Sporozoite movement prior (*red projection*) and after (*green projection*) invasion of a lymph vessel. Maximum intensity projections of other sporozoites in the field of view are shown in different colors with numbers indicating times in seconds of the images used for maximum intensity projections. **b** Single frames from the areas indicated with *white boxes* in **a**. Numbers correspond to the respectively pseudocolored frames. Note the forward movement of sporozoites during gliding (early time points) and the drifting during later time points. **c** The decrease of speed after the sporozoite entered a lymph vessel

a vacuole (Mota et al. 2001). Two studies aimed at visualizing the process of liver infection in vivo. This was achieved by either allowing hundreds of mosquitoes to bite an operated-on mouse placed on a microscope stage (Frevert et al. 2005), or by intravenous injection of isolated salivary gland sporozoites into the tail vein of the mouse (R. Amino, F. Frischknecht, R. Menard, S. Shorte and S. Thiberge, unpublished results). While the first approach delivered sporozoites in the natural way, it also limited the numbers of sporozoites that could be observed. The second approach injected large numbers of sporozoites directly into the bloodstream, which consequently allowed the observation of a large number of sporozoites in the liver. To expose the liver, an incision was made along the rib cage and the left liver lobe was placed onto a cover glass and immobilized with superglue or tissue glue. This allowed the use of a variety of oil or water immersion objectives for imaging. To avoid cooling of the mouse, a thermal blanket or a heated incubator can be used, with the latter keeping both the mouse and the microscope at 37°C, resulting in a stable setup for long-term observation. A general problem with imaging the liver of a living mouse is the continuous breathing and the beating heart, which can cause unwanted vibrations, similar to those experienced when imaging ookinetes in living mosquitoes. This results in a shifting focus, so the parasites cannot be localized appropriately in the tissue, and thus severely limits or prevents the use of such movies for quantitative analysis. However, stable preparations show that a combination of fast confocal 4D imaging with red fluorescent sporozoites and recombinant mice expressing GFP in a variety of cell types are likely to let us gain more interesting insights into the interaction of sporozoites with the liver. In addition a number of recombinant parasites are available that fail at different points to successfully develop into liver stages (Bhanot et al. 2005; Ishino et al. 2004, 2004; Mueller et al. 2005a, b; van Dijk et al. 2005). Clearly, expressing fluorescent proteins in these mutants will give additional insight into host–sporozoite interactions. The creation of such mutants is possible owing to the existence of a second selectable marker that allows the generation of a gene-disruption mutant in a line that already expresses GFP (de Koning-Ward et al. 2000). This has been further simplified by recent improvements in the efficiency of *P. berghei* transfection that have allowed the use of GFP as a selectable marker to either create a fluorescent line that may serve as the template for the subsequent selection of gene disruptions by conventional means (Janse et al. 2006) or directly select for gene-disruption mutants that may be immediately used in imaging studies (Franke-Fayard, Janse and Waters, unpublished results). Lastly, fluorescent *P. yoelii* parasites have also been established which will help to probe results obtained from experiments using *P. berghei* (Tarun et al. 2006).

Imaging developing parasites in the liver will be a future challenge as it would require animals to be kept anesthetized and stable for many hours. A first step should be the development of quantitative imaging assays of parasite development in liver cells in vitro. Imaging the release of parasites from fully developed liver parasites took advantage of the stable liver imaging preparations described above (Sturm et al. 2006). This revealed that merozoites are released in bags termed "merosomes" from the hepatic parasite (Fig. 12.7) containing only a few or as many as several thousand merozoites. The complex interactions between parasite and host cell are partially revealed by the modulation of liver cell death, which seems fundamentally different from apoptosis, and is inhibited during early stages but is induced just before merosome release (Sturm et al. 2006; van de Sand et al. 2005). What the exact signaling pathways are that are subverted in this complex host cell–parasite interaction are not known but, clearly, imaging will play a major part in deciphering them (Münter et al. 2006).

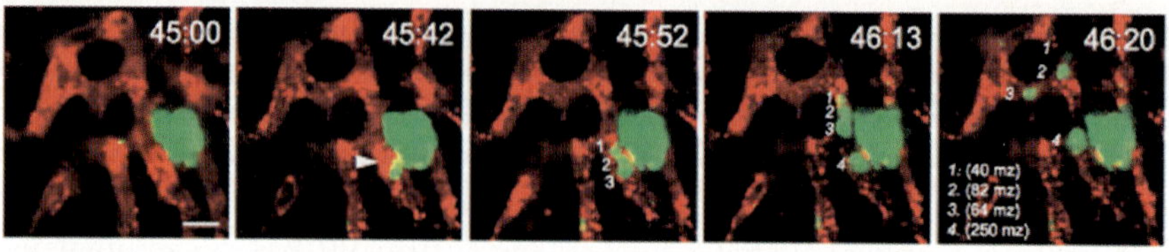

Fig. 12.7 Release of malaria parasites from infected hepatocytes. Release of malaria parasites from the liver. An intrahepatic parasite (*green*) releases merosomes (small numbers) of various sizes containing different numbers of merozoites (*mz*) into the liver sinusoids (*red*). Liver sinusoids were visualized by intravenous injection of fluorescently labeled bovine serum albumin. Numbers indicate time in hours and minutes after infection. (From Sturm et al. 2006, reprinted with permission from the American Association for the Advancement of Science)

12.8 Towards Molecular Imaging in Vivo

Once knockout parasites can be imaged in vivo the next step would be to also image a parasite expressing the protein of interest as a GFP fusion. Clearly, a limiting factor of such an approach is the amount of GFP-fusion protein expressed in a cell, especially when imaging of fast-moving parasites limits the exposure time to the subsecond range. Another caveat is that some GFP-fusion proteins cannot be expressed at the necessary high levels as the GFP moiety might interfere with the function of the protein. Nevertheless, GFP-tagging of parasite proteins has revealed a number of interesting insights when infections were performed in cultured cells in vitro (Amino et al. 2005). For example, secreted proteins or food vacuole resident enzymes were followed and the dynamics of major parasite organelles were investigated (Klemba et al. 2004; van Dooren et al. 2005; Wickham et al. 2001). Furthermore, the use of dyes to investigate the pH and calcium concentrations in parasites and infected cells allowed a number of unexpected observations to be made, including the sensitivity of intraerythrocytic parasites to light (Gazarini et al. 2003; Rohrbach et al. 2005, 2006; Wissing et al. 2002). The use of both intravital dyes and parasites expressing fluorescent fusion proteins might also be feasible for in vivo studies. For example, the observation that sporozoites are able to migrate through cells was known (Vanderberg et al. 1990); however, only the availability of intravital dyes has allowed the indirect observation of this phenomenon in mouse liver (Frevert et al. 2005; Mota et al. 2001). Fluorescently coupled dextran was injected into the blood and accumulated in liver cells that contained no sporozoites, indicating that the dye was taken up by the damaged cell after a sporozoite ruptured the plasma membrane. However, the cells might have also been damaged in other ways that led to an uptake of the dye. With imaging approaches in the liver allowing the observation of sporozoites expressing high levels of GFP or red fluorescent protein, one might now be able to use these in addition to intravital dyes to clearly demonstrate when a parasite is rupturing/passing through a cell in real time. In addition, this question could be answered by imaging parasites expressing GFP fusions with proteins that are specifically secreted during gliding, cell disruption or cell invasion.

12.9 A Look at Other Parasites

The most devastating parasitic disease, malaria, is only one of a large panoply of parasite-caused debilities that haunt more humans today than at any time in history. Trypanosomes are the causative agents of sleeping sickness in Africa (*Trypanosoma brucei*) and Chagas disease (*T. cruzi*) in South and Central America, respectively, and together are responsible for over 60,000 deaths per year. *T. brucei* parasites are transmitted by both sexes of the bloodsucking tsetse fly. Once *T. brucei* parasites are taken up by a fly, they first establish an infection in the midgut, undergo complex cycles of differentiation and multiplication and end up in the salivary glands. A first study aiming at imaging the events in the gut used green fluorescent *T. brucei* parasites and showed their motile behavior in the blood meal of dissected fly guts using epifluorescence microscopy (Gibson and Bailey 2003). The parasites were present for several days and quantification confirmed that the parasites multiplied exponentially during the first 3 days but little detail was revealed on how the parasites proceed within the tsetse fly or how they float back into the host. A similar study on *Leishmania donovani*, the causative agent of Old World Kala-azar, revealed that GFP parasites could establish an infection in the *Lutzomyia* sandflies, which normally host only New World *Leishmania* parasites (Guevara et al. 2001). Again, much more remains to be discovered, especially as *Leishmania* parasites are deposited within the skin, where *Leishmania major* promasitoges can establish an infection by first invading neutrophils, which are then taken up by macrophages (Frischknecht 2007, Laskay et al. 2003). *T. cruzi* parasites are transmitted by the bloodsucking *Triatoma* or *Rhodnius* kissing bugs. Curiously, the parasites are located within the feces of the insect and are subsequently rubbed into the wound by the unsuspecting victim. How they proceed through the skin is currently not known and one might expect that an in vivo imaging approach using already available fluorescent *T. cruzi* (Guevara et al. 2005) could yield new insights as with malaria parasites.

A recent elegant study investigating the motile behavior of a parasitic amoeba made full use of fluorescent parasites, genetic manipulation and two-photon microscopy. *Entamoeba histolitica* causes amoebiasis in 50 million people annually, with some 100,000 deaths. The parasite, which is taken up with contaminated food or water, invades the intestinal epithelium and disseminates via the blood circulation to cause abscesses in the liver. Central to its pathogenicity is its ability to migrate. Employing two-photon laser scanning microscopy to visualize the path of the parasites on gut epithelium and within the liver of experimentally infected hamsters, the authors used wild type parasites as well as parasites over-expressing a dominant negative myosin II or a major adhesive lectin (Coudrier et al. 2005). This demonstrated that while myosin II is important for migration through intestinal monolayers and the liver, the lectin is exclusively involved in liver invasion. Similar to the observations in *Plasmodium* sporozoites, it was also shown, using specifically developed tracking programs (Zimmer et al. 2002), that the motile behavior in vitro does not reflect the complexity of motility in vivo.

12.10 Conclusion

In conclusion, we hope that our brief outline has shown how readily available and easy-to-handle imaging techniques have cast a new light on the interactions between malaria and other parasites with their hosts and vectors. Although much remains to be discovered, it should now be possible to visualize even rare (but potentially important) events in host–parasite and vector–parasite interactions using wild-type or mutant parasites. The next challenge will then constitute the imaging of protein function in vivo.

Acknowledgements We thank Volker Heussler, Kai Matuschewski, Markus Meissner and Sylvia Münter for reading the manuscript. The authors are supported by grants from the European Union, the Howard Hughes Medical Institute, the BioFuture Program of the German Federal Ministry for Education and Research (BMBF) and the German Research Foundation (SFB 544, SPP 1128).

References

Akaki M, Dvorak JA (2005) A chemotactic response facilitates mosquito salivary gland infection by malaria sporozoites. J Exp Biol 208:3211–3218

Aly AS, Matuschewski K (2005) A malarial cysteine protease is necessary for Plasmodium sporozoite egress from oocysts. J Exp Med 202:225–230

Amino R, Menard R, Frischknecht F (2005) In vivo imaging of malaria parasites – recent advances and future directions. Curr Opin Microbiol 8:407–414

Amino R, Thiberge S, Martin B, Celli S, Shorte S, Frischknecht F, Menard R (2006) Quantitative imaging of Plasmodium transmission from mosquito to mammal. Nat Med 12:220–224

Baer K, Roosevelt M, Clarkson AB Jr, van Rooijen N, Schnieder T, Frevert U (2007) Kupffer cells are obligatory for Plasmodium yoelii sporozoite infection of the liver. Cell Microbiol 9:397–412

Baldacci P, Menard R (2004) The elusive malaria sporozoite in the mammalian host. Mol Microbiol 54:298–306

Baton LA, Ranford-Cartwright LC (2005) How do malaria ookinetes cross the mosquito midgut wall? Trends Parasitol 21:22–28

Baum J, Papernfuss AT, Baum B, Speed TP, Cowman AF (2006) Regulation of apicomplexan actin-based motility. Nat Rev Microbiol 4:621–628

Beier JC (1993) Malaria sporozoites: survival, transmission and disease control. Parasitol Today 9:210–215

Beier JC (1998) Malaria parasite development in mosquitoes. Annu Rev Entomol 43:519–543

Bhanot P, Schauer K, Coppens I, Nussenzweig V (2005) A surface phospholipase is involved in the migration of plasmodium sporozoites through cells. J Biol Chem 280:6752–6760

Billker O, Lindo V, Panico M, Etienne AE, Paxton T, Dell A, Rogers M, Sinden RE, Morris HR (1998) Identification of xanthurenic acid as the putative inducer of malaria development in the mosquito. Nature 392:289–292

Blandin S, Levashina EA (2004) Mosquito immune responses against malaria parasites. Curr Opin Immunol 16:16–20

Coudrier E, Amblard F, Zimmer C, Roux P, Olivo-Marin JC, Rigothier MC, Guillen N (2005) Myosin II and the Gal-GalNAc lectin play a crucial role in tissue invasion by Entamoeba histolytica. Cell Microbiol 7:19–27

Danielli A, Barillas-Mury C, Kumar S, Kafatos FC, Loukeris TG (2005) Overexpression and altered nucleocytoplasmic distribution of Anopheles ovalbumin-like SRPN10 serpins in Plasmodium-infected midgut cells. Cell Microbiol 7:181–190

de Koning-Ward TF, Thomas AW, Waters AP, Janse CJ (1998) Stable expression of green fluorescent protein in blood and mosquito stages of Plasmodium berghei. Mol Biochem Parasitol 97:247–252

de Koning-Ward TF, Fidock DA, Thathy V, Menard R, van Spaendonk RM, Waters AP, Janse CJ (2000) The selectable marker human dihydrofolate reductase enables sequential genetic manipulation of the Plasmodium berghei genome. Mol Biochem Parasitol 106:199–212

Dessens JT, Beetsma AL, Dimopoulos G, Wengelnik K, Crisanti A, Kafatos FC, Sinden RE (1999) CTRP is essential for mosquito infection by malaria ookinetes. EMBO J 18:6221–6227

Doyle TC, Burns SM, Contag CH (2004) In vivo bioluminescence imaging for integrated studies of infection. Cell Microbiol 6:303–317

Enninga J, Mounier J, Sansonetti P, Tran Van Nhieu G (2005) Secretion of type III effectors into host cells in real time. Nat Methods 2:959–965

Franke-Fayard B, Trueman H, Ramesar J, Mendoza J, van der Keur M, van der Linden R, Sinden RE, Waters AP, Janse CJ (2004) A Plasmodium berghei reference line that constitutively expresses GFP at a high level throughout the complete life cycle. Mol Biochem Parasitol 137:23–33

Franke-Fayard B, Janse CJ, Cunha-Rodrigues M, Ramesar J, Buscher P, Que I, Lowik C, Voshol PJ, den Boer MA, van Duinen SG, Febbraio M, Mota MM, Waters AP (2005) Murine malaria parasite sequestration: CD36 is the major receptor, but cerebral pathology is unlinked to sequestration. Proc Natl Acad Sci USA 102:11468–11473

Frevert U (2004) Sneaking in through the back entrance: the biology of malaria liver stages. Trends Parasitol 20:417–424

Frevert U, Engelmann S, Zougbede S, Stange J, Ng B, Matuschewski K, Liebes L, Yee H (2005) Intravital observation of Plasmodium berghei sporozoite infection of the liver. PLoS Biol 3:e192

Frevert U, Usynin I, Baer K, Klotz C (2006) Nomadic or sessile: can Kupffer cells function as portals for malaria sporozoites to the liver? Cell Microbiol 8:1537–1546

Freyvogel TA (1966) Shape, movement in situ and locomotion of plasmodial ookinetes. Acta Trop 23:201–222

Frischknecht F, Baldacci P, Martin B, Zimmer C, Thiberge S, Olivo-Marin JC, Shorte SL, Menard R (2004) Imaging movement of malaria parasites during transmission by Anopheles mosquitoes. Cell Microbiol 6:687–694

Frischknecht F, Martin B, Thiery I, Bourgouin C, Menard R (2006a) Using green fluorescent malaria parasites to screen for permissive vector mosquitoes. Malar J 5:23

Frischknecht F, Renaud O, Shorte SL (2006b) Imaging today's infectious animalcules. Curr Opin Microbiol 9:297–306

Frischknecht F (2007) The skin as interface in the transmission of arthropod-borne pathogens. Cell Microbiol 9:1630–1640

Gazarini ML, Thomas AP, Pozzan T, Garcia CR (2003) Calcium signaling in a low calcium environment: how the intracellular malaria parasite solves the problem. J Cell Biol 161:103–110

Gibson W, Bailey M (2003) The development of Trypanosoma brucei within the tsetse fly midgut observed using green fluorescent trypanosomes. Kinetoplastid Biol Dis 2:1

Greber UF, Way M (2006) A superhighway to virus infection. Cell 124:741–754

Guevara P, Pinto-Santini D, Rojas A, Crisante G, Anez N, Ramirez JL (2001) Green fluorescent protein-tagged Leishmania in phlebotomine sand flies. J Med Entomol 38:39–43

Guevara P, Dias M, Rojas A, Crisante G, Abreu-Blanco MT, Umezawa E, Vazquez M, Levin M, Anez N, Ramirez JL (2005) Expression of fluorescent genes in Trypanosoma cruzi and Trypanosoma rangeli (Kinetoplastida: Trypanosomatidae): its application to parasite-vector biology. J Med Entomol 42:48–56

Hardy J, Edinger M, Bachmann MH, Negrin RS, Fathman CG, Contag CH (2001) Bioluminescence imaging of lymphocyte trafficking in vivo. Exp Hematol 29:1353–1360

Hitziger N, Dellacasa I, Albiger B, Barragan A (2005) Dissemination of Toxoplasma gondii to immunoprivileged organs and role of Toll/interleukin-1 receptor signalling for host resistance assessed by in vivo bioluminescence imaging. Cell Microbiol 7:837–848

Ho M, Hickey MJ, Murray AG, Andonegui G, Kubes P (2000) Visualization of Plasmodium falciparum-endothelium interactions in human microvasculature: mimicry of leukocyte recruitment. J Exp Med 192:1205–1211

Hunt NH, Grau GE (2003) Cytokines: accelerators and brakes in the pathogenesis of cerebral malaria. Trends Immunol 24:491–499

Ishino T, Yano K, Chinzei Y, Yuda M (2004) Cell-passage activity is required for the malarial parasite to cross the liver sinusoidal cell layer. PLoS Biol 2:E4

Ishino T, Chinzei Y, Yuda M (2005) Two proteins with 6-cys motifs are required for malarial parasites to commit to infection of the hepatocyte. Mol Microbiol 58:1264–1275

Janse CJ, Franke-Fayard B, Mair GR, Ramesar J, Thiel C, Engelmann S, Matuschewski K, van Gemert GJ, Sauerwein RW, Waters AP (2006) High efficiency transfection of Plasmodium berghei facilitates novel selection procedures. Mol Biochem Parasitol 145:60–70

Kappe SH, Buscaglia CA, Nussenzweig V (2004) Plasmodium sporozoite molecular cell biology. Annu Rev Cell Dev Biol 20:29–59

Kaul DK, Liu XD, Nagel RL, Shear HL (1998) Microvascular hemodynamics and in vivo evidence for the role of intercellular adhesion molecule-1 in the sequestration of infected red blood cells in a mouse model of lethal malaria. Am J Trop Med Hyg 58:240–247

Khan SM, Franke-Fayard B, Mair GR, Lasonder E, Janse CJ, Mann M, Waters AP (2005) Proteome analysis of separated male and female gametocytes reveals novel sex-specific Plasmodium biology. Cell 121:675–687

Klemba M, Beatty W, Gluzman I, Goldberg DE (2004) Trafficking of plasmepsin II to the food vacuole of the malaria parasite Plasmodium falciparum. J Cell Biol 164:47–56

Lang T, Goyard S, Lebastard M, Milon G (2005) Bioluminescent Leishmania expressing luciferase for rapid and high throughput screening of drugs acting on amastigote-harbouring macrophages and for quantitative real-time monitoring of parasitism features in living mice. Cell Microbiol 7:383–392

Larson DR, Johnson MC, Webb WW, Vogt VM (2005) Visualization of retrovirus budding with correlated light and electron microscopy. Proc Natl Acad Sci USA 102:15453–15458

Larson DR, Ma YM, Vogt VM, Webb WW (2003) Direct measurement of Gag-Gag interaction during retrovirus assembly with FRET and fluorescence correlation spectroscopy. J Cell Biol 162:1233–1244

Laskay T, van Zandbergen G, Solbach W (2003) Neutrophil granulocytes – Trojan horses for Leishmania major and other intracellular microbes? Trends Microbiol 11:210–214

Luker GD, Prior JL, Song J, Pica CM, Leib DA (2003) Bioluminescence imaging reveals systemic dissemination of herpes simplex virus type 1 in the absence of interferon receptors. J Virol 77:11082–11093

Luker KE, Hutchens M, Schultz T, Pekosz A, Luker GD (2005) Bioluminescence imaging of vaccinia virus: effects of interferon on viral replication and spread. Virology 341: 284–300

Medica DL, Sinnis P (2005) Quantitative dynamics of Plasmodium yoelii sporozoite transmission by infected anopheline mosquitoes. Infect Immun 73:4363–4369

Mota MM, Pradel G, Vanderberg JP, Hafalla JC, Frevert U, Nussenzweig RS, Nussenzweig V, Rodriguez A (2001) Migration of Plasmodium sporozoites through cells before infection. Science 291:141–144

Mueller AK, Camargo N, Kaiser K, Andorfer C, Frevert U, Matuschewski K, Kappe SH (2005a) Plasmodium liver stage developmental arrest by depletion of a protein at the parasite-host interface. Proc Natl Acad Sci USA 102:3022–3027

Mueller AK, Labaied M, Kappe SH, Matuschewski K (2005b) Genetically modified Plasmodium parasites as a protective experimental malaria vaccine. Nature 433:164–167

Münter S, Way M, Frischknecht F (2006) Signaling during pathogen infection. Sci STKE 2006:re5

Natarajan R, Thathy V, Mota MM, Hafalla JC, Menard R, Vernick KD (2001) Fluorescent Plasmodium berghei sporozoites and pre-erythrocytic stages: a new tool to study mosquito and mammalian host interactions with malaria parasites. Cell Microbiol 3:371–379

Osta MA, Christophides GK, Vlachou D, Kafatos FC (2004) Innate immunity in the malaria vector Anopheles gambiae: comparative and functional genomics. J Exp Biol 207:2551–2563

Pelkmans L, Fava E, Grabner H, Hannus M, Habermann B, Krausz E, Zerial M (2005) Genome-wide analysis of human kinases in clathrin- and caveolae/raft-mediated endocytosis. Nature 436:78–86

Pimenta PF, Touray M, Miller LH (1994) The journey of malaria sporozoites in the mosquito salivary gland. J Euk Microbiol 41:608–624

Rietdorf J, Ploubidou A, Reckmann I, Holmstrom A, Frischknecht F, Zettl M, Zimmermann T, Way M (2001) Kinesin-dependent movement on microtubules precedes actin-based motility of vaccinia virus. Nat Cell Biol 3:992–1000

Rohrbach P, Friedrich O, Hentschel J, Plattner H, Fink RH, Lanzer M (2005) Quantitative calcium measurements in subcellular compartments of Plasmodium falciparum-infected erythrocytes. J Biol Chem 280:27960–27969

Rohrbach P, Sanchez CP, Hayton K, Friedrich O, Patel J, Sidhu AB, Ferdig MT, Fidock DA, Lanzer M (2006) Genetic linkage of pfmdr1 with food vacuolar solute import in Plasmodium falciparum. EMBO J 25:3000–3011

Rust MJ, Lakadamyali M, Zhang F, Zhuang X (2004) Assembly of endocytic machinery around individual influenza viruses during viral entry. Nat Struct Mol Biol 11:567–573

Saeij JP, Boyle JP, Grigg ME, Arrizabalaga G, Boothroyd JC (2005) Bioluminescence imaging of Toxoplasma gondii infection in living mice reveals dramatic differences between strains. Infect Immun 73:695–702

Schlumberger MC, Muller AJ, Ehrbar K, Winnen B, Duss I, Stecher B, Hardt WD (2005) Real-time imaging of type III secretion: Salmonella SipA injection into host cells. Proc Natl Acad Sci U S A 102:12548–12553

Schüler H, Matuschewski K (2006) Regulation of Apicomplexan microfilament dynamics by a minimal set of actin-binding proteins. Traffic 7:1433–1439

Sidjanski S, Vanderberg JP (1997) Delayed migration of Plasmodium sporozoites from the mosquito bite site to the blood. Am J Trop Med Hyg 57:426–429

Srinivasan A, McSorley SJ (2004) Visualizing the immune response to pathogens. Curr Opin Immunol 16:494–498

Sterling CR, Aikawa M, Vanderberg JP (1973) The passage of Plasmodium berghei sporozoites through the salivary glands of Anopheles stephensi: an electron microscope study. J Parasitol 59:593–605

Sturm A, Amino R, van de Sand C, Regen T, Retzlaff S, Rennenberg A, Krueger A, Pollok JM, Menard R, Heussler VT (2006) Manipulation of host hepatocytes by the malaria parasite for delivery into liver sinusoids. Science 313:1287–1290

Sultan AA, Thathy V, Frevert U, Robson KJ, Crisanti A, Nussenzweig V, Nussenzweig RS, Menard R (1997) TRAP is necessary for gliding motility and infectivity of plasmodium sporozoites. Cell 90:511–522

Sultan AA, Thathy V, Nussenzweig V, Menard R (1999) Green fluorescent protein as a marker in Plasmodium berghei transformation. Infect Immun 67:2602–2606

Tarun AS, Baer K, Dumpit RF, Gray S, Lejarcegui N, Frevert U, Kappe SH (2006) Quantitative isolation and in vivo imaging of malaria parasite liver stages. Int J Parasitol 36:1283–1293

Templeton TJ, Kaslow DC, Fidock DA (2000) Developmental arrest of the human malaria parasite Plasmodium falciparum within the mosquito midgut via CTRP gene disruption. Mol Microbiol 36:1–9

Turner GD, Morrison H, Jones M, Davis TM, Looareesuwan S, Buley ID, Gatter KC, Newbold CI, Pukritayakamee S, Nagachinta B et al. (1994) An immunohistochemical study of the pathology of fatal malaria. Evidence for widespread endothelial activation and a potential role for intercellular adhesion molecule-1 in cerebral sequestration. Am J Pathol 145:1057–1069

Vanderberg JP (1974) Studies on the motility of Plasmodium sporozoites. J Protozool 21:527–537

Vanderberg JP, Frevert U (2004) Intravital microscopy demonstrating antibody-mediated immobilisation of Plasmodium berghei sporozoites injected into skin by mosquitoes. Int J Parasitol 34:991–996

Vanderberg JP, Chew S, Stewart MJ (1990) Plasmodium sporozoite interactions with macrophages in vitro: a videomicroscopic analysis. J Protozool 37:528–536

van de Sand C, Horstmann S, Schmidt A, Sturm A, Bolte S, Krueger A, Lutgehetmann M, Pollok JM, Libert C, Heussler VT (2005) The liver stage of Plasmodium berghei inhibits host cell apoptosis. Mol Microbiol 58:731–742

van Dijk MR, Douradinha B, Franke-Fayard B, Heussler V, van Dooren MW, van Schaijk B, van Gemert GJ, Sauerwein RW, Mota MM, Waters AP, Janse CJ (2005) Genetically attenuated, P36p-deficient malarial sporozoites induce protective immunity and apoptosis of infected liver cells. Proc Natl Acad Sci USA 102:12194–12199

van Dooren GG, Marti M, Tonkin CJ, Stimmler LM, Cowman AF, McFadden GI (2005) Development of the endoplasmic reticulum, mitochondrion and apicoplast during the asexual life cycle of Plasmodium falciparum. Mol Microbiol 57:405–419

Vlachou D, Kafatos FC (2005) The complex interplay between mosquito positive and negative regulators of Plasmodium development. Curr Opin Microbiol 8:415–421

Vlachou D, Zimmermann T, Cantera R, Janse CJ, Waters AP, Kafatos FC (2004) Real-time, in vivo analysis of malaria ookinete locomotion and mosquito midgut invasion. Cell Microbiol 6:671–685

Wang Q, Fujioka H, Nussenzweig V (2005) Mutational analysis of the GPI-anchor addition sequence from the circumsporozoite protein of Plasmodium. Cell Microbiol 7:1616–1626

Wickham ME, Rug M, Ralph SA, Klonis N, McFadden GI, Tilley L, Cowman AF (2001) Trafficking and assembly of the cytoadherence complex in Plasmodium falciparum-infected human erythrocytes. EMBO J 20:5636–5649

Wissing F, Sanchez CP, Rohrbach P, Ricken S, Lanzer M (2002) Illumination of the malaria parasite Plasmodium falciparum alters intracellular pH. Implications for live cell imaging. J Biol Chem 277:37747–37755

Yu J, Wu J, Francis KP, Purchio TF, Kadurugamuwa JL (2005) Monitoring in vivo fitness of rifampicin-resistant Staphylococcus aureus mutants in a mouse biofilm infection model. J Antimicrob Chemother 55:528–534

Yuda M, Sakaida H, Chinzei Y (1999) Targeted disruption of the plasmodium berghei CTRP gene reveals its essential role in malaria infection of the vector mosquito. J Exp Med 190:1711–1716

Zieler H, Dvorak JA (2000) Invasion in vitro of mosquito midgut cells by the malaria parasite proceeds by a conserved mechanism and results in death of the invaded midgut cells. Proc Natl Acad Sci USA 97:11516–11521

Zimmer C (2000) Parasite rex: inside the bizarre word of nature's most dangerous creatures. Free, New York

Zimmer C, Labruyere E, Meas-Yedid V, Guillen N, Olivo-Marin JC (2002) Segmentation and tracking of migrating cells in videomicroscopy with parametric active contours: a tool for cell-based drug testing. IEEE Trans Med Imaging 21:1212–1221

13 Computer-Assisted Systems for Dynamic 3D Reconstruction and Motion Analysis of Living Cells

David R. Soll, Edward Voss, Deborah Wessels, and Spencer Kuhl

Abstract Cell motility is an essential component of the life history of cells ranging in complexity from soil amoebae to human white blood cells. Because the methods we standardly employ to view moving cells are two-dimensional and rarely quantitative, we have developed an incomplete and in some cases inaccurate view of motile behavior. A solution to these problems is the development of a computer-assisted system and software program for the three-dimensional reconstruction and motion analysis of live, migrating cells. Strategies are described for obtaining optical sections of migrating cells at short time intervals, outlining and reconstruction that provide three-dimensional representations of the transparent cell surface and internal architecture. Because outlines of the cell surface and cell compartments are converted in these systems to β-spline models, every aspect of cell behavior and the dynamic changes in internal architecture can be quantitated in space and time. Methods have been refined to reconstruct in three dimensions the filopodia of migrating cells, as well as to reconstruct and motion-analyze every cell and nucleus in a developing embryo. In the future, this technology can be used to reconstruct and motion-analyze fluorescent components of a cell, including cytoskeletal localization. It is argued that without such four-dimensional analyses, inaccurate descriptions of normal cellular locomotion will persist, and aberrant, mutant and metastatic cell behavior will have no valid contextual framework for interpretation.

13.1 Introduction

Migrating animal cells are three-dimensional (3D). Yet, we tend to think of them in two dimensions, primarily because that is how we normally view them through a microscope. Furthermore, most migrating cells also change shape, and yet we tend to think of them as morphologically constant. There are two reasons for this. First, because we commonly fix migrating cells to assess their cytoskeletal organization or the distribution of other molecules, we tend to think that the snapshot represents the migrating cell. Second, because cells move slowly, we tend to miss changes in

S.L. Shorte and F. Frischknecht (eds.), *Imaging Cellular and Molecular Biological Functions.*

shape when viewing them in real time. In casual observation, we lack a precise mechanism for storing a cell image at an earlier time point in order to compare it with an image at a later time point. Time-lapse and continuous recordings of cells have helped in appreciating both motility and dynamic morphology, but these presentations again are primarily two-dimensional (2D). Finally, most cells probably move along 3D routes in a natural setting, whether it is in the wild or in a human body. This may be through complex environments in the soil, in ponds and streams, or in extracellular matrices or tissue in a multicellular organism. However, we tend to think that cells move in two dimensions, because we usually view their motility in vitro on the flat (2D) surface of a transparent material like glass or plastic. A partial solution to this myopic view of cell behavior is the development of computer-assisted software for the 3D reconstruction and motion analysis of live, migrating cells.

Here, we will first describe the computer-assisted 3D reconstruction and motion analysis system 3D Dynamic Image Analysis System (3D-DIAS; Soll 1999; Soll et al. 2000; Wessels et al. 1998a, 2006), which was developed specifically for reconstructing and motion-analyzing individual animal cells in three dimensions. This 3D system evolved from 2D systems, which began to be developed at the University of Iowa in the late 1980s (Soll et al. 1988; Soll 1995, 1988; Soll and Voss 1998). Then we will describe a customized version of 3D-DIAS, 3D-DIASemb, developed for the 3D reconstruction and motion analysis of live cells in living embryos in three dimensions (Heid et al. 2002; Ulrich et al. 2003). Finally, we will review new technology for reconstructing and motion-analyzing filopodia in three dimensions. Our intent will not be to review the history of 3D reconstruction and analysis, but rather to describe specifically the advanced systems we have developed as prototypes for the development of future 3D systems. We will try to point out both the caveats and the potentials of these systems. Hopefully this chapter will stimulate an interest in scientists from a wide range of disciplines to continue to develop and apply such dynamic 3D systems.

13.2 Approaches to 3D Reconstruction and Motion Analysis

The challenge of dynamic 3D reconstruction is to generate an accurate 3D reconstruction of a living cell at short time intervals during cell migration. To reconstruct a moving cell in three dimensions, one must obtain all optical sections within a time period short enough to ensure that movement of the cell between the first and the last optical section does not cause a significant reconstruction artifact. The optical sections in this short time period are then used to generate a 3D reconstruction. This process of optical sectioning must be repeated at short enough time intervals to generate a time sequence of reconstructions. The simplest 3D reconstruction is that of the original stack of optical sections. While much has been made of the simple retrieval and storage of optical sections for future analysis, it represents only the first and least complicated of the steps in computer-assisted reconstruction and motion analysis. For a cell 20 μm in length moving at a speed of 10 μm/min, the distance moved in 2 s is 0.33 μm. The movement error between the bottom and the top section collected in a 2-s period is approximately 2% (Shutt et al. 1995a). For a pseudopod 4 μm in height that is growing at an average rate of 30 μm/min, the time of optical sectioning is 0.8 s and the movement error is 10%. Straightforward methods have been developed to obtain a set of optical sections of an unstained living cell using differential interference contrast (DIC) optics, or a vitally stained cell using fluorescence or laser scanning confocal microscopy (LSCM). The former method involves the use of customized, computer-regulated stepper motors with character generators for identifying each frame temporally and spatially (Heid et al. 2002). The latter method is usually too slow for reconstructing cells translocating at velocities above 5 μm/min, since the scan rates for high-resolution imaging are too slow. This results in an unacceptably large movement error. In addition, the edges of the cell perimeters and internal architecture are both of low contrast in optical sections obtained through DIC microscopy, and the cytosol and intranuclear regions are refractive. Therefore, stacked images lose internal 3D architecture that may be readily identifiable in single sections. Images of vitally stained cells can also be stacked, but the lack of transparency of stained regions, especially external surfaces, hides internal architecture. Side views of such stacks are latticed. More importantly, such reconstructions do not provide the kind of pixel data that can be used to measure motility and dynamic morphology parameters of the cell or intracellular components. While the collection of such images can be highly informative, especially in the analysis of embryos and cell lineages (Thomas and White 1998; Schnabel et al. 1997), it must be emphasized that it represents only the first step in a computer-assisted reconstruction and motion analysis system.

To extend the preceding procedures, which basically represent the stacking of raw data, cells and internal architecture must be reconstructed as computer-assisted mathematical representations, based upon the outlines of the cell and internal architecture. In this approach, optical images obtained using DIC or other optical methods are collected, and the perimeter (edge) of the cell and internal architecture (e.g., nuclear perimeter, pseudopod, vacuole perimeter, etc.) outlined in each optical section. The outlines are converted to β-spline representations, stacked and linked in the *z*-axis (encapsulated) to generate a faceted representation of the cell and internal architecture. Such faceted images of the cell body, nucleus, pseudopod and vacuole are generated in independent trace slots, so each can be reconstructed at short time intervals either alone or in combination by merging with another component or all components of the cell. Because the representations of the cell's surface and cell components are individual and mathematical, one can quantitate a variety of 3D parameters as a function of time. One can also compute 2D parameters for any single section.

The faceted images generated by the above method can then be made semitransparent, like chicken-wire fencing, or nontransparent. For instance, representation of the surface of the cell body can be made transparent, while the surfaces of the nuclear representation and/or pseudopod representation can be made nontransparent. In addition, other components such as cytoplasmic particles, stained molecular complexes or even filopodia can be made nontransparent. Both transparent and nontransparent components can then be color-coded.

The preceding computer-assisted approach to 3D reconstruction and motion analysis can be applied not only to a single cell crawling on a substratum, but also to two or more cells interacting on a substratum, a single cell crawling in an acellular matrix, a single cell crawling amongst other cells in a tissue, several cells interacting in a tissue, a pathogen penetrating a cell or tissue, individual cells in a developing embryo, or all cells in a developing embryo.

13.3 Obtaining Optical Sections for 3D Reconstruction

3D-DIAS was developed at the University of Iowa in the 1990s for the 3D reconstruction of live, migrating cells at short time intervals. With use of a computer-regulated stepper motor (Heid et al. 2002) attached to the focusing knob of a microscope equipped with DIC optics and a character generator that notes temporal and spatial parameters, optical sections through a cell are collected in the *z*-axis in a 1- or 2-s period (30 or 60 sections, respectively), and the process is repeated every 5 s, on average. A time series of optical sections can be acquired in this manner at 30 frames per second using iMovie on a Macintosh G3 or higher-performance computer in conjunction with an analog-to-digital converter. The digital movie is then exported into QuickTime® format, opened in DIAS and saved using the DIAS compression format. Image processing features available in DIAS can be used to facilitate automated or manual outlining for subsequent reconstruction.

13.4 Outlining

3D outlining of a cell migrating along a substratum results in a very large number of optical sections. For 30 optical sections collected in a 1-s interval, then repeated at 5-s intervals over 20 min of cell migration, one collects 7,200 optical sections. Automatic outlining has, therefore, been an important objective in the development of 3D-DIAS software. Here we describe the general approach we have most recently taken in the development of a Java-based 3D DIAS program.

Our objective was to reconstruct and motion-analyze in three dimensions the hierarchical boundaries of a cell and its subcellular components. For a cell, this could potentially include the entire outer cell boundary, pseudopodia, nucleus, vacuole, cytoplasmic particles and filopodia. Outlining involves detecting the edges of the object of interest. The edges may be detected directly or *inferred* from detail that implies the existence of an edge. The final outline may be stored as a list of pixels making up the outline, a set of control points with connecting *β-splines* (Barsky and Beatty 1983) or a region consisting of all pixels interior to the outline. The last method is appropriate when the outline is more complex than a simple collection of closed curves, such as a fibrous network of microtubules or intermediate filaments. The following represents select situations for outlining. We will assume images are in 8-bit grayscale. That is, each pixel has an intensity ranging from 0 (black) to 255(white). In the solution that follows, black pixels are considered unwanted background and white pixels are desired detail. Images can be inverted if the opposite is desired. Color images may be broken into their red, green and blue (RGB) components, with each component represented by an 8-bit gray image. These may be outlined separately and then reassembled, a useful solution to multicolor fluorescent dyes.

The most common automatic outlining method is *thresholding*. This relies on differences in pixel intensity between the inside and outside of an object. Both phase-contrast and fluorescence microscopy have proven amenable to this method. There are two thresholding methods: *direct cutoff thresholding* and *gradient thresholding*. In the direct cutoff method, the user supplies an intensity threshold level, T, between 0 and 255 that determines the set S of desired pixels (I is the entire original image):

$$S = \text{direct cuttoff } (I, T) = \{\text{all points } P \in I \mid \text{intensity } (P) = T\}.$$

Thus, pixels that have intensities less than T are discarded.
If the result is complex (i.e., if the image contains holes, fibrous networks or other very thin processes), the entire set of white pixels is saved as a *quadtree* for efficiency (Pavlidis 1998). When an outline is desired, the resulting white pixels are dilated, eroded and traced. Dilation (also called *closing*; Russ,1992) fills in gaps caused by artifacts, while erosion reduces the dilated figure back to its original size. They are defined by

Dilation $(S) = I \cup \{\text{all points } P \mid \text{there exists an 8-neighbor } Q \text{ of } P \text{ such that } Q\ \hat{\text{I}}\ S\}$
and
Erosion $(S) = \{\text{all points } P\ \hat{\text{I}}\ S \mid \text{all 8-neighbors of } P\ \hat{\text{I}}\ S\}$,

where S is the original set of white pixels, and an "8-neighbor" of P refers to a pixel that is horizontally, vertically or diagonally adjacent to P.

Dilation and/or erosion may be skipped if not needed. The result is traced in a counterclockwise direction as follows. First, the pixels are grouped into their connected components. The centroid (center of area) of each group is computed. A horizontal line is extended through the centroid until it reaches the leftmost pixel in the group. A trace is then generated by starting at this point and moving along the edge, keeping the pixels within the group always to the right, until the starting point is reattained. When all the pixel groups have been traced, a collection of closed outlines is obtained (Fig. 13.1). Any holes hidden within an object will be ignored. The resulting traces may now be stored as pixel arrays or alternatively stored as β-splines. If β-splines are desired, the user selects *control points* by specifying a pixel increment. For example, every third point might be designated as a control point. Those points are then connected by mathematical curves called β-splines. The default splines are defined as follows (Barsky and Beatty 1983).

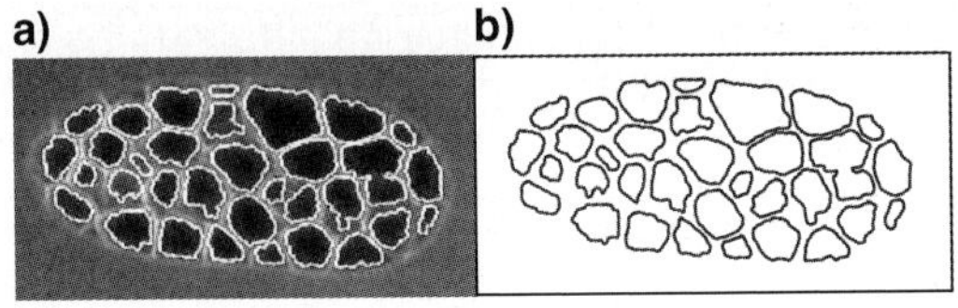

Fig. 13.1 Outlining by thresholding with multiple groups. In this case, the original image, provided by W. Mohler of the University of Connecticut School of Medicine, was a *Caenorhabditis elegans* embryo vitally stained with a membrane dye. **a** Original outlining. **b** Reversal of white. The method works even better with single cells. Dilation of 1 and erosion of 2 was used

Let $V_i = (x_i, y_i)$ be the control points.

The spline curve $[X_i(u), Y_i(u)]$ for 0 £ u £ 1 connecting the control points V_i and V_{i+1} is defined by

$$X_i(u) = \left[b_{-2}(u)x_{i-2} + b_{-1}(u)x_{i-1} + b_0(u)x_i + b_1(u)x_{i+1}\right]/6$$

and

$$Y_i(u) = \left[b_{-2}(u)y_{i-2} + b_{-1}(u)y_{i-1} + b_0(u)y_i + b_1(u)y_{i+1}\right]/6,$$

where

$b_1(u) = u^3,$
$b_0(u) = 1 + 3u + 3u^2 - 3u^3,$
$b_{-1}(u) = 4 - 6u^2 + 3u^3$
and
$b_{-2}(u) = 1 - 3u + 3u^2 - u^3.$

Optionally *bias* (β_1) and *tension* (β_2) may be used to further control the splines. The 15 equations used to define such splines can be found in Barsky and Beatty (1983). The control points are then saved (along with the bias and tension, if needed). This has the result of smoothing the outline and reducing artifacts due to "zigzagging" pixels.

The second variant is gradient thresholding. Here the user selects a minimal pixel intensity difference, D. The set of selected pixels S from the entire image I is

$$S = \text{gradient}\,(I,T) = \{\text{all points } P \in I \mid \text{there exits an 8-neighboour } Q \text{ of } P \text{ such that} \mid \text{interior}\}$$

Thus, any pixel that has a neighboring pixel with an intensity difference greater than D is selected. The edge that is determined in this way may have gaps. Dilation is used to "heal" the outline. The resulting pixels are then traced and β-splined just as with the threshold cutoff method. The gradient method is useful when the edges are sharp but the microscope lighting is uneven.

For DIC optics, the edges do not have clear intensity differences and there is a characteristic DIC "shadow," which makes the threshold method impractical. Interior in-focus detail, however, is quite clear. In this case the edges are best "inferred" from the enclosed detail. This is the essence of the complexity method. The set of pixels S selected from the entire image I is

$$S = \text{complexity}\ (I, K, T) = \{\text{all points } P \in I | \text{standard deviation } (B_{K,P}) \geq T\},$$

where

$$B_{K,P} = \{\text{intensity } (Q) | Q \in \text{the box of size } K \times K \text{ centered at } P\}$$

and K is the kernel size (usually 3 or 5) and T is the threshold.

That is, pixels selected by the complexity method have a high standard deviation of intensities in a local neighborhood of that pixel. The resulting set S consists of a number of clumps of detail and smaller specks of noise. To remove noise, S is filtered by removing clumps containing fewer than N pixels (N=5 by default: the user may vary this number as needed). Dilation is repeated (typically three times) to join the clumps into coherent traceable groups. The result is then eroded the same number of times to reduce the object to its original size. Then it is traced and splined just as in the threshold method (Fig. 13.2). The user chooses the kernel size, threshold, noise filter cutoff, dilation and erosion interactively, allowing the user to see the immediate effects of choices.

This same method can be used to detect smooth regions such as nuclei by reversing the threshold inequality:

$$S = \text{anticomplexity}\ (I, K, T) = \{\text{all points } P \in I | \text{standard deviation } (B_{K,P}) \leq T\}.$$

Here the selected pixels have *low* standard deviations in a local neighborhood. Dilation and erosion are not usually needed in this case (Fig. 13.3).

In the case of complex fibrous structures, often the *negative space* (that is, the pixels not in the desired object) is easier to trace. The image is inverted by reversing the intensity of each pixel via

New pixel intensity = 255–original pixel intensity.

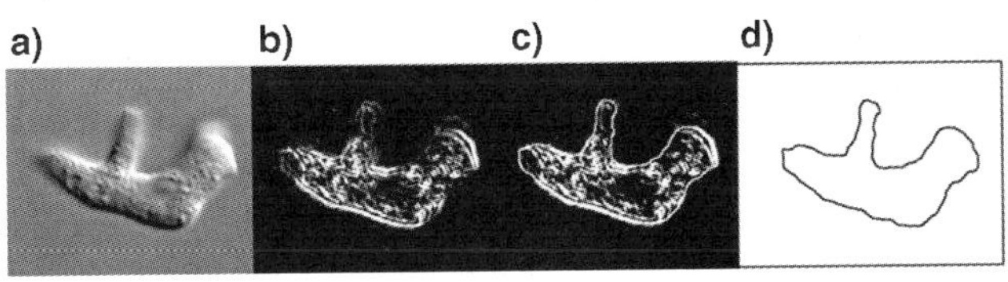

Fig. 13.2 Outlining by complexity. In this case, the original image of a *Dictyostelium discoideum* amoeba was obtained with differential interference contrast (DIC) microscopy. **a** Original image. **b** Initial complexity outlining. **c** Final outline. **d** Outer edge with internal architecture subtracted. Dilation of 3 and erosion of 3 was used

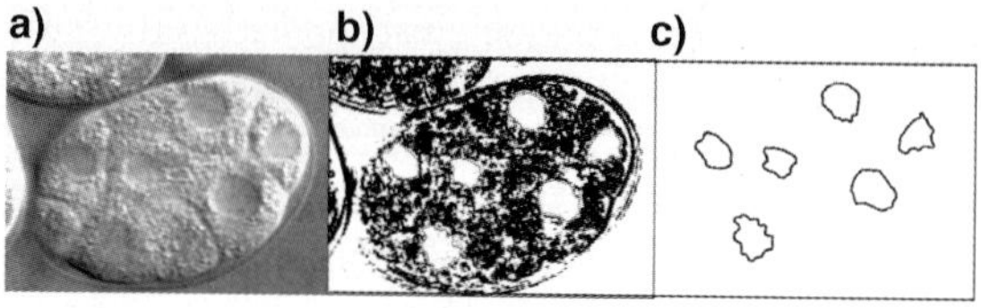

Fig. 13.3 Outlining by anticomplexity. In this case, the object was to outline the nuclei of cells in an early *C. elegans* embryo. The regions of low standard deviation are shown as *white*. **a** Original image obtained with DIC optics. **b** Final image showing nuclei as *white holes*. **c** Nuclei alone

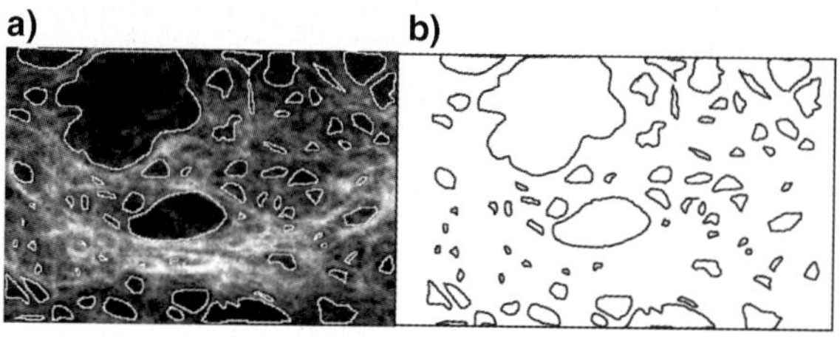

Fig. 13.4 Negative space outlining. **a** Thresholding was used to trace the dark background regions in a fibrous network image, in this case an image of an intermediate filament network. **b** Reversal of white

That is, black and white are reversed. The resulting image is outlined by any of the above methods. Thus the set of selected pixels S represents the *background* of the desired detail. The complement of S, S^c, is then taken:

$$S^c = \text{complement}\ (S) = \{\text{all points}\ P \in I | \mathit{not}\ P \in S\}.$$

S^c is then traced as in the above methods (Fig. 13.4).

Sometimes parts of cells must be traced manually. Direct unassisted tracing gives a "jittery" line that, even when dilated, eroded and splined at completion, is often not what the user intended. A better result is obtained when the processing is continuously applied during the tracing process – giving the user continuous feedback. It is much easier to "back up" a few pixels than to retrace the entire object. To do this, the trace in progress must be temporarily closed. The current trace position is connected to the starting point and the resulting closed region is dilated, eroded and β-splined just as with the automatic tracing methods described above. The resulting trace (minus the line that connects the current trace position with the starting point) is continuously drawn. The result is that the user has a much greater sense of outline stability than is otherwise possible. In complex tracing (such as dendritic processes) this "healing" process begins anew whenever the user lets go of the mouse button and begins with a new starting position.

13.5 Reconstructing 3D Faceted Images and Internal Architecture

The β-spline representations of optical sections are stacked in the *z*-axis to generate a contour map. This is done independently for the cell perimeter, the cell nucleus, the vacuole and the pseudopod of a migrating cell. These stacked contours are then wrapped to generate a faceted image which may be transparent or nontransparent. The details of the wrapping method we use can be found in Soll et al. (2000). In brief, β-spline representations of outlined perimeters of the cell surface are stacked, generating a 3D contour map. A triangular net is projected over the top contours and a second is projected in the reverse direction over the bottom contours. The nets are trimmed at their junction and joined by triangular sections in the intervening spaces. The cell surface, nucleus and pseudopods are independently wrapped and can be viewed together (color-coded) (Fig. 13.5a, b) or independently (Fig. 13.5c–d) over time. Together or individually, the reconstructions can be viewed from any angle. This method has been applied to the study of vacuolar inheritance (Barelle et al. 2003) and mating (Lockhart et al. 2003) in the pathogenic yeast *Candida albicans*, and cytokinesis in fibroblasts (Li et al. 2003).

13.6 Quantitative Analyses of Behavior

Converting images to 3D mathematical models facilitates quantitation of dynamic shape and motility parameters. 2D-DIAS provides over 30 parameters computed at intervals as short as 1/30th of a second (Soll 1995; Soll and Voss 1998; Wessels et al. 2006). Dynamic shape parameters are computed from the contour of the cell and motility parameters from either the dynamics of the centroid or contour (Soll 1995; Soll and Voss 1998; Wessels et al. 2006). Generating a 3D reconstruction increases the number of parameters to well over 100. These parameters are selected from a menu and can be presented in tabular form or plotted as a function of time. Time plots can be smoothed by applying Tukey windows and multiple parameters can be coplotted. Correlations of changes in different parameters can be quite revealing. For instance, plotting the instantaneous velocity of migrating *Dictyostelium* amoebae (Wessels et al. 1994), human polymorphonuclear leukocytes (Murray et al. 1992), human T cells (Sylwester et al. 1995), and sperm from *Ascaris suum* (Royal et al. 1995) and *Caenorhabditis elegans* (Royal et al. 1997) revealed a velocity cycle in each, in which cells sped up, then slowed down, every 1–2 min in the case of *Dictyostelium* and human cells, or 0.5 min in the sperm. When directional change was coplotted with velocity as a function of time, it was immediately evident that each time the cell slowed down, its frequency of turning increased (Murray et al. 1992). Subsequently, it was discovered that each time cells slowed down, the front end of the cell extended in the *z*-axis (Wessels et al. 1994). Indeed, plots of height

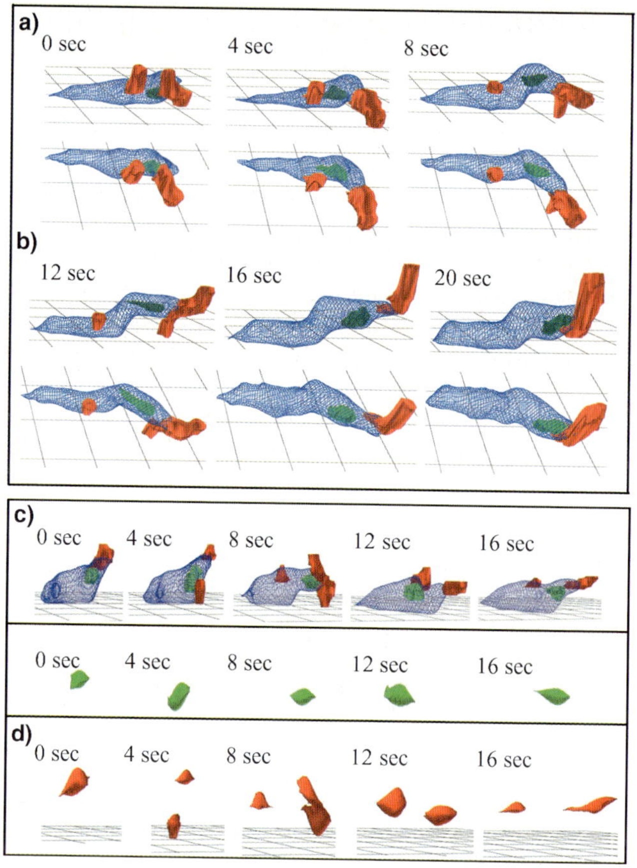

Fig. 13.5 3D-DIAS reconstructions of a translocating *D. discoideum* amoeba. **a** Complete cell reconstruction with cell surface a *transparent net*, nucleus *nontransparent green* and pseudopod *nontransparent red*, viewed from a top angle. **b** A second complete cell reconstruction. **c** Nuclei alone reconstructed for the cell in **b**. **d** Pseudopods alone reconstructed for the cell in **b**

showed peaks coinciding with velocity troughs. In addition, the dynamics of a portion of the cell body, such as the uropod, can be windowed and analyzed over time. Since the nucleus and pseudopod are traced independently of the entire cell surface, they can be independently reconstructed and motion-analyzed (Fig. 13.5c, d, respectively). Their behavior can be compared with that of the main cell body. The quantitation of 3D parameters during cell motility has been woefully underutilized. It is clear that many of the dynamics of cellular behavior can be described and used to model cellular translocation in 3D by analyzing the time plots of 3D-DIAS parameters. For cells like *Dictyostelium discoideum* amoebae and human polymorphonuclear leukocytes, 2D reconstructions are inadequate.

13.7 3D-DIASemb

Upon development of a functional 3D-DIAS system, it became clear that by modifying the program, it could be used to reconstruct and motion-analyze every cell and nucleus in an early developing embryo. Early *C. elegans* embryogenesis was selected as the experimental system used to generate 3D-DIASemb because the cells in the embryo are relatively transparent, embryogenesis is rapid, and embryos can be readily stabilized. Hence, the developing embryo could be optically sectioned with DIC optics. Two major hurdles first had to be overcome in customizing the 3D-DIAS program (Heid et al. 2002). First, the number of independent trace slots had to be increased from ten to 10,000. The solution involved compressing 1,024 traces at a time and bundling them into individual resource forks, the mini-database in each file in the Macintosh. Second, the video file had to be compressed, since 1 h of embryo recording, in which 60 optical sections are collected every 5 s, results in 43,200 optical sections. The scheme, based on the discrete cosine transform, achieved 20:1 compression and resulted in file sizes between 600 and 700 MB for 1 h of image capture. Finally, 3D-DIAS had to be reconfigured to handle the increase in the number of trace slots, the 3D viewer had to be reconfigured for reconstructing multicellular complexes (i.e., embryos), and finally algorithms had to be developed for rapidly playing back complex faceted images of embryos that were compatible with the 3D workstation employed. While the software in 3D-DIAS outlined the outer surface of the entire embryo automatically, it did not accurately outline in DIC images the in-focus edges of individual cells and nuclei within the developing embryo. Manual outlining of the latter two was, therefore, necessary to reconstruct in three dimensions the *C. elegans* embryo through the 28-cell stage (Heid et al. 2002). Recently, we analyzed images of embryos vitally stained with a membrane dye, provided to us by W. Mohler of the University of Connecticut School of Medicine. The laser scanning confocal images proved amenable to automatic outlining, as demonstrated in Fig. 13.1. New software also has proven effective in automatically outlining nuclei, as demonstrated in Fig. 13.3.

In 3D-DIASemb, the facets of each cell body and nucleus are individually stored in a facet data file. The capacity of 3D-DIASemb is 10,000 facets per reconstruction. Information on each cell and nucleus is stored at the resource fork of each facet file. The multiple cell and nuclear reconstructions are then merged.

While 3D-DIASemb is far from a complete, automated system, its present output is still quite remarkable. In Fig. 13.6, reconstructions are presented of a single, live, developing embryo from the four-cell to eight-cell stage, in this case between 28.5 and 38.5 min of embryogenesis. The embryo is viewed at a 20° angle (Fig. 13.6a) and at a 70° angle (Fig. 13.6b). The surface of each cell in the embryo is presented as a transparent blue faceted net. Each surface of each nucleus is presented in nontransparent red during interphase or nontransparent yellow when undergoing mitosis. Note that the exact relationships of each cell to the other, and the positions and division phases of each nucleus can be assessed. In addition, the timing of events can be ascertained to within 5-s intervals if desired. The sequence presented for the

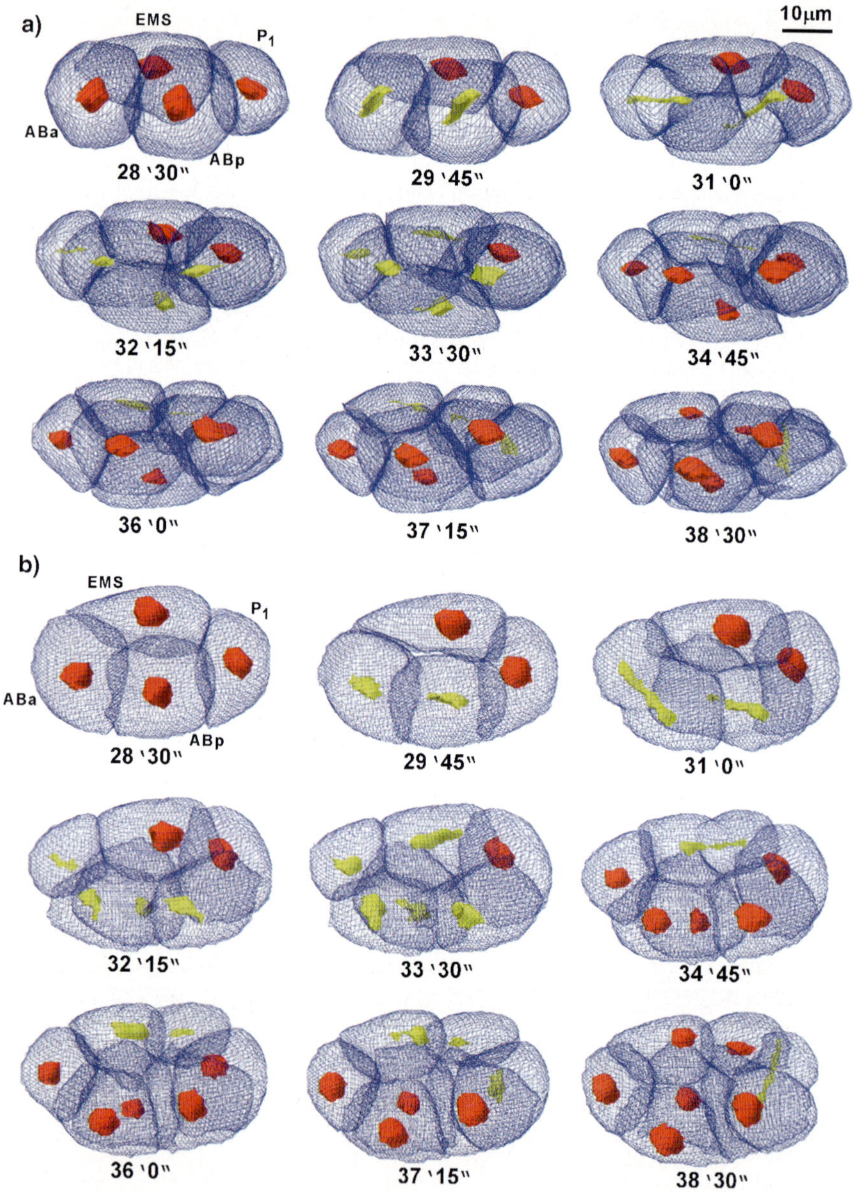

Fig. 13.6 3D-DIASemb reconstruction of a developing embryo at 15-s intervals between the four- and eight-cell stages. Cell surfaces are presented as *blue faceted images*, interphase nuclei as *nontransparent red faceted images* and mitosing nuclei as *nontransparent yellow faceted images*. The views in **a** are at 20° and in **b** at 70° angles. The initial cell types are noted at 28.5 min

embryo between 28.5 and 38.5 min includes only select reconstructions from those generated every 5 s during this period. A total of 1,152 reconstructions were generated through the 28-cell stage that can be analyzed for dynamics.

Because each reconstructed entity (the surface and nucleus of each cell) is individually reconstructed, one can also reconstruct cells in a single lineage or only nuclei. In Fig. 13.7, nuclei alone are reconstructed between the two-cell and 28-cell stage. Note that the positions, relationships and shapes of the nuclei are accurate. Note also that nuclear lineages can be color-coded and followed, in this case malignant schwannoma derived nuclei. The capacity to isolate a cell or nucleus and follow its lineage reveals events that cannot be ascertained in full reconstructions. For instance, by reconstructing only the C cell lineage, it was revealed that this cell extends itself around another cell just prior to nuclear division of the latter (Heid et al. 2002). Because one can plot 3D parameters of each cell as a function of time, detailed temporal and spatial relationships can be identified, for instance, the exact time of cytokinesis of each cell, unequal divisions and changes in nuclear position. In this case, the resolution is the interval of 5 s between reconstructions.

3D-DIASemb can also map cytoplasmic flow in two dimensions by generating vector flow plots of cytoplasmic particles (Heid et al. 2002). In this method, vector arrows are constructed that indicate the direction and extent of movement for hundreds of particles in the cytoplasm. This feature has been described but it has not been fully utilized to map the flow of cytoplasm during cell divisions. Software to extend vector flow plots to three dimensions should eventually be developed. 3D-DIASemb can also be used on later embryos, in which case windowed or boxed areas of the embryo can be analyzed. It can also be used to study cellular events in tissues. It should also be feasible to study in three dimensions how pathogens invade tissue.

3D-DIASemb has been used to analyze cell movements and extensions during gastrulation in zebrafish embryos. In Fig. 13.8, a series of faceted 3D reconstructions of hypoblast cells (light gray) and overlying epiblast cells (dark gray) is presented at 15° and 90° for a wild-type zebrafish embryo at the onset of gastrulation. Blastomeres in the embryo were fluorescently labeled and scans acquired using a Bio-Rad Radiance 2000 multiphoton confocal microscope as described in Ulrich et al. (2003). It can be seen that cells in the epiblast and hypoblast move past each other in opposing directions in the wild-type zebrafish embryo. Furthermore, it was determined that hypoblast cells in *slb*-mutant embryos were selectively defective in their movements during this stage of development (Ulrich et al. 2003).

13.8 Resolving Filopodia

If a subcellular component can be visualized over time in or on a living cell by any microscopic method available, it can be digitized, reconstructed at time intervals and motion-analyzed. One of the most challenging cellular components we encountered was the filopodia. Most, if not all, motile cells form these hair-like projections

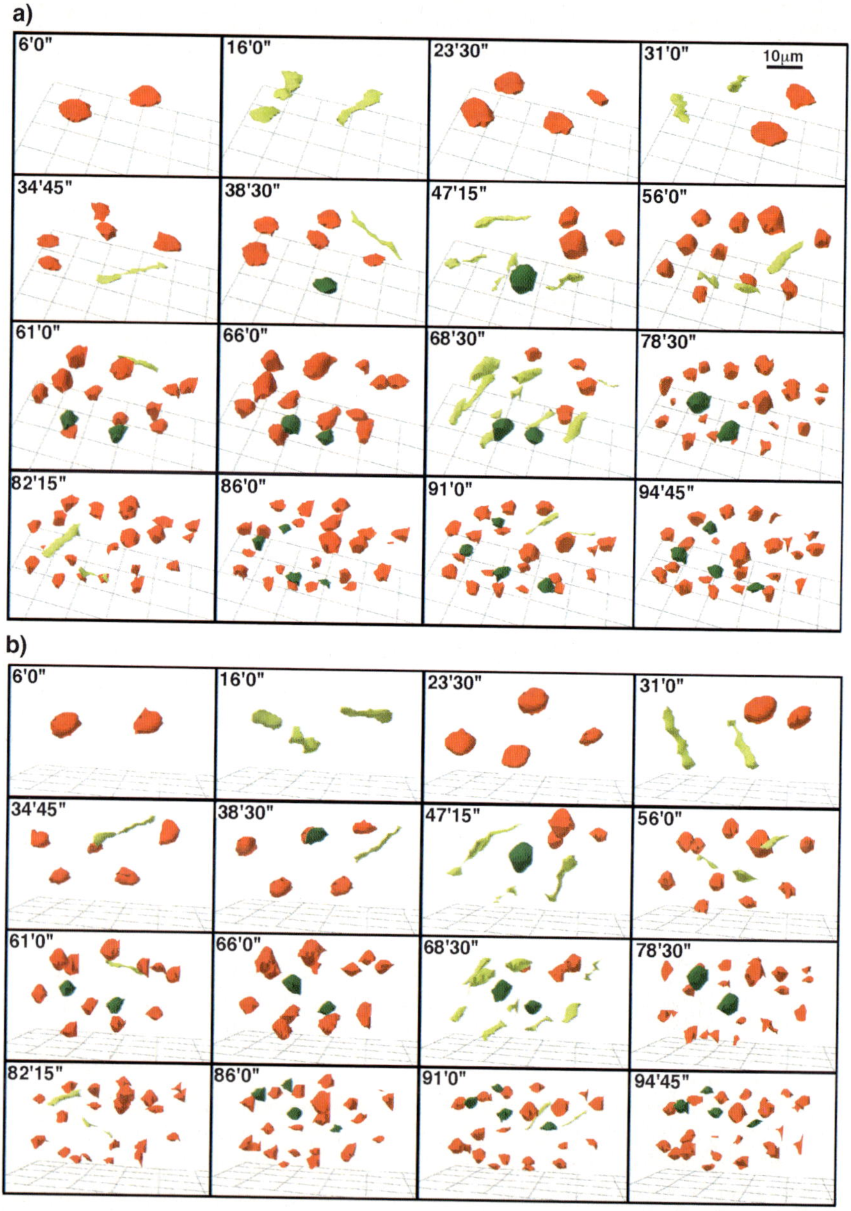

Fig. 13.7 3D-DIASemb allows presentation of nuclei alone. **a, b** Two views at different angles (see grid for references). Interphase nuclei are color-coded *red*, mitosing nuclei *yellow* and malignant schwannoma derived nuclei *green*

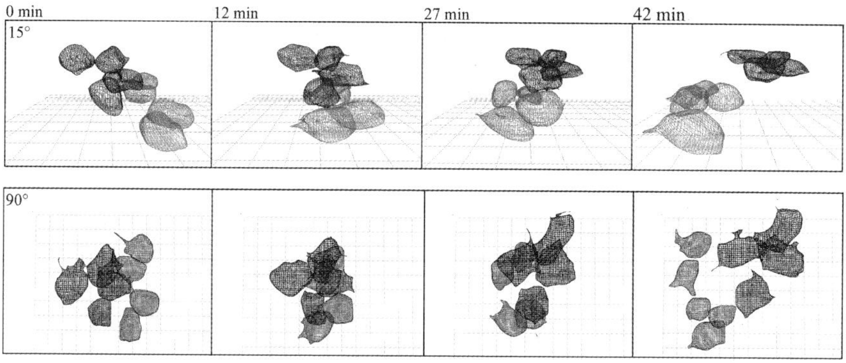

Fig. 13.8 3D-DIAS was used to reconstruct over time the migration of hypoblast cells (*light gray*) and epiblast cells (*dark gray*) in zebrafish embryos. Cells were fluorescently labeled (Ulrich et al. 2003)

that can extend over long distances. These projections are barely visible under normal light microscope settings since they are commonly 0.02–0.1 μm in diameter and can reach lengths equivalent to the cell diameter, or even greater, for cells 15-μm long and 6-μm wide. The 'hairs' are enriched in F-actin, which extends along the filopodia as an unbranched bundle, which emanates from the cross-linked cortical F-actin network (Lewis and Bridgmen 1992; Small et al. 2002; Svitkina and Borisy 1999; Svitkina et al. 2003). Because they are so difficult to see, they have been largely neglected in most cellular studies. For most cell types, their function(s) remains elusive. When a cell, such as a *D. discoideum* amoeba or a human polymorphonuclear leukocyte, is optically sectioned using DIC optics, filopodia are visible; hence, they can be manually outlined (Heid et al. 2005). For 3D reconstruction, filopodia are traced as a series of short line segments. Because of the difficulty of identifying all segments of each filopodia, a customized program has been developed to dilate the segments in order to achieve continuity, and then to erode the reconstruction in order to restore the outline to a realistic, but not necessarily accurate, width. The filopodia are then color-coded to distinguish them from other cell components (Fig. 13.9). In the first application of this software, the dynamics of filopodial extension, retraction and localization were mapped in time as *D. discoideum* amoebae crawled in buffer or towards aggregation streams in the process of chemotaxis. The average number and length of filopodia were demonstrated to differ in myosin heavy chain phosphorylation mutants. This behavioral study revealed that rather than a tactile role, filopodia appear to play a role in stabilizing pseudopodia to the substratum in *D. discoideum*. There have also been suggestions that filopodia play a role in sensing a cell's environment, either physically or chemotactically (Niell et al. 2004; Gallo and Letourneau 2004; De Joussineau et al. 2003; Koleske 2003; Lee and Goldstein 2003; Dalby et al. 2004). The availability of 3D-DIASemb should help reveal the exact role these ubiquitous structures play.

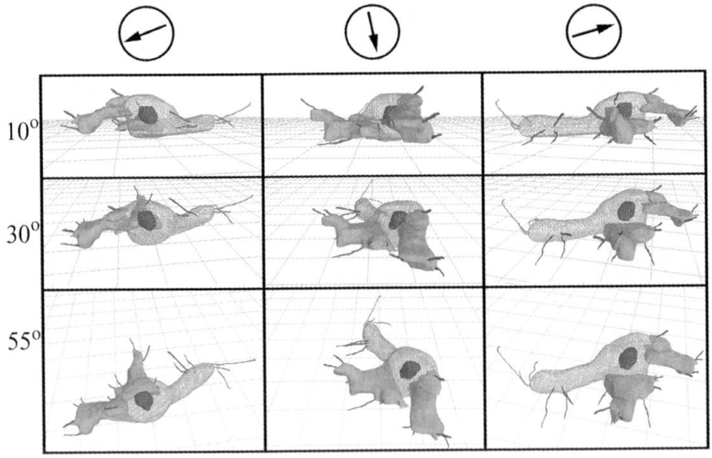

Fig. 13.9 Software has been developed that reconstructs filopodia on the surface of a migrating cell. The cell surface is presented as a transparent gray net, the nucleus as nontransparent dark gray, and the pseudopodia and filopodia as nontransparent gray. The orientation of the cell is diagrammed at the top by an *arrow* and the tilt is noted in degrees on the *left*

13.9 The Combined Use of LSCM and 3D-DIAS

The next challenge in the application of 3D-DIAS will be to combine it with laser scanning confocal microscopy (LSCM) in order to map in space and time the changes in cytoskeletal organization during cellular translocation and chemotaxis. There are two ways in which this can be accomplished. First, one can combine DIC and LSCM reconstructions of cells expressing a fluorescently tagged protein (e.g., green fluorescent protein tagged myosin heavy chain). There are two limitations to this approach. First, the LSCM scans are slow for the necessary resolution, so the artifact due to cellular movement will usually be too great for accurate reconstructions. Second, the DIC images obtained simultaneously with fluorescent images in laser scanning confocal microscopes do not provide the necessary detail for 3D-DIAS reconstructions of pseudopods. There are compromises that can facilitate reconstructions and the limits of these compromises are now being explored to optimize this approach. For example, one can take fewer LSCM scans or decrease the pixel per line resolution to increase acquisition speed, but we have found that the loss in 3D detail is unacceptable under these conditions. The second way 3D-DIAS and LSCM can be combined to map the localization of the cytoskeleton is to take an end-point "snapshot." In this approach, a living cell is reconstructed at short time intervals in two dimensions or three dimensions, and when it performs a particular behavior, it is fixed. The fixed cell is then optically sectioned using DIC optics for high-resolution reconstruction of cellular compartments, and then optically sectioned by LSCM for optimum resolution of the fluorescently tagged molecule.

The DIC reconstruction and LSCM reconstruction are then merged. This latter scheme forces the deduction of temporal events from different fixed cells, but provides the necessary detail for refined intracellular localization.

13.10 Reasons for 3D Dynamic Image Reconstruction Analysis

It is surprising how scientists continue to view cell behavior as a 2D phenomenon. There are several reasons. The first reason for this is the way cells are viewed microscopically. Whether from above or below, we tend to focus on the plane that gives us the largest cell image, or we employ optics and light that provide a thick focal plane and hence a full view of a cell that lacks depth and is virtually 2D. The second reason is the increased level of difficulty in analyzing a living, crawling cell in three dimensions. This is best demonstrated in a comparison of 2D-DIAS and 3D-DIAS applications. In 2D-DIAS, the behavior of a cell at a single focal plane is continuously recorded. Collection of a 2D movie in this manner every 2 s for 10 min results in 300 frames of images that are readily outlined and motion-analyzed. For 3D-DIAS, one must have a stepper motor synchronized to a character generator, shutter or other mechanism that permits the software to subsequently distinguish between up and down scans. One must then use complex software for 3D reconstruction. For 3D reconstructions, if one reconstructs a cell every 2 s for 10 min, one accumulates 9,000 frames. That is roughly 30 times the number for a comparable 2D analysis. The third reason we view cell behavior as 2D stems from the natural versus experimental substrate. While we examine cell migration on a flat plastic or glass surface in vitro, natural settings frequently include a 3D substrate.

Studies of the cell behavior of *D. discoideum* mutants, using side views of cells or 3D-DIAS analysis, have revealed how important 3D information can be. By microscopically viewing cells sideways on flat surfaces by rotating preparations 90°, Shelden and Knecht (1996) found that cells devoid of myosin II heavy chain were unable to undergo normal *z*-axis extension. Previous analyses never dealt with the 3D consequences of the mutation (Wessels et al. 1988b). When Shutt et al (1995b) used 2D-DIAS to reconstruct and analyze cell behavior, they found no significant defects in *D. discoideum* amoebae lacking ponticulin (Luna et al. 1990) but when they used 3D-DIAS, they found a 3D defect in pseudopod stability. In ponticulin-minus cells, lateral pseudopods that formed off the substratum abnormally floated across the surface of cells, whereas in wild-type cells, lateral pseudopods formed off the substratum were positionally fixed in relation to the substratum and did not migrate. Wessels et al. (1994) using 3D-DIAS described for the first time differences in the fate of pseudopods formed on versus off the substratum. Wessels et al. (1996) further demonstrated that deletion of a myosin I gene affected the frequency of pseudopods formed on but not off the substratum. These are just a few examples demonstrating the importance of 3D analyses. An immediate goal of mutant analysis is to first define the behavioral roles of genes involved in motility and chemotaxis. For the cytoskeletal proteins, most deletions are not lethal, and in the majority of cases, casual or nonquantitative observations suggest no aberrant consequence. However, computer-assisted 2D, and in a growing number of cases computer-assisted 3D, reconstruction has revealed abnormalities in shape and/or translocation and/or chemotaxis in these mutants. Most molecules associated with shape, translocation, pseudopod dynamics and chemotaxis are not essential for the basic behavior, but are necessary for fine-tuning the behavior. Most of these mutants exhibit a decrease in the efficiency of the behavior. It is therefore essential that in analyzing cell behavior associated with a mutation, neoplasia, environmental change or drug treatment, one obtains an exact, quantitative behavioral phenotype, and that may require computer-assisted 3D reconstruction and quantitative motion analysis.

Acknowledgements The most recent work described in this review article was supported by NIH grant HD-18577 and the Developmental Studies Hybridoma Bank at Iowa.

References

Barelle C, Bohula E, Kron S, Wessels D, Soll DR, Shafer A, Brown A, Gow NAR (2003) Asynchronous cell cycle and asymmetric vacuolar inheritance in true hyphae of *Candida albicans*. Eukaryot Cell 2:398–410

Barsky BA, Beatty JC (1983) Local control of bias and tension in beta splines. ACM Trans Graph 2:109–134

Dalby MJ, Gadegaard N, Riehle MO, Wilkinson CD, Curtis AS (2004) Investigating filopodia sensing using arrays of defined nano-pits down to 35 nm diameter in size. Int J Biochem Cell Biol 36: 2015–2025

De Joussineau C, Soule J, Martin M, Anguille C, Montcourrier P, Alexandre D (2003) Delta-promoted filopodia mediate long-range lateral inhibition in Drosophila. Nature 426:555–559

Gallo G, Letourneau PC (2004) Regulation of growth cone actin filaments by guidance cues. J Neurobiol 58:92–102

Heid PJ, Geiger J, Wessels D, Voss E, Soll DR (2005) Computer-assisted analysis of filopod formation and the role of myosin II heavy chain phosphorylation in *Dictyostelium*. J Cell Sci 118:2225–2237

Heid P, Voss E, Soll DR (2002) 3D-DIASemb: a computer-assisted system for reconstructing and motion analyzing in 4D every cell and nucleus in a developing embryo. Develop Biol 245:329–347

Koleske AJ (2003) Do filopodia enable the growth cone to find its way? Sci STKE 183:20

Lee JY, Goldstein B (2003) Mechanisms of cell positioning during *C. elegans* gastrulation. Development 130:307–320

Lewis AK, Bridgman PC (1992) Nerve growth cone lamellipodia contain two populations of actin filaments that differ in organization and polarity. J Cell Biol 119:1219–1243

Li Y, Wessels D, Wang T, Lin JL-C, Soll DR, Lin JJ-C (2003) Regulation of caldesmon activity by Cdc2 kinase plays an important role in maintaining membrane cortex integrity during cell division. Cell Mol Life Sci 60:198–211

Lockhart S, Daniels K, Zhao R, Wessels D, Soll DR (2003) Cell biology of mating in *Candida albicans*. Eukaryot Cell 2:49–61

Luna E, Wuestehube J, Chia C, Shariff A, Hitt A, Ingalls M (1990) Ponticulin, a developmentally regulated plasma membrane glycoprotein, mediates actin binding and nucleation. Dev Genet 11:354–361

Murray J, Vawter-Hugart H, Voss E, Soll DR (1992) Three-dimensional motility cycle in leukocytes. Cell Motil Cytoskel 22:211–223

Niell CM, Meyer MP, Smith SJ (2004) *In vivo* imaging of synapse formation on a growing dendritic arbor. Nat Neurosci 7:254–260

Pavlidis T (1998) Motion picture restoration. Springer, New York, pp 13–45

Royal D, Royal M, Italiano J, Roberts T, Soll DR (1995) A computer-assisted analysis of *Ascaris* sperm motility and MSP fiber dynamics. Cell Motil Cytoskel 22:241–253

Royal D, Royal M, Wessels D, L'Hernault S, Soll DR (1997) Quantitative analysis of *Caenorhabditis elegans* sperm motility and how it is affected by mutants *spe11* and *unc54*. Cell Motil Cytoskel 37:98–110

Russ JC (1992) The image processing handbook. CRC, London, pp 294–299

Schnabel R, Hutter H, Moerman D, Schnabel H (1997) Assessing normal embryogenesis in *Caenorhabditis elegans* using a 4D microscope: variability of development and regional specification. Dev Biol 184:234–265

Sheldon E, Knecht D (1996) *Dictyostelium* cell shape generation requires myosin II. Cell Motil Cytoskel 35:59–67

Shutt D. Stapleton J, Kennedy R, Soll DR (1995a) HIV-induced syncytia of peripheral blood cell cultures crawl by extending giant pseudopods. Cell Immun 166:261–274

Shutt D, Wessels D, Wagenknecht K, Chandrasekhar A, Hitt A, Luna E, Soll DR (1995b) Ponticulin plays a role in the positional stabilization of pseudopods. J Cell Biol 131:1495–1506

Small JV, Stradal T, Vignal E, Rottner K (2002) The lamellipodium: where motility begins. Trends Cell Biol 12:112–120

Soll DR (1995) The use of computers in understanding how animal cells crawl. Int Rev Cytol 163:43–104

Soll DR (1988) "DMS", a computer-assisted system for quantitating motility, the dynamics of cytoplasmic flow and pseudopod formation: its application to *Dictyostelium* chemotaxis. Cell Motil Cytoskel 10(Suppl):91–106

Soll DR (1999) Computer-assisted three-dimensional reconstruction and motion analysis of living, crawling cells. Comput Med Imaging Graph 23:3–14

Soll DR, Voss E (1998) Two and three dimensional computer systems for analyzing how cells crawl. In: Soll DR, Wessels D (eds) Motion analysis of living cells. Wiley, New York, pp 25–52

Soll DR, Voss E, Varnum-Finney B, Wessels D (1988) "Dynamic Morphology System": a method for quantitating changes in shape, pseudopod formation and motion in normal mutant amoebae of *Dictyostelium discoideum*. J Cell Biochem 37:177–192

Soll DR, Voss E, Johnson O, Wessels DJ (2000) Three-dimensional reconstruction and motion analysis of living crawling cells. Scanning 22:249–257

Svitkina TM, Borisy GG (1999) Progress in protrusion: the tell-tale scar. Trends Biochem Sci 24:432–436

Svitkina TM, Bulanova EA, Chaga OY, Vignjevic DM, Kojima S-I, Vasiliev JM, Borisy GG (2003) Mechanism of filopodia initiation by reorganization of a dendritic network. J Cell Biol 160:409–421

Sylwester A, Shutt D, Wessels D, Stapleton JT, Stites J, Kennedy R, Soll DR (1995) T cells and HIV-induced T cell syncytia exhibit the same motility cycle. J Leukoc Biol 57:643–650

Thomas CF, White JG (1998) Four-dimensional imaging: the exploration of space and time. Trends Biotechnol 16:175–182

Ulrich F, Concha M, Heid P, Voss E, Witzel S, Roehl H, Tada M, Wilson S, Adams R, Soll DR, Heisenberg C-P (2003) Slb/Wnt11 controls hypoblast cell migration and morphogenesis at the onset of zebrafish gastrulation. Dev 130:5375–5384

Wessels D, Vawter-Hugart H, Murray J, Soll DR (1994) Three dimensional dynamics of pseudopod formation and the regulation of turning during the motility cycle of *Dictyostelium*. Cell Motil Cytoskel 27:1–12

Wessels D, Titus M, Soll DR (1996) A *Dictyostelium* myosin I plays a crucial role in regulating the frequency of pseudopods formed on the substratum. Cell Motil Cytoskel 33:64–79

Wessels D, Voss E, Von Bergen N, Burns R, Stites J, Soll DR (1998a) A computer-assisted system for reconstructing and interpreting the dynamic three-dimensional relationships of the outer surface, nucleus and pseudopods of crawling cells. Cell Motil Cytoskel 41:225–246
Wessels D, Soll DR, Knecht D, Loomis WF, DeLozanne A, Spudich J (1988b) Cell motility and chemotaxis in *Dictyostelium* amoebae lacking myosin heavy chain. Develop Biol 128:164–177
Wessels D, Kuhl S, Soll DR (2006) Application of 2D and 3D DIAS to motion analysis of live cells in transmission and confocal microscopy imaging. In: Eichinger L, Rivero F (eds) Methods in molecular biology. Humana, Totowa, pp 261–279

14 High-Throughput/High-Content Automated Image Acquisition and Analysis

Gabriele Gradl, Chris Hinnah, Achim Kirsch, Jürgen Müller, Dana Nojima, and Julian Wölcke

Abstract In this chapter we describe why and how automated imaging of cellular events was established in the pharmaceutical and biotech industry with the goal of discovering new drugs and approaches to fight against disease. There is increasing demand for such methods already in use today predominantly in the areas of oncology, neurology, in vitro toxicology and immunology and they are termed "high-content screening." We describe how high-throughput/high-content screening is approached technically and which technologies are relevant for improving and expanding its use. We focus on the application of confocal imaging in high-throughput screening, its advantages over wide-field microscopy and how the speed necessary to investigate processes in live cells with high throughput and for several in parallel color is achieved. We also address the specific requirements for image analysis software and how the ultrafast data processing necessary for on-line image analysis in a screening campaign can be realized. Last but not least we provide examples for high-content screening assays.

14.1 The Driving Forces for High-Throughput/High-Content Automated Imaging

There is an increasing demand for automated microscopy in high-content screening (HCS) – a term generally understood to mean the analysis of cellular responses in a high number of different samples by automated (fluorescence) microscopy, by multiplexing and multiparametric analysis and most of the time by subcellular imaging at a high resolution (Comley and Fox 2004). The greatest interest for HCS today is in oncology, neurology, in vitro toxicology and immunology. While HCS assays already have widespread use in target identification, secondary screening, lead optimization and compound profiling, growth areas of the future are primary screening and especially ADME Tox/preclinical research (Comley 2005).

S.L. Shorte and F. Frischknecht (eds.), *Imaging Cellular and Molecular Biological Functions.*

High-throughput screening today is generally perceived as analyzing the effect of 100,000 or more compounds in 24 h in a given assay with a single data point per compound. "HCS is used in a different context with an emphasis on higher information content for a single data point than just a single value. Such multiparametric analysis is essential, e.g., when screening against a set of targets at the same time. It is also required for multiplexed readout of several cellular functions in order to provide a more comprehensive picture of drug activity and action, in pathway screening and also in screening for unwanted side effects.

The desired throughput of high-content cellular assays varies. Depending on the application, a few or a high number of data points per compound or reagent under investigation are required. For example, hits which are generated in a primary non-high-content screen are tested in triplicate and in five to seven concentrations generating dose–response curves in order to rank the compounds as highly active, active, and low-activity compounds and false negatives or false positives. An ultra-high-throughput screen of one million compounds with a hit rate of 1.5% will generate 15,000 compounds for follow-up testing. Follow-up testing generating IC_{50} curves consisting of the said seven concentrations in triplicate means measurement of a minimum of 315,000 data points plus controls. Kinetic measurements might also be required. Measurement of a rare event in a cell population may require analyzing tens of thousands of cells. In order to cover the entire area of one well of a 96-well plate, 150 image fields have to be generated when using a ×20 lens and a camera chip size of 1,280×1,024 pixels. In general, HCS imaging assays have to handle between 10,000 and 100,000 data points per day, each comprising one or multiple images from a sample.

14.2 Confocal Imaging in High Throughput – The Principles Available

Whereas for some cellular assay applications wide-field imaging may be sufficient, adequate identification of cellular compartments such as plasma-membrane areas, lipid rafts, clathrin-coated pits, vesicles and intracellular organelles, endosomes, mitochondria, actin filaments, etc. requires an increased spatial resolution in the imaging technique as found in confocal imaging. One key advantage of confocal imaging versus wide-field imaging is the suppression of background light and thus removal of interference from out-of focus features. When high-throughput assays are the goal, background suppression is essential for homogeneous (mix and read) assays. The assay protocol and the steps required for the assay setup are less complex when excess reagents are used, e.g., a fluorescently labeled ligand or antibody does not have to be removed before measurement. In addition, liquid handling errors are reduced. Confocal laser scanning microscopy is therefore the ultimate method to obtain high-resolution imaging data but conventional laser scanning microscopes are by far too slow to achieve sufficiently high sample throughput, i.e., up to 100,000 analyses per day. In addition, the local laser light intensity is very high within the sample and is not well suited for live-cell imaging (Stephens and Allan 2003). Two other techniques have emerged which are used for achieving fast scanning rates: the line scanning approach and the Nipkow (pinhole) disc scanning approach (see also Chap. 10 by Kaestner and Lipp).

Line scanners illuminate the sample along a narrow line instead of a pointlike spot and image the illuminated region onto a linear array of detector elements, resulting in parallel acquisition of image signals along the line. The sample is scanned sequentially in the transverse direction. Today CCD cameras with 2,000–3,000 pixels are available for line scanners, enabling high lateral resolution. A typical scanner line width is 600 nm. This makes resolution in the direction of scanning comparable with that of confocal microscopy. However, crosstalk between adjacent pixels along the line of excitation is not suppressed, and lateral resolution along this axis is not increased. Since light from out-of-focus object planes can also reach the detector along beam trajectories through this line, the optical sectioning achieved by line scanners is compromised compared with point scanners. Such an approach is also called "semiconfocal" because confocal resolution is not reached in all three dimensions, but only in one. With line scanners, higher-magnification lenses have to be used in order to achieve the same spatial resolution as point-scanning confocal microscopes. The area of scanning can be easily increased by scanning more lines, partly compensating for the disadvantage of the smaller field of view, but at the cost of longer acquisition times.

Point scanning can be parallelized, in order to significantly reduce the scanning time for a complete image, by using a so-called Nipkow spinning disc. By projecting an array of points into the sample via an array of pinholes, and imaging the sample onto a camera through the same pinhole array, one can observe on the order of 1,000 points confocally in parallel. By arranging the pinhole array on a rotating disc, this grid of points is rapidly scanned through the field of view to obtain a complete confocal image. Crosstalk and out-of-focus contributions from the local neighborhood of each observed pixel are effectively masked in all spatial dimensions, resulting in image quality comparable with that of single-point laser scanning. Only for thick samples, when sample thickness approaches the lateral distance of the spots projected into the sample, some crosstalk over large distances becomes noticeable and reduces optical sectioning to some extent. (Sandison and Webb 1994). Illuminating the sample via a conventional Nipkow disc will, however, introduce a significant loss in illumination intensity. To achieve spot sizes close to the diffraction limit, and adequate separation between the spots, typical Nipkow discs have a sparse array of small pinholes, with a total transmission of 1–2% of the incident light, only. The transmission can be improved dramatically by combining the Nipkow pinhole disc with a microlens array. By precisely matching the lateral spacing and the focal plane of the microlenses with the pinhole array, and rotating both arrays in synchrony during scanning, one can selectively focus the incident light on the pinholes. Total transmission values of 40% are achieved for such microlens/pinhole arrays (Taanami et al. 2002; Wang et al. 2005). We have implemented parallel point scanning using such a microlens-enhanced Nipkow spinning disc in our high-throughput microscopy reader Opera™ (Fig. 14.1). This approach has

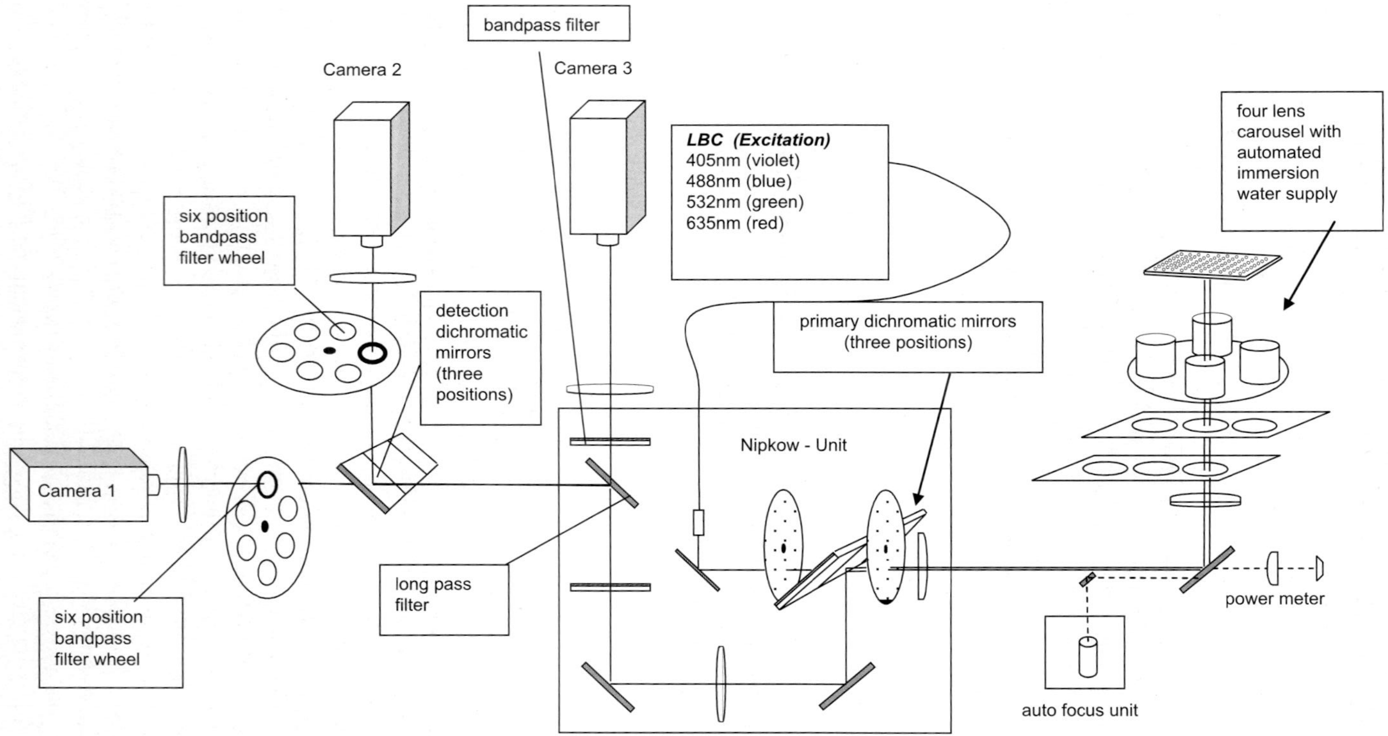

Fig. 14.1 Optical path of Opera (nonconfocal path not included)

several advantages. It is much faster than a single-point scanner, but with no significant compromise in resolution. There is less crosstalk between neighboring points in a point scanner compared with line scanning instruments and therefore better rejection of out-of-focus light as well as higher signal-to-background ratio. This in combination with the fast sample scanning rates makes fluorescence imaging using a point scanner much less prone to phototoxicity than any other method. The Nipkow disc is already renowned for inducing fewer photodamage effects than traditional confocal laser scanning microscopes. Altogether this approach is superior to line scanning owing to a higher optical quality and background suppression.

Another very powerful approach for high-resolution imaging is multiphoton excitation, parallelized approaches of which being facilitated by multifocal multiphoton microscopy (Straub et al. 2000; Hell and Andresen 2001). Multiphoton excitation is based on the simultaneous absorption of two or more low-energy photons by a fluorophore. Absorption of two photons of twice the wavelength of the absorption peak of the fluorophore leads to its efficient excitation. This nonlinear optical effect ensures that excitation is confined to the focal region, thus providing confocality without the need for a pinhole. The excitation in out-of-focus areas is nearly eliminated, reducing photodamage of the cell sample. It also has the potential to deeply penetrate into a tissue section. The need for expensive high-power lasers has to date prevented the widespread use of multiphoton microscopy in automated imaging, but this may change in the future.

14.3 Resolution and Sensitivity

Microscope-based systems use high-magnification lenses to image subcellular structures with high resolution. While the type of lens used has a dominant contribution to resolution and sensitivity, the intermediate optics of an instrument between the objective and the detector also influences the results. For example, for clathrin-coated pit quantification the ×20 lens with high numerical aperture in one instrument has been shown to be sufficient to deliver results of comparable quality to a ×40 lens in another instrument. However, not all experiments can be performed with the same objective lens and magnification and they should be chosen according to the assay used. While colocalization studies in some cases may require ×40 or even ×60 magnifications – at the expense of image field and the numbers of cells imaged in a single image field – other assays might be run with only ×10 magnification, thus increasing the number of cells imaged and improving the statistical significance of the results, especially in a high-throughput scenario. If the number of cells is not high enough, one can simply acquire several image fields within a well (Fig. 14.2).

In order to get high-resolution images the image has to be digitized using a pixel size at or below optical resolution (see also Chap. 1 by Hazelwood et al.). Today many high-resolution cameras are available; however, there is always a trade-off between resolution and sensitivity. If the light coming from the sample is very dim, the exposure time will be long. A possibility to reduce exposure time is binning, but

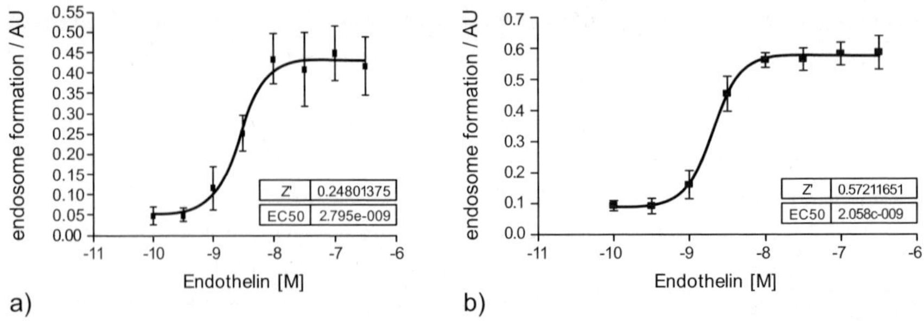

Fig. 14.2 Endothelin receptor activation in U2OS cells, measured upon addition of the ligand endothelin-1. Dose–response curves for measurements in one field per well (**a**) and in six fields per well (**b**) using a ×20 lens. The average number of cells for data acquisition is approximately 100 cells for **a** and 600 cells for **b**. The *Z′* value increases from 0.25 to 0.57 on increasing the number of fields measured per well sixfold.

this means that signals from binned pixels can no longer be differentiated, i.e., resolution will be decreased. We use Peltier-cooled CCD cameras for simultaneous detection. The resolution for cameras currently used for high-throughput screening is 1.3 megapixels and 12 bit with a CCD field of 1,360 (rows)×1,040 (lines) pixels; intensity values are represented by 12 bits, resulting in a dynamic range of 4,000:1. The pixel size for a ×20 objective lens is 0.35 μm, and for a ×40 lens it is 0.15 μm, which are both well below optical resolution. A 2×2 binning can still be used in order to increase the signal-to-noise ratio without a significant loss of resolution.

When it is the final goal to colocalize molecules and/or organelles, each labeled with a different chromophore, the optical effect of chromatic aberration is an issue. The reason is that refraction of light is dependent upon its wavelength, ultimately leading to focus differences of the different wavelengths. There are two types of chromatic aberration in microscopy: transverse and axial. The transverse aberration leads to an apparent difference in the magnification of the optical system. A similar difference can also occur owing to tolerances in the lenses used for relaying the image onto the different cameras. For this type of aberration we have included specific correction routines which compensate for this error with a precision only limited by the pixel resolution of the camera. The axial chromatic aberration of the microscope results in different *z*-positions of focal planes depending on the wavelength of the light being used. While some laser scanning microscopes correct for this effect by positioning separate pinholes at different distances from an ideal image plane, multispot point scanning systems and line scanning systems usually employ a single confocal diaphragm for all wavelengths used. Thus, the correction for the chromatic aberration with these systems cannot be performed by using pinholes in different positions. The *z*-shift of the focus caused by the chromatic aberration predominantly depends on the wavelengths of the light used and the objective lens. There are lenses available which correct for two or three different specific wavelengths. Should more colors be required a correction mechanism has to be employed. One method is to sequentially acquire each color and to reposition in z for each color according to the respective aberration. It is possible to furnish suitable calibration data for wavelength/lens combinations. One would first determine the z-shift between the focal positions of wavelengths used either by calculations or by adjustment measurements and toggle between these positions for the subsequent images at the different wavelengths. The z-displacement can be reduced to zero this way so that colocalization of two different probes can be measured with a precision down to the ultimate optical resolution. Opera can carry out such measurements without significant extension of the cycle time. Since four cameras are available there is no overhead time required for switching of filters. There is no overhead time for xy-table movement and autofocus measurements for the subsequent colors because there is no movement and autofocusing is not necessary. The exposure times for the colors add up but not the overhead time for positioning, which dominates the cycle time in most cases. Only at very low fluorescence intensities does an increase of the exposure time lead to a significant reduction in throughput.

The sensitivity limit of the detector used depends on the detector's quantum efficiency for registering incoming photons, the dark (background) signal produced internally in the detector, and the readout noise.

The quantum efficiency specifies what fraction of the photons impinging on the detector will actually cause a photoelectron to be registered and accumulated in the detector. This factor will directly influence the sensitivity of the detector. Noise contributions can be quantified as "photo electrons" (i.e., compared with the number of photoelectrons actually registered by the detector) or as digital "counts" (i.e., compared with the intensity scale delivered from the detector to the computer, after analogue-to-digital conversion). Noise results in a reduction of the dynamic range of the camera system, i.e., the range of separate intensity levels that can be resolved. The dynamic range is often expressed in "bit," i.e., the base 2 logarithm of the number of resolvable intensity levels. Noise components are (the values currently used for our camera are given in parentheses):

- Dark signal of the CCD chip, i.e., spurious electrons registered without the incidence of photons. While the average dark signal can be measured separately and subtracted from acquired images, the dark signal also introduces statistical fluctuations, which cannot be separated. For Opera's standard and ultra-high-sensitivity cameras this contribution can be neglected. In the standard camera this is less than 1.5 counts per second and can be neglected in the typical exposure times of less than 0.5 s (total dark noise 1.8 bit).
- Readout noise, associated with the transport of charges inside the detector matrix and their conversion into digital data, as specified by the camera manufacturer (six counts).
- Signal statistics of the fluorescence photons themselves, i.e., random variations of the number of photons counted when observing a weak fluorescence signal. This is a fundamental noise contribution inherent in the photon stream, rather than a technical limitation of a specific detector. Typically, the variation amounts to the square root of the number of photoelectrons detected.

14.4 Measurements

In multicolor principle, one can use either a continuous-spectrum light source such as a xenon arc lamp or lasers for confocal excitation in combination with a conventional Nipkow spinning disc. The continuous-spectrum light source provides more flexibility for the excitation wavelengths but a lamp-based solution has inherent limitations. Less than 5% of the illumination energy passes a standard pinhole disc. Large pinholes have to be used (e.g., 70 μm in diameter) in order to generate illumination intensity within the focal plane which is sufficient for a short exposure time. The resolution of the system is determined by the pinhole size; therefore, the use of small pinholes with ×20 and ×40 lenses is essential.

Microlens-enhanced Nipkow discs allow for the use of smaller pinholes, while still achieving high excitation intensities and hence short exposure times. However, they are more demanding with respect to the excitation sources: The light incident onto the microlens array must essentially be a plane wave to provide a grid of foci in the exact plane of the pinhole array. Lasers can provide radiation with this well-defined geometry as well as a precisely defined wavelength for highly selective excitation of fluorescent labels.

We use four solid-state lasers in Opera: 405, 488, 532 or 561 and 635 nm. Up to three of the laser lines can be used for simultaneous confocal excitation and detection using three CCD cameras in parallel. For addressing the UV range and other important wavelengths, such as the *Discosoma* sp. red fluorescent protein (DsRed; BD Biosciences) or Texas Red excitation, there is an additional nonconfocal light path employing a xenon arc lamp and an independent fourth CCD camera. The available spectral range of the cameras is 420–750 nm with quantum efficiency of greater than 40% and maximum quantum efficiency (more than 60%) at 500–600 nm. Exposure times can be as fast as 50 ms for kinetic measurements. We have implemented primary and camera dichromatic mirror sets to provide maximum flexibility for dual, triple and quadruple color simultaneous measurement (Fig. 14.1). Useful excitation combinations are, e.g., 488/635 nm or 405/532/635 nm for confocal measurement since one can find dye combinations with well-separated excitation and emission spectra for these excitation lines. The nonconfocal channel can be used in parallel, e.g., for a UV dye.

The individual channels of the four cameras have to be spatially aligned, which is done by software. The images acquired have a position offset because mechanical alignment of cameras is not accurate enough for color matching in the submicron range. An image analysis procedure creates skew and crop adjustment parameters for alignment of all subsequent images based on reference images of beads labeled with multiple fluorophores. Such alignment is dependent on the filter settings and is also performed individually for, e.g., each combination of dichroic mirrors. Other systematic errors of the optics and cameras, such as uneven brightness distribution, are also eliminated by reference measurements. In this case, dye solutions are used which provide a homogeneous fluorophore distribution in the entire field of view and allow for calibration of the intensity profile.

Fluorescence emission crosstalk can be an issue when performing simultaneous acquisition of multiple colors in order to achieve the highest throughput. Since this effect is sample-dependent it cannot be compensated via measurements on reference objects. The DNA-staining dyes 4′,6-diamidino-2-phenylindole (DAPI) and Hoechst 33342, for example, are prone to bleed through into other channels because of their very broad emission spectra. A workaround in cases where narrowing the emission filter bands does not solve the problem and correct quantification of signals is hampered by fluorescence crosstalk is to sacrifice some throughput by sequential acquisition including controls for background subtraction in the respective channels.

14.5 Where Is the Signal and How To Focus?

Confocal imaging systems record signals from within a single plane in a cell layer. Thickness of this layer depends on imaging magnification. For example, the use of a ×20 lens, 0.7 NA, will lead to fluorescence excitation and fluorescence emission collection from a section of 3 μm thickness. It is mandatory to adjust the focus for every position where measurements are taken because the thickness of a cell layer is only a few micrometers and is well below the mechanical tolerances of the well plates within the area of subsequent exposures. The first automated focus systems available used an image-based approach by taking a *z*-stack image sequence and determining the *z*-position of maximum intensity or contrast. This approach is very slow and not suitable for high throughput since it requires multiple-exposure, camera readout and image processing cycles – a full autofocus sequence can easily take an order of magnitude longer than the actual exposure time needed for the measurement. In addition, it may fail when, e.g., features or movements at well-defined object planes have to be resolved in an extended sample, which may even exhibit dominant structures in another object plane. The better option for achieving high speed and a well-defined observation plane is to determine in advance the imaging plane within the cell layer providing the best signal-to-noise ratio for a given assay – defined by its *z*-position relative to the well bottom – and adjust the focus to this distance automatically before each measurement. This can be achieved by an active autofocus system, which uses a dedicated light source and detector to observe a reflection from the well bottom. We have developed such a system employing an infrared laser and confocal optics. A diffraction-limited spot is projected onto the upper interface of the sample carrier's bottom glass or foil, and is observed via a matched confocal pinhole. At the correct *z*-position of the interface, the observed intensity is maximized. The autofocus beam is projected through the same optical system used for image acquisition. Hence, it measures the well bottom position directly at the imaging field of view, and with the full optical *z*-resolution provided by the objective currently in use. Autofocusing and adjusting the focal plane is part of each automated measurement after changing *xy*-position and takes less than 200 ms. Finding the optimal focal plane for a high-throughput measurement set is one task of assay development. One must be aware that it might not be the plane of highest fluorescence intensity or in which cells appear to be well focused visually (Fig. 14.3).

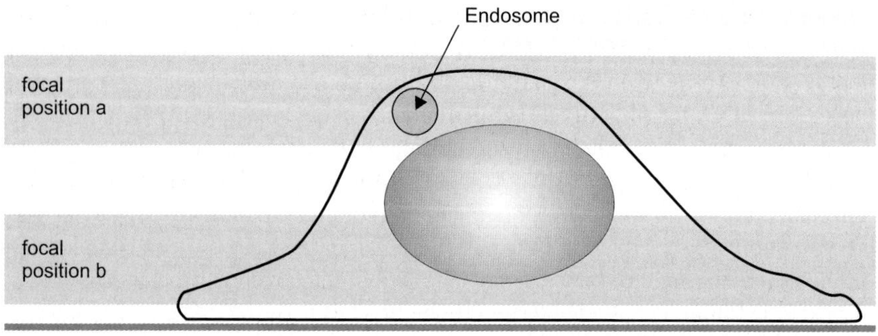

Fig. 14.3 Finding the focal plane in a receptor activation assay based on endosome formation. For *b* the cell appears to be well focused, e.g., when looking at fluorescent protein tagged nonactivated receptors, but most endosomes will be out of focus. For *a* the endosomes are focused and this focal position will provide a reliable readout.

14.6 Plates and Lenses

The first plate parameter to consider in a high-throughput assay setup is the plate format. While the format most commonly used for high-throughput screening is 384 wells, many cell assays are still run in a 96-well format – sometimes owing to the necessity to acquire images from a large number of cells in order to detect rare events, and sometimes just for the easy adaptation process from assay development to a screening situation. Miniaturization of the well format is of importance in high-throughput screening. The main drivers for this are compound and reagent savings but also the shorter time required per data point because of reduced overhead time for plate changing and movement from well to well for imaging. Homogeneous cell assays can easily be set up in 1,536-well plates at 2–3-μl assay volume.

A large number of different assay plates are available. These plates differ in their suitability for growing adherent cells in them and also in their optical quality which can influence their performance in an imaging reader drastically. Selecting the most suitable cell imaging assay plate requires finding a compromise between cell compatibility and optical resolution. Important aspects to take into account are the material of the plate bottom, the thickness of the plate bottom, the plate bottom geometry, and the flatness of the plate. Other plate properties such as surface treatment of the wells (e.g., polylysine coating), plate wall material or color (white, black, transparent) have not been found to influence the optical performance of a plate in a confocal setup. Glass is the gold standard as a plate bottom material for superior image quality. The best results for short exposure times and high resolution can be achieved with this material. However, without a coating very few cells are capable of attaching and spreading on glass surfaces and if they do, they come off in washing procedures more easily than they do in tissue culture plates. In addition, coated plates are expensive; therefore, glass-bottom plates are very rarely used for adherent cell screening. Still, they are a good choice when using nonadherent cells. Organic polymers such as polystyrene are much better suited for adherent cells. This material can be “tissue culture treated," i.e., modified physically to enhance cell attachment and spreading. Conventional plastic plates for cell culture are usually not well suited for imaging, especially when low-intensity fluorescence signals have to be analyzed. Their plate bottom thickness is 0.5 mm or more, leading to significant signal loss and the necessity to use long working distance objective lenses with low numerical apertures. Recently plates have become available with very thin and optically high grade plastic bottoms for cell culture. They are well suited for high-speed image acquisition using high numerical apereture lenses, but lack the mechanical stability to create a flat plate bottom in large wells (96-well plates).

It is evident that in order to achieve the highest throughput and to avoid artifacts one has to carefully select the optimal plate and lens combination for the assay. For the lens, the smallest features to be resolved and the number of cells to be imaged per sample define the magnification required. But not only magnification is important for high resolution and high speed. The numerical aperture and lens type are just as important. Air lenses lead to substantial reflection of the fluorescence light at the air/glass interface of the lens and the plastic/air interface at the plate bottom. Such reflection is greatly reduced with water immersion lenses at the water/glass and the water/plastic interface. For refractive indices n_1 for the lens or plate bottom glass (n_1~1.52 typically) and n_2 for the medium in the lens/sample gap (n_2=1 for air, n_2=1.33 for water), reflection losses amount to

$$R=(n_1-n_2)^2/(n_1+n_2)^2$$

for beams close to the optical axis. Losses become even larger for off-axis beams. As a consequence, exposures are longer for air lenses compared with water immersion lenses. In addition, total reflection limits the optical acceptance angle (numerical aperture) of air objective lenses. We have fitted Opera with an automated immersion water supply for water immersion lenses. The advantage of immersion objective lenses is a higher numerical aperture and therefore higher fluorescence detection efficiency and better lateral resolution when compared with air lenses. Another advantage of high numerical aperture lenses is the better rejection of out-of-focus light in confocal setups. This results in a thinner focal plane and hence better axial resolution when compared with lower numerical aperture systems.

14.7 Image Analysis

The tasks for the analysis of cellular images are manifold and their number is growing fast (Table 14.1). In most cases there are numerous readouts for a given analysis class and which one will work best in a screening scenario can be determined by using the Z' factor from the test data set. Not only does every assay scenario require a dedicated image analysis algorithm, but a modification of the assay, e.g., changing cell type (Fig. 14.4), magnification or schedule, may make adaptation of the algorithm necessary. Basic algorithms for tasks frequently used are helpful, but algorithm adaptation, algorithm combination and new algorithm development are an essential step of assay development.

Table 14.1 Some sample image analysis tasks of cellular imaging assays

Analysis class (detection of)	Application examples	Measurement parameters
Whole-cell fluorescence intensity	Cell viability	Esterase activity based on live-cell stains
	Cytotoxicity	Membrane permeability based on dead-cell stains
	Calcium flux	Ca^{2+} influx via activation of G-protein coupled receptors assessed with a calcium-sensitive dye
	Cell-membrane transporter assays	Transporter activity assessed by influx/efflux of a fluorogenic substance cleaved by intracellular esterases
	Kinases	Binding of fluorescently labeled antibodies to phosphorylated proteins
	Cell/substrate interaction	Cell spreading using live-cell stain
Nucleus	Transcription factor activation/translocation	FP-tagged proteins accumulating in the nucleus. Accumulation of nuclear proteins assessed with fluorescently labeled antibodies
	Apoptosis/cytotoxicity	Nuclear condensation assessed with DNA-binding dyes
Membrane	Receptor/ligand binding and competition assays	Binding of fluorescently labeled ligand to membrane receptor
	Receptor expression	Expression of FP-tagged membrane receptor
	Early apoptosis	Fluorescent annexin-V binding
Cytoplasm	Signal transduction events	Phosphorylation assessed by fluorescent antibody detection
	Fatty acid uptake	Uptake of fluorescently labeled fatty acids
Subcellular structure, e.g., single large spots	Translocation of FP-tagged receptors into single endosomal compartment	Receptor internalization
Subcellular structure, e.g., numerous small spots	Coated pits/vesicle formation after activation of G-protein coupled receptors	Recruitment of β-arrestin carrying a fluorescent tag. Internalization of receptors carrying a fluorescent tag.
		Detection of internalized receptors in endosomes, antibodies with pH sensitive tag
	Hepatosteatosis	Fluorescently labeled lipid droplets
Redistribution events	Cytoplasm-to-nucleus redistribution, FP-based or antibody-based	Transcription factor activation
	Cytoplasm-to-membrane redistribution	Signal-molecule recruitment
Nucleus and whole cell	Angiogenesis	Cell assembly assessed by actin staining plus nuclear dye
	Mitosis	Nuclear shape/intensity based on binding of specific probes to nuclei

(continued)

Table 14.1 (continued)

Analysis class (detection of)	Application examples	Measurement parameters
	Cytokinesis	Number of nuclei based on actin staining plus nuclear dye
Cytoskeleton structures	Cell division/cell cycle	Differentiation between cell cycle phases based on tubulin and nuclear staining
Cellular extensions	Neurite outgrowth	Recognition of axons and dendrites

FP fluorescent protein

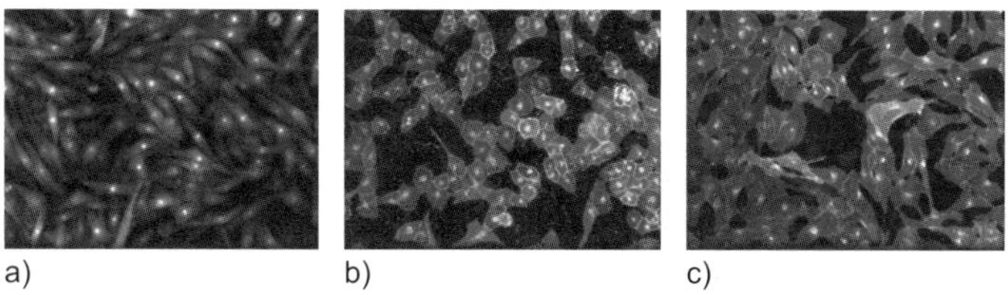

Fig. 14.4 Fluorescently tagged endothelin receptors, internalized into endosomes in CHO cells (**a**), HEK293 cells (**b**) and U2OS cells (**c**). Adaptation of an evaluation algorithm is useful in order to account for the difference in cell size, cell shape and clustering behavior

High-throughput image analysis relies on software with fast data processing capability. We have developed a script-based runtime system called Acapella™ which manages fully automated on-line analysis of images during the screening process. It is composed of a modular library of routines for algorithm development. A number of basic algorithms for many standard applications are already implemented. New algorithms – so-called scripts – can be generated by combining image analysis modules. All computation-intensive tasks are performed in such compiled modules. Modules implement image processing functionality (threshold calculation, image filters), data manipulation (statistical measures, numerical fitting), control structures (if/else structures, loops), and the interface for user interaction (input/output parameters). They can also implement the preliminary data interface, such as displaying an image showing an intermediate result. The script only directs the data flow and is generally not very time consuming itself.

In order to simplify the generation of new algorithms we clustered image analysis tasks which are used in many applications in "procedures." As examples of such generic image processing procedures we describe how single cells are identified and how regions of interest within these cells where signals occur can be detected.

The task of identification and analysis of single cells in a complex image is most reliably solved when a reference image is available for determining cell positions which is independent of assay conditions. A nuclear stain is well suited for such reference images and is often used for single-cell identification, even when the nucleus has no further meaning for the specific cellular event under analysis.

Images of nuclei are more suitable than images of other cell organelles in this respect as they have a very consistent and simple shape. There is usually only one nucleus per cell and the nuclei are also well separated from each other. Different cell types, staining procedures and cell constitution (passage number, cell cycle, etc.) result in different visual appearance of the nuclei. We developed a set of algorithms for identifying nuclei, depending on their visual appearance. Nuclei detection results depend on various image properties such as contrast between nuclei and background, contrast between nuclei and cytoplasm, spatial noise properties in the nuclear regions and in the background regions, etc. The result can be used as a search mask for detection of nuclear events.

Some cellular events are expected to occur within the nucleus, others in one of the other major cellular compartments, the cytosol or plasma membrane. Expression of a G-protein coupled receptor (GPCR), a receptor protein tyrosine kinase or ion channel proteins or binding of a ligand to such receptors are events localized typically in the plasma membrane. Other events will occur as traffic between one compartment and the other. Identification of such regions of interest is one way of analyzing these events. Our Cytoplasm Detection Library provides a set of routines based on nuclei detection and a cytoplasmic stain. For these routines the starting point is identification of the nucleus. The outer border of the cell is defined by an intensity threshold or by the cytoplasm of a neighboring cell. Besides relatively simple homogeneous staining of the whole cytoplasm, there are several structures within the cytoplasm area (e.g., vesicles, actin fibers, etc.) which can be fluorescently labeled and used as an indicator of the position and extension of the cell body. However, these are typically more difficult to use for cytoplasm detection, as they are often highly structured and inhomogeneous and do not always correlate with the cytoplasm. Most nuclear-staining dyes give rise to spurious staining of other cellular components within the cytoplasm. This spurious staining and/or cellular autofluorescence in the nuclear channel enables also the detection of the cytoplasmic area (Fig. 14.5a). Spurious staining generally occurs with nuclear dyes which have excitation and emission wavelengths in the visible range, such as DRAQ5™ red fluorescent nucleic acid stain (BioStatus) or HCS Cell Mask ™ Red 59 red fluorescent nucleic acid stain (Invitrogen), whereas autofluorescence can be observed using UV excitation.

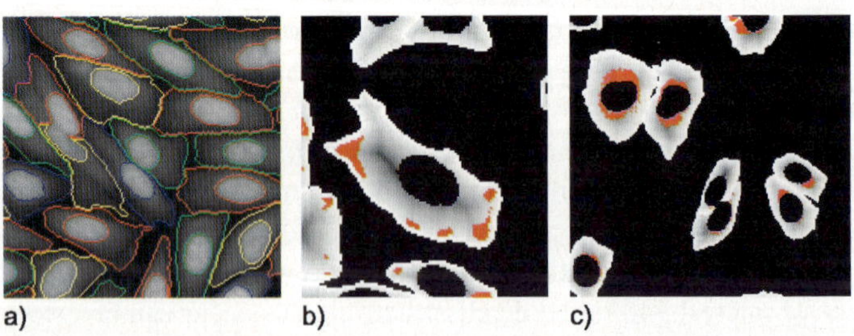

Fig. 14.5 **a** Detection of cell borders and nuclei, based on a nuclear stain and spurious cytoplasm labeling. The individual colors correspond to individual cells. **b** Detection of membrane-bound fluorescence (pixels marked in *red*), based on a weighted evaluation of brightest pixels with a long distance from the nucleus. **c** Detection of cytosolic fluorescence (pixels marked in *red*), based on weighted evaluation of the brightest pixels with a short distance from the nucleus

More sophisticated routines allow signal quantification independent of identification of a compartment/region of interest, but taking into account its localization. The task of identifying a signal in the plasma membrane is an example. The optics of confocal imaging have the power to distinguish between a plasma membrane and a cytosolic region. The use of a membrane dye in order to label membrane regions would mean the introduction of a further color, consuming another channel in multicolor imaging. For most cases we do not use such dyes, but use spatial parameters in combination with intensity parameters in order to distinguish between a cytosolic and a plasma-membrane localization of a signal. The cell is divided into radial regions starting at the interface between the nucleus and the cytosol and ending at the interface between the cytosol and background based on cytosolic staining. These radial regions are used to obtain a weighting function for the pixel locations. The weighting function has higher values further away from the nucleus/cytosol border. Pixels are then sorted according to their fluorescence intensity and a certain percentage of the brightest pixels are selected. For these selected pixels, the values of the weighting function are determined. Higher average values indicate a higher degree of membrane localization of the signal (Fig. 14.5b); lower values indicate a higher degree of cytoplasmic localization (Fig. 14.5c). This method is also suited for search-mask-independent quantification of local cellular events and translocation processes.

Detection and quantification of the specific event of interest is the next step in image analysis. Finally, various correction algorithms can be implemented, such as background correction, detection of autofluorescent or quenching compounds and filtering of contaminations (e.g., precipitated compounds or dye molecules). Advanced filtering is required in order to identify and discriminate subpopulations of cells such as apoptotic cells, degraded cells or nontransfected cells in the case of transient transfections.

14.8 Throughput: How to Acquire and Analyze Data Rapidly

When we designed our HCS hardware and software we had the goal to achieve a throughput of 100,000 multicolor data points per day in a 24 h per day, 7 days per week situation. This meant that we would have less than 1 s for each data point in a scenario of sequential imaging of samples with all colors acquired simultaneously. The process of image acquisition includes table movement from one well to another, focusing and exposure time. Table movement requires approximately 230 ms when using 1,536-well plates. The laser-based autofocus process is very fast and is finished within 120 ms. This leaves approximately 500 ms for exposures which works well for applications using bright dyes and high laser power. Image analysis can be done while the next acquisitions are made, but must not consume more time than the acquisition itself. The throughput of image analysis depends mainly on computing power. A complex image analysis algorithm can easily take 5 s or more for evaluation of one multicolor image set when using a standard PC. It can take much longer if an image series, e.g., from a whole set of positions within the same well, has to be analyzed. One possibility to speed up the process is to use high-speed data processors; however, this was not our preferred choice since it is only possible at very high cost and still runs as a serial process for a single processor.

We have achieved high-throughput image analysis matching the speed of image acquisition by a distributed computing concept. A solution was established in which the Opera control computer is connected to a cluster of PCs which perform the analysis tasks. For most applications the cluster consists of three nodes and is capable of storing several terabytes of data but is easily expandable. Data transfer is realized via gigabit LAN. A data analysis rate of one data point per second in a three-color application, for example, requires data transfer rates of at least 8 MB/s. The Opera control computer itself does not require or reserve a prespecified amount of disc space in order to execute a batch of experiments consisting of more than one plate. However, buffering disc space is reserved for the data generated from an experiment on a plate-by-plate basis so that in the event of network breakdown the measurement of one plate can still be finished. The concept ensures that data analysis runs at a speed as fast as image acquisition. Data evaluation is done in parallel to the measurement within the evaluation cluster. Run-time management makes the solution both reliable and fast. The size of the cluster is scalable and can be adjusted in order to keep track with the image acquisition. The data evaluation cluster can work simultaneously for several image acquisition stations.

The first step in the process is to define an imaging task by defining data acquisition parameters, referencing and an image analysis algorithm (script) to be performed on a set of related images. Acquisitions from the cell sample at one or more positions within a well using one or more different excitation and emission wavelengths belong to the task as well as a set of reference images for channel alignment and illumination gradients. The control computer stores this information and evaluation commands in containers of specific data transfer format. The automated process is started and for every well all respective images together with the task are sent to the image analysis cluster within a container. This means that image data are moved to the evaluation cluster during measurement. The size of the container is determined primarily by the size of the images. The "commands" in the container are as follows: correct the image for spatial alignment and illumination gradients (see above), save and evaluate them. Raw or corrected images are saved on the image evaluation cluster or in network attached storage. Information about file locations and the numerical results themselves are sent back to the control computer. The results are one file per plate containing the analysis results and the requested metadata in XML format. The images themselves can be stored in a TIFF-compatible data format in 8 or 16 bit. The image files can also be compressed using, e.g., the LuraWave® (LuraTech, Berlin, Germany) wavelet algorithm. Data file export to a networked drive is done automatically.

An example for the throughput achieved in a screening situation is given in Table 14.2. It describes a two-color assay using a reference dye for cell identification and a fluorescent tag to identify a specific signal. The cells were seeded into 96-well plates and a comparison was made for analyzing one and five image fields per well. The durations of the processes of image acquisition and image analysis

Table 14.2 Example for the high throughput achieved in a screen

	Time required for individual steps	
	Image acquisition	Image analysis
One region per well		
Data acquisition time per image pair	0.4 s	1/3×1.2 s (runs on 3 parallel PCs)
Overhead time per image pair (table movement, autofocus)	0.6 s	–
Total time per well	1 s	0.4 s
Time per 96-well plate	1.6 min	0.64 min
Total time per 96-well plate		1.6 min
Five regions per well		
Data acquisition time per image pair	0.4 s	1/3×1.2 s (runs on 3 parallel PCs)
Overhead time per image pair (table movement, autofocus)	0.6 s	–
Total time per well	5 s	2 s
Time per 96-well plate	8 min	3.2 min
Total time per 96-well plate		8 min

are listed. In this case the duration of image acquisition determines the speed of the overall process; image analysis using a cluster of three evaluation computers is much faster than acquisition. In higher-density plates such as 384- or 1,536-well plates the overhead time required for table movement between acquisitions can be reduced, thus speeding up the whole process.

14.9 Screening Examples

Imaging-based automated quantification of pharmacologically relevant cellular effects were described several years ago (Giuliano et al. 1997, Ding et al. 1998, Conway et al. 1999). The first of those papers describes measurement of human glucocorticoid receptor activation according to its transfer from cytosol into the nucleus. It further describes quantification of the apoptotic effect of a compound with respect to nuclear morphology, actin polymerization and mitochondrial membrane potential. Another example for cytosol to nucleus translocation is described by Ding et al. (1998). Here, the target NF-κB was under investigation. Conway et al. (1999) studied activation of parathyroid hormone receptor by monitoring internalization of a green fluorescent protein tagged receptor. These examples provided dose–response curves for one or a few effectors.

Among the first imaging-based high-throughput cell screening assays were a cell viability and a P-glycoprotein pump-inhibitor assay (Jäger et al. 2003). The readout parameter was whole-cell fluorescence in both cases. The use of one channel only in the case of the P-glycoprotein pump assay and of two channels for

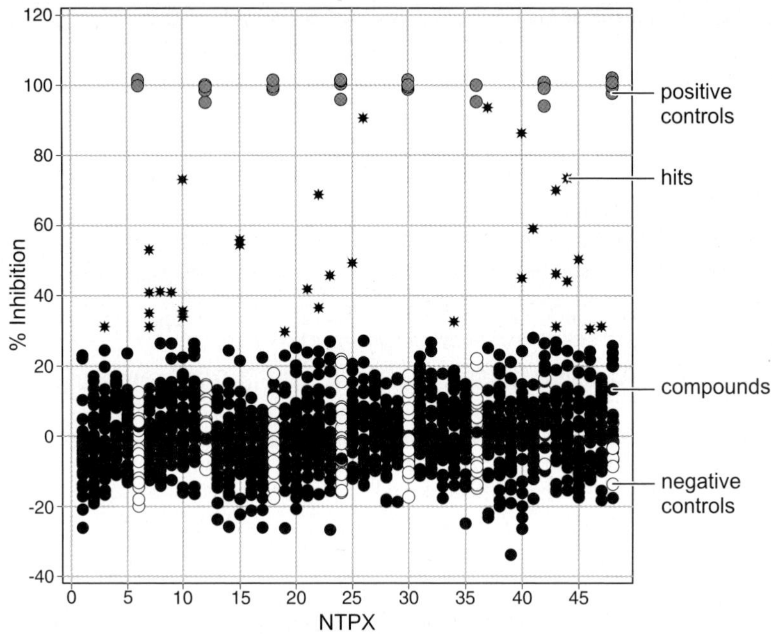

Fig. 14.6 Endothelin A receptor high-throughput screening run result in 1,536-well format; result from one screening plate example. The *x*-axis shows the columns of the 1,536-well nanotiter plate (*NTPX*). Readout was percentage inhibition of endothelin A receptor activation

cell viability shows the potential of running those types of assay in parallel to further functional assays in the future making use of additional wavelengths. Both HCS campaigns of more than 100,000 data points revealed high assay robustness expressed by mean *Z′* factors of higher than 0.6 (Zhang et al. 1999), assay sensitivity and reproducibility of data.

High-throughput screening for inhibitors of the endothelin A receptor is a further example application established by Evotec (Fig. 14.6). Receptor activation was measured using CHO cells stably expressing fluorescently tagged endothelin A receptors. The readout was endosome formation which was diminished in the presence of endothelin inhibitors. BQ-123, a selective endothelin A receptor antagonist, and the nonselective PD-142893 were used as standard inhibitors to monitor the assay sensitivity throughout the HCS run. A selection of highly diverse compounds from Evotec's Lead Discovery Library were screened in the 1,536-well format in a total assay volume of 9 μl per well and at a compound concentration of 5 μM. A total of 120,000 data points were generated at an average throughput of 50,000 data points per 24 h. The mean Z′ factor of 0.73 documented excellent data quality and robustness of the assay. 1,379 compounds were identified as primary hits on the basis of a threshold set at 3 times the standard deviation from the mean value of the negative controls at 32%. False-positive hits originating from low densities of viable cells in a well could easily be identified and sorted out at the stage of primary HCS. Sixty-five percent of the primary hits could be confirmed by testing them again in triplicates under the same conditions as in the primary screen. Five hundred and an three confirmed hits were then selected for duplicate IC_{50} titrations at 11 concentrations ranging from 0.7 to 20 µM. Among those, 94 highly active compounds with IC_{50} values lower than 1 µM, 337 compounds with IC_{50} values between 1 and 10 µM, and 38 compounds with IC_{50} values between 10 and 50 µM with respect to inhibition of the endothelin A receptor were identified.

The high-throughput assays described so far were live-cell assays and did not require washing, cell fixation or multistep staining procedures, making them easy to automate even for low-volume plates. The majority of high-throughput cell imaging assays described today are fixed-cell assays using specific fluorescently labeled antibodies. A general drug profiling according to cell phenotype has been described for a set of test compounds (Perlman et al. 2004). Parameters such as influence of a compound on cell cycle, on cytoskeleton, on calcium regulation, on energy metabolism, on several kinases and many more were measured using a DNA stain and a set of two fluorescently labeled antibodies in each cell sample.

The first genomewide analysis screen was described by the HT – Technology Development Studio Dresden (Pelkmans et al. 2005; Pelkmans and Zerial 2005). The role of human kinases in clathrin- and calveolae/raft-mediated endocytosis was revealed in a phenotypic profiling using small interfering RNA based silencing of kinases. Endocytosis was monitored either in a viral uptake process by detection of the large T antigen of SV40 and a fluorescently labeled antibody or by uptake of fluorescently labeled transferrin in the case of clathrin-mediated endocytosis (Fig. 14.7).

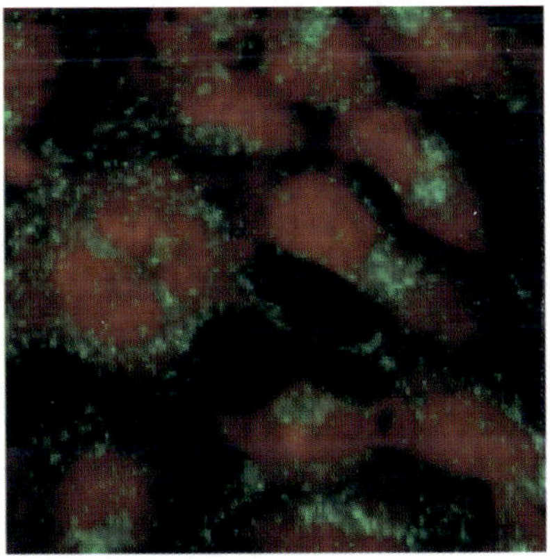

a)

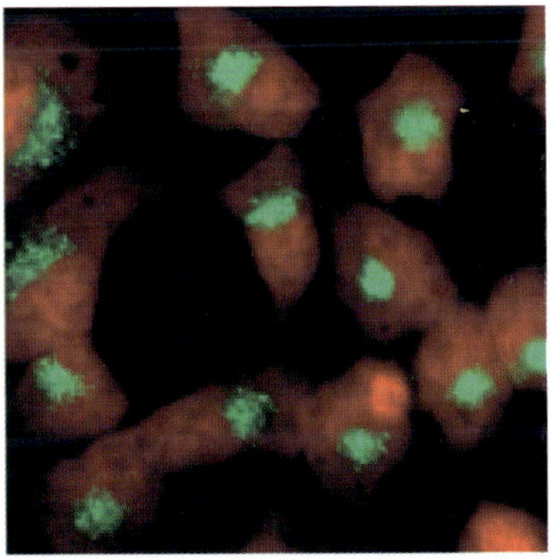

b)

Fig. 14.7 Images of endocytosis screening based on uptake of AlexaFluor488-labeled transferrin (Pelkmans et al. 2005). **a** Negative control showing internalization of transferrin into HeLa cells treated with small interfering RNA (siRNA) against firefly luciferase (*green spots*). **b** Perinuclear accumulation of internalized transferrin into HeLa cells treated with siRNA against the kinase EPHA7 indicating a role of this kinase in inhibition of endocytosis. Nuclei were stained using DRAQ5™. (Images were kindly provided by the HT – Technology Development Studio, Dresden, Germany)

The final high-throughput high-content assay example is that of screening for modulators of an orphan GPCR based on β-arrestin recruitment to activated receptors and the formation of clathin-coated pits/vesicles (Transfluor® technology of Molecular Devices; Garippa et al. 2006). Screening for multiple cellular events and extending screening dimensions to kinetic data will be one of the challenges of the future. Simpson (2005) has described how this could support drug discovery for CNS diseases.

Altogether high-throughput high-content automated imaging is an essential enabling tool for cell analysis tasks in drug discovery, in biomedical research for drug development, in genome analysis and in other fields of cellular systems biology. Pioneer cell-based assay applications were developed in academic research and in the pharmaceutical industry. It will find widespread use in the future, allowing systematic phenotypic and physiological assay approaches where high volumes of high quality data points are required.

References

Comley J (2005) High content screening, emerging importance of novel reagents/probes and pathway analysis. Drug Discov World Summer 31–53

Comley J, Fox S (2004) Growing market for high content analysis tools. Drug Discov World Spring 25–34

Conway BR, Minor LK, Xu JZ, Gunnet JW, DeBiasio R, D'Andrea MR, Rubin R, DeBiasio R, Giuliano K, Zhou L, Demarest KT (1999) Quantification of G-protein coupled receptor interalization using G-protein coupled receptor-green fluorescent protein conjugates with the ArrayScan™ high-content screening system. J Biomol Screen 4:75–86

Ding GJF, Fischer PA, Boltz RC, Schmidt JA, Colaianne JJ, Gough A, Rubin RA, Miller DK (1998) Characterization and quantitation of NF-κB nuclear translocation induced by interleukin-1 and tumor necrosis factor-α. J Biol Chem 273:28897–28905

Garippa RJ, Hoffman AF, Gradl G, Kirsch A (2006) High-throughput confocal microscopy for beta arrestin green fluorescent protein translocation G-protein coupled Receptor assays using the Evotec Opera. In: Inglese J (ed) Methods in enzymology, vol 414. Measuring biological responses with automated microscopy. Elsevier, Amsterdam, pp 99–120

Giuliano KA, DeBiasio R, Dunlay T, Gough A, Volosky JM, Zock J, Pavlakis GN, Taylor DL (1997) High-content screening: a new approach to easing key bottlenecks in the drug discovery process. J Biomol Screen 2:249–259

Hell SW, Andresen V (2001) Space-multiplexed multifocal nonlinear microscopy. J Microsc 202:457–63

Jäger S, Garbow N, Kirsch A, Preckel H, Gandenberger FU, Herrenknecht K, Rüdiger M, Hutchinson JP, Bingham RP, Ramon F, Bardera A, Martin J (2003) A modular, fully integrated ultra-high-throughput screening system based on confocal fluorescence analysis techniques. J Biomol Screen 8:648–659

Perlman ZE, Slack MD, Feng Y, Mitchison TJ, Wu LF, Altschuler SJ (2004) Multidimensional drug profiling by automated microscopy. Science 306:1194–1198

Pelkmans L, Zerial M (2005) Kinase regulated quantal assemblies and kiss-and-run recycling of caveolae. Nature 463:128–133

Pelkmans L, Fava E, Grabner H, Hannus M, Habermann B, Krausz E, Zerial M (2005) Genome-wide analysis of human kinases in clathrin- and caveolae/raft-mediated endocytosis. Nature 436:78–86

Sandison DR, Webb WW (1994) Background rejection and signal-to-noise optimization in confocal and alternative fluorescence microscopes. Appl Opt 33:603–615

Simpson PB (2005) Getting a handle on neuronal behaviour in culture. Eur Pharm Rev 2:56–62

Stephens DJ, Allan VJ (2003) Light microscopy techniques for live cell imaging. Science 300:82–86

Straub M, Lodemann P, Holroyd P, Jahn R, Hell SW (2000) Live cell imaging by multifocal multiphoton microscopy. Eur J Cell Biol 79:726–734

Tanaami T, Otsuki S, Tomosada N, Kosugi Y, Shimizu M, Ishida H (2002) High-speed 1-frame/ms scanning confocal microscope with a microlens and Nipkow discs. Appl Opt 41:4704–4708

Wang E, Babbey CM, Dunn KW (2005) Performance comparison between the high-speed Yokogawa spinning disc confocal system and single-point scanning confocal systems. J Microsc 218:148–159

Zhang JH, Chung TD, Oldenburg, KR (1999) A simple statistical parameter for use in evaluation and validation of high throughput screening assays. J Biomol Screen 4:67–73

15
Cognition Network Technology – A Novel Multimodal Image Analysis Technique for Automatic Identification and Quantification of Biological Image Contents

Maria Athelogou, Günter Schmidt, Arno Schäpe, Martin Baatz, and Gerd Binnig

Abstract Detailed knowledge about the morphology of a biological system gives valuable and precious information about its functions and dynamics. Improvements in imaging technologies enable users to acquire thousands of images of different modalities with different resolutions from biological systems. Such images show subcellular structures, cells, cell groups, tissue, organs and organisms. On the other hand, image data are of high value only if they can be transformed into valuable knowledge. Therefore, the problem of automatic information extraction from such images has become a prime priority in academic and industrial biomedical research and development. The Definiens Cognition Network Technology (CNT) solves that problem by simulating human cognition processes using knowledge-based and context-dependent processing. It is represented in its entirety by the image data, image processing methods, image derived layers, and image objects and their definitions in a unified model. CNT incorporates elements from semantic networks, description logics and functional programming. We applied this technology to imagery of different magnification, resolution and different modalities such as electron micrographs, optical microscopy and modalities in the area of radiology. Using a unified approach with a scale- and problem-invariant processing and knowledge model is a prerequisite for studying complex hierarchical systems such as biological systems.

15.1 Introduction

Today, medical scientists and biologists have the necessity and the means to acquire high-content images in high-throughput scenarios, routinely generating tens of thousands of images per day. Likewise, even "low-throughput" imaging modalities, for example, multidimensional fluorescence imaging situated in a busy multiuser imaging laboratory, can easily generate thousands of images a day. In turn, the need to systematically extract information from these images is becoming exceedingly important, and especially where it concerns the need to make efficient routine imaging that should benefit the statistical advantages levied by such

S.L. Shorte and F. Frischknecht (eds.), *Imaging Cellular and Molecular Biological Functions.*

enormous data flow. Definiens has developed the so-called *Cognition Network Technology* (CNT), a general framework aimed to facilitate and automate the analysis process using Cognition Network Language (CNL), a high-level semantic computer language that provides users with an image analysis "developer environment," where solutions may be designed efficiently.

Interactive graphical representations of CNL scripting, a rich variety of classification and segmentation algorithms, as well as variables and control structures are the main components of CNT, which make it a very powerful tool. The CNT–CNL solution is extremely flexible, inasmuch as it can work on any image type and delivers by the rule of thumb "…if the user can see the object of interest, then the CNT–CNL can too…". The history of the development of the underlying algorithms, custom language, and user-interface semantics includes a broad gamut of scientific and exploratory imaging modalities and applications, including light and fluorescence microscopy, magnetic resonance imaging (MRI), computer tomography (CT), photoacoustic tomography, positron emission tomography (PET), X-ray, ultrasound, and infrared imaging. Each of these image acquisition methods represents one "image modality" with their respective characteristics (e.g., contrast, noise, background, etc; see Chap. 1 by Hazelwood et al.), which can limit conventional analysis methods. During conventional image analysis, the objects of interest are extracted through a series of more or less sophisticated filters, for example, using information concerning intensity thresholds, proximity, gradients, and edges. Such mathematical filters are generally applied to a whole image or to regions of interest of those images. These filters evaluate the pixels or pixel fields and relate them to the neighboring pixels to produce new images with different appearances. The goal is the transformation of regions of interest in the images to objects of interest that can be extracted by a simple threshold. This process of creating the objects of interest is called segmentation and is usually followed by a classification process that depends on the properties of those objects of interest. In contrast to conventional approaches, when CNT is applied it not only creates the final product, i.e., "objects of interest," but also creates a large number of intermediate steps and intermediate objects that are involved in the process. In CNT this process evolves rather than being predefined. Thus, through classification of all objects including the intermediate objects semantic relations are generated. Knowledge of the objects gathered *during* the course of segmentation and their relations can then be used effectively for future purposes. Therefore, image analysis by CNT is not only a segmentation and classification process, but also an evolutionary process with an alternation of classification and segmentation that can be visualized in the form of a "spiral" (Fig. 15.1). The "spiral" represents the global process of analysis for an image consisting of all the local processes driven by the local properties and the local contexts. On the other hand, each local process might not only consist of a single process but also of a stepwise evolution. This local subevolution might again contain subevolutions within it. Thus, a simple spiral turns into a self-similar spiral consisting of subspirals and subsubspirals (not shown in Fig. 15.1). The bigger spiral can continue with the next turn when all local subspiral turns are completed. The key advantage is, that CNT works across

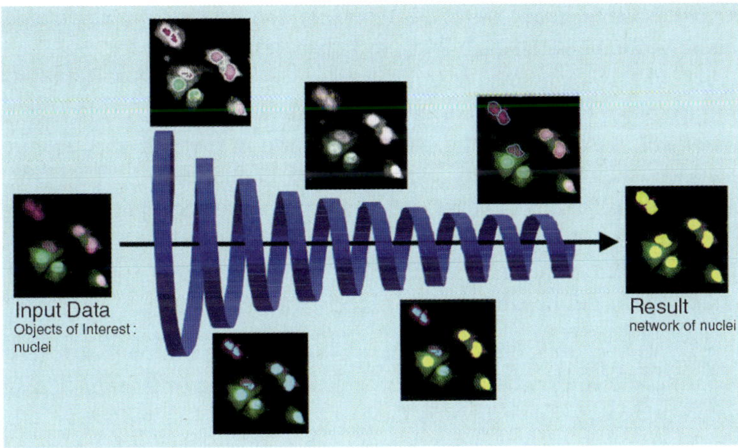

Fig. 15.1 Cognition Network Technology (CNT) allows interplay between domain-based segmentation (intermediate results *above the spiral*) and classification (intermediate results *below the spiral*) in an iterative way for extracting and quantifying objects of interest. The example shows the automatic extraction of nuclei in a fluorescence image. (Image courtesy of BioImage A/S, Denmark)

all imaging modalities and multiplexed, multidimensional high-content series in a manner that is feature-based and mostly robust to systematic variables. Here we outline how CNT–CNL works and give an idea of the unique and immediate impact we believe this sort of approach may have on biological and biomedical imaging.

15.2 Cognition Network Technology and Cognition Network Language

15.2.1 Cognition Networks

A program written in CNL is able to extract, represent, and store information from complex inputs like images or texts. As a result of an image being processed, a hierarchical network of classified objects is produced. This network represents the extracted information. Once the objects of interest are found, defined, and classified by the program, measuring their features is relatively fast and easy. Some of the features are represented by the context of the objects. Relations to other objects might be relevant for the interpretation of their state and meaning. This leads to a meaningful network of objects. As some of the relations are hierarchical, the network has a hierarchical form too. Typical hierarchical relations are, for example, a cell has two nuclei or a nucleus being part of a cell covers 30% of the cross section of the complete cell. To introduce these sorts of semantic definitions CNT–CNL uses five types of software operations:

1. *Data objects* (or "image objects") represent concrete data input (images with pixels or vectors, characters, tables, metadata, texts etc.).
2. *Instance objects* are created as a result of processing data objects and they represent groups of data objects or groups of instance objects.
3. *Class objects* describe potential instance objects in terms of what features they might have and what relations to other objects might occur.
4. *Process objects* mainly process data and instance objects but in principle are able to modify all kinds of objects and their cross-linking. Their order also defines the order of the analysis steps.
5. *Domain objects* define which set of objects will be processed.

Using these five operators, one can define all kinds of target through structured hierarchical form, defined by what may be considered a unique *millefeuille* derived from interrelated operators. For example, there is usually a class hierarchy, a process hierarchy, an instance object hierarchy, a data object hierarchy, and a domain hierarchy. As these hierarchies are linked to each other they form one hierarchical object network, the cognition network. Thus, to begin, "process objects" creates a hierarchy of instance objects by grouping data objects using specific segmentation methods. In turn, instance objects could be linked to class objects using a classification step (i.e., process objects), wherein the method and parameters used could in turn be stored in class objects, adding another level to the hierarchical network. The state of this network as defined by each instance, class, and process is called a *network situation*, which is modified iteratively until a final state is reached. This final network situation represents the information extracted from the input data (Baatz et al. 1999; Binnig et al. 1999; Schaepe et al. 2005).

15.2.2 Input Data and Image Object Hierarchy

CNT recognizes a data object, i.e., images, which may have an arbitrary number of layers, and each layer may have different properties, e.g., 24-bit RGB, 8- or 16-bit signed/unsigned integer, or 32-bit single precision floating point number (see Chap. 1 by Hazelwood et al.). Images and image layers are treated as data objects within the instance network and the pixels are described using an "*is-part-of*" ontology that functions as an operator defining the connectedness of multiple data objects (i.e., among complete multiple images, and subregions therein) (Baatz et al. 1999; Binnig et al. 1999; Schaepe et al. 2005). Image data objects are also described by their neighbor relationships (*neighbur link objects*). Neighbor link objects use local partitioning at the *image object level*. For example, the output of any segmentation algorithm could be interpreted as a valid image object level. Each part of this segmentation result defines the associated region of an image object. Two trivial and definitive image object levels are (1) the partition of the image into pixels (the *pixel level*) and (2) where only one object covers the entire image, the *scene level*. Image object levels (Fig. 15.2) are structured in an *image object hierarchy*. The image object levels of the hierarchy are ordered according to inclusion. The image objects

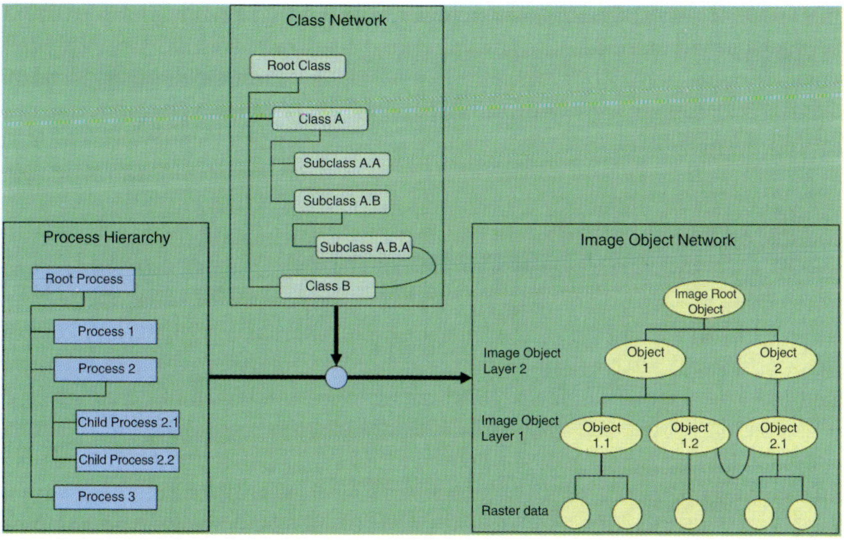

Fig. 15.2 Representation of the CNT: a Cognition Network Language (CNL) script (rule set) executes a hierarchy of processes (*left*). A network of image objects is generated from input raster data (*right*). A predefined network of image object classes (*middle*) allows a process-controlled classification of the image objects as well as a classification-controlled segmentation. All processes are executed in the context of the actual classifications

of any level are restricted to be completely included (according to their associated image regions) in some image objects on any "higher-order" image object level. The image object hierarchy together with the image forms the instance object network that is generated from the input data.

15.2.3 Features and Variables

Features are numbers, which are computed by a well-defined algorithm from the current network situation (for an example of a typical work space, see Fig. 15.3) – reading a data entry is also considered as a computation. There are two major types of features: *object features*, which are linked to an object in the cognition network, and *global features*, which could be any kind of other information. Object features measure properties of the individual image data objects. Since regions in the image provide much more information than single pixels, there are a large number of different image data object features for measuring color, shape, and texture of the associated regions. Additional information might be extracted by taking into account the network structure and the classification of the image objects. Important examples of this type of feature are "relative border to neighboring

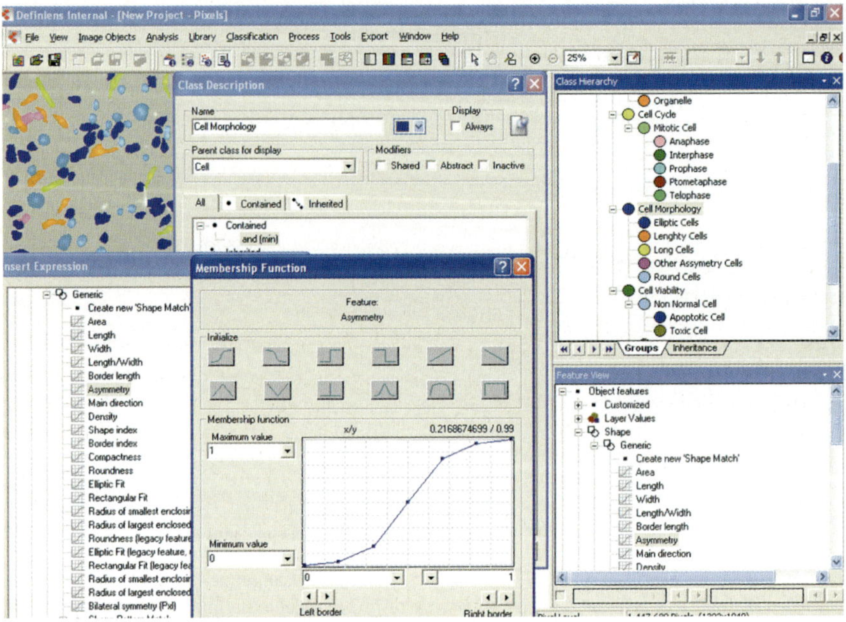

Fig. 15.3 Screenshot of the Definiens eCognition image analysis software with focus on object features (right dialog) and class descriptions using fuzzy membership functions (center dialog)

objects of a given class" and "number of subobjects of a given class." On the other hand, global features describe the current network situation in general. Examples include "mean value of a given image channel," "number of levels in the image object hierarchy," and "number of objects classified as a given class." Global features may also be used for metadata content as an additional part of the input data and include, for example, the type of tissue sample used in an experiment. All numerical values of an image object can be used for statistical analysis. CNL provides a fully automated functionality which extracts and then exports this information into a wide variety of formats (file types).

Process variables are stored data values. These values (numbers, texts) may be used and modified by process objects. Variables may be efficiently managed using *parameter sets*, containing variable names and values stored within or retrieved from the file system. A parameter set may be applied to a given network situation, which means that all variable values in the current cognition network are set to the values stored in the parameter set. There are global and local variables. Global variables are conventional variables used for processing. Local variables are kinds of object attributes as they are linked to image objects. Local variables represent in a way the history of an object. A feature can be calculated once, stored as an attribute, and used again later many times. The feature of the object might, however, have changed in the meantime. On the first glance historical object information seems not to make much sense. It might, however, be valuable to know that during the course of evolution of a nucleus its classification as a nucleus at a given time is much worse than it had been in an earlier state. One might want to reset its state to the earlier one which might have been stored in some form also in the form of an attribute.

15.2.4 Classes and Classification

Class objects are used to assign a semantic description to other objects in the cognition network (Figs. 15.2, 15.4). Classes can be linked by *inheritance links* to inherit class descriptions and by *group links* to group different classes together to a group class. Class descriptions are created via a fuzzy logic based system. The classes form a structured subnetwork of the cognition network called the *class hierarchy* (Baatz et al. 1999; Binnig et al. 1999; Schaepe et al. 2005). Image objects are linked to class objects by *classification link objects*. Each classification link stores the fuzzy membership value of the image object to the linked class. An image object may have an arbitrary number of classification links and the class with the highest membership value for the image object is called the *current class* of the image object. For example, the *image objects of class* C are the set of all image

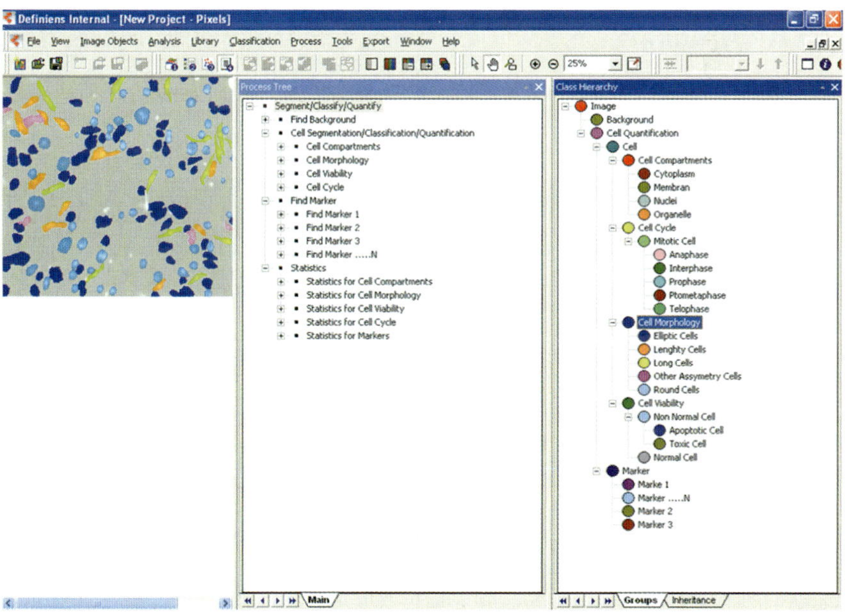

Fig. 15.4 Example of a CNL script (rule set): classes and processes with modules for segmentation, classification, and quantification of cells

objects with current class C. Classification could be performed with any classification algorithm as long as the results could be translated into fuzzy membership values. The current implementation uses fuzzy membership functions on image object features and a *nearest-neighbor classifier*. Since process variables could be used for class descriptions, these could be modified by the cognition network itself during the image analysis.

15.2.5 Processes

The CNT is modified by *process* objects, where processes (Figs. 15.2, 15.4) are linked by *flow-control* objects to describe their order of execution in time (Baatz et al. 1999; Binnig et al. 1999; Schaepe et al. 2005). During process execution, each process holds a temporary *execution context* object that stores all information related to the process-execution state. A process is the combination of an *algorithm* and an *image object domain*, and may have an arbitrary number of subprocesses. The algorithm describes *what* the process will do, for example, classification, creating image objects (segmentation), or image object modifications like merging, splitting of an image object, or rearranging subobjects of an image object. Other important algorithms are computing and modifying image object attributes and process variables and exporting results.

15.2.6 Domains

The image object domain describes *where* the algorithm and the subprocesses of the process will be executed in the image object hierarchy. The present implementation allows the defining of domains by a hierarchical navigation through the class and object hierarchy. This navigation starts from basic sets of objects that are defined by selecting the image object level, the object classes, and any additional condition. An example could be "go to the cell level, to all objects classified as cells that have a length shorter than 120 pixels." This way a set of objects is already selected and could serve as the domain. In this case all those particular cells will be modified by the related process. The story, however, could continue in a form of navigation: "from those cells go to all subobjects classified as a nucleus and from there to all neighbors that are classified as spots and that are brighter than 2,000". In such a case the domain would not be represented by all particular cells but rather by all bright spots that are neighbored to nuclei and that are part of short cells. Since during process execution image objects of a domain are treated one after the other, image object domains could be defined relative to the *current image object of the parent process* (PPO), e.g., the subobjects or the neighboring objects of the PPO.

15.2.7 Using CNT-CNL for Image Analysis

To exemplify the utility of CNT, we assumed that a large number of fluorescence images are produced from a cell-based assay (see also Chap. 14 by Gradl et al. and Chap. 16 by Fava et al.) in a high-throughput/high-content screening and a fully automated, fast, reliable, and detailed morphological quantification for each individual cell is required. However, owing to the inherent property of the cells in cell culture to adhere to and to overlap with each other, a very rigorous and precise segmentation is needed to quantify each cell. Therefore, a more sophisticated algorithm is required for detailed and accurate segmentation than the traditional pixel, threshold-based cell segmentation. Ideally, a segmentation algorithm has to have the ability to extract cells, separate them, and to extract and separate cell compartments from each other in order to quantify a single object. Furthermore, all these created objects have to be linked through semantic relations that have biological relevance in the context of the actual screening. The corresponding image analysis solution has to be comprehensive, fast, robust, reliable, and reusable. Image analysis applications based on traditional pixel-based algorithms are not sufficient to accomplish all these tasks in a high-throughput/high-content scenario. Most of these traditional algorithms use segmentation methods based on color and intensity thresholds (see also Chap. 2 by Meijering and van Cappellen) and have limitations in dealing with context information and semantic relations that are biologically relevant. Therefore, the information gained using a pixel-based approach loses valuable knowledge pertaining to biologically relevant image objects and their mutual relations.

Figure 15.5 represents a schematic creation of a CNT network of an image using CNL. A first hierarchical level (L0) with objects representing "cells" and an object representing the image "background" is created using an initial segmentation and classification algorithm. Since CNT provides well-defined interfaces to external algorithms, it is possible to include any known segmentation and classification algorithm into this framework. For background extraction a threshold method based on local contrast produces excellent results. The classification can be efficiently done using a fuzzy set description of the object properties and their relations. For example, to calculate the morphological relations of cell cytoplasm and nuclei, and at the same time the relations of each cell to its cytoplasm and nucleus, a new hierarchical level (L0–1, below L0) is created. Through local segmentation of "cell" objects, the "nuclei" and "cytoplasm" subobjects can be extracted. To identify cell groups and their parameters, a new hierarchical level (L0+1, above L0) is created. To create another level the algorithms "copy image object level" and "merge image object with same classification" can be applied. Such analysis of cell group properties could provide, for example, important information on tissue formation (Classen et al. 2005). As a result of the iterative segmentation and classification processes, all objects in all hierarchical levels (L0, L0+1, and L0–1) are linked to each other, forming an image object network. Objects or mutual relations between objects, for example, morphological properties of cells included in a cell group, can be quantified automatically and in detail.

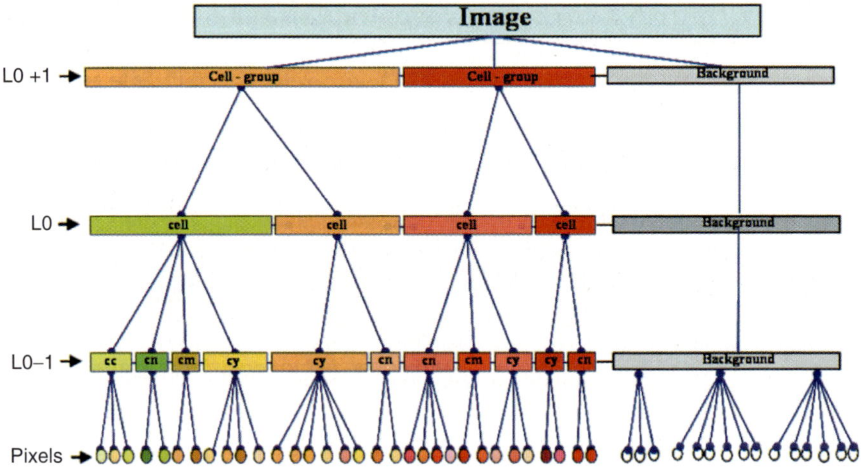

Fig. 15.5 Representation of a hierarchical network of biologically relevant objects extracted from image data. The CNT network consists of three hierarchical levels. Each object on the lowest level (L0–1) represents unspecified cell area (*cc*), nucleus (*cn*), membrane (*cm*), or cytoplasm (*cy*). Each object on level L0 represents a cell. Each object on level L0+1 represent a cell group

Each cell group of level L0+1 contains information about its morphology, its position, and the relative morphological properties and positions of its cells. At the same time detailed information about each cell of level L0 can be extracted. The image object network enables the generation of object statistics with the ability to summarize according to classes and levels or with respect to superobjects on a higher hierarchical level.

CNL solutions for image analysis are applicable for both cell and tissue images independent of the assay (Abraham et al. 2005; Athelogou et al. 2004; Baatz et al. 2004; Constans 2004; Hannon 2002). For example, a CNL-script is able to extract and quantify nuclei at all stages of a cell cycle from normal, mitotic (prophase, prometaphase, metaphase, anaphase, telophase, cytogenese), toxic, and apoptotic cells. Similar solutions can be developed for cell organelles (e.g., mitochondria), for marker detection and cell membrane analysis. Biological phenomena such as translocation, colocalization, *z*-stack image analysis and quantification, particle detection and tracking, and neuron tracing and quantification have been developed successfully. Thus, a high-quality detailed quantification of cell- and tissue-based assays can be accomplished with higher accuracy and greater detail based on a method that uses the contextual relevance of regions of interest extracted from the image. Finally, and as will be elaborated by examples in the next section, CNL is independent of any image modality. For example, one can use CNL to create robust, reproducible analysis of images recorded by modalities as diverse as transmission or epifluorescence optical microscopy, electron microscopy, CT, MRI, ultrasound, PET, and X-ray imaging. So, CNL is unique inasmuch as it offers one solution applicable to all conditions from functional molecular imaging to basic light microscopy.

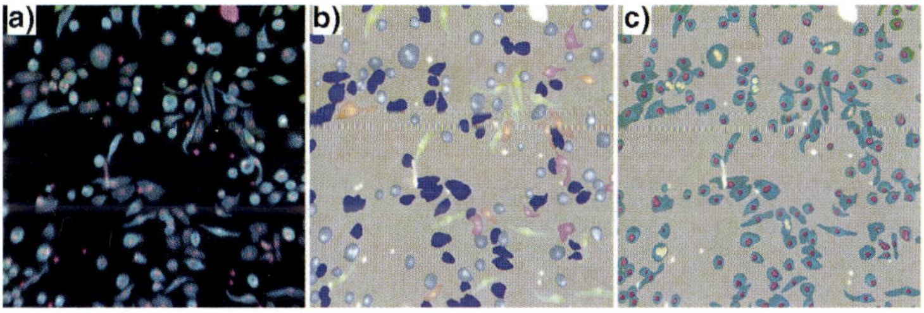

Fig. 15.6 High-content image analysis. **a** Input data overlay. **b** Cells are segmented and classified according to their morphology; different colors represent different cell shape classes. **c** Cytoplasm (*green*), mitotic nuclei (*yellow*), single nuclei (*magenta*). (Image courtesy of Cenix Bioscience GmbH, Dresden, Germany)

15.2.8 Application Notes

Using CNL it is possible to extract knowledge about the same and/or different objects of interest by combining context information contained in different image layers and/or in different images simultaneously. Figure 15.6 represents such an example. The input image is a representative example for a high-throughput and high-content assay (for a definition see Chap. 14 by Gradl et al.) imaging process. This specific example is taken from a RNA interference assay, a gene-silencing mechanism originally elucidated in plants, *Caenorhabditis elegans*, and *Drosophila* (Bernstein et al. 2001; Hannon 2002). As gene silencing might have an influence on cell morphology this CNL solution quantifies in detail the potential morphological cell characteristics following the script below:

- Extract nuclei and cells as objects of interest
- Separate nuclei, assign nuclei to the cells
- Separate cells, classify cells according to the cell cycle using the number of nuclei they contain
- Reclassify cells according to their morphology, i.e., long, round, elliptic, or alternatively formed cells
- Quantify the hierarchical mutual relations between cells, nuclei, and cytoplasm
- Quantify the mutual relations between cells on the same hierarchical level

Using CNL it is also possible to extract knowledge from objects of interest on different levels of resolution, using different imaging techniques and image modalities. This knowledge provides the user with the possibility to connect multimodal and multiscale information contained within such data and to acquire knowledge on a higher semantic level. Such an object of interest or a system might be cell assays, tissue, organ, organism, a number of organisms, or plants, etc. This kind of system analysis and quantification is useful for modeling and simulations in biology (e.g. systems biology, and molecular dynamics; see also Chap. 9 by Jaqaman et al.). Liver, for example, might be imaged on the subcellular level using electron microscopy (Fig. 15.7), on the cell assay level using fluorescence microscopy (Figs. 15.8, 15.9),

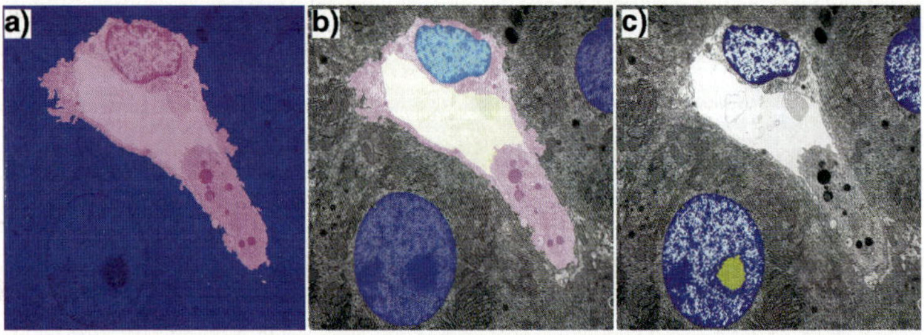

Fig. 15.7 Analysis of electron micrographs representing liver tissue. **a** Hepatocytes (*blue*), sinusoid (*pink*), **b** lumen of sinusoid (*white*), Nuclei of hepatocyte (*blue*), nuclei of the endothelial cell (*azure*). **c** Chromatin (*blue and white*), nucleolus (*yellow*). (Image courtesy of Institute of Surgical Research of the Ludwig Maximilians University Munich, Germany)

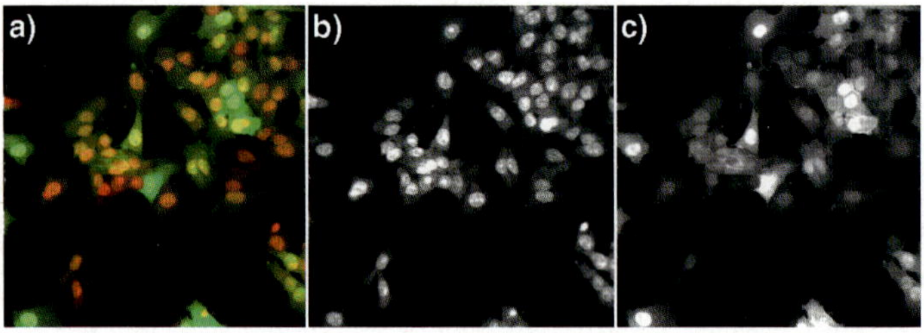

Fig. 15.8 Sample fluorescence microscopy data of a liver cell line with DRAQ5 staining. The picture shows HepG2 cells, stably expressing green fluorescent protein (GFP)–Akt1 grown on collagen-coated plastic, and exposed to 400 nM insulin-like growth factor 1 for 2 min. A GFP marker is in the membrane. **a** Pseudocolor overlay, **b** DRAQ5 layer, **c** GFP layer. (Image courtesy of BioImage A/S, Denmark)

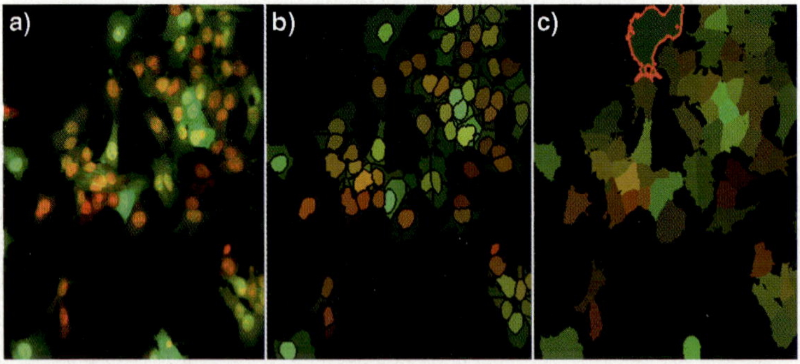

Fig. 15.9 Input data and representative image analysis results for fluorescence microscopy of a liver cell line (see also raw input channels in Fig. 15.8). **a** Original image. **b** Nuclei segmentation in first image object level. **c** Cell segmentation in second image object level. (Image courtesy BioImage A/S, Denmark)

on the tissue level using optical microscopy (Fig. 15.10), or on the whole-organ level embedded within the human or animal body using, for example, CT (Fig. 15.11).

Figure 15.7 shows an electron micrograph of the rat liver. The images represent a rat liver section at a magnification of 3,000. The aim of the solution was to extract from the images selected objects of interest such as the nuclei, classify them

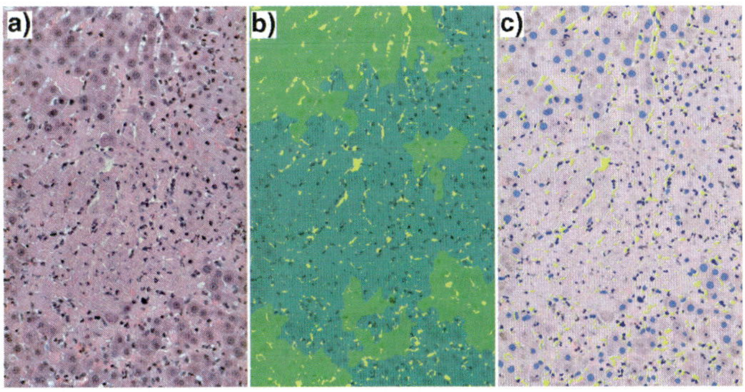

Fig. 15.10 Image analysis of mouse liver tissue. **a** Input data. **b** Healthy tissue areas (*light green*) in contrast to necrotic tissue areas (*dark green*), white spaces (*yellow*). **c** "Normal," round nuclei (*light blue*), "nonnormal," shrunk nuclei (*deep blue*). (Image courtesy of Institute for Pathology, Ludwig Maximilians University Munich, Germany)

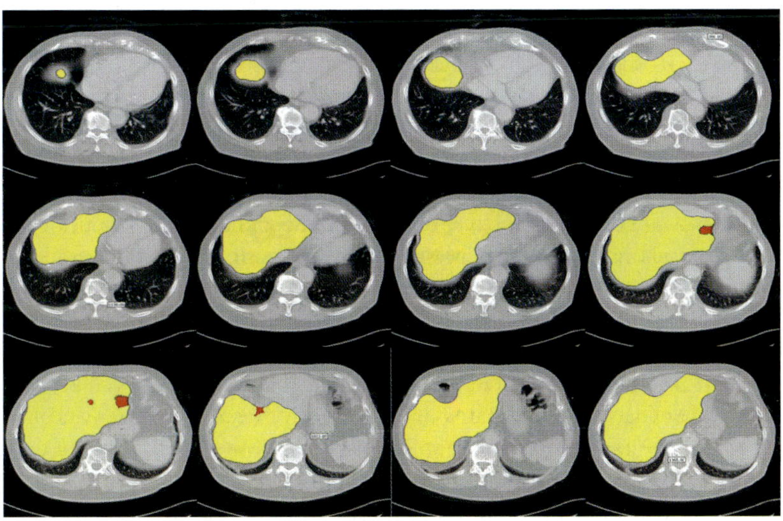

Fig. 15.11 Example of computer tomography slides of human liver (*yellow*) with tumor (*red*). The solution uses the same CNL script for analyzing each of the images in a fully automated workflow. (Image courtesy of Institute for Radiology, Ludwig Maximilians University Munich, Germany)

according to the kind of cells they belong to, extract the contents of nuclei and classify them in different classes, e.g., chromatin and nucleolus. There are two types of nuclei in the image: nuclei belonging to the hepatocytes and those belonging to the endothelial cells. Any nucleus object conveys information about its neighborhood and the mutual relations to each of its neighbors, the neighbors of its neighbors, and the whole image. In the same way it is possible to quantify each cell by extracting organelles like mitochondria, peroxisomes, or lysosomes. According to these kinds of analysis it is possible to extract relevant features concerning cells, sinusoid, and their contents for quantifying damage to or improvements of treated or nontreated liver in different ways, for example to quantify relevant morphological features characterizing liver damage following ischemia (Biberthaler et al. 2003; Urbani 2004).

Figure 15.9 (see also Fig. 15.8) shows input data and image analysis results for fluorescence microscopy of a liver cell line (HepG2) stained with DRAQ5 (red channel in merge image). DRAQ5 is a novel far-red fluorescent DNA dye that can be used in live cells and in this case reveals cell nuclei. The cell line is modified to stably express green fluorescent protein Akt1, and in this example the cells grown on collagen-coated plastic were stimulated by exposure to 400 nM insulin-like growth factor 1 for 2 min expected to cause a relocation of Akt1 from plasma membrane/cytoplasm to nucleus. The rule set extracts nuclei and classifies nuclei in two classes: those containing marker and those not. Nuclei are characterized by the surrounding cytoplasm. A first process extracts the background. Nuclei and cytoplasm objects created on a hierarchical image object level can be grouped together to create cell objects in an elevated hierarchical level. Cells can be classified also according to their subobjects, which means nuclei that contain marker, or those that do not. This analysis demonstrates the ability of the eCognition software to handle cell based assay type datasets.

Figure 15.10 shows images of necrotic and nonnecrotic mouse liver tissue. The aim of the image analysis solution was to extract nuclei, classify them as normal and nonnormal according to morphological and context criteria, and to separate necrotic from nonnecrotic tissue areas. After extraction of the nuclei, a detailed separation of necrotic from nonnecrotic regions within the images was performed. CNT allows a single-cell, objective, fully automatic and detailed quantification and characterization of morphological tissue properties. Such a solution is applicable to large numbers of tissue slides, that today can be handled by automated histology slide imaging systems, and which can now realistically be quantified within a short time using Definiens CNT.

Figure 15.11 shows CT images of a human liver. The image analysis solution separates fully automatically the liver from the other organs and the surrounding tissue, and tumor from nontumor tissue in the liver. Thus, in this case CNL allows to distinguish and quantify detailed morphological properties of the liver and the surrounding organs and to extract morphological features of a disease.

15.3 Discussion

The application of CNT to image analysis permits the development of powerful, automated, user-tailored, flexible solutions for image analysis. These solutions are applicable to different kinds of image content and different imaging modalities. The adaptive processing, which uses the domain concept, aims to provide the user with the ability to limit the analysis to "objects of interest." CNT provides integration of additional approaches, algorithms, and methods of image analysis, such as those based on pixels. The benefits of CNT are most visible if this technology is integrated in an automated or semiautomated workflow like in a screening or imaging facility. For a screening facility, which develops and uses cell-based assays, eCognition high-throughput/high-content applications might be developed to support and accompany the whole workflow from the idea to the production to data storage and downstream data analysis. Aside from the basic ability to measure any image object of interest, the user may develop sophisticated ideas and methods on the conceptual level allowing the combination of digital images, assay development, experimental workflow design, and implementation in order to achieve maximal knowledge exploitation. CNT–CNL provides a practical means towards these ends.

Acknowledgements We thank R. Leiderer, K. Messmer, H. Meissner, M. Urbani, and R. Schönmeyer for data and image contributions as well as heuristic recommendations and J.D. May and D. Neil for editing and content support.

References

Abraham K, Fritz P, McClellan M, Hauptvogel P, Athelogou M, Brauch H (2005) Prevalence of CD44+/CD24–/low cells in breast cancer may not be associated with CD44+ clinical outcome but may favour distant metastasis. Clin Cancer Res 11:1154–1159

Athelogou M, Schaepe A, von Büren E, Hummel M, Stein H, Binnig G (2004) Vollautomatische, detaillierte Quantifizierung des Ki-67 Indizes bei Brustkrebs (Fully automated, detailed quantification of Ki-67 indices for breast cancer). BioSpektrum 6:778–799

Baatz M, Schäpe A, Schmidt G (1999) Method for processing data structures German Patent Appl DE19960372.3, 14 Sept 1999

Baatz M, Schäpe A, Athelogou M (2004) Automatisierung durch objektorientierte Bildanalyse. Analyse von Strukturen in Zell- und Gewebebildern (Automation by object-oriented image analysis. Analysis from structures in cell and tissue images). Laborwelt 6(6):17–20

Bernstein E, Denli AM, Hannon GJ (2001) The rest is silence. RNA 7:1509–1521

Biberthaler P, Athelogou M, Langer S, Luchting B, Leiderer R, Messmer K (2003) Evaluation of murine liver transmission electron micrographs by an innovative object-based quantitative image analysis system (Cellenger). Eur J Med Res 8:257–282

Binnig G, Schmidt G, Athelogou M et al (1999) N-th order fractal network for handling complex structures. German Patent Appl DE10945555.0, 2 Oct 1998, and DE19908204.9, 25 Feb 1999; US Patent Appl 09/806,727, 24 Sept 1999 and 9 July 2001

Classen AK, Anderson KI, Marois E, Eaton S (2005) Hexagonal packing of drosophila wing epithelial cells by the planar cell polarity pathway, Dev Cell 9:1–13

Constans A (2004) A brainy twist to image analysis. Definiens' Cellenger offers object-oriented software for the high-content imaging market. Scientist 18(11):44

Hannon GJ (2002) RNA interference. Nature 418:244–251

Schaepe A, Athelogou M, Benz U, Krug C, Binnig G (2005) Extracting information from input data using a semantic cognition network. Patent no EP1552437; international G06F17/30, G06N5/02, G06T5/00; Appl no EP20030808849 20031015; priority nos DE20021048013 20021015, WO2003IB06431 20031015

Urbani M (2004) Computer supported and automated analysis from transmission electron micrographs of the liver. Doctoral thesis, Ludwig Maximilians University Munich

16
High-Content Phenotypic Cell-Based Assays

Eugenio Fava, Eberhard Krausz, Rico Barsacchi, Ivan Baines, and Marino Zerial

Abstract The use of high-content screening (HCS) as a medium to high-throughput technology for the analysis of microscopy-based cellular assays is a reality. Automated cell handling, microscopy and quantitative image analysis are all essential components of HCS. However, the key core of HCS is the use of sophisticated high-content phenotypic cell-based assays (HC-PCAs), which allow large-scale analysis of cellular phenotypes to be applied to genetic and compound screening. In this chapter we describe the process of of HC-PCA development for HCS, including the different phases from assay design to assay optimization, validation and troubleshooting.

16.1 A New Tool for Biological Research and Drug Discovery

Despite almost four centuries of debate, the inventor of the microscope remains disputed (Galileo Galilei or Hans and Zachariah Janssen). This is mainly due to the enormous impact this instrument has brought to biology, beginning with the discovery of the cell (Hooke 1665), establishment of followed by the foundations of "the cell theory" (formulated by Schwann 1839 and Schleiden 1839). More recently, the discovery of green fluorescent protein (GFP) (Shimomura et al. 1962) enabling fluorescent tagging of proteins has brought about what may reasonably be considered the most important advance since the microscope's invention (Prasher et al. 1992). The application of fluorescence microscopy to study protein dynamics within living cells provides an alternative to end-point assays on fixed cells. Multidimensional fluorescence-based imaging over time, although still somewhat technically challenging, is now a routine method in biological research laboratories and has rapidly (in less than two decades) promoted microscopy from being merely a visualization tool, to becoming a fully fledged quantitative method. As such, what was until recently considered an "operator-dependent" technology, relying on the scientist's subjective expertise to interpret manually images obtained from biological samples to arrive at a qualitative interpretation has been radically changed. Automated

S.L. Shorte and F. Frischknecht (eds.), *Imaging Cellular and Molecular Biological Functions.*

microscopy and sophisticated image processing and analyse (Baatz et al. 2006; Giuliano et al. 2003; Liebel et al. 2003) is currently the "state-of-the-art" approach for the so-called high-throughput "visual" screening experiments. This approach is capable of not only extracting complex quantitative parameters from cell-based experiments, but also resolving them in a statistically robust manner in a single step (see also Chap. 14 by Gradl et al.). This chapter considers the methodological basis to the process of assay development in high-content phenotypic cell-basesd assays (HC-PCAs) whereby high-content imaging is upscaled, in a manner that aims to measure the phenotypic features that are the target readout in biological analysis. In contrast to cell-based homogenous assays, HC-PCAs open the possibility to quantitatively relate multiple parameters on a single cell, or even at the subcellular level, to each other. This can be done within a large cell population either in an end-point assays or a time-lapse setting. For example, it is possible to study the cell cycle in combination with receptor signaling and cytotoxicity, generating multiparametric "high-content" information. Thus, HC-PCAs allow efficient and cost-effective analysis of parameters relevant to biological discovery. Here we describe the process of HC-PCA development, including the different phases from assay design to assay optimization, validation and some potential pitfalls.

16.2 What Is High-Content Screening and How Can Biologists Use It?

The failure of high-throughput screening campaigns based on homogenous assays to find new drug targets or new active molecules (Dove 2003) has pushed pharmaceutical and biotech companies to look again at the cell as a tool for discovery (Giuliano et al. 2003; Hoffman and Garippa 2007). High-content screening (HCS) is rapidly emerging as a promising tool to overcome the bottleneck in target discovery. HCS can be defined as the automated process of detecting cellular and/or intracellular events in arrayed cells by using multiple markers (e.g., nuclei, mitochondria, endosome, etc.) whose features are summarized by multiparametric descriptors. However, it should be mentioned that this approach is substitute not for bench-scale experiments that – as we learned from the past – enable significant scientific breakthrough. HCS has its value in allowing the scientist to use a systematic approach to discovery, such as the use of genomic libraries (complementary DNA (cDNA), RNA interference (RNAi), etc.), large chemical collections or multivariate experiments where multiple variables are systematically changed (i.e., cell type, small molecules, concentration of substances, etc.). HCS has therefore the possibility to generate not only morphological, phenotypic or genotypic data, but also to generate functional data in the relevant biological context of a cell. The three main application areas for HCS are (1) functional genomics, (2) chemical genomics and (3) biological variable matrix: each of these applications use the same workflow and experimental design, consisting of (1) automated sample preparation, (2) automated image acquisition, (3) archiving, (4) automated image analysis, (5) data extraction and analysis and (6) bioinformatics

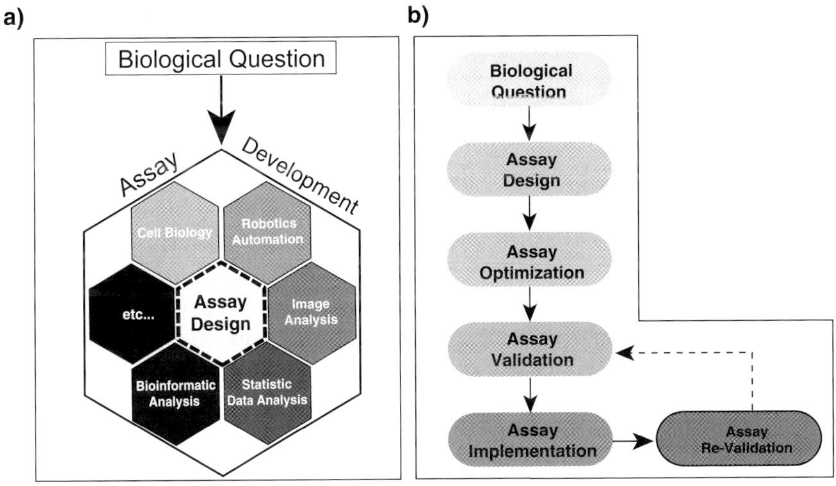

Fig. 16.1 Assay development. **a** As a result of their complexity and multi-technological method, the development of high-content phenotypic assays require a multidisciplinary approach. The biological question is the driving force of the entire process of assay development. During assay development, different expertise must be gathered together to contribute to both assay design and assay development. In particular, image analysis and robotics are involved from the very early phases of assay development. **b** Assay development consists of different phases. After the assay design has been defined by the biological question, the assay will be optimized and then validated before the final implementation. Most importantly, data obtained during assay implementation must be used to revalidate the assay, thus increasing assay quality.

analysis (Fig. 16.1). However, before performing the aforementioned tasks, there is a fundamental step that precedes and enables the remainder: assay development. In the next section we will discuss how to approach assay development for HCS and explain the main criteria to be taken into consideration.

16.3 Assay Design: First Think, Then Act

Designing a HC-PCA for HCS is a very time- and energy-consuming activity and it is one of the limiting factors for the more extensive use of HCS in both industry and academia. Hence, the need for more efficient HC-PCA design is of utmost importance. Assay development consists of different phases (Fig. 16.1b), described as (1) assay design, (2) assay optimization, (3) assay validation, (4) assay implementation and (5) assay re-validation.

Assay design is the most important step of HC-PCA development. The assay design originates from a relevant biological question aimed at solving a scientific problem. However, it is important to consider from the very beginning that HCS assays require a multi-disciplinary approach. HCS assay development require the joint action of several disciplines including automation, informatics, biology etc. (Fig. 16.1a). Therefore, it is essential that all persons responsible for each area are involved in the assay design from a very early stage. This allows for comprehensive action to be taken across all faccets of assay development, resulting in a more accurate assay. The biological question is the basis of the design and therefore requires a thorough understanding of the scientific field. The choice of cells, cellular markers, readout parameters, positive and negative controls, microscope type, magnification, etc. should be carefully considered.

Assay development for HCS must take into consideration that the assay is applied systematically to an array of variables such as chemical genomic libraries. Although the assay has to be reliable and robust enough to give a stable and repeatable readout it should never sacrifice biological complexity. Notably, the value of HCS is more in the output than in the throughput.

16.4 Assay Optimization

Assay optimization is the phase following assay design. During assay optimization, the analytical process defined by assay design is elaborated and optimized, resulting in a robust and reproducible assay. The assay optimization phase requires a continuous evaluation, clearly separate from the validation phase. The final goal of assay optimization is to choose its optimal format and to find and eliminate (or limit) the factors that may reduce assay performance. In particular, optimization for HCS assays has to take into account the integrated final format normally composed of (1) biology, (2) automation and (3) image analysis. With this intended use in mind, the assay developer should delineate appropriate performance characteristics. This section aims to discuss some of the main factors that are relevant for HC-PCA optimization. However, it should be noted that this is not an exhaustive list of the possible factors influencing specific assays. Further discussions on assay characteristics are available from the International Conference for Harmonization (ICH 1994) and the NCBI *Assay Guidance Manual* (Inglese 2006). Although these documents were not developed specifically for HC-PCAs, they are still a valuable resource for the assay developer. Note however, we believe that HC-PCAs will necessitate a revision of such documents in view of the different and more complex nature of these assays.

16.5 Cell Culture

The cells used for HC-PCAs are typically of three types: (1) cell lines, (2) cell lines transfected with a gene of interest or (3) primary cells originating from explants. There are a number of different distributors offering a large selection of mammalian cells derived from different tissues, with different characteristics. Therefore, the assay developer has the possibility to choose one or more cell lines for optimal modeling of the biological process to be investigated.

Naturally, selection of the appropriate cell line is decided with respect to the biological question, e.g., if the user is interested in studying macrophage physiology, a macrophage cell line will be chosen. However, it is also possible to choose a cell line to use as workhorse for several primary screens. The advantage of this choice is the consistency through different screening campaigns and the possibility to analyze and cross-check the outcome of different screens. Adherent cell lines, such as HeLa or SAOS, which form epithelial monolayers, are normally preferred for microscopy applications.

Once a specific cell line has been chosen, it is very important to monitor its performance and stability in the desired assay. Cell quality and contamination represent the major concerns in the stability of an assay. Careful daily routine and good sterile practice are necessary to avoid contamination. Cells must be certified as mycoplasma-free before starting assay development. Large stocks of cells frozen in aliquots should be prepared beforehand and ideally are used for both assay development and screening. After freezing, an aliquot should be thawed and checked for mycoplasma to ensure the quality of all the stock. For the daily use of cells, the medium should be checked routinely for contamination, by simply plating medium on agar plates.

Cell count variability and cell aging are two aspects of cell culture that may generate problems. These can be standardized across assays by the use of automatic or semiautomatic devices to count and seed cells and by using a fixed number of cell passages. As rule of thumb, 15 passages are suggested, meaning 5 weeks with three passages a week. However, one should check the stability of the cells as this may vary for different cell types. Assessable parameters for cell stability are growth rate, spontaneous apoptosis and/or the presence of micronuclei. As the number of passages may effect the performance of the assay, it is important to carefully test the assay performance at different cell passages.

In order to increase cell seeding quality it is useful to preincubate the plate at room temperature after cell seeding until the cells are attached, before moving the plates to the incubator (Lundholt et al. 2003). This allows for constant results across different experiments and homogenous sampling during image acquisition.

If an assay is designed to include long-term incubation before readout (i.e., 48 h or longer), evaporation is also an issue. Wells on the outer border of the plate (first and last columns/rows) lose medium, resulting in the concentration of the solute, which in turn causes osmotic distress for cells. Potentially, these samples may generate artifacts in the readout, jeopardizing the stability of the assay. Possible solutions to minimize this effect are to seal the plates with breathable foils that allow CO_2 exchange but impede water evaporation; an easier and cheaper solution is to add a higher volume of medium or to add sterile water to the cells after a determined incubation time. In this case one must check if the addition of water is affecting the assay. If the two solutions mentioned above are not applicable to the assay, it is possible to simply not use the outer wells for the assay or use statistical correction for the values obtained from the affected wells. However, note that statistical correction should be used only if all other solutions are not applicable. In this case it is strongly recommended to contact a biostatistician to ensure the correct algorithms are used for the specific assay.

16.6 Cell Vessels

HC-PCAs rely on microscopy detection and therefore the vessels for the cells must have a clear-bottom surface suitable for imaging with an inverted microscope. There are two main substrates amenable for microscopy: glass and optical-grade plastic. Many laboratory material suppliers offer optical-grade plastic plates for cell culture in different formats (96 wells, 384 wells, etc.). Normally cells adhere relatively well to cell culture plastic, while adhesion to glass can be more problematic. However, as different cells may adhere to the same substrate in a different way, it is important to check the specific conditions for each individual assay. Importantly, conditions for manual cell handling might be completely different from the conditions for automated cell processing. For this reason it is essential to coordinate test activities with the automation team to check that assay steps can be performed without disturbing cell adherence. Additionally, cell treatment, incubation time or fixation can alter cell adherence and influence the final number of cells available during the readout. It is therefore necessary to check adherence by including all possible conditions (i.e., positive and negative controls, solvent, etc.) and all steps (medium exchange, fixation, etc.) that will be performed in the assay. The risk of losing cells is especially high during medium exchange or washing steps. Adding gelatin (1%) or other colloidal particles to increase solution viscosity can partly circumvent this.

16.7 Cellular Imaging

A key component of HC-PCAs is cellular imaging. Staining and image acquisition are a relevant source of artifacts and introduce problems that must be taken into consideration during HC-PCA development. Cellular imaging is one of the most important steps in the assay development process, hence particular care must be taken for a successful final result.

There are several different microscopy techniques that can be used for cellular imaging, however the most widely used in HCS is fluorescence microscopy. Fluorescence microscopy (Herman and Tanke 1998; Ploem and Tanke 1987) can be divided into two main areas: epifluorescence and confocal microscopy (Pawley 2006). Both techniques have advantages and disadvantages, hence the investigator must carefully choose which will be the most suitable for the specific assay.

Fluorescence microscopy relies on fluorophores; the choice of appropriate probes is of particular importance for assay developers. Fluorescent probes can be divided into two general categories: (1) small molecules and (2) fluorescent proteins. Indeed, fluorescent proteins have already been extensively used in HC-PCAs (Baatz et al. 2006; Laketa et al. 2007; Neumann et al. 2006); however, the use of fluorescent proteins requires the time-consuming generation of cell lines stably expressing the respective gene. Additionally, it should be taken into consideration that the use of fluorescent proteins may require a license. The main area of use of fluorescent proteins is kinetic assays. Time-resolved image analysis remains a challenge for HCS at this time, but fluorescent proteins will be at the center of future development for HC-PCAs.

Small fluorescent molecules have seen an impressive expansion in the last 20 years in both numbers and quality. The use of small fluorescent molecules can be divided into two types: antibody/protein tagging and direct organelle/compartment tagging (e.g., DNA or mitochondria staining). Both techniques are used in HC-PCAs and are amenable for HCS. The disadvantage of the use of small fluorescent molecules is photobleaching; in the case of protein tagging it is cost. However, the latest generation of small fluorescent molecules is performing excellently with regard to photobleaching. Additionally, the use of "photobleaching friendly" microscopes such as the Nipkow disc (Graf et al. 2005) will further reduce this problem. The issue of cost in the use of tagged antibodies can be addressed either by assay miniaturization and consequent reduction in the use of reagents or by obtaining a cheap and reliable source of antibody (e.g., monoclonal production, phage display, etc.).

HC-PCAs developed for HCS campaigns rely on the use of automated microscopy to acquire images in a user-independent fashion: in turn, automated microscopes rely on routines for autofocus, exposure and all the other functions normally executed by the operator. Although these routines have improved extremely in quality and reliability, the researcher should be aware of possible problems arising in these steps that could compromise the assay readout. The most common problems are out-of-focus or soft-focus images. The consequence of this problem is obviously the loss of images to be analyzed. The problem can be reduced by an accurate choice of the right vessels during assay development. Where possible, it is suggested to always use a combination of hardware with an object-based autofocus routine. This will result in a slightly increased acquisition time but also an increase in image quality, and hence a better assay.

HC-PCAs rely on the detection of cellular or intracellular phenotypes differing from a control phenotype. However, during assay development it is almost impossible to forecast all the possible phenotypes that can be encountered during a screening campaign. This can result in the generation of artifacts or even assay failure. Therefore, it is good practice during the advanced phase of assay development to investigate a large number of positive and negative controls (or a subset of the library to be screened) in order to assess phenotype variability and robustness of the image analysis algorithm. If unexpected phenotypes are may be necessary to implement a new image analysis algorithm that can correctly analyze the images. This monitoring process should also be carried out during the entire screening campaign and should continuously feed information back to the image analysis routine. Thus, assay development is a continuous activity during the screening campaign.

16.8 Autofluorescence

The large majority of HC-PCAs are based on fluorescence detection of selective fluorochromes chosen by the user to mark a specific cell compartment or protein. However, autofluorescence can impair the readout of the fluorochrome. Common sources of autofluorescence include natural and fixative-induced autofluorescence. Natural (endogenous) autofluorescence is due to the fact that cells contain molecules which become fluorescent when excited by UV and visible light. The majority of natural autofluorescence is generated by aromatic amino acids, lipopigments and largely by pyridinic (NADPH) and flavin coenzymes. Natural autofluorescence is normally localized in mitochondria or acidic organelles, but it can also be diffused throughout the cytoplasm. Fixative-induced autofluorescence is due to the reaction of a fixative with biomolecules that generate an autofluorescent substrate. In particular, aldehyde fixatives react with amines and protein to generate fluorescent products.

It is necessary before starting the development of a HC-PCA to check the level of autofluorescence for the selected setting of the experiments at all the wavelengths that will be used for the HC-PCA optimization. Autofluorescence can also be generated by substrates or hydrogels used for coating or transfection agents. Although one can check for autofluorescence levels, it is not always possible to avoid it. Nonetheless, there are some techniques that can help: (1) autofluorescence filtering during image acquisition; (2) chemically removing autofluorescence.

The cell line can be exchanged if it is the source of the problem. To avoid aldehyde-based fixative autofluorescence, a non-aldehyde fixative or a lower concentration of aldehyde can be used. An alternative to formaldehyde and glutaraldehyde is dimethyl suberimidate (Davies and Stark 1970; Dodson 2000).

To filter autofluorescence during acquisition, strict band-pass filters can be used. However, autofluorescence generally has a broad spectrum of emission compared with the spectra of dyes used for signal detection. This can make it difficult to separate wanted from unwanted fluorescence by using conventional filter techniques. An interesting alternative solution to the problem is to use software filtering calibrated on autofluorescence spectral characteristics. In particular, linear unmixing (Dickinson et al. 2001; Zimmermann 2005) is useful for this purpose. Unfortunately, linear unmixing requires microscopes able to detect either excitation-based or emission-based spectra, which is currently not possible with HCS microscopes. Lastly, there are many protocols to chemically reduce autofluorescence but analysis of this aspect goes beyond the scope of this chapter (Billinton and Knight 2001; Mosiman et al. 1997; Neumann and Gabel 2002).

16.9 Image Analysis

Automated image analysis constitutes the readout of the assay and thus is central to assay development (see also Chap. 14 by Gradl et al.). There are three main approaches for image analysis in HC-PCAs: (1) the use of image analysis software that runs predefined algorithms to identify objects, (2) the use of open image analysis platforms that enable the construction of personalized algorithms making use of primitive functions and (3) to develop software for a specific use. As image analysis will be an essential part of the assay, a largely defined algorithm to detect the readout parameters must be available from the beginning of assay development. During assay design it is essential to identify cell components for measurement and determine the number of channels and what resolution will be used. Compiling a list of objects and parameters that must be extracted from the images will help in the choice of the best image analysis solution and will drive the development of image analysis. Finally, as early in the process as possible, some images of control cells should be made using the entire set of markers and given to the person responsible for image analysis development. This will enable programming and troubleshooting of the core algorithm that will be further improved during the assay development steps (Fig. 16.1b).

16.10 Transfection Optimization for RNAi-Based Assays

HCS-based studies aimed at functional genomics screening use "gain of function" and "loss of function" assays. Gain of function can be achieved by expression of cDNAs, while loss of function can be achieved using RNAi technologies. In both cases it is necessary to deliver nucleic acid with high efficacy to obtain the desired effect. For both cDNA and RNAi, large arrayed (genome-wide) libraries are available. As a thorough discussion of functional genomics screening is beyond the scope this section, we will focus specifically on transfection issues related to assay development for RNAi screens.

Recently, the use of HC-PCAs in functional genomics applications has seen a rapid increase. Indeed, RNAi has revolutionized the role of HC-PCAs in functional genomics applications, including HCS. In the past few years, there has been exponential growth in the application of RNAi technologies, ranging from small-scale experiments to genome-wide screens (Carpenter and Sabatini 2004; Neumann et al. 2006; Pelkmans et al. 2005). Approaching a high-throughput RNAi screening campaign involves a complex setup that goes beyond assay development (for a comprehensive review see Echeverri and Perrimon (2006 and Rines et al. (2006).

However, some considerations should be taken into account during assay development if the HC-PCA will be used for RNAi HCS. In particular, transfection of double-stranded RNAs (dsRNAs) may have a strong impact on the stability of the HC-PCA in question. To transfer RNAi to cells there are three types of technique: (1) lipofection, (2) cationic polymers and (3) electroporation. In addition to these conventional methods, the use of peptides (Li et al. 2006) and the use of viral vectors as alternative methods are possible.

During assay design it should be carefully evaluated which of these methods is amenable for the assay. It is of particular importance to consider if and what kinds of artifacts transfection reagents generate. The most diffuse method to deliver nucleic acids to the cells is lipofection (Li et al. 2006; Simoes et al. 2005). Lipofection can be performed using different strategies, such as forward transfection, reverse transfection (Ovcharenko et al. 2005) and solid-phase transfection (Ziauddin and Sabatini 2001). For most of the cell lines commonly used, small interfering RNAs (siRNAs) can be efficiently transiently transfected using commercial reagents (Table 16.1). However, careful optimization of the transfection protocol is necessary in the preparation of a HC-PCA to be used in RNAi screens. There are different sources of problems generated by transfection. Those most interest to this discussion are (1) "escapers," (2) silencing efficiency, (3) toxicity and (4) off-target effects, or unspecific reactions.

16.11 Escapers and Silencing Efficiency

The duration of gene silencing due to RNAi can last up to 5–6 days (Song et al. 2003), although the use of chemically modified RNAi significantly extends the silencing period (Elmen et al. 2005). The efficiency of transfection will therefore have a strong influence on the silencing duration. For each cell line used in a HC-PCA it is important to calculate the number of cells that "escape" transfection. Escapers are defined as the cells that are not transfected or that did not downregulate the gene of interest to an extent able to generate a desired phenotype. It is also recommended to calculate the doubling time of the specific cell lines in use as this directly affects the duration of silencing. The shorter the doubling times, the greater the risk of the phenotypes in the HC-PCA being masked by escapers. Escapers are a relevant problem especially in end-point RNAi-based screens or HC-PCAs where the sampling is limited to a small number of cells (e.g., time lapse or high magnification). In order to observe a phenotype it is generally necessary to incubate cells with the RNAi of interest for at least 48 h. This is generally the length of time required to obtain a significant downregulation of messenger RNA and proteins in RNAi transfected cells (Song et al. 2003), although in many cases screens for RNAi are extended to 72 h or longer. It is conceivable that if the transfection protocol allows a high number of escapers this population might mask the phenotype. One method to accurately identify escapers is to use RNAi for

Table 16.1 Selected transfection reagents commercially available and suited for transfection of small interfering RNA (*siRNA*). Transfection reagents can be divided into (1) cationic lipids, (2) polymers and (3) penetrating peptides and nanomaterials. In some cases, companies offer a combination of one or more of these components. It is important to note that different reagents may behave differently with different cells

Company	System name	Web address	Comments
Ambion	siPORT NeoFX	www.ambion.com	Lipid-based formulation specially developed for reverse transfection, serum-compatible
Biontex Laboratories	METAFECTENE	www.biontex.com	Serum compatible, polycationic transfection reagent
Bio-Rad	siLentFect Lipid Reagent	www.bio-rad.com	Cationic lipid
Dharmacon	DharmaFeCT siRNA Transfection Reagents 1, 2, 3 and 4	www.dharmacon.com	
Genlantis	GeneSilencer siRNA Transfection Reagent	www.genlantis.com/RNAi/GeneSilencer	Cationic lipid formulation, serum-compatible, but efficiency might be increased under serum-free conditions
Invitrogen	OligofectAMINE	www.invitrogen.com	Serum-free conditions recommended
Invitrogen	Lipofectamine RNAiMAX	www.invitrogen.com	Serum-compatible, no antibiotics recommended
Mirus	TransIT-TKO	www.mirusbio.com	Serum compatible
Mirus	TransIT-siQUEST	www.mirusbio.com	Unique lipopolyplex formulation offers advantages in select cell types, serum-compatible
orbigen	RNAi-Shuttle siRNA Transfection Reagent	www.orbigen.com	Cationic lipid based
OZ Biosciences	Lullaby siRNA Transfection Reagent	www.ozbiosciences.com	Cationic lipids formulation that triggers endosomal escape
PAA	Nanofectin-siRNA	www.paa.com	2 components: positively charged polymer to bind nucleotides and porous nanoparticles on which the polymers are coiled
Panomics	DeliverX siRNA Transfection Kit	www.panomics.com	Virus-derived amphipathic MPG peptides
PolyPlus	INTERFERin siRNA Transfection Reagent	www.polyplus-transfection.com	Compatible with serum and antibiotics

(continued)

Table 16.1 (continued)

Company	System name	Web address	Comments
QIAGEN	HiPerFect Transfection Reagent	www.qiagen.com	
Roche	X-tremeGENE siRNA Transfection Reagent	www.roche-applied-science.com	Blend of lipids and other components
Synvolux Therapeutics	SAINT-RED siRNA/ RNAi-Delivery System	www.synvolux.nl	Synthetic amphiphilic (nonliposomal) delivery systems based on the cationic pyridinium head group SAINT-18 with two C18 tails combined with dioleoylphosphatidylethanolamine
Targeting Systems	Targefect-siRNA transfection kit	www.targetingsystems.com	Combination of novel reagents

Table 16.2 Genes recommended as controls during transfection optimization. The genes are divided into (1) cell cycle related genes, (2) organelle or structural genes and (3) nonessential genes. Transfection efficiency for all classes can be assessed by reverse-transcription PCR and/or western blot. For the first two classes, it is possible to determine the visual phenotype and they can therefore be used in combination with image analysis

Gene name	Accession number	Phenotype detection	Essential	Function
KIF11	NM_004523	Mitotic arrest, cell number decrease	+	Cell cycle
INCENP	NM_020238[a]	Mitotic arrest, cell number decrease	+	Cell cycle
AURKA/B/C	NM_198433[a]	Mitotic arrest, cell number decrease	+	Cell cycle
COPB2	NM_004766	Apoptotic	+	Golgi trafficking
NUMA	NM_006185	Apoptotic	+	Nuclear
Lamin B1	NM_005573	Apoptotic	+	Nuclear
Emerin	NM_000117	–	–	
Zyxin	NM_003461[a]	–	–	
Lamin A/C	NM_170707[a]	–	–	Nuclear

Accession number is given for human protein

[a]More than one transcript is known

genes involved in the cell cycle in mitotic cell lines or genes that are essential for cell survival. Silencing of these genes generates a mitotic or toxic phenotype, hindering the proliferation of the cells and allowing estimation of the number of escapers to the particular transfection protocol. In both cases the vitality of the cells measured either by microscopy or by alternative methods (e.g., methyltetrazolium reduction assay, alamar blue assay, etc.) will give the investigator a reference for the efficacy of the transfection protocol. A list of suggested genes is shown in table 16.2.

It is also recommended that after establishing of a reliable transfection protocol, the investigator should carry out more extensive experiments with genes specifically related to the HC-PCAs in development and characterize the phenotype effectiveness for the conditions chosen.

16.12 Toxicity

Although transfection optimization for the highest silencing efficiency is highly sought after, in order to obtain an excellent assay for RNAi use, it is essential to avoid toxic effects. In many cases optimal transfection efficiency comes at the price of higher unspecific toxicity or alterations of physiological functions that may compromise assay validity. The assay developer must tackle this problem by including all the appropriate controls and taking actions in order to eliminate or reduce such phenomena. The goal of the assay developer in this case is to obtain maximum silencing efficiency in healthy cells.

Different cell lines react differently to different transfection reagents. Conditions that are optimal for one cell line (e.g., HeLa) can be extremely toxic or completely inefficient for another cell line (e.g., A431). Therefore, before starting optimization of the transfection protocol it is worthwhile testing different reagents on the selected cell line. A suggested scheme is indicated in Fig. 16.2a and b. The transfection reagent tested alone and in combination with different nontargeting RNAi duplexes will indicate the best reagent for the selected cell line. As shown in Fig. 16.2c, reagent B shows unspecific toxicity not visible in the sample treated with reagent A. However, silencing potency is similar in the two reagents. In this case, simply changing transfection reagents increases the quality of the assay.

Other possible sources of toxicity can be contamination in the dsRNA oligos (Denise Kensky, personal communication). In the case of siRNA, different purification grades can lead to different results. It is important therefore to compare the same oligos with two different purification grades in order to exclude possible toxic effects due to contaminantion. Unfortunately, in many cases siRNA libraries to be used for screens come with only a "crude" purification of the siRNA, limiting the possibility to increase the quality of the assay – and hence the screen – by using cleaner siRNA.

Another source of toxicity or change in physiological response is the interferon response (Sledz et al. 2003) due to cellular self-defense against exogenous RNA or DNA. However, it has been shown that the interferon response is not activated when RNAi duplexes shorter than 30 base pairs are used in mammalian cells (Elbashir et al. 2001). Nonetheless, it is strongly recommeded to check for an interferon response when a new cell line is to be used.

Although toxicity is a relevant issue in transfection optimization, loss or alteration of physiological function due to transfection conditions is far more dangerous and difficult to identify. This problem must not be underestimated during assay development and particular care should be dedicated to identify and troubleshoot it. The best way to address this problem is to define a "gold standard." In this case a gold standard is defined as one or more genes known to give a specific phenotype in the specific assay.

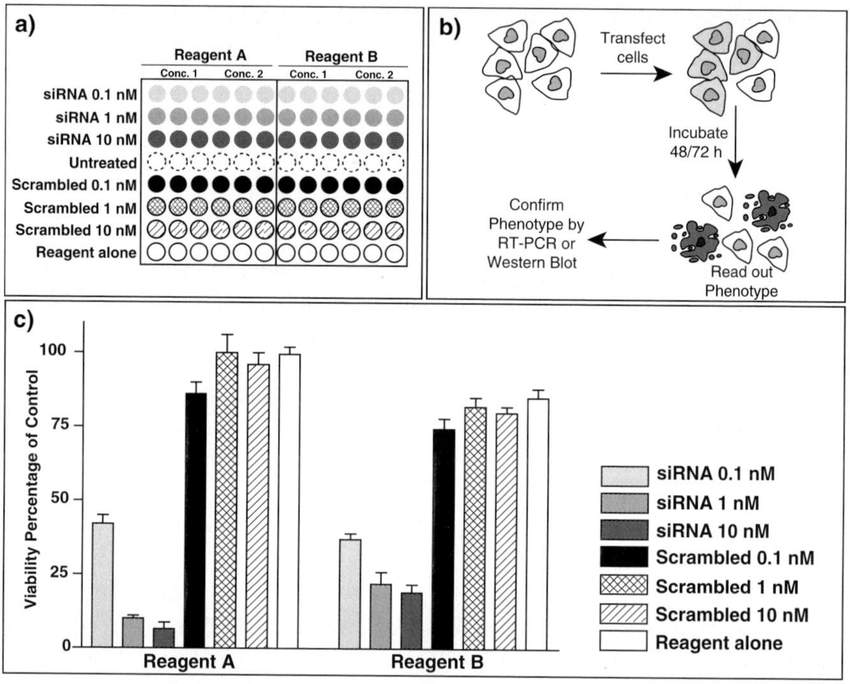

Fig. 16.2 Assay optimization. **a** To optimize the transfection protocol it is useful to design a master plate containing different positive and negative controls, possibly at different concentrations. Different transfection reagents could be tested on the same plate to assess transfection efficiency and toxicity. **b** The master plate is applied to the assay in question. Evaluation of the phenotype will be determined by phenotypic characterization and/or RT-PCR western blot. **c** Reagent B shows an unspecific toxicity in both scrambled sequences and transfection reagent alone. In contrast, reagent A does not show any toxicity under the same condition. Additionally transfection efficiency is not affected as shown by the results for the different siRNA concentrations. By adopting reagent A, the assay will gain in stability without sacrificing transfection efficiency. (*RT-PCR* reverse-transcription PCR, *siRNA* small interfering RNA)

The assay developer must then identify specific RNAi molecules for this gene(s), validate them by reverse-transcription PCR and/or western blot and finally use them with the transfection conditions chosen. Experiments must be carried out in triplicate and repeated on a minimum of three different days using RNAi-treated and RNAi-untreated cells. The assay developer must then analyse the results and verify the occurrence of the expected phenotype. Discrepancy from the expected phenotype will indicate a loss or alteration of physiological function due to transfection.

16.13 Off-Target or Unspecific Reactions

RNAi is a recent discovery in science (Elbashir et al. 2001; Fire et al. 1998) however it has been adopted in many research fields with extraordinary speed. Although not yet fully understood, the molecular mechanisms that govern RNAi include an off-target effect of dsRNA. This complicates interpretation of the data and can increase the number of false positives. Obviously, this phenomenon hinders assay development, (the aim of which is to design the most robust and reliable assay) in obtaining the highest accuracy in the interpretation of hits. At the moment, understanding of the off-target phenomenon is still lacking and unfortunately very little can be done during assay development to avoid off-target issues. It was recently shown that chemical modification of the siRNA significantly reduces off-target effects (De Paula et al. 2007; Jackson et al. 2006) and the use of enzymatically produced siRNA is less prone to off-target effects (Kittler et al. 2007). What the assay developer must consider is that lowering the concentration of siRNA as much as possible reduces unspecific effects of siRNA. Additionally, the assay developer could apply secondary assays to test the off-target effect, such as a rescue assay (Kittler et al. 2005).

16.14 Assay Quality

How should one determine assay quality? What tools do we have to decide if an assay is suitable for our purpose or not? The statistical methods to determine assay quality are not well defined and hence the assay developer must decide which statistical tools are more suitable for the assay in question. A good reference on the use of statistical tools during assay development is Derzko (2005).

An intuitive criterion to determine assay quality is the size of the screening window! Simply stated, the screening window is represented by the difference between the positive and negative control. The bigger the difference, the better the possibility to assign a hit to one side or the other. However, measurements of empirical data coming from the same sample are subject to a certain degree of variability. The data variability is determined mathematically by using the standard deviation. The assay developer must take data variability into account to determine the real validity of the assay. An assay with a large screening window, but a high standard deviation can be worse than an assay with a small screening window but excellent standard deviation. However, for normally distributed data sets (Gaussian distribution) there is a formula to calculate the quality of an assay. The Z' factor (Zhang et al. 1999) is often referred to as a marker of assay quality. This coefficient is dimensionless and reflects both the dynamic range of the assay as well as the variation of data measurement. The formula for Z' is

$$Z' = 1 - \frac{3(\sigma_{\mathrm{pc}} + \sigma_{\mathrm{nc}})}{|\mu_{\mathrm{pc}} - \mu_{\mathrm{nc}}|},$$

where σ is the standard deviation and μ is the average of, positive (pc) and negative (nc) controls respectively. The numerator represents the variability of data measurements and the denominator is the dynamic range of the assay. Values close to 1 indicate an excellent assay, while values below 0.5 indicate a poor assay (see Table 16.3 for a list of values and relative indications). In Fig. 16.3 different situations

Table 16.3 The *Z′* factor is a mathematical evaluation of assay quality. For an assay to be used in a screening campaign, values above 0.5 are accepted as satisfactory. Assays with a Z′ factor between 0 and 0.5 are normally described as a yes/no assay. In this case the user must determine if the particular assay is suitable for the screening condition. Note that the Z′ factor applies only to data sets that are normally distributed. (SD is Standard Deviation)

Z′ factor	Assay	Outcome
1	SD = 0 or dynamic range = ∞	Optimal assay
1>*Z′*>0.5	Separation band is large	Excellent assay
0.5>*Z′*>0	Separation band is small	Double assay
0	No separation band, sample signal and control signal variation touch	Yes/no assay
<0	No separation band, sample signal and control signal variation overlap	Unusable assay

are summarized to describe the different possible conditions. The *Z′* factor has been used extensively in assay development in high-throughput screening campaigns and it is a valuable tool for assay developers. However, the limitation of the *Z′* factor lies in the fact that it is applicable only to normally distributed data. Numerical parameters derived from HC-PCAs are frequently not normally distributed, hence rendering the *Z′* factor useless for the determination of assay quality. In this case different statistical tools need to be used instead of the *Z′* factor. The Kolgomorov–Smirnov test, also known as the KS test (Chakravarti et al. 1967), is a nonparametric and distribution-free test that is used to determine if two datasets are significantly different. Basically, the KS test measures the vertical distance of two empirical cumulative distribution functions and calculates if they are significantly different. The KS test can be used on the positive and negative controls to determine understand if the assay is sensitive enough to distinguish data coming from the two populations. However, the KS test should be used with caution, as it is most sensitive around the median of the values and less sensitive on the tails of the data set. Variations of the KS test that are more sensitive on the tails of the data set are the Anderson–Darling test (Stephens 1974) and the Shapiro–Wilk test (Stephens 1974). A deeper explanation of the statistical tools for assay optimization and screening data analysis is beyond the scope of this chapter. A comprehensive and accessible review (Malo et al. 2006) may help to better understand the main statistical concepts needed to optimize HC-PCAs.

16.15 Assay Validation

In contrast to assay development, the statistical methods for assay validation are well defined. A comprehensive list of parameters to be validated has been described by the International Conference of Harmonization (ICH Geneva) (ICH 1994).

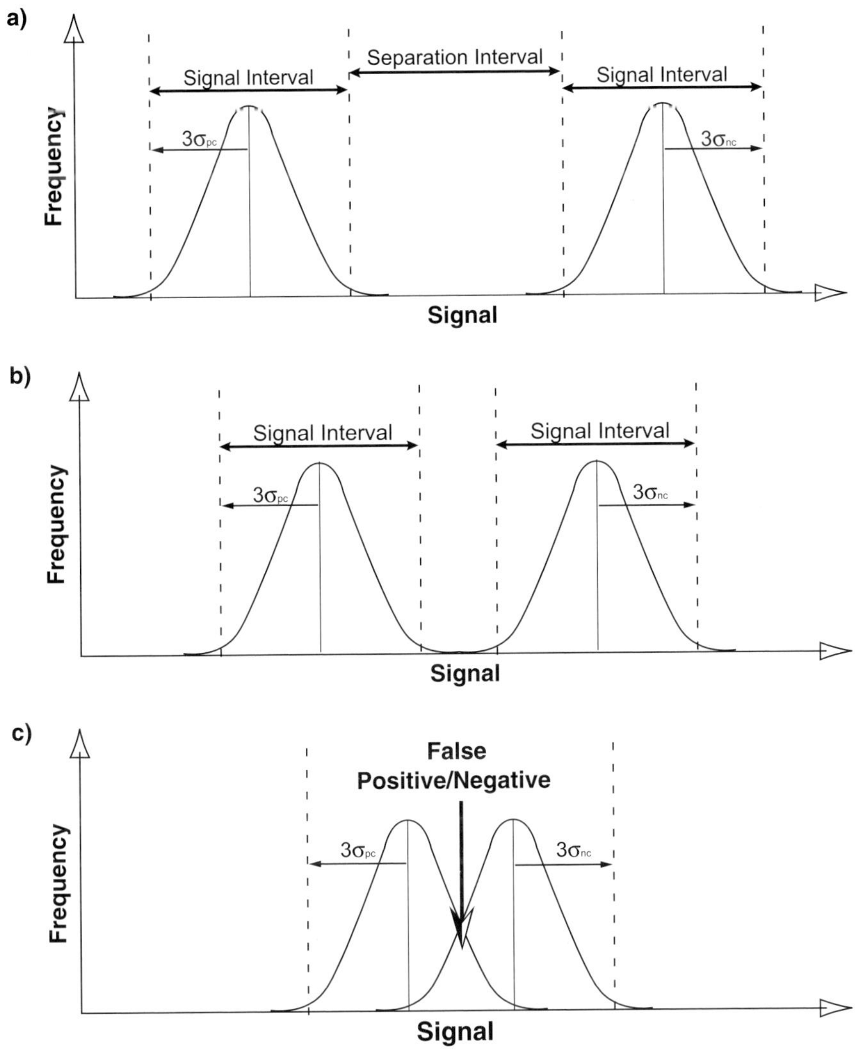

Fig. 16.3 The Z' factor defines the quality of an assay with normally distributed data. **a** Optimal conditions with either no standard deviation or infinite dynamic range (Z'=1). **b** Good separation interval with clear separation of the positive and negative controls (0>Z'>0.5); the assay is suitable for screening. **c** Positive and negative controls overlap, indicating either extreme variability in the data set or scarce dynamic range. Under this condition it is not possible to separate false positive and false negative; the assay is not suitable for screening.

This document provides an exhaustive explanation of the parameters to be validated. However, these criteria have not been designed for HC-PCA and thus the assay developer should apply direction when deciding which parameters should apply discretion for HC-PCA validation. See Ritter et al. (2001) for a more extensive discussion.

16.16 Conclusion and Outlook

The use of HC-PCAs and their application in HCS is clearly only just beginning. However, HC-PCAs have already shown their power in both basic research and target discovery (Gasparri et al. 2004; Pelkmans et al. 2005). It is possible to forecast a future development in the automation of more specialized microscopy techniques that will open new frontiers to the cell biologist. Similarly, image analysis software is becoming more and more powerful and user-friendly. In this field, the use of parallel computing or cluster CPUs will open the possibility to analyze a greater number of images with the extraction of multiparametric descriptors. This will increase the precision of phenotype annotation and consequently improve the quality of HCS campaigns.

Regarding cell biology, it is possible to predict an increased use of HC-PCAs to study intracellular events. Application of confocal imaging and acquisition of 3D and 4D data sets will enable the development of HC-PCAs dedicated to specific intracellular targets. On the other hand, application of HC-PCAs to primary cells (e.g., neurons, hepatocytes, etc.), and to model organisms, will help to develop more physiologically relevant HC-PCAs. Time-lapse and kinetic HC-PCAs will also become an essential tool in HCS.

In closing, it is foreseeable that the correlation of results of different screening campaigns performed using diverse HC-PCAs and analysis of this information with bioinformatics tools, will unveil protein functions in multiple pathways. Connecting these pathways will lead to a more thorough evaluation of a protein target within the complex network of cell physiology and possibly shorten the efforts to understand its function in the whole organism as well as during infection or disease. In time, HC-PCAs and their application in HCS will demonstrate their validity and will take their place as routine tools in the laboratory.

Acknowledgements We thank B.M. Simon, M. Bickle and Melissa Thomas for critical reading of the manuscript and for many helpful discussions. We also would like to thank all members of the Technology Development Studio (TDS) at the Max Planck Institute of Molecular Cell Biology in Dresden. Without their work, the "exploration" of the high-content assay would not have been possible.

References

Baatz M, Arini N, Schäpe A, Binnig G, Linssen B (2006) Object-oriented image analysis for high content screening: detailed quantification of cells and sub cellular structures with the cellenger software. Cytometry A 69:652–658

Billinton N, Knight AW (2001) Seeing the wood through the trees: a review of techniques for distinguishing green fluorescent protein from endogenous autofluorescence. Anal Biochem 291:175–197

Carpenter AE, Sabatini DM (2004) Systematic genome-wide screens of gene function. Nat Rev Genet 5:11–22

Chakravarti IM, Laha RG, Roy J (1967) Handbook of methods of applied statistics. Wiley, New York

Davies GE, Stark GR (1970) Use of dimethyl suberimidate, a cross-linking reagent, in studying the subunit structure of oligomeric proteins. Proc Natl Acad Sci USA 66:651–656

De Paula D, Bentley MV, Mahato RI (2007) Hydrophobization and bioconjugation for enhanced siRNA delivery and targeting. RNA 13:431–456

Derzko AN (2005) Statistical practice in assay development and validation. IVD Technol. http://www.devicelink.com/ivdt/archive/05/03/002.html

Dickinson ME, Bearman G, Tille S, Lansford R, Fraser SE (2001) Multi-spectral imaging and linear unmixing add a whole new dimension to laser scanning fluorescence microscopy. Biotechniques 31:1272, 1274–1276, 1278

Dodson MS (2000) Dimethyl suberimidate cross-linking of oligo(dt) to DNA-binding proteins. Bioconjug Chem 11:876–879

Dove A (2003) Screening for content – the evolution of high throughput. Nat Biotechnol 21:859–864

Echeverri CJ, Perrimon N (2006) High-throughput RNAi screening in cultured cells: a user's guide. Nat Rev Genet **7**:373–384

Elbashir SM, Harborth J, Lendeckel W, Yalcin A, Weber K, Tuschl T (2001) Duplexes of 21-nucleotide RNAs mediate RNA interference in cultured mammalian cells. Nature 411:494–498

Elmen J, Thonberg H, Ljungberg K, Frieden M, Westergaard M, Xu Y, Wahren B, Liang Z, Orum H, Koch T, Wahlestedt C (2005) Locked nucleic acid (LNA) mediated improvements in siRNA stability and functionality. Nucleic Acids Res 33:439–447

Fire A, Xu S, Montgomery MK, Kostas SA, Driver SE, Mello CC (1998) Potent and specific genetic interference by double-stranded RNA in Caenorhabditis elegans. Nature 391:806–811

Gasparri F, Mariani M, Sola F, Galvani A (2004) Quantification of the proliferation index of human dermal fibroblast cultures with the arrayscan high-content screening reader. J Biomol Screen 9:232–243

Giuliano KA, Haskins JR, Taylor DL (2003) Advances in high content screening for drug discovery. Assay Drug Dev Technol 1:565–577

Graf R, Rietdorf J, Zimmermann T (2005) Live cell spinning disc microscopy. Adv Biochem Eng Biotechnol 95:57–75

Herman B, Tanke HJ (1998) Fluorescence microscopy. Springer, Berlin

Hoffman AF, Garippa RJ (2007) A pharmaceutical company user's perspective on the potential of high content screening in drug discovery. Methods Mol Biol 356:19–31

Hooke (1665) Micrographia: or some physiological descriptions of minute bodies made by magnifying glasses. Royal Society, London

ICH (1994) Q2(r1): validation of analytical procedures: text and methodology. International Conference of Harmonization

Inglese J (2006) Assay guidance manual. Eli Lilly and the National Institutes of Health Chemical Genomic Center

Jackson AL, Burchard J, Leake D, Reynolds A, Schelter J, Guo J, Johnson JM, Lim L, Karpilow J, Nichols K, Marshall W, Khvorova A, Linsley PS (2006) Position-specific chemical modification of siRNAs reduces "off-target" transcript silencing. RNA 12:1197–1205

Kittler R, Pelletier L, Ma C, Poser I, Fischer S, Hyman AA, Buchholz F (2005) RNA interference rescue by bacterial artificial chromosome transgenesis in mammalian tissue culture cells. Proc Natl Acad Sci USA 102:2396–2401

Kittler R, Surendranath V, Heninger AK, Slabicki M, Theis M, Putz G, Franke K, Caldarelli A, Grabner H, Kozak K, Wagner J, Rees E, Korn B, Frenzel C, Sachse C, Sonnichsen B, Guo J, Schelter J, Burchard J, Linsley PS, Jackson AL, Habermann B, Buchholz F (2007) Genome-wide resources of endoribonuclease-prepared short interfering RNAs for specific loss-of-function studies. Nat Methods 4:337–344

Laketa V, Simpson JC, Bechtel S, Wiemann S, Pepperkok R (2007) High-content microscopy identifies new neurite outgrowth regulators. Mol Biol Cell 18:242–252

Li CX, Parker A, Menocal E, Xiang S, Borodyansky L, Fruehauf JH (2006) Delivery of RNA interference. Cell Cycle 5:2103–2109

Liebel U, Starkuviene V, Erfle H, Simpson JC, Poustka A, Wiemann S, Pepperkok R (2003) A microscope-based screening platform for large-scale functional protein analysis in intact cells. FEBS Lett 554:394–398

Lundholt BK, Scudder KM, Pagliaro L (2003) A simple technique for reducing edge effect in cell-based assays. J Biomol Screen 8:566–570

Malo N, Hanley JA, Cerquozzi S, Pelletier J, Nadon R (2006) Statistical practice in high-throughput screening data analysis. Nat Biotechnol 24:167–175

Mosiman VL, Patterson BK, Canterero L, Goolsby CL (1997) Reducing cellular autofluorescence in flow cytometry: an in situ method. Cytometry 30:151–156

Neumann B, Held M, Liebel U, Erfle H, Rogers P, Pepperkok R, Ellenberg J (2006) High-throughput RNAi screening by time-lapse imaging of live human cells. Nat Methods 3:385–390

Neumann M, Gabel D (2002) Simple method for reduction of autofluorescence in fluorescence microscopy. J Histochem Cytochem 50:437–439

Ovcharenko D, Jarvis R, Hunicke-Smith S, Kelnar K, Brown D (2005) High-throughput RNAi screening in vitro: from cell lines to primary cells. RNA 11:985–993

Pawley JB (2006) Handbook of biological confocal microscopy, 2nd edn. Springer, New York

Pelkmans L, Fava E, Grabner H, Hannus M, Habermann B, Krausz E, Zerial M (2005a) Genome-wide analysis of human kinases in clathrin- and caveolae/raft-mediated endocytosis. Nature 436:78–86

Ploem JS, Tanke HJ (1987) Fluorescence microscopy. BIOS, Oxford

Prasher DC, Eckenrode VK, Ward WW, Prendergast FG, Cormier MJ (1992) Primary structure of the Aequorea victoria green-fluorescent protein. Gene 111:229–233

Rines DR, Tu B, Miraglia L, Welch GL, Zhang J, Hull MV, Orth AP, Chanda SK (2006) High-content screening of functional genomic libraries. Methods Enzymol 414:530–565

Ritter NM, Hayes T, Dougherty J (2001) Analytical laboratory quality: part II. Analytical method validation. J Biomol Tech 12:11–15

Schleiden MJ, Schwann T, Schultze MJS, Ilse J (1839) Klassische Schriften zur Zellenlehre. Ostwalds Klassiker der exakten Wissenschaften, vol. 275. Sander, Berlin, p 166

Schwann T (1839) Mikroskopische Untersuchungen über die Übereinstimmung in der Struktur und dem Wachsthum der Thiere und Pflanzen. Ostwalds Klassiker der exakten Wissenschaften, vol 176. Sander, Berlin

Shimomura O, Johnson FH, Saiga Y (1962) Extraction, purification and properties of aequorin, a bioluminescent protein from the luminous hydromedusan, Aequorea. J Cell Comp Physiol 59:223–239

Simoes S, Filipe A, Faneca H, Mano M, Penacho N, Duzgunes N, de Lima MP (2005) Cationic liposomes for gene delivery. Expert Opin Drug Deliv 2:237–254

Sledz CA, Holko M, de Veer MJ, Silverman RH, Williams BRG (2003) Activation of the interferon system by short-interfering RNAs. Nat Cell Biol 5:834–839

Song E, Lee SK, Dykxhoorn DM, Novina C, Zhang D, Crawford K, Cerny J, Sharp PA, Lieberman J, Manjunath N, Shankar P (2003) Sustained small interfering rna-mediated human immunodeficiency virus type 1 inhibition in primary macrophages. J Virol **77**:7174–7181

Stephens MA (1974) Edf statistic for goodness of fit and some comparison. J Am Stat Assoc 69:730–737

Zhang JH, Chung TD, Oldenburg KR (1999) A simple statistical parameter for use in evaluation and validation of high throughput screening assays. J Biomol Screen 4:67–73

Ziauddin J, Sabatini DM (2001) Microarrays of cells expressing defined cDNAs. Nature 411:107–110

Zimmermann T (2005) Spectral imaging and linear unmixing in light microscopy. Adv Biochem Eng Biotechnol 95:245–265

Index

Printing: Krips bv, Meppel, The Netherlands
Binding: Stürtz, Würzburg, Germany